CONTRACTILE PROTEINS IN MUSCLE AND NON–MUSCLE CELL SYSTEMS

CONTRACTILE PROTEINS IN MUSCLE AND NON–MUSCLE CELL SYSTEMS

Biochemistry, Physiology, and Pathology

Edited by

Emanuele E. Alia
Nicolò Arena
Matteo A. Russo

PRAEGER

PRAEGER SPECIAL STUDIES • PRAEGER SCIENTIFIC

New York • Philadelphia • Eastbourne, UK
Toronto • Hong Kong • Tokyo • Sydney

Library of Congress Cataloging-in-Publication Data
Main entry under title:

Contractile proteins in muscle and non-muscle cell systems.

Based on the First Symposium on Biochemistry, Physiology, and Pathology of Contractile Proteins in Muscle and Nonmuscle Cell Systems, held in Sassari, Sardinia, Italy, Oct. 1-5, 1983; organized by the Institute of Histology and General Embryology of the University of Sassari.
1. Actomyosin – Congresses. 2. Muscle proteins – Congresses. 3. Membrane proteins – Congresses. I. Alia, Emanuele E. II. Arena, Nicolò. III. Russo, Matteo A. IV. Symposium on Biochemistry, Physiology, and Pathology of Contractile Proteins in Muscle and Nonmuscle Cell Systems (1st: 1983: Sassari, Sardinia) V. Università degli studi di Sassari. Institute of Histology and General Embryology. [DNLM: 1. Contractile Proteins – congresses. 2. Muscle Proteins – congresses. QU 55 C764 1983]
QP552.A32C66 1985 591.1'852 85-12274
ISBN 0-03-071654-3 (alk. paper)

Published and Distributed by the
Praeger Publishers Division
(ISBN Prefix 0-275)
of Greenwood Press, Inc.,
Westport, Connecticut

Published in 1985 by Praeger Publishers
CBS Educational and Professional Publishing, a Division of CBS Inc.
521 Fifth Avenue, New York, NY 10175 USA

56789 052 987654321

Printed in the United States of America on acid-free paper

Preface

This book is the fruit of the First Symposium on Biochemistry, Physiology, and Pathology of Contractile Proteins in Muscle and Nonmuscle Cell Systems, which was held in Roccaruja, Sassari (Sardinia, Italy) in October 1983, and which was organized by the Institute of Histology and General Embryology of the University of Sassari.

Contractile and cytoskeletal proteins have represented one of the most exciting topics in cell biology and pathology during the last decade. Therefore, the Organizing Committee combined the attractions of this rapidly evolving topic with the splendid Sardinian landscape in order to bring together a number of scientists who are at the frontier of cytoskeletal research.

We thank, first all the contributors who have made this publication possible, containing, as it does, almost all of the papers presented at the Symposium. We are also pleased to express our acknowledgments to all the sponsors and public and private institutions that have generously sustained and helped the Organizing Committee. In particular, we thank Hon. Sandro Pertini, the President of Italy, for giving his high patronage, Hon. A. Rujck, the President of the Sardinian Government, and all the sponsors listed below:

Italian Minister of Public Education
Italian Minister of Public Health
Sardinian Regional Government
Sardinian Regional Agency for Public Health
District Government of Sassari
District Tourist Office of Sassari
Autonomous Tourist Agency of Sassari
City Government of Sassari
University of Sassari
Cocco's Real Estate
Alingros, Sassari
Commerbus, Sassari
Prodifarm, Sassari
Degros, Sassari
Garau Arredamenti, Sassari
UTET Publishing Company, Sassari
Libreria Scientifica Internazionale, Sassari
Banco di Sardegna
Tecnolab, Cagliari

Lyons Club, Sassari
City Tourist Agencies of Alghero, Arzachena, and La Maddalena
Cantina Sociale di S.M. La Palma
Cantina Sociale di Sorso
Sella & Mosca, Alghero

Matteo A. Russo gives thanks to the Assicurazione Generali Foundation for supporting his work in Philadelphia during 1983 and 1984.

Finally, we would like to express our appreciation to Prof. Cappuccinelli for the warm and helpful advice, to Miss Maria Papaspyropoulos for the organizing work and to Miss Maggie Murray of the Praeger Publishing Company for her superb efficiency in editing the book.

Sassari, November 1983

Philadelphia, December 1983

Emanuele E. Alia
Nicolo' Arena
Matteo A. Russo

Introduction

The study of the molecular basis of contractility has shown a fascinating inversion. In an evolutionary sense, we have been going in reverse. Such study was begun with the skeletal muscles of higher animals, which was quite appropriate from an experimental point of view. Not only is skeletal muscle that tissue which is most highly specialized for contractility, but its remarkably organized fibrillar structure as seen in the microscope seemed to be an important clue to the way muscle worked. Indeed, the isolation and characterization of skeletal muscle actin and myosin, and the elegant demonstration of the sliding filament mechanism of contractility, relied on these special features of the tissue. However, as has become clear over the last 30 years, skeletal muscle is a relatively recent evolutionary development of a very ancient contractile protein system. Proteins closely homologous to skeletal muscle actin and myosin go back to the earliest forms of eukaryotic cells. In fact, actin is one of the most abundant cytoplasmic proteins in almost all cells exdept the prokaryotes. The evolutionarily primitive forms of myosin show similar biochemical and enzymic properties to those of skeletal muscle, and it is highly likely that in conjunction with actin, these myosins are capable of generating force by a sliding filament mechanism analogous to that of skeletal muscle. Now that these evolutionary relationships have been recognized, they have been purposefully exploited in detailed comparative studies of the actins, myosins, and functionally related proteins of lower organisms, studies which bid fair to illuminate many of the molecular properties of these proteins in higher organisms.

The area of muscle contractility has expanded in many directions in the last two decades. While the sliding filament mechanism has provided the conceptual framework of skeletal muscle activity, many of the details of that mechanism are still not completely understood at the molecular level. Many biochemists and biophysicists are engaged in such finer detail studies. The regulation of contractility is also a highly active field for molecular exploration. Furthermore, there is a great deal left to learn about what protein components and structures are responsible for the highly organized myofibrillar arrangement of skeletal muscle.

In addition to such developments, however, there has been a great expansion in our awareness of the variety of contractile and force-generating systems inside cells. The behavior of actin itself, apart from its interactions with myosin, has revealed a wide range of new mechanochemical activities. The reversible polymerization of G-actin to F-actin microfilaments, the capping and severing of these filaments, the bundling of F-actin into fibers, the reversible gelation-solation of microfilament aggregates, and the attachment of actin filaments and bundles to cell membranes and intracellular structures, are all activities that are related to force generation, which appear to be of particular importance in cells that have a low molar ratio of myosin to actin; i.e., all eukaryotic nonjuscle cells. There is a growing body of evidence that such actin-mediated force-generating activities may provide the mechanisms for cytokinesis, cell motility, and similar very basic and universal cellular phenomena. A truly explosive development of the last few years has been the discovery of a large and growing number of distinctive cellular proteins that interact with actin to induce and control these polymerization, capping, severing, bundling, etc., activities. How these actin-related processes are all integrated and regulated in a given cell has become a key problem in cell biology.

Apart from actomyosin and actin-related components, two other major intracellular filamentous systems have been recognized and characterized: microtubules and intermediate filaments. The critical role played by microtubules as force-generating systems in mitosis and in ciliary movement has been much investigated, and the participation of microtubules in the movements of intracellular organelles has been well documented. The intermediate filaments, the most recently discovered filamentous system, have rapidly emerged as prominent and highly diversified elements of higher eukaryotic cells. Although their relative abundance in such cells suggests that they have important functions, it is still not entirely clear what these functions are. No general force-generating or contractile activity has as yet been assigned to intermediate filaments, although by reversible interaction and bundling with microtubules, for example, they could participate in such activity. Nor does a strictly structural or cytoskeletal role for intermediate filaments seem commensurate with the diversity and heterogeneity of their constituent subunits, or with the fact that one type of subunit often undergoes replacement by another during developmental diversification of cells.

One of the crucial conclusions that has arisen from the studies of these contractile and structural filamentous systems, firmly established by appropriately designed specimen preparation techniques for electron microscopy, is that all eukaryotic cells possess a dense intracellular matrix of filamentous structures in which the cellular organelles and macromolecules are embedded. This matrix has come to be called the cytoskeleton. The cytoskeleton is not all that different in amorphous cells such as amebae and in highly structured cells such as skeletal muscle; it is mainly the degree of order imposed on this matrix that is different.

As might be expected, this wealth of new information about contractile and structural systems inside cells has begun to find interest and applications in pathology. We may safely anticipate that much new insight into the molecular basis for many disorders of contractile or mechanochemical functions of cells and tissues will eventually derive from knowledge of the basic molecular and cell biology of these systems.

Many of the problems of contractility, and of the molecular systems that have become recognized as participating in contractile activities, are addressed in this book, which consists of the proceedings of an "International Symposium on Contractile Proteins in Muscle and Nonmuscle Cell Systems and their Morpho-Physio-Pathology" that was held at Sassari, Italy, October 1-5, 1983. A very broad spectrum of subjects, ranging from the detailed biophysics of myosin cross-bridge dynamics during muscle contraction to molecular and structural alterations in the cytoskeleton in a variety of developmental and pathological changes, were addressed in a remarkably diversified and yet coherent manner, under ideal scientific and environmental conditions. This book clearly demonstrates the great breadth of molecular and cellular studies related to contractility, as well as the intense excitement attending the plethora of new discoveries and new insights into cell biology that characterize this area of biological science.

S. J. Singer
La Jolla, California
October 1983

The following table indicates the percent reduction by which illustration magnifications must be adjusted.

Figure	Page	Adjustment
4.1	32	.67
4.2	33	.67
4.3	34	.67
5.1	41	.90
5.2	42	.90
7.1	53	.75
7.2	53	.63
7.3	54	.70
7.4	54	.67
7.5	55	.67
13.1	92	.60
13.2	92	.60
13.3	94	.60
13.4	95	.60
14.1	102	.67
15.1	106	.66
15.2	107	.66
15.3	108	.66
16.1	113	.62
16.2	113	.62
16.3	114	.62
22.3	176	.75
23.5	186	.95
23.9	189	.52
23.10	190	.52
23.11	190	.50
23.12	191	.43
23.13	191	.37
36.6	329	.67
36.8	332	.71
36.9	333	.75
37.1	345	.75
37.2	346	.75

Figure	Page	Adjustment
37.3	347	.75
37.4	349	.74
37.5	349	.74
37.6	350	.75
37.7	350	.75
39.1	368	.72
39.2	370	.50
39.3	371	.50
39.4	372	.50
39.5	374	.50
39.6	375	.50
44.2	410	.67
45.1	419	.67
49.2	449	.67
49.4	450	.67
51.3	477	.80
51.4	478	.75
51.6	481	.66
59.1	552	.75
59.2	552	.75
59.3	552	.75
59.4	552	.75
59.5	554	.75
59.6	554	.75
59.7	554	.75
59.8	554	.75
62.1	576	.47
62.2	577	.47
62.3	577	.47
62.4	578	.47
62.5	579	.47
62.6	579	.47
79.1	728	.63
79.2	729	.62

Contents

PART I

STRUCTURE AND ORGANIZATION OF CONTRACTILE AND CYTOSKELETAL PROTEINS

1 The Chicken Erythrocyte Cytoskeleton: Structure, Synthesis, and Assembly in Development

Elias Lazarides

Over the past several months we have succeeded in establishing the nucleated chicken erythrocyte as a model system to study cell structure and its assembly during development. In addition to an equatorially disposed marginal band of microrubules this cell contains a subcortical filamentous network analogous to the mammalian erythrocyte spectrin-actin network. Chicken erythrocyte spectrin is composed of two nonidentical subunits, α and β, and is associated with the plasma membrane in a manner presumed to be analogous to what has been described for the human erythrocyte (1). These cells contain a protein which is biochemically and immunologically related to human erythrocyte ankyrin and termed globin. Additionally, they contain a polypeptide analogous to the human erythrocyte anion transporter. Association of spectrin with the anion transporter is indirect through ankyrin which simultaneously associates with -spectrin and a subset of the anion transporters (1). The chicken erythrocyte also contains a system of intermediate filaments which form a three-dimensional network perpendicular to the marginal band of microtubules and which interconnects the nucleus and the spectrin-actin network on opposite sides of the cell thus forming a transcytoplasmic network interlinking the nucleus with the plasma membrane (2). Extraction studies have established that in these cells the intermediate filament network associates with the spectrin-actin network (2,3), but the molecular details of this interaction have not yet been established. Biochemical and immunoelectron microscopic studies have established that the intermediate filament network is composed of vimentin as its major subunit in adult cells and is crosslinked at periodic intervals by a 230,000 dalton protein termed synemin. The synemin periodicity is approximately 180 nm in adult cells and 230 nm in cells from embryos suggesting a fundamental change in the structure of the filaments during chicken development (4). To study the synthesis and assembly of the spectrin and intermediate filament networks we made use of the observation that both spectrin subunits as well as vimentin are quantitatively retained in a Triton X-100 (TX-100) or hypotonic cytoskeletal fraction of erythrocytes (3,5). Continuous labeling studies as well as pulse-chase studies have revealed that assembly of the two filament networks occurs independently of each other. We have observed that newly synthesized vimentin first enters a TX-100 or hypotonically soluble pool before being assembled into filaments. Quantitation by immunoprecipitation or two-dimensional gel electrophoresis show that the vimentin soluble pool saturates rapidly while the radioactivity in cytoskeletal vimentin increases linearly. Pulse-chase studies reveal that soluble vimentin is converted to cytoskeletal vimentin with a half-life of approximately 7 min and with little evidence of turnover (6). The rapid saturation of the vimentin soluble pool suggests that the amount of soluble vimentin is not in great excess of cytoskeletal binding sites; hence, newly synthesized vimentin is rapidly incorporated into the cytoskeleton. These observations establish for the first time that vimentin containing intermediate filaments assemble posttranslationally from a soluble precursor pool and refute previous suggestions that vimentin nascent polypeptide chains assemble cotranslationally. The kinetics of synemin assembly are more complex than those of vimentin. They suggest that synemin first enters a soluble pool which initially exceeds the available cytoskeletal binding sites and accumulates in the cytoskeletal fraction only after a considerable lag of time and with little evidence of turnover. This evidence collectively suggests that both synemin and vimentin assemble posttranslationally onto filaments from soluble precursors and that the rate of vimentin assembly limits the rate of synemin assembly onto filaments (7). Immunoprecipitation studies indicate that cytoskeletal synemin and vimentin form a tight complex and coimmunoprecipitate, while soluble vimentin and synemin do not. This and other supportive evidence allows us to hypothesize that the posttranslational binding of synemin onto elongating vimentin filaments may require the generation of synemin binding sites as a result of vimentin filament elongation and a higher order change in vimentin structure (7).

Further characterization of intermediate filaments from embryonic erythroid cells has revealed that a 70,000 dalton polypeptide is a component of embryonic erythroid cell intermediate filaments but not adult erythrocyte intermediate filaments.

Iodopeptide mapping, immunoautoradiography and immunofluorescence with antibodies raised against the neurofilament 70,000 dalton subunit (NF70) have unambiguously established that the erythroid cell protein is highly homologous, if not identical to NF70 (8). This is the first time that any neurofilament subunit has been detected in a nonneuronal cell. Immunoelectron microscopy and immunofluorescence have suggested the following:

(1) E70 (NF70) and vimentin randomly copolymerize into intermediate filaments and that modulation of the ratio of these two subunits in a given erythrocyte is a function of their differential synthesis.

(2) The relative decline in the incorporation of E70 during development is responsible for the reduction in the linear periodicity of the association of synemin with the core filament; this would imply that addition of an E70 subunit to a vimentin polymer would lengthen the filament more than the addition of another vimentin subunit, thereby increasing the distance between successive synemin binding sites.

(3) Individual erythroid cells do not modulate their vimentin/E70 ratio during differentiation, but that this radio changes as the cell population changes during development and growth of the organism, culminating in the virtual absence of E70 from most cells in the adult.

These results emphasize that regulation of the variable ratios of intermediate filaments during development and in certain adult cells is regulated predominantly, if not exclusively, at the transcriptional and mRNA abundance level. Furthermore, since assembly occurs from a soluble pool for each subunit, a variable ratio of a copolymer can be achieved by different pool sizes of each subunit.

Similar studies on the mechanism by which chicken embryo erythroid cells achieve the stoichiometric assembly of α- and β-spectrin with their membrane-cytoskeleton during erythrophoresis have revealed the following: α-spectrin is synthesized in a threefold excess over β-spectrin, but only a small percentage of the newly synthesized α-spectrin molecules are incorporated into the membrane cytoskeleton. On the other hand, most of the newly synthesized β-spectrin is incorporated into the membrane cytoskeleton early in erythroid development, with progressively lesser proportions later in erythroid development (5). The results support a scheme of spectrin assembly in which the amount of β-spectrin synthesized and assembled in the cytoskeleton regulates the rate of assembly of stoichiometric amounts of α-spectrin. However, the stable assembly of stoichiometric amounts of α- and β-spectrin (and globin) with the membrane cytoskeleton is regulated by the availability of membrane receptor binding sites, most likely the anion transporter. Unassembled α-spectrin cannot stabilize unassembled β-spectrin and both are degraded coordinately. Thus assembly of both filamentous systems appears to occur posttranslationally from soluble precursor pools and to follow different regulatory pathways; in the case of intermediate filaments it is predominantly transcriptional, while in the case of spectrin it is posttranslational.

REFERENCES

1. Branton, D., Cohen, C. M., and Tyler, J. (1981) Cell 24, 24-32.
2. Granger, B. L., Repasky, E. A., and Lazarides, E. (1982) J. Cell Biol. 92, 299-312.
3. Repasky, E. A., Granger, B. L., and Lazarides, E. (1982) Cell 29, 821-33.
4. Granger, B. L. and Lazarides, E. (1982) Cell 30, 263-75.
5. Blikstad, I., Nelson, W. J., Moon, R. T., and Lazarides, E. (1983) Cell 32, 1081-91.
6. Blikstad, I. and Lazarides, E. (1983) J. Cell Biol. (In press).
7. Moon, R. T. and Lazarides, E. (1983) Proc. Natl. Acad. Sci. USA (in press).
8. Granger, B. L. and Lazarides, E. (1983) Science (in press).

2 Expression of Myofibrillar M-line Proteins during Differentiation

J. C. Perriard, B. K. Grove, L. Cerny,
M. Eppenberger, E. E. Strehler,
H. M. Eppenberger

The process of biogenesis of myofibrils is still elusive and, in order to gain more insight into myofibrillogenesis, it appears to be essential to learn more about the building blocks of the contractile apparatus. Thus the identification and possible characterization of the myofibrillar proteins that are incorporated into myofibrils are necessary. In addition, more information has to be gathered on the regulation of the synthesis and on the tissue specificity of myofibrillar proteins. Many of the most abundant myofibrillar proteins have been characterized thoroughly and the study of their interactions are well documented at the biochemical and structural level. Little, however, is known about the ontogeny of the contractile structure and what "minor" proteins are required to order these elements into an efficient contractile apparatus. Within the myofibril, there are two zones which are potentially of major importance for maintaining the order and ensuring the functioning of myofibrils. On one hand there is the Z-line, where thin filaments insert and where lateral connections to neighboring myofibrils and other organelles and to the cellular envelope are effected; on the other hand the M-line structure may have important tasks in keeping the thick filaments in register during the contraction cycle and ensuring the flux and regeneration of energy rich compounds (Eppenberger et al., 1983; Wallimann et al., 1984). Another role of the minor myofibrillar proteins which contribute to the Z-line and M-line may be their involvement in myofibril assembly as possible "nucleation centers" during muscle differentiation.

The M-region structure of cross-striated muscle has been thoroughly studied by electron microscopy and some of the transverse substriations have been postulated to be the site of primary and secondary crossbridges linking thick filaments to each other (Luther et al., 1981; Thornell and Sjöström, 1981). The M_4 M_4' cross-striations have been identified as part of the electron opaque M-line made up by MM-CK (Strehler et al., 1983; Wallimann et al., 1983). This bound MM-CK can rephosporylate "in vitro" ADP generated by Ca^{++} stimulated myosin ATPase activity in the presence of creatine phosphate (Eppenberger et al., 1983; Wallimann et al., 1984). Another protein was shown to be localized by immunohistochemistry in the M-line, the M-line protein with MW of 165,000 (Masaki and Takaiti, 1974). Polyclonal antibodies were generated against such an M-line protein fraction and immunological experiments demonstrated that the antibody bound to differentiated muscle cells but not to nonmuscle or undifferentiated myogenic cells (Eppenberger et al., 1981). They also served to localize muscle tissue in young chicken embryos in the somites and later in the limb buds, the future muscle masses being outlined by the staining pattern with M-line protein antibodies (Chiquet et al., 1981).

Monoclonal antibodies (mAb) were generated primarily to study more precisely the interaction of isolated thick filaments and M-line proteins. The immunogen used was a fraction of isolated M-line protein from chicken breast muscle purified to apparent homogeneity with an MW of 165,000 (Strehler et al., 1980). The resulting clones were screened with an enzyme linked immune sorbent assay (ELISA) and a number of positive clones were identified, seven of them were more thoroughly characterized (Grove et al., 1984). The monoclonal antibodies thus generated all bound to isolated breast muscle myofibrils within the M-line. No staining of the Z-line occurred (often seen with not affinity purified polyclonal antibody). Interestingly, antibodies from some clones required prior fixation and SDS treatment of the myofibrils in order to yield positive staining. Low ionic strength extraction of blycerinated myofibrils for extended periods of time resulted in drastically diminished staining of the M-line. If the extracted proteins were subjected to immunoblotting procedures and stained with the various antibodies, two distinct staining patterns could be observed. One group of antibodies A6, A5, A1, C2 bound to a protein band of the expected MW of 165,000 (165 K); however, another group of antibodies, namely B5, B4, A2, bound to a band of MW of 185,000 (185 K) exclusively. Using these new reagents and a refined column chromatography protocol with various precautions to inhibit proteolysis, the two species could also be enriched for on DE-52 ion exchange resin in two separate column peaks. Furthermore, immunocompetition experiments in which the binding of biosynthetically labeled monoclonal antibody to the immobilized antigen was challenged by unlabeled antibody, no competition was found between the mAb's against the MW 165 K and those against the 185 K proteins. In addition, the peptides generated by limited endogenous proteolysis subjected to immunoblotting showed distinctly different patterns after staining with the two sets of antibodies. All the evidence, therefore, demonstrates that mAb's binding to the MW 165,000 band bind to the M-protein previously described (Masaki and Takaiti, 1974), but the group of mAb's binding to the protein with an MW 185,000 recognized a new element so far not recognized as an M-line component which we called Myomesin (Eppenberger, 1981; Grove et al., 1983; 1984).

The question of the involvement of the two proteins Myomesin and M-protein in early events of myofibrillogenesis was addressed using first the technique of immunohistochemical staining of cryosections of breast muscle anlagen of 7- to 18-day-old chicken embryos. While the mAb directed against Myomesin showed distinct staining in a cross-striated pattern throughout all stages, the mAb against M-protein gave only a reaction with the tissue from 11 day old or older embryos (Grove et al., 1983). The qualitative analysis was then supplemented by data using quantitative ELISA assays, where it appears that M-protein is also detectable at earlier stages of development. It remains to be seen whether Myomesin is integrated earlier into myofibrils than M-protein and is therefore more readily detectable at these early stages. Myomesin and M-protein distribution was also studied in differentiating myogenic cell cultures of different stages. Generally it can be stated that Myomesin is stained more readily at the different stages of myogenic differentiation (Fig. 2.1a,c). M-protein is not easily if at all detectable in young myotubes (Fib. 2.1b) but both antigens are found in cells from pectoralis cultured for 5 days (Fig. 2.1c,d) and heart muscle (not shown). It had been found before that with polyclonal anti-M-line protein antibody shown to react with Myomesin and M-protein even mononucleated differentiating cells can be stained showing a pattern of cross-striated fluorescence (Eppenberger et al., 1981). A possible interpretation of this result could be that the protein detected in cells of early stages of myofibrillogenesis with the polyclonal antibody is probably also mainly Myomesin that is already incorporated into M-lines and thus is detectable more easily than M-protein. This may indicate the indispensability of Myomesin during early myofibrillogenesis for M-line formation. The M-protein, however, appears to be also specific

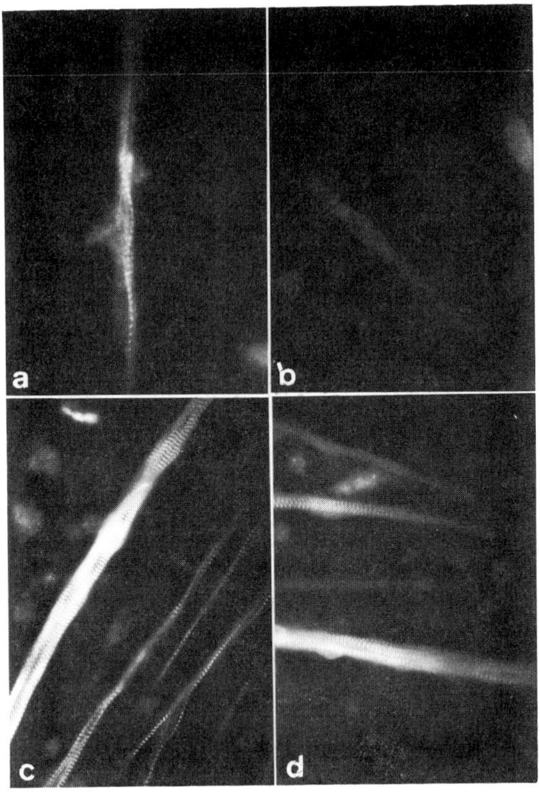

Fig. 2.1. Indirect immunofluorescence of cultured breast muscle cells stained with monoclonal antibodies against Myomesin and M-protein. Cultures of myogenic cells derived from 11 d embryonic breast muscle anlagen were fixed (0.5 min) with 20% ethanol, 1% acetic acid, 0.72% formaldehyde and permeabilized with a short (0.5 min) treatment with 0.2% Triton X-100 in PBS. Incubation with primary antibody (undiluted culture supernatant containing mAb) was carried out at room temperature for 45 min. After extensive washing with PBS, bound antibody was visualized with rhodamine labeled sheep antimouse immune globulin at a dilution of 1:100. In one day old cultures few cells showed cross-striations with mAb B5 (anti-Myomesin) but no striations with mAb (C2) (anti-M-protein). First striations with mAb C2 were apparent only in cells cultured for at least three days. (a,b): 24 h cultures; (c,d): 120 h cultures; (a,c): stained with mAb B5 (anti-Myomesin 185 K); (b,a): stained with mAb C2 (anti-M-protein 165 K).

for cross-striated muscle but might have a role different from the one of Myomesin.

If this hypothesis is correct then one would expect Myomesin to be present in the M-lines of all cross-striated muscle types. M-protein, however, would be expected in many but not necessarily all muscle types. ELISA measurements were made on extracts from various tissues for the presence of Myomesin and M-protein and relative amounts of the two proteins are compiled in Table 2.1. Myomesin and M-protein are present

Table 2.1 Relative amounts of Myomesin and M-protein in various chicken tissues.

Total extract from	Myomesin (185 K)	M-protein (165 K)
breast muscle	100%	100%
PLD	118%	118%
ALD	233%	2%
heart	93%	16%
thymus	7.8%	5.3%
gut	not detectable	not detectable
gizzard	not detectable	not detectable
liver	not detectable	not detectable

Note: Tissues were homogenized and solubilized in PBS containing 2% SDS and 1% 2-mercaptoethanol and adjusted to identical protein concentration and brought to 0.2% SDS, 0.1% 2-mercaptoethanol, 2% Triton X-100 and 10 mg/ml BSA. Myomesin and M-protein was determined in these extracts with a modified ELISA assay. Microtiter plates were coated with purified polyclonal rabbit antibody to chicken M-line protein fraction (reacts with both Myomesin and M-protein). Increasing amounts of extracts were incubated in the wells overnight, the wells were then washed and bound antigen was reacted with a mixture of mAbs B4 and B5 (for Myomesin) or with mAbs A5 and A6 (for M-protein). Bound mAb was detected by incubation with peroxidase labeled antimouse antibody as in regular ELISA. The resulting curves were standardized against breast muscle which was defined as 100% for both antigens.

in striated muscle containing tissues like breast muscle, leg muscle, posterior latissimus dorsii (PLD), and heart, but no significant amounts were found in smooth muscle containing tissue like gizzard and stomach or nonmuscle tissues (liver, brain, spleen). There is, however, the exception of a typical slow muscle of chicken, like the anterior latissimus dorsi (ALD), where the analysis shows a different distribution. The amount of Myomesin is higher than that found in fast cross-striated muscle but a significantly lower amount of M-protein is found. Immunohistological analysis of cross-sectioned material was carried out in collaboration with Dr. L. E. Thornell (University of Umeå, Sweden) and demonstrated the distribution of Myomesin in all ALD fibers, but M-protein was found only in a few fibers that had also a different ATPase, indicating the presence of fast fibers. The results with heart extracts resemble those of the slow muscle, but M-protein is distributed throughout the tissue. Thus, in all types of cross-striated muscle so far tested, Myomesin appears to be present.

If one assumes Myomesin to be indispensable for proper myofibrillogenesis and functioning of the contractile apparatus then the same or a similar protein would be expected to occur in tissue containing myofibrils from different species. Binding studies of various mAb to isolated myofibrils were carried out and showed the antibody secreted by clone B4 to stain the M-line of vertebrate myofibrils but not those of drosophila or crayfish. Similar results have been obtained for M-protein which was found in M-lines of vertebrate myofibrils. In competition ELISA assays binding of mAb B4 to the chicken antigen was challenged by muscle extracts of various sources including those from mammalian, bird, reptile, amphibian, fish and crayfish muscle. All extracts except fish and crayfish extracts inhibited B4 binding to the chicken antigen with very similar kinetics as shown in Fig. 2.2. This may indicate that mAb B4 recognizes a strongly conserved antigenic determinant present on most vertebrate Myomesins. Fish and crayfish extracts do not compete in these assays. The fish protein has probably a much lower affinity for B4 and thus cannot compete with the chicken antigen. It is not clear if invertebrate myofibrils have an M-line organization like that of vertebrate M-line and the question remains which proteins, if any of this type exist, take the place of Myomesin and MM-CK. Arginine kinase, for instance, the phosphagen kinase of invertebrates, could not be found in the M-line of drosophila myofibrils (Lang et al., 1980).

The mAb B4 can be used to trace the development and the stability of the M-line region in myogenic cells from hamster

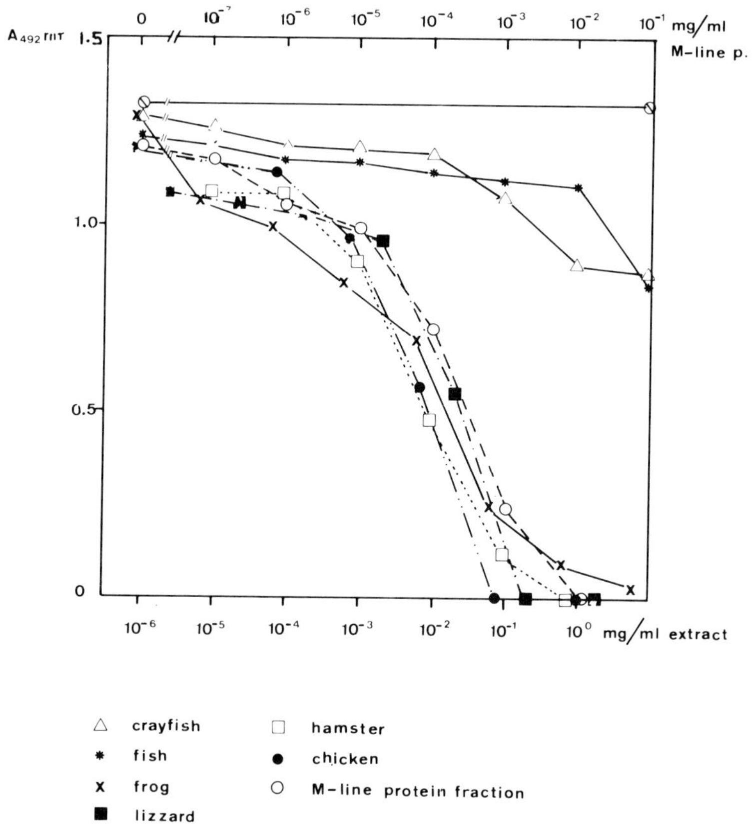

Fig. 2.2. Competition of mAb B4 (anti-Myomesin) in ELISA assays by various vertebrate and crayfish muscle extracts. Microwells of ELISA assay plates were coated with 10 μg of purified M-line protein fraction containing both Myomesin and M-protein. 50 μl mAb B4 (diluted 1,000x in PBS) was pre-incubated with increasing amounts of extract (indicated on lower abscissa) or with M-line protein fraction (upper abscissa) and then transferred to the coated microwells. The antibody not complexed bound to antigen coated on the wells. After extensive washing antibody bound to wells was revealed by incubation with horse raddish peroxidase conjugated rabbit antimouse immunoglobulin diluted 1:1000. Bound peroxidase activity was determined after incubation with o-phenylenediamine and H_2O_2 at 492 nm in a Dynatech ELISA reader.

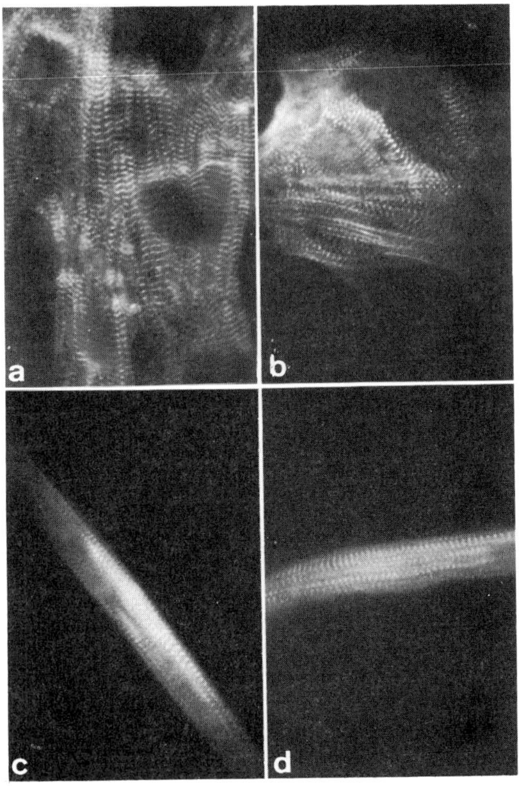

Fig. 2.3. Indirect immunofluorescence of primary hamster muscle cell cultures stained with mAb B4 (anti-Myomesin). Skeletal muscle cells were fixed in 3% paraformaldehyde in PBS containing 0.1 mM $CaCl_2$, 1.0 mM $MgCl_2$, washed in PBS and quenched overnight in PBS with 0.1 M glycine and finally permeabilized with 0.2% Triton X-100 in PBS for 5 min. Cultures were stained with undiluted mAb B4 culture supernatant. Bound antibody was then revealed by incubation with appropriately diluted secondary rhodamine labeled rabbit antimouse immunoglobulin and viewed with epifluorescence optics. (a): 6 d old heart culture from control animals. (b): 6 d old heart culture from dystrophic animals (Bio 14.6). (c): 6 d old skeletal muscle culture from control animals. (d): 6 d old skeletal muscle culture from dystrophic animals (Bio 14.6).

carrying an autosomal defect leading to muscular dystrophy in heart and skeletal muscle (Homburger et al., 1962; 1966). As shown in Fig. 2.3 the M-lines in cell cultures from controls and from diseased animals are clearly stained in heart as well as in cells from skeletal muscle (M. Eppenberger et al., 1984). There was, however, no indication of a difference in alignment and morphological appearance of myofibrils in cultured cells from heart or skeletal muscle if controls were compared to cells from diseased animals. It is possible that cultivation periods will have to be extended in order to see the dystrophic phenotype to appear.

The function of these proteins still cannot be defined and besides the studies on their developmental behaviour more information will be required on the exact structural localization within the M-line. Using polyclonal antibody against M-line proteins decoration of the M6, M6' substriations as well as longitudinal decoration was observed in the M-region (Strehler et al., 1983). So far none of the known M-line proteins seem to be obviously associated with the central substriation M1. In addition, there might exist other minor components not yet discovered that may be important for the organization of thick filaments in the M-region.

ACKNOWLEDGMENTS

This work was supported by SNF-grant No. 3.707-0.80 and by a grant from the MDA.

REFERENCES

Chiquet, M., Eppenberger, H. M., and Turner, D. C. (1981) Dev. Biol. 88, 220.

Eppenberger, H. M., Perriard, J. C., Wallimann, T. (1983) In "Isozymes," Current Topics in Medical and Biological Research. Vol. 7: Molecular Structure and Regulation. p. 19. Alan R. Liss, New York.

Eppenberger, H. M., Perriard, J. C., Rosenberg, U., and Strehler, E. E. (1981) J. Cell Biol. 89, 185-93.

Eppenberger, M. E., Schoenenberger, R., and Eppenberger, H. M. (1984) Muscle and Nerve (in press).

Grove, B. K., Kurer, V., Lehner, C., Doetschman, T. C., Perriard, J. C., and Eppenberger, H. M. (1984) J. Cell Biol. (in press).

Grove, B. K., Kurer, V., Lehner, C., Perriard, J. C. and Eppenberger, H. M. (1983). In: Exptl. Biol. Med. Vol. 9 (in press), S. Karger, Basel.

Homburger, F., Nixon, C. W., Eppenberger, M., Baker, J. R. (1966). Ann. N.Y. Acad. Sci. 138, 14-27.

Homburger, F., Baker, J. R., Nixon, C. W., Whitney, R. (1962) Med. Exp. 6, 339-45.

Lang, A. B., Wyss, C., and Eppenberger, H. M. (1980) J. Muscle. Res. Cell Mot. 1, 147-61.

Luther, P. K., Munro, P. M., and Squire, J. M. (1981). J. Mol. Biol. 151, 703-30.

Masaki, T., and Takaiti, O. (1974). J. Biochem. (Tokyo) 75, 367.

Strehler, E. E., Carlson, E., Eppenberger, H. M., and Thornell, L. E. (1983). J. Mol. Biol. 160, 141-58.

Strehler, E. E., Pelloni, G., Heizmann, C. W., and Eppenberger, H. M. (1980). J. Cell Biol. 86, 775-83.

Thornell, L. E., and Sjöström, M. (1981). J. Microsc. 104, 263-69.

Wallimann, T., Schlösser, T., Eppenberger, H. M. (1984). J. Biol. Chem. (in press).

Wallimann, T., Doetschman, T. C., and Eppenberger, H. M. (1983). J. Cell Biol. 96, 1771-79.

3 Structure and Function of the Erythrocyte Membrane Skeleton

Vincent T. Marchesi

Red blood cells have an elaborately engineered multiprotein complex that performs a variety of functions. This protein ensemble, generally referred to as the membrane skeleton (1), confers stability on the lipid bilayer, is responsible for plastic deformation, regulates topography of surface receptors via attachment to their cytoplasmic segments, and probably also plays an important role in transmembrane signaling (2-8).

The membrane skeleton is composed of three distinct but interrelated components; these include: (1) supporting network, (2) linking-proteins, and (3) attachment sites on transmembrane glycoproteins. The proteins that make up these structures are shown in Figure 3.1.

THE SUPPORTING NETWORK

A supporting network underlies the erythrocyte membrane that is made up of spectrin and actin. The major component on a mass basis is spectrin, a water-soluble protein easily extracted from red cell membranes in low ionic strength buffers that appears as an elongated, noodlelike form when examined by low angle shadowing and electron microscopy (9). An example of this is shown in Figure 3.2. Analysis of the spectrin molecule by SDS gel electrophoresis shows that it is composed of two similar but nonidentical subunits. Both subunits are extremely large, the α subunit is 240 KD and the β subunit is approximately 225 KD. Domains of each of the spectrin subunits have been defined by restricted proteolytic cleavage experiments and analyzed by high-resolution peptide mapping techniques (10-12). These domains, designated α-I, II, III, IV, and V and β-I, II, III, and IV are structurally unique

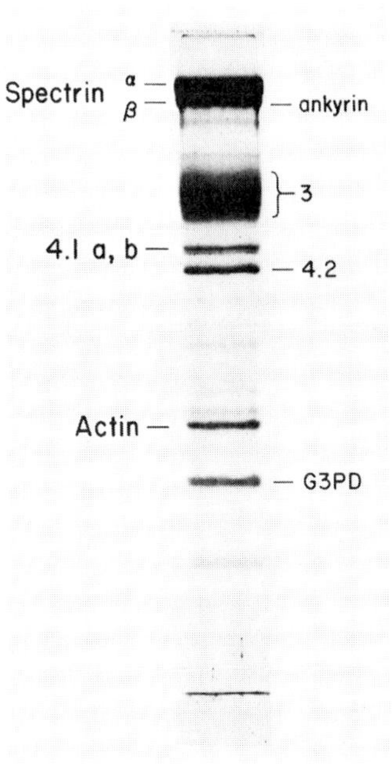

Fig. 3.1. SDS PAGE of the major polypeptides of the human red cell membrane.

segments, and some also carry specific functional sites as described below.

Spectrin Has Remarkable Powers of Self-Association

Purified spectrin molecules exist in solution in many different forms. The individual subunits, generally referred to as monomers, rarely exist as such under nondenaturing conditions but instead form extremely stable dimers. Spectrin dimers have the capacity to self-associate to form tetramers (13-17) and multiple higher oligomeric forms (18,19). The capacity to self-associate into oligomeric units is a property inherent in the spectrin molecule and can occur without the collaboration of any other protein. The formation of oligomeric units of spectrin is a mass action driven process that appears to proceed through a single common pathway. Spectrin dimers form tetrameric units by joining together at one specific end through noncovalent interactions between the terminal portions of each subunit; each α subunit links to a corresponding β unit.

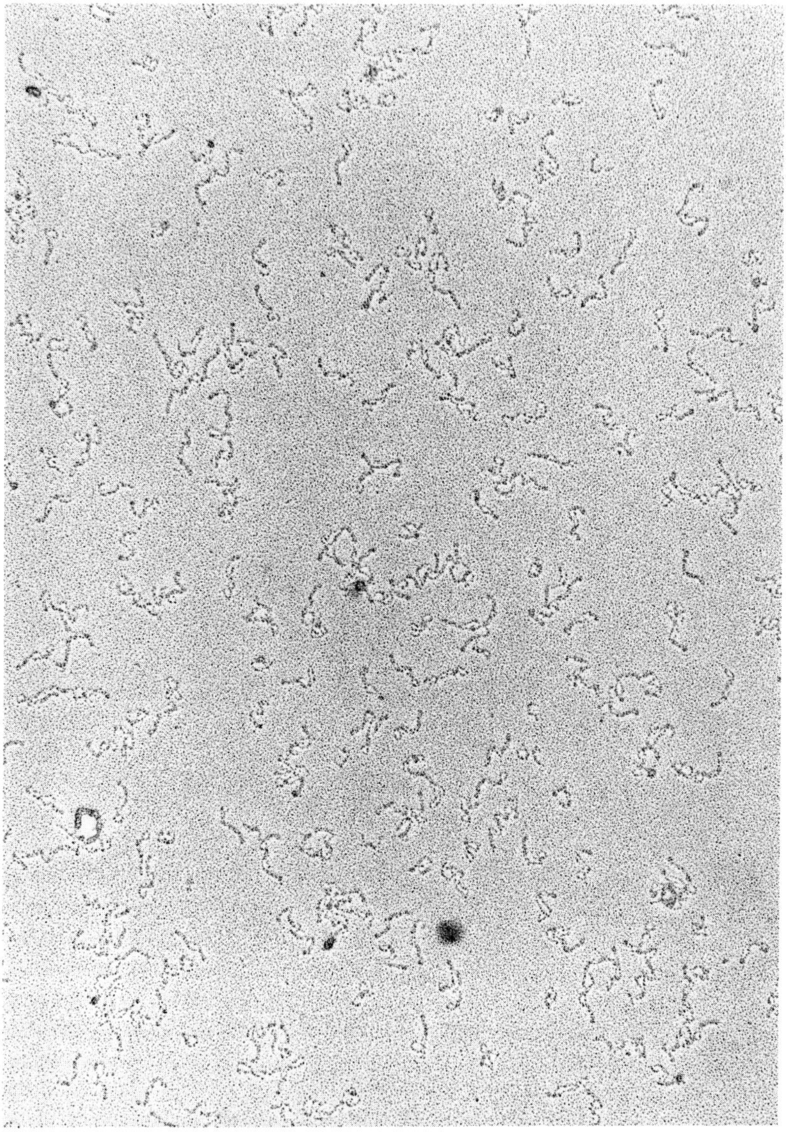

Fig. 3.2. Electron micrograph of human erythrocyte spectrin prepared by low angle shadowing.

Spectrin Has Multiple Functional Sites

The spectrin molecule has multiple functional sites that are distributed at specific positions along each of the subunits (20,21). The self-assembly process described above seems to be mediated through interactions between a terminal segment of an α subunit with a counterpart segment of the β subunit. Spectrin tetramers also have binding sites for at least two linking-proteins, to be described below, and in addition sites for the binding of calmodulin, a possible regulatory protein (22,23). A highly schematic and very provisional model of these functional sites is illustrated in Figure 3.3. Sites of intrinsic phosphorylation of the spectrin molecule have also been localized to the terminal end of the β chain proximal to the self-assembly site (24,25).

Amino acid sequencing studies of different parts of both spectrin subunits have revealed some interesting insights into the functional anatomy of the molecule. The amino acid sequences of both the α-I and α-II domains have been completed (26,27), and predictions of their conformation suggest that each domain is composed of multiple α helical loops. The simplest interpretation of these findings is that each spectrin subunit is composed of multiple linear arrays of repeating triple helices. The amino acid sequence results also reveal another remarkable finding: both spectrin subunits seem to be composed of multiple copies of a single homologous segment of 106 amino acids. Preliminary evidence obtained from sampling all nine domains suggests that this may be a characteristic feature of the entire spectrin molecule.

This remarkable finding suggests that perhaps the present-day erythrocyte spectrin molecule may be the product of multiple gene duplications. Based on this idea one might speculate that the 106 amino acid repeat, which has been duplicated multiple times, folds into a structural unit that is particularly well adapted to form a supporting network for the red cell membrane. Since there is abundant evidence that spectrins also exist in nonerythroid cells (28-33), it is likely that they will share this same characteristic structural feature.

Actin May Also Exist in the Form of Short Oligomers

Although it has been known for many years that actin is associated with human red blood cell membranes (34), the size and state of the membrane-bound form is still unclear. Recent

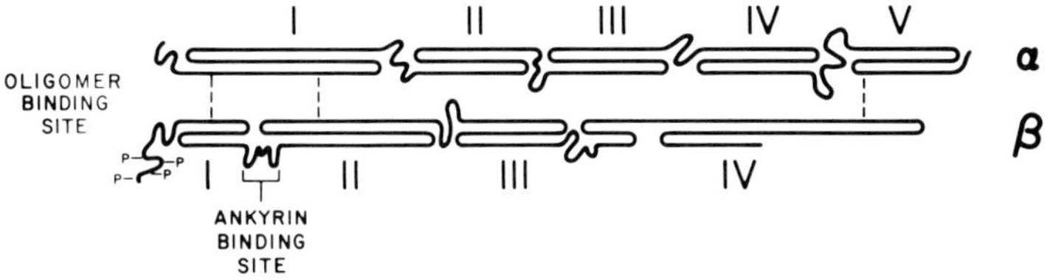

Fig. 3.3. A schematic diagram illustrating the domain structure of human erythrocyte spectrin and some specific functional sites.

evidence from several laboratories suggests that this actin may exist in the form of short oligomeric units (35-38). When actin is extracted from phalloidin-stabilized red cell ghosts that have been digested with proteolytic enzymes it appears to be approximately 1,000 Å, about the length of a spectrin dimer (37). Spectrin tetramers attach to filamentous forms of muscle actin at their free ends (39). This observation has led to the suggestion that spectrin and actin interact through noncovalent associations between ends of the spectrin molecules that are opposite from the self-assembly sites. If erythrocyte spectrin binds to erythrocyte actin in the same way such spectrin-actin complexes could take on many forms. One possibility is that large spectrin oligomers could be tethered together by short actin oligomers. Spectrin-actin complexes could be further modified by the action of a recently described actin bundling protein (40), the so-called band 4.9, or by the presence of an erythrocyte form of tropomyosin (41).

The bulk of the evidence favors the idea that spectrin and actin join together to form the supporting framework of the membrane skeleton. How these two proteins actually do this is still quite unclear, as is the role that other proteins may play in regulating their actions. These are critical questions that deserve more intensive study.

THE LINKING-PROTEINS 2.1 AND 4.1 PLAY A PIVOTAL ROLE

Proteins 2.1 and 4.1 bind to the spectrin-actin network, and by so doing link this network to the overlying cell membrane. Protein 2.1, first described by Bennett and coworkers and named Ankyrin (42,43) binds to the β subunit of spectrin at a region that spans the first and second domains (20,44). Ankyrin is a 210 KD polypeptide that is also composed of multiple domains that can be defined by limited proteolytic cleavage and high resolution peptide mapping (45).

Ankyrin binds to the β subunit of spectrin through a high-affinity link with the end of the molecule containing the 55 KD peptide. This segment of ankyrin is also phosphorylated, and it can be fractionated further into a 32 KD segment that is phosphorylated and has the spectrin-binding site.

Ankyrin links spectrin to the cytoplasmic segment of band 3 (46-48), the most abundant transmembrane glycoprotein of the human red cell and the site of anion exchange across the membrane. This function has been localized to the 82 KD peptide illustrated above. These results indicate that ankyrin is a multidomain protein with at least two structurally distinct specialized sites that are required for its interaction between spectrin and band 3. The peptide analysis described above also confirms that ankyrin is composed of a single polypeptide chain rather than a family of closely related molecules. Since ankyrinlike molecules have been described in many cell types in addition to erythrocytes (49), it is likely that ankyrinlike molecules represent a class of linking proteins that are active components of all membrane skeletal complexes.

The human red blood cell membrane skeleton has a second linking protein, operationally referred to as band 4.1, and this molecule has a number of interesting features. Protein 4.1 is composed of two similar subunits (50) molecular weights 80 and 78 KD), and this molecule has the potential to link the spectrin-actin network to the inner surface of the membrane via noncovalent associations with two different transmembrane glycoproteins. Limited proteolytic digestion experiments and specific chemical cleavages were applied to the study of protein 4.1, and a series of intermediate-sized peptides were generated that were found to represent specific structural and functional domains of the original molecule (51).

The 4.1 molecule displays unusual polarity in that each subunit contains a 30 KD peptide that is relatively basic and appears to be resistant to further digestion by trypsin or chymotrypsin. Peptides derived from the remainder of the 4.1 subunit are more acidic and are readily degraded into smaller fragments. Peptides derived from the acidic end of protein 4.1 appear as doublets and have identical peptide maps, suggesting that the two subunits of 4.1 differ from each other by approximately 2,000 daltons close to the terminus of the acidic end. The 30 KD basic peptide of protein 4.1 is particularly interesting since it is both rich in cysteine and contains a spectrin-binding site. The binding of this 30 KD peptide of 4.1 to spectrin is salt dependent but of relatively low affinity, and is inhibitable by mercurials. Protein 4.1 is also phosphorlated, and sites of phosphorylation have been identified.

Earlier studies had suggested that protein 4.1 binds to the inner surface of the human red cell membrane through associations with glycophorin A, the major sialic acid-containing transmembrane glycoprotein of the human red cell (52). Recently we have found that soluble glycophorin, isolated by the Lis-phenol method, also binds protein 4.1 (53). The specific association between protein 4.1 and isolated blycophorin is abolished if glycophorin preparations are first extracted with acidified chloroform-methanol under conditions that remove all tightly bound lipids including polyphosphoinositides. Earlier studies indicated that both phosphatidylinositol-4,5-diphosphate (TPI) and phosphatidylinositol-4-phosphate (DPI) coisolated with glycophorin prepared by the Lis-phenol method (54). The capacity of delipidated glycophorin to bind protein 4.1 is restored if such blycophorin preparations are reconstituted with either DPI or TPI. On closer examination it appears that glycophorin reconstituted with TPI binds protein 4.1 with higher affinity than preparations reconstituted with DPI, suggesting that both monoester phosphates contribute to the activation of the blycophorin molecule. Protein 4.1 does not associate with either TPI or DPI in the absence of glycophorin.

Many previous studies reported by others suggest that protein 4.1 plays an important role in linking the spectrin-actin complex to the inner surface of the cell membrane (55-57), and it is likely that the polyphosphoinositides that are bound to glycophorin modulate the association of 4.1 with the overlying membrane, possibly through their states of phosphorylation. A recent report suggests that TPI increases the mobility of transmembrane glycoproteins when added to erythrocyte membranes (58). Such an effect could be related in part to the effects of TPI on protein 4.1 functions.

ATTACHMENT SITES ON TRANSMEMBRANE GLYCOPROTEINS

The spectrin-actin supporting network is linked indirectly to the overlying lipid bilayer by the linking proteins which bind to the cytoplasmic segments of both band 3 and the glycophorins. Ankyrin links the β subunit of spectrin to the 43 K peptide of band 3, but neither the site on band 3 nor the properties of this association have yet been defined. The ankyrin-band 3 connection is disrupted by high salt treatments, suggesting that electrostatic interactions play an important role.

The association between protein 4.1 and glycophorin is particularly interesting since it now appears that the 4.1 "receptor" is a lipid-protein complex composed of the cytoplasmic segment of glycophorin and equimolar amounts of polyphosphoinositides. Both TPI and DPI "activate" glycophorin (TPI-glycophorin provides higher affinity binding) but phosphatidyl inositol (PI) and other phospholipids (PE, PS, PC) are inactive. The TPI bound to glycophorin is associated primarily with a peptide fragment of glycophorin that includes the intramembranous hydrophobic domain and a short segment of the C-terminal part of the molecule that is rich in basic amino acids (54).

This clustering of basic amino acids seems to be ideally placed to interact with the negatively charged head groups of the polyphosphoinositides while also allowing the hydrophobic fatty acid chains of the inositides to associate with the nonpolar segments of the transmembrane proteins. The widespread occurrence of this relatively unique structural feature suggests that its function may be shared by a number of different transmembrane glycoproteins. It would be particularly exciting if transmembrane glycoproteins that share this structural feature also bind polyphosphoinositides and become receptors for membrane skeletal proteins.

Studies now in progress provide another wrinkle of complexity: Protein 4.1 also has the capacity to bind to the cytoplasmic segment of band 3 (59). The apparent affinity of the binding of 4.1 to band 3 is weaker than that of 4.1 to the glycophorin PPI complex, thus 4.1 binds preferentially to glycophorin but only if the latter is complexed with either TPI or DPI. These results raise the possibility that 4.1 may oscillate between these two different binding sites depending upon the metabolic state of the cell.

Activation of many different physiological receptors by their appropriate ligands causes striking changes in the degree of phosphorylation of the phosphoinositides (60-66), and this is believed to be an important link in the transmembrane signaling cascade that occurs in response to specific ligands. Since dephosphorylation of TPI markedly reduces the capacity of glycophorin to bind 4.1, such changes in phospholipid metabolism could cause significant shifts in the attachment of the membrane skeletal complex to the membrane. This may be one of the mechanisms responsible for the ligand-induced shape change in red cells reported earlier (67), and it is conceivable that a similar mechanism may be operative in all cells. The factors that regulate the linkage of the membrane skeletal complex to the overlying cell membrane clearly represent a fruitful area for further investigation.

REFERENCES

1. Marchesi, V. T. (1983) Blood 61:1.
2. Branton, D., Cohen, C. M., and Tyler, J. (1981) Cell 24:24.
3. Bennett, V. (1982) J. Cell Biochem. 18:49.
4. Lux, S. E. 1979) Semin. Hematol. 16:21.
5. Sheetz, M. P. (1979) Biochim. Biophys. Acta 557:122.
6. Tsukita, S., Tsukita, S., and Ishihawa, H. (1980) J. Cell Biol. 85:567.
7. Hainfeld, J. F. and Steck, T. J. (1977) J. Supramol. Structr. 6:301.
8. Goodman, S. R. and Shiffer, K. (1983) Am. J. Physiol. 244:C121.
9. Shotton, D. M., Burke, B. E., and Branton, D. (1979) J. Mol. Biol. 131:303.
10. Speicher, D. W., Morrow, J. S., Knowles, W. J., and Marchesi, V. T. (1980) Proc. Natl. Acad. Sci. 77:5673.
11. Speicher, D. W. and Marchesi, V. T. (1982) J. Cell Biochem. 18:479.
12. Speicher, D. W., Morrow, J. S., Knowles, W. J., and Marchesi, V. T. (1982) J. Biol. Chem. 257:9093.
13. Shotton, D., Burke, B., and Branton, D. (1978) Biochim. Biophys. Acta 536:313.
14. Ungewickell, E. and Gratzer, W. (1978) Eur. J. Biochem. 88:379.
15. Goodman, S. R. and Weidner, S. A. (1980) J. Biol. Chem. 255:8082.
16. Ji, T. J., Kiehm, D. J., and Middaugh, C. R. (1980) J. Biol. Chem. 255:2990.
17. Liu, S-C. and Palek, J. (1980) Nature 285:586.
18. Morrow, J. S. and Marchesi, V. T. (1981) J. Cell Biol. 88:463.
19. Morrow, J. S., Haigh, W. B., and Marchesi, V. T. (1981) J. Supramol. Structr. 17:275.
20. Morrow, J. S., Speicher, D. W., Knowles, W. J., Hsu, C. J., and Marchesi, V. T. (1980) Proc. Natl. Acad. Sci. USA 77:6592.

21. Knowles, W. J., Marchesi, S. L., and Marchesi, V. T. (1983) Sem. Hem. 20:159.
22. Sobue, K., Muramoto, Y., Fujita, M., and Kakiuchi, S. (1981) Biochem. Biophys. Res. Commun. 100:1063.
23. Sears, D., Morrow, J., Speicher, D., and Marchesi, V. T. Manuscript in preparation.
24. Harris, H. W. and Lux, S. E. (1980) J. Biol. Chem. 255:11512.
25. Speicher, D. W., Morrow, J. S., Knowles, W. J., and Marchesi, V. T. (1982) J. Biol. Chem. 257:9093.
26. Speicher, D. W., Davis, G., Yurchenco, P. D., and Marchesi, V. T. (1983) J. Biol. Chem. (In press).
27. Speicher, D. W., Davis, G., and Marchesi, V. T. (1983) J. Bio. Chem. (In press).
28. Goodman, S. R., Zagon, I. S., and Kulikowski, R. R. (1981) Proc. Natl. Acad. Sci. USA 78:7570.
29. Bennett, V., Davis, J., and Fowler, W. E. (1982) Nature 299:126.
30. Repasky, E. A., Granger, B. L., and Lazarides, E. (1982) Cell 29:821.
31. Glenney, J. R., Glenney, P., and Weber, K. (1982) J. Biol. Chem. 257:9781.
32. Glenney, J. R., Glenney, P., Osborn, M., and Weber, K. (1982) Cell 28:843.
33. Lazarides, E. and Nelson, W. J. (1983) Science 220:1295.
34. Oshnishi, T. (1962) J. Biochem. (Tokyo) 52:307.
35. Tilney, L. G. and Detmers, P. (1975) J. Cell Biol. 66:508.
36. Brenner, S. L. and Korn, E. D. (1980) J. Biol. Chem. 255:1670.
37. Atkinson, M. A. L., Morrow, J. S., and Marchesi, V. T. (1982) J. Cell Biochem. 18:493.
38. Pinder, J. C. and Gratzer, W. B. (1983) J. Cell Biol. 96:768.
39. Cohen, C. M., Tyler, J. M., and Branton, D. (1980) Cell 21:875.
40. Siegel, D. L. and Branton, D. (1982) J. Cell Biol. 95:265a.
41. Fowler, V. and Bennett, V. Personal communication.
42. Bennett, V. and Stenbuck, P. J. (1979) J. Biol. Chem. 254:2533.
43. Bennett, V. and Stenbuck, P. J. (1980) J. Biol. Chem. 255:2540.
44. Tyler, J. M., Reinhardt, B. N., and Branton, D. (1980) J. Biol. Chem. 255:7034.

45. Weaver, D. C. and Marchesi, V. T. Manuscript in preparation.
46. Bennett, V. and Stenbuck, P. J. (1979) Nature 280:468.
47. Hargreaves, W. R., Giedd, K. N., Verkleij, A., and Branton, D. (1980) J. Biol. Chem. 255:11965.
48. Nigg, E. A. and Cherry, R. J. (1980) Proc. Natl. Acad. Sci. USA 77:4702.
49. Bennett, V. (1979) Nature 281:597.
50. Goodman, S. R., Yu, J., Whitfield, C. F., Culp, E. N., and Posnak, E. J. (1982) J. Biol. Chem. 257:4564.
51. Leto, T. and Marchesi, V. T. Manuscript in preparation.
52. Anderson, R. A. and Lovrien, R. (1982) Fed. Proc. 41: 513.
53. Anderson, R. A. and Marchesi, V. T. Manuscript in preparation.
54. Armitage, I. M., Shapiro, D. L., Furthmayr, H., and Marchesi, V. T. (1977) Biochem. 16:1317.
55. Ungewickell, E., Bennett, P. M., Calvert, R., Ohanian, V., and Gratzer, W. B. (1979) Nature 280:811.
56. Cohen, C. M. and Korsgren, C. (1980) Biochem. Biophys. Res. Commun. 97:1429.
57. Fowler, V. and Tayler, D. L. (1980) J. Cell Biol. 85:361.
58. Sheetz, M. P., Febbroriello, P., and Koppel, D. E. (1982) Nature 296:91.
59. Pasternack, G., Leto, T., Anderson, R. A., and Marchesi, V. T. Manuscript in preparation.
60. Billah, M. M. and Lapetina, E. G. (1982) J. Biol. Chem. 257:12705.
61. Imai, A., Nakashima, S., and Nozawa, Y. (1983) Biochem. Biophys. Res. Commun. 110:108.
62. Rhodes, D., Prpic, V., Exton, J. H., and Blackmore, P. F. (1983) J. Biol. Chem. 258:2770.
63. Thomas, A. P., Marks, J. S., Coll, K. E., and Williamson, J. R. (1983) J. Biol. Chem. 258:5716.
64. Agranoff, B. W., Murthy, P., and Seguin, E. B. (1983) J. Biol. Chem. 258:2076.
65. Harrington, C. A. and Eichberg, J. (1983) J. Biol. Chem. 258:2087.
66. Billah, M. M. and Lapetina, E. G. (1983) Proc Natl. Acad. Sci. USA 80:965.
67. Anderson, R. A. and Lovrien, R. (1980) J. Cell Biol. 85:534.

4 Microfilaments in Theca Interna Cells of Mouse Ovary: Immunological and Ultrastructural Observations

Carlo Cavallotti, Guiseppe Familiari,
Elisabetta Simongini

INTRODUCTION

Three different cell types are recognized in the theca interna of secondary and graafian follicles of mammals; these are fibroblastlike cells, steroid-secreting cells, and transitional cells. Their ultrastructure has been extensively studied, but the presence of microfilaments in these three types of cells was not functionally investigated (Hiura and Fujita, 1977; O'Shea et al., 1978; Familiari and Mota, 1979; Familiari et al., 1981).

Several studies have demonstrated the presence of contractile proteins such as myosin, actin, or an actomyosin complex in several nonmuscular cell types (Hitchock, 1977). In addition the presence of contractile proteins in the ovary has been revealed in the rabbit (Cavallotti et al., 1975a and b) and in the rat (Amsterdam et al., 1977). Furthermore, in the last years there has been a renewal of interest in the contention that ovarian contractions are necessary for mammalian ovulation (Espey, 1978).

The purpose of the present report has been to study the presence of microfilaments and myosinlike protein in theca interna cells of mouse ovarian graafian follicles. The immunochemical and ultrastructural results were compared and the possible significance of these findings was discussed.

MATERIALS AND METHODS

Mature mice, housed in air conditioned rooms at 20°C with natural lighting and with lab chow and water available ad libitum, were used in this study. Each animal was anesthetized with ether. The ovaries were removed rapidly and the animals subsequently were killed by an overdose of ether.

Electron Microscopy

Ovaries were fixed in 3.0% glutaraldehyde plus 1.0% paraformaldehyde in 0.1 M Cacodylate buffer at pH 7.4.

The specimens were then washed and postfixed in 1.0% osmium tetroxide in the same buffer. The tissues were dehydrated in a graded acetone series and enbedded in EPON 812.

Thin sections were cut on an LKB ultramicrotome and stained with uranyl acetate and lead citrate (Reynolds, 1963) before viewing in a Zeiss EM9A electron microscope.

Immunochemistry

The specimens were washed in an isotonic sodium sulphate solution, frozen in isopentane, cooled in liquid nitrogen, and cut in several sections of 4-6 μm in thickness on a cryostat at a temperature of -20°C. The sections were placed on slides and air-dried. Some sections were fixed in chloroform-methanol at 4°C and washed in phosphate buffered saline (PBS, pH 7.4). Antimyosinlike protein antibodies (AMA) were obtained and purified in our laboratory as previously described (Cavallotti et al., 1975a and b; Melis et al., 1977).

AMA were bound with C14 cystein (Amersham, England), in an irreversible way, in the presence of 8 M urea (Porcelli et al., 1973) or conjugated with fluoresceinisothiocyanate (Sigma). Fixed and unfixed sections were incubated in a moist chamber at 25°C for 30 min in 1/40 labeled (or fluorescent) AMA dissolved in PBS.

The specificity of the reaction was tested by the treatment of the sections with nonlabeled (or nonfluorescent) AMA and their subsequent exposition to labeled (or fluorescent) AMA, or by the treatment of the sections with labeled (or fluorescent) AMA denaturated or previously adsorbed with the myosinlike protein. For the autoradiographic treatment the sections were coated by dipping with Kodak NTB 2 emulsion diluted 2:1. After 20 days of exposure these autoradiographs were developed in D19 Kodak, stained with toluidine blue and examined in the light microscope Zeiss.

The fluorescence observations were carried out on a Zeiss UV photomicroscope with UG1 or UG5 excitor filters and Zeiss

41-47/65 barrier filter using Ilford HP5 black and white panchromatic safety film.

OBSERVATIONS

Electron Microscopy

The theca interna of mouse graafian follicles is characterized by three cellular types: fibroblastlike cells, steroid-secreting cells, and intermediate cells. The fibroblastlike cells have rough endoplasmic reticulum, Golgi apparatus, mitochondria with lamellar cristae, few lipid droplets, and free ribosomes. On the contrary the theca gland cells are provided with large ellipsoidal nuclei, abundant lipid droplets, mitochondria with tubular cristae, and smooth endoplasmic reticulum. The transitional cells have oval nuclei, many lipid droplets, mitochondria with tubular or lamellar cristae, and small amounts of smooth endoplasmic reticulum (Fig. 4.1a, b, c).

All three thecal cellular types of mouse graafian follicles contain a considerable number of microfilaments measuring 40 to 70 Å in diameter. The microfilaments appear particularly concentrated in discrete parallel bundles in the cortical areas of the cytoplasm and within microvilli and ameboid evaginations of the cells (Fig. 1a, b, c).

Immunochemistry

(a) Fluorescence microscopy

A specific fluorescence has been observed in the follicular wall. The fluorescence is localized both at the level of the thecal cells and in the corresponding level of the granulosa layer of developing follicles. The intensity of fluorescence is decreased in theca interna cells compared with granulosa cells (Fig. 4.2a). At higher magnification theca interna cells show clearly the presence of fine diffuse fluorescence interpreted as corresponding to microfilaments observed in their cytoplasm with the electron microscope (Fig. 4.2b).

(b) Autohistoradiography

The distribution of silver grains exhibited the same pattern as the fluorescence microscopy. A diffuse autoradiographic reaction has been observed on the follicular wall

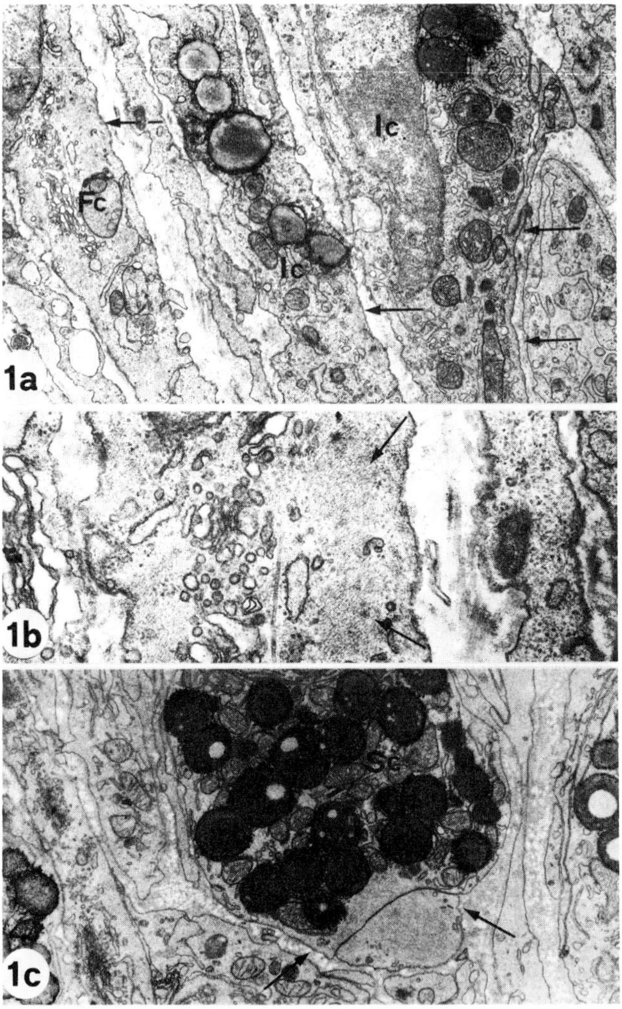

Fig. 4.1. Electron microscopy. Theca interna cells.

(a) Fibroblastlike cells (Fc). Intermediate cells (Ic). Note the presence of microfilaments in these cells (arrows) (12,000×).

(b) Bundles of microfilaments are located in the cytoplasm of fibroblastlike cells (arrows) (36,000×).

(c) Steroid-secreting cell (Sc). Microfilaments are present in the cytoplasm of this type of cell (arrows) (12,000×).

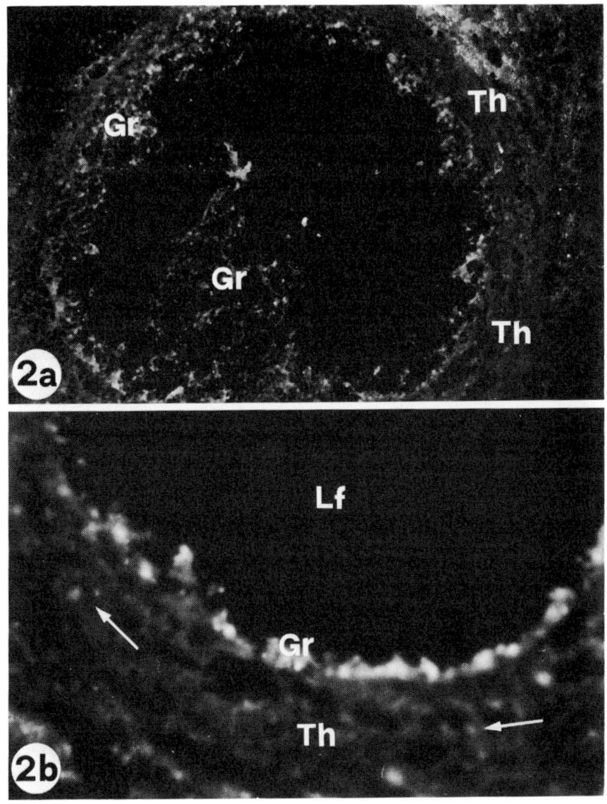

Fig. 4.2. Immunofluorescence. Forzen section of mouse ovary treated with fluorescent AMA.
(a) Cavitary follicle showing a discrete fluorescent activity in granulosa (Gr) and theca interna cells (Th) (180×).
(b) Note the strong fluorescent layer on granulosa cells (Gr), and the fine, diffuse fluorescence (arrows) in theca interna cells (Th) (470×).

(Fig. 4.3a). Silver grains are less numerous on theca interna cells if compared to those observed in granulosa cells (Fig. 4.3b).

The immunofluorescence and the autoradiographic reaction are less intense in chloroform-methanol fixed ovaries. Control incubations are negative.

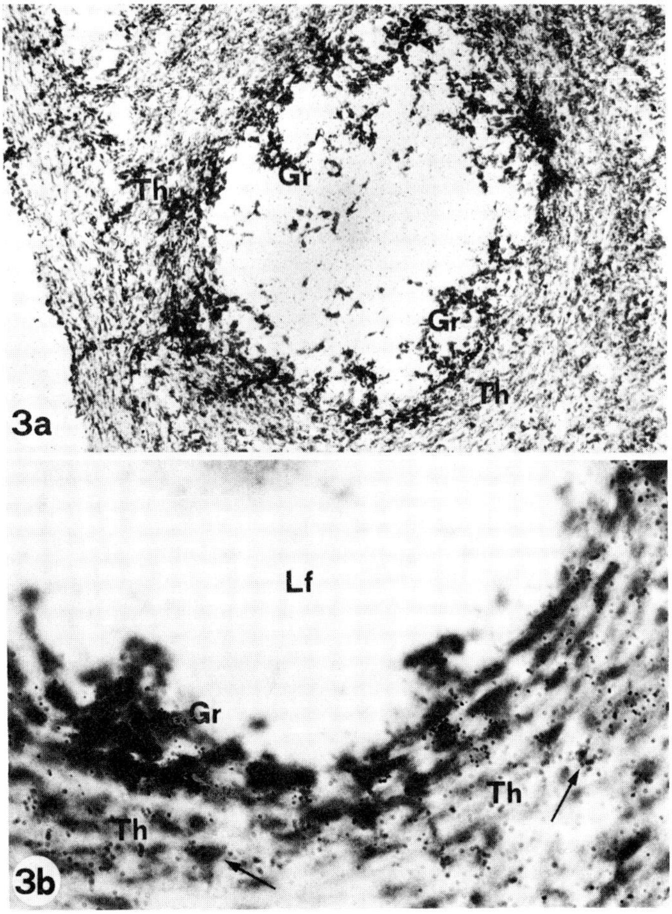

Fig. 4.3. Autoradiography. Frozen section of mouse ovary treated with C14 AMA.

(a) Cavitary follicle. Silver grains are localized on the granulosa cells (Gr) and thecal cells (Th) (180×).

(b) Silver grains (arrows) are less numerous on the theca interna cells (Th) if compared to those observed on granulosa cells (Gr) (470×).

DISCUSSION

Present results provide direct evidence that both steroidogenic and fibroblastlike theca interna cells are kinetically active and capable of producing in vivo discrete cellular movements for their content of microfilaments and contractile myosinlike proteins.

Some studies have analyzed the distribution of contractile proteins in the mammalian ovaries. Typical smooth muscle cells have been demonstrated by electron microscope in the theca externa of the mammalian ovaries (DiDio et al., 1980; Familiari and Motta, 1981). In addition, studies of Amsterdam et al. (1977) provided direct evidence for the existence of actin and smooth muscle myosin in high concentration in theca externa cells of the rat and justified the designation of external thecal cells as smooth muscle cells that reach their maximal development in the mature graafian follicle.

Furthermore, thin microfilaments and myosinlike proteins have been demonstrated in the granulosa cell layer of rabbit and mouse developing follicles (Motta and DiDio, 1974; Cavallotti et al., 1975a and b; Amenta and Cavallotti, 1980; Moscarini and Amenta, 1980). The correspondence between these immunochemical and ultrastructural findings suggested that the ovarian follicles were not passive organs, but, containing contractile granulosa cells, were capable of producing in vivo discrete movements, that might be used for morphogenetic phenomena such as the detachment of the cumulus ooforus from the granulosa wall preparing the oocyte to be extruded from the ruptured follicle (Motta and DiDio, 1974).

There are only fibroblastlike cells around the primordial follicle. On the contrary three different cell types are recognized around the graafian follicles; these are fibroblastlike cells, steroid-secreting cells and transitional cells (Hiura and Fujita, 1977). During the process of follicular atresia, steroid-secreting cells of the theca interna survive to the degenerative process, constituting, intermixed with several fibroblasts, the pool of steroidogenic elements called "interstitial gland of thecal origin."

Present studies demonstrate the presence of the same quantity of microfilaments in all cell types recognized in the theca interna, irrespective of their degree of maturation. Moreover, our observations show that theca interna cells bind antimyosinlike protein antibodies. In fact a fine diffuse fluorescence and a diffuse autoradiographic reaction was observed between theca interna cells.

It seems obvious to conclude therefore that microfilaments contain a myosinlike protein, and it is then possible to suggest that theca interna cells are capable of producing in vivo movements that could facilitate the reorganization of theca cells in order to form groups of cells in the ovary stroma (interstitial gland of the ovary).

SUMMARY

The present observations demonstrated that the theca interna cells of developing follicles in the mouse ovary bind antimyosinlike antibodies (AMA). The same theca interna cells studied with the electron microscope showed bundles of microfilaments in the cortical areas of the cytoplasm.

The immunochemical and ultrastructural results suggest that theca interna cells are capable of producing in vivo movements involved in morphogenetic phenomena.

ACKNOWLEDGMENTS

We are grateful to Dr. Wade Collier, who revised the manuscript, and to Prof. Pietro M. Motta for their suggestions and criticisms.

REFERENCES

Amenta F. and Cavallotti C. Immunohistochemical demonstration of a myosin-like protein in the oocyte. Acta Histochem. Cytochem. 13, 619-22. 1980.

Amsterdam A., Linder H.R. and Groschel-Stewart U. Localization of actin and myosin in the rat oocyte and follicular wall by immunofluorescence. Anat. Rec. 187, 311-28. 1977.

Cavallotti C, DiDio L. J. A., Familiari G., Fumagalli G. and Motta P. Microfilaments in granulosa cells of rabbit ovary: immunological and ultrastructural observations. Acta Histochem. 52, 253-56. 1975a.

Cavallotti C., Familiari G., Fumagalli G., DiDio L. J. A. and Motta P. Contractile microfilaments in granulosa cells. An untrastructural and immunochemical study. Arch. Histol. Jap. 38, 171-75. 1975b.

DiDio L. J. A., Allen D. J., Correr S. and Motta P. Smooth musculature in the ovary; in: Biology of the Ovary, Motta and Hafez eds. pp. 106-118, Martinus Nijhoff, The Hague, 1980.

Espey L. L. Ovarian contractility and its relationship to ovulation. A review. Biol. Reprod., 19, 540-51. 1978.

Familiari G. and Motta P. Gap communicating junctions in theca interna cells of mouse ovarian follicles. In: Endocrinological cancer and ovarian function and disease, H. Adlercreutz, R. D. Bulbrook, H. J. Van Der Molen, A. Vermeulen and F. Sciarra eds. pp. 216-19, Excerpta Medica, Amsterdam, 1980.

Familiari G., Correr S. and Motta P. Gap junctions in theca interna cells of developing and atretic follicles. In: Advances in the morphology of cells and tissues, E. Galina and A. Vidrio eds. pp. 337-48, Alan Liss New York, 1981.

Familiari G. and Motta P. Occurrence of a contractile tissue in the theca externa of atretic follicles in the mouse ovary. Acta Anat. 109, 103-114. 1981.

Hitchock S. E. Regulation of motility in non muscle cells. J. Cell Biol., 74, 1-15. 1977.

Hiura M. and Fujita H. Electron microscopy of the cytodifferentiation of the theca cell in the mouse ovary. Arch. Istol. Jap., 40, 95-105. 1977.

Melis M., Carpino F., Familiari G. and Cavallotti C. The contractile nature of the tubular-like cells of the parietal layer of the Bowman's capsule in male mice. Acta Anat. 98, 31-38. 1977.

Moscarini M. and Amenta F. Myosin-like protein in mouse ovary. Immunohistochemical and autohistoradiographic study. Cellular&Molecular Biol., 26, 275-79. 1980.

Motta P. and DiDio L. J. A. Microfilaments in granulosa cells during the follicular development and transformation in corpus luteum in the rabbit ovary. J. Submicr. Cytol. 6, 15-27. 1974.

O'Shea J. D., Cran D. G., Hay M. F. and Moor R. M. Ultrastructure of the theca of ovarian follicles in sheep. Cell Tiss. Res. 187, 457-72. 1978.

Porcelli G., Bettolo G. B., DiIorio M. and Ranieri M. Preliminary observations of labeled C14 protein with C14 cystein. J. Biochem., 22, 141-47. 1973.

Reynolds E. S. The use of lead citrate at high pH as an electron opaque stain in electron microscopy. J. Cell Biol. 17, 208-212. 1963.

5 High Resolution Autoradiographic Localization of Myosinlike Immunoreactivity within Rat Brain

Carlo Cavallotti, Francesco Amenta
Maria Caterina Mione

The actomyosinlike system is the contractile system most widely diffused in the nervous system (see Berl, 1975). While the actinlike protein is thought to be the constituent of the microfilamentous structures which lie close to the cell membrane no data are yet available concerning the precise location of the myosin molecules within nerve cells or nerve fibers (see Berl, 1975; Unsicker et al., 1978). In view of this, the subcellular location of the myosinlike protein was studied using a high resolution autoradiography technique.

MATERIALS AND METHODS

Antimyosinlike protein antibodies were obtained and purified in our laboratory as described in earlier papers (Cavallotti et al., 1975a,b; Melis et al., 1977). The specificity of the antibodies for the myosinlike protein was tested by blocking studies (absorption and immunodiffusion). The antimyosinlike antibodies were bound with [^{14}C]-cystein in an irreversible way in the presence of 8 M urea as described by Porcelli et al. (1973). In brief 0.1 μCi [^{14}C]-cystein were incubated with 10 mg of antimyosinlike antibodies and dissolved in 2.5 ml of 0.01 M acetic acid containing 8 M urea. After 20 hr of incubation at 4°C unbound labelled cystein was separated by conjugated antibodies by gel filtration or dialysis against distilled water for 48 h at 4°C.

Rats of the Wistar strain were anesthetized and perfused with warm (35°-38°C) physiological saline followed by 1% paraformaldehyde solution buffered to pH 7.4 with 0.1 M sodium phosphate buffer (PBS) at 4°C. The brain was then removed and cut into slices of 50-100 μm using a Vibratome at a temperature of 4°C. The slices were postfixed for 3-4 hr with a mixture of formalin-picric acid (Stefanini's fluid) at 4°C. The slices were then washed for 6-12 hr in PBS containing 10-25% sucrose and exposed to radiolabelled antimyosinlike antibodies (mc 10^{-7}M, 25 Ci/m-mole, see Moscarini and Amenta, 1980) diluted 1/60 in PBS. The incubation was done in a moist chamber, at room temperature, for 40 min. After the exposure to radiolabelled antibodies the slices were washed in PBS to remove unbound antibodies and air-dried. The specificity of the reaction was verified by treating parallel sections with an excess of nonlabelled antibodies before [^{14}C]-cystein conjugated antimyosinlike antibodies.

The slices were exposed to osmic acid vapors, dehydrated in graded ethanol and propylene oxide and embedded in a mixture of araldite and epon.

Semithin sections were processed for light microscope autoradiography by dipping into Ilford L4 emulsion (diluted 1:1). After 3-4 weeks of exposure the sections were developed, stained with toluidine blue and observed under light microscope. Thin sections (pale-gold) obtained from blocks exhibiting satisfactory labelling at the light microscope level were collected on celloidin-coated slides, stained with uranyl and lead, lightly carbonated and dipped into Ilford L4 emulsion (diluted 1:4). After 15-20 weeks of exposure at 4°C the autoradiographies were developed with Kodak Microd 1× and examined on a Zeiss EM 9 electron microscope.

RESULTS AND CONCLUSIONS

Under light microscope radiolabelled antimyosinlike antibodies were seen to be bound primarily by brain white matter and in smaller amounts by gray matter; electron microscopic studies showed that the radiolabelled antibodies could easily be recognized at the level of neuronal structures and more infrequently close to the cell membrane of glial elements (Fig. 5.1). At the neuronal level the antibodies were bound by axonal microtubular-microfilamentous structures, by axon terminals containing synaptic vesicles, and, more infrequently, by mitochondria as well as by the cellular membrane of nerve cells (Fig. 5.2). No radioactivity was found at the level of nuclear structures both of nerve or glial cells. The microtubular-microfilamentous structures and axon terminals containing synaptic vesicles are the neuronal components that more remarkably bound antimyosinlike protein antibodies.

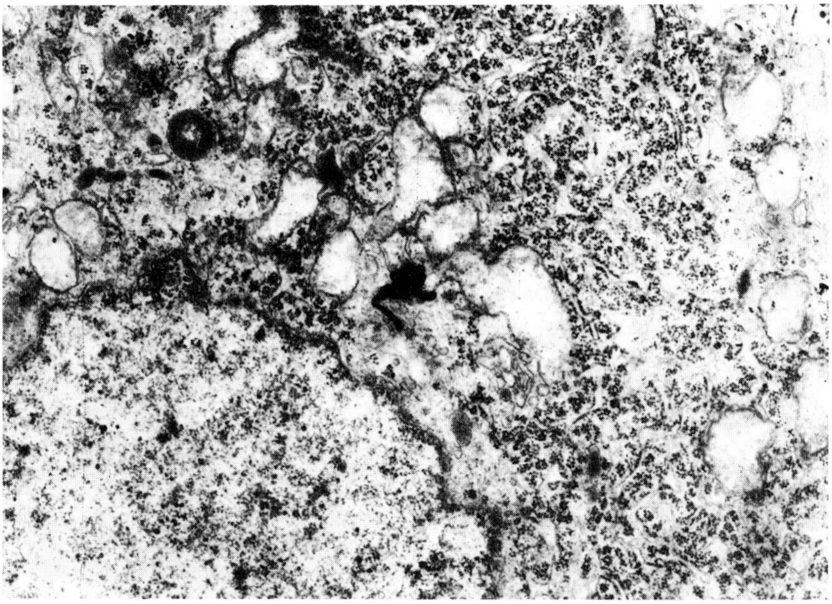

Fig. 5.1. Rat cerebral cortex. Autoradiographic demonstration of myosinlike immunoreactivity. The autoradiographic reaction appears to be localized at the level of cellular membrane of a glial cell. ×20,000

On the basis of our morphological data we are not able to establish the meaning of the myosinlike system within nervous system. The high density of myosinlike immunoreactivity within axon terminals as well as in close apposition to the synaptic vesicles lends support to the hypothesis that the myosinlike system may be involved in the release of neurotransmitter from synaptic vesicles (see Berl, 1975). However, the location of radiolabelled antibodies within nerve or glial cell membrane as well as within the microtubular-microfilamentous system and mitochondria let us hypothesize that brain myosinlike protein may play a role in the contractile activity of cell membranes and in the composition of microtubular-microfilamentous system. Further studies are in progress in our laboratory to deepen the problem.

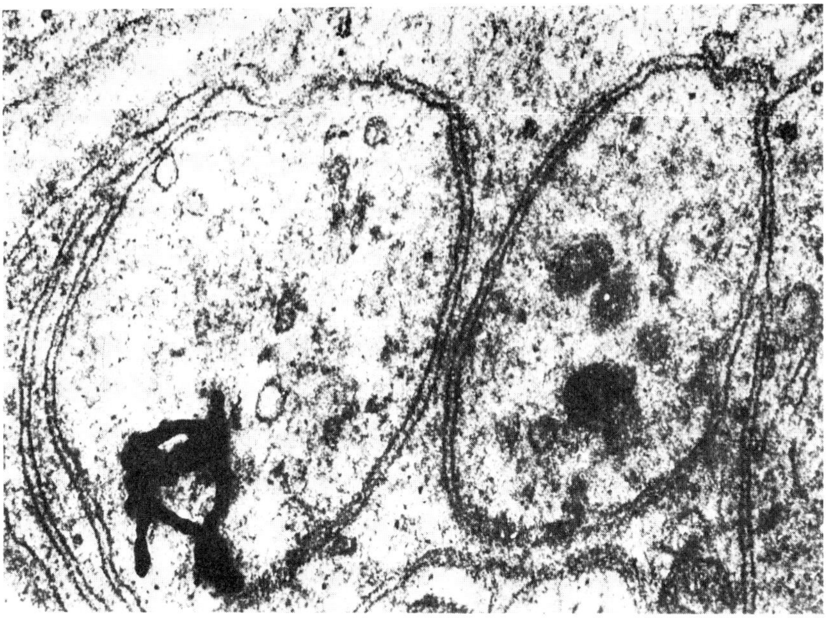

Fig. 5.2. Rat cerebral cortex. Autoradiographic demonstration of myosinlike immunoreactivity. The label is clearly localized at the level of axonal varicosities. ×75,000

REFERENCES

Berl, S., The actomyosin-like system in nervous tissue. In: The Nervous System (Tower D. B. Ed.), Vol. 1, Raven Press, New York, pp. 565-73, 1975.

Cavallotti, C., Di Dio, L. J. A., Familiari, G., Motta, P., Microfilaments in granulosa cells of rabbit ovary: immunological and ultrastructural observations. Acta Histochme. 52, 253-56, 1975a.

Cavallotti, C., Familiari, G., Fumagalli, G., Di Dio, L. J. A., Motta, P., Contractile filaments in granulosa cells. An ultrastructural and immunochemical study. Arch Histol. Jap. 38, 171-75, 1975b.

Melis, M., Carpino, F., Familiari, G., Cavallotti, C., The contractile nature of tubular-like cells of the parietal layer of the Bowman's capsule in male mice. Acta Anat. (Basel) 98, 31-38, 1977.

Moscarini, M., Amenta, F., Myosin-like protein in mouse ovary: immunohistochemical and autohistoradiographic study. Cellular & Molecular Biology 26, 275-79, 1980.

Porcelli, G., Bettolo, G. B., Di Iorio, M., Ranieri, M., Preliminary observations of labeled ^{14}C protein with ^{14}C cystein. J. Biochem. 22, 141-47, 1973.

Unsicker, K., Drenckhahn, D., Groschel-Stewart, U., Further immunofluorescence-microscopic evidence for myosin in various peripheral nerves. Cell Tiss. Res. 188, 341-44, 1978.

6 Effect of Various Fixatives and Embedding Methods on Myosinlike Immunoreactivity within Central and Peripheral Nervous System

Maria Caterina Mione, Francesco Amenta,
Wade L. Collier, Carlo Cavallotti

In the central nervous system as well as in other nonmuscular tissues (see Hitchock, 1977), the existence of an actomyosin contractile system that dissociates into actinlike and myosinlike components of the molecular weight of 45,000-47,000 and 220,000, respectively, has been described (Berl and Puszkin, 1970; Puszkin and Berl, 1972; Puszkin et al., 1968). The actinlike protein is thought to be the constituent of the microfilamentous structures which lie close to the cell membrane. We know even less about how and where the myosinlike protein is situated (Berl, 1975). The morphologic data available on this subject are limited to immunofluorescence localization of myosin within central and peripheral nerve fibers, while no information concerning the subcellular localization of the myosin in the central gray matter is yet available (Isenberg et al., 1977; Puszkin et al., 1972).

In view of this we have studied the effects of various fixatives and embedding methods on myosinlike immunoreactivity within the central and peripheral nervous systems. The present study was carried out in order to standardize a technique of fixation and embedding of nerve tissue in order to preserve the best myosinlike immunoreactivity.

MATERIALS AND METHODS

Rats of the Wistar strain were used in the present study. The animals were anesthetized and perfused with 300-500 ml of warm (35°-38°C) physiological saline followed by one of the solutions mentioned here: (a) 3.5% glutaraldehyde in Millonig buffer; (b) 4% paraformaldehyde solution buffered to pH 7.4 with 0.1 M sodium phosphate; (c) 1% paraformaldehyde solution buffered to pH 7.4 with 0.1 M sodium phosphate. All solutions were maintained at 4°C and each animal was perfused for 40-60 min. with 1,000-1,200 ml of fixative solution. The brain and the ischiatic nerve were then removed and cut into slices of 50-250 μm using a Vibratome. The slices were then postfixed for 60-240 min. in the same perfusion fixative with the exception of sections perfused as indicated in point c, which were postfixed for the period mentioned above using a mixture of formalin-picric acid (Stefanini's fluid). At the end of the postfixation the slices were washed for 6-12 hr at 4°C by rinsing in the same buffer used in the fixative mixture with the addition of 10-25% sucrose. Several slices were then exposed to osmic acid vapors at room temperature for 1 hr, dehydrated in graded ethanol and propylene oxide, embedded in epon or in a mixture of araldite and epon. Other sections were incubated with antimyosinlike antibodies as described later, washed, and then exposed to osmic acid vapors, dehydrated and embedded as described above. Sections of 1-2 μm were mounted on microscope slides.

Immunohistochemistry

Antimyosinlike antibodies were obtained, purified, and tested in our laboratory as previously described (Cavallotti et al., 1975a,b; 1977) and conjugated with fluorescein isothiocyanate (see Cavallotti et al., 1975a). Unembedded slices were exposed to fluorescent antimyosinlike antibodies diluted 1/60 in 0.1 M phosphate buffered saline (PBS, pH 7.2) for 40 min. at room temperature in a moist chamber. Plastic embedding media were removed using sodium methoxide according to Mayor et al. (1961); the sections were washed in PBS and then exposed to fluorescent antimyosinlike antibodies as above. After exposure to fluorescent antibodies unembedded slices and the other sections were rinsed in PBS to remove unbound antibodies, and air-dried.

The specificity of the reaction was verified by treating parallel sections with nonfluorescent antimyosinlike antibodies (see Cavallotti et al., 1975a). The sections were observed and photographed using a fluorescence microscope equipped with an epi-illumination system. The intensity of fluorescence developed after exposure to antibodies of differently treated sections was measured microfluorometrically. The slides were then examined at a "Fluoval photometric" microfluorometer at 520 nm. The instrument was calibrated considering "zero" the value of the background tissue autofluorescence. After the calibration single central and peripheral fibers were examined using a 63× objective and an 8× ocular. The area measured was delimited by a circular spot of 10 μm in diameter that was delineated by a diaphragm. The fluorescence was expressed in arbitrary units proportional to the binding of fluorescent antibodies to the tissue sections.

RESULTS AND CONCLUSIONS

A yellow fluorescence showing the spectral characteristics of fluorescein isothiocyanate (see Pearse, 1974) developed in sections incubated with fluorescent antimyosinlike antibodies, while only a faint tissue autofluorescence was found in sections incubated with nonfluorescent or denatured antibodies.

The effects of used fixative mixture and embedding media are seen in Figure 6.1. As can be seen, the glutaraldehyde and the 4% paraformaldehyde solutions caused a loss of specific fluorescence of 70% and 48% respectively, while a 1% paraformaldehyde perfusion followed by formalin-picric acid postfixation caused a loss of about 30% fluorescence compared with the intensity of fluorescence developed in frozen unfixed sections of rat brain used as a marker of intensity of fluorescence.

A comparison between the intensity of fluorescence in sections incubated with antimyosinlike antibodies before and after embedding of slices in araldite-spon and epon alone showed that the pre-embedding step causes a very remarkable loss of fluorescence (Fig. 6.2).

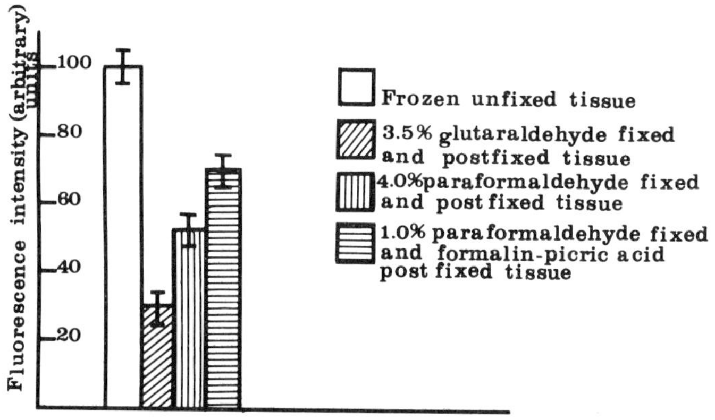

Fig. 6.1. Effect of different fixatives used on myosinlike immunoreactivity within rat ischiatic nerve.

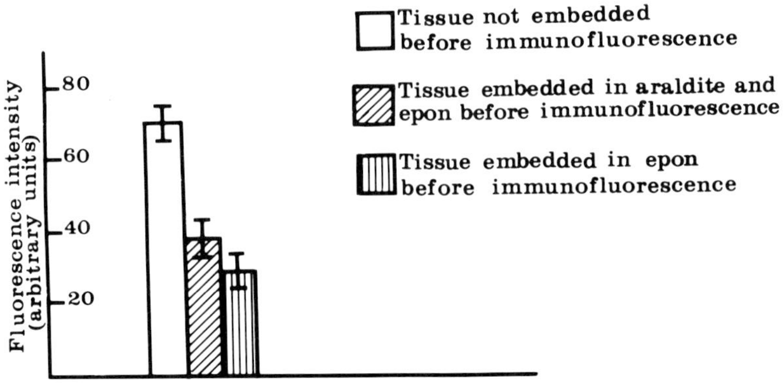

Fig. 6.2. Effect of pre-embedding on myosinlike-immunoreactivity within rat ischiatic nerve. (Tissue fixed with 1.0% paraformaldehyde and postfixed with formalin-picric acid.)

In summary, the perfusion with low concentrations of paraformaldehyde, the use of Vibratome-obtained nervous sections, the postfixation with the formalin-picric acid mixture, the exposure of sections to labelled antibodies, and the subsequent embedding in araldite-epon gave the best preservation of nervous tissue myosinlike immunoreactivity.

The standardization of this experimental protocol may represent a first stage for further studies aimed at analyzing the subcellular localization of contractile proteins at the level of the central nervous system.

REFERENCES

Berl, S., The actomyosin-like system in nervous tissue. In: The Nervous System (Tower D. B. Ed.), Vol. 1, Raven Press, New York, pp. 565-73, 1975.

Berl, S., Puszkin, S., Mg^{2+}-Ca^{2+}-activated adenosine triphosphatase system isolated from mammalian brain. Biochemistry 9, 2058-67, 1970.

Cavallotti, C., Carpino, F., Familiari, G., Re, M., Vicari, A., Myosin-like microfilaments in the human normal and pathological testis. Acta Anat. 99, 220-27, 1977.

Cavallotti, C., Di Dio, L. J. A., Familiari, G., Motta, P., Microfilaments in granular cells of rabbit ovary: Immunological and ultrastructural observations. Acta Histochem. 52, 253-56, 1975a.

Cavallotti, C., Familiari, G., Fumagalli, G., Di Dio, L. J. A., Motta, P., Contractile filaments in granulosa cells. An ultrastructural and immunochemical study. Arch. Histol. Jap. 38, 171-75, 1975b.

Hitchock, S. E., Regulation of motility in nonmuscle cells. J. Cell Biol. 74, 1-15, 1977.

Isenberg, G., Rieske, E., Kreutzberg, G. W., Distribution of actin and tubulin in neuroblastoma cells. Cytobiologic 15, 382-89, 1977.

Mayor, H. D., Hampton, J. C., Rosario, B. A simple method for removing the resin from epoxy embedded tissue. J. Cell Biol 9:909-910, 1961.

Pearse, A. G. E., Histochemistry, theoretical and applied. Little, Brown and Co., Boston, 1974.

Puszkin, S., Berl, S., Actomyosin-like protein from brain: Separation and characterization of the actin-like component. Biochem. Biophys. Acta 256, 695-709, 1972.

Puszkin, S., Berl, S., Puszkin, E., Clarke, D. D., Actomyosin-like protein isolated from mammalian brain. Science 161, 120-21, 1968.

Puszkin, S., Nicklas, W. J., Berl, S., Actomyosin-like protein in brain: subcellular distribution. J. Neurochem. 19, 1319-33, 1972.

Unsicker, K., Drenckhahn, D., Groschel-Stewart, U., Further immunofluorescence-microscopic evidence for myosin in various peripheral nerves. Cell Tiss. Res. 188, 341-44, 1978.

7 Actin Filament Meshworks and Microtubules in Dictyostelium Discoideum

S. Rubino, J. V. Small, M. Claviez,
C. Sellitto, P. Cappuccinelli

INTRODUCTION

Despite the considerable progress that has been made on the biochemistry of contractile proteins of amoeboid cells (e.g., Spudich and Spudich, 1982) the structural interactions of these proteins, leading to movement in vivo, remain virtually unknown. This is explained by the dynamic activity and associated lability of the motile apparatus of amoebae that has constituted a major stumbling block in attempts to define ultrastructural organizations by microscopic methods.

The present report concerns itself with the motile apparatus of the cellular slime mold Dictyostelium discoideum. Spreading and locomotion in this amoebe is mediated via the extension of thin veils of cytoplasm, lamellipodia, as well as of filopodia, not unlike those seen on locomoting vertebrate cells (Cooke et al., 1976). These regions have been shown by immunofluorescent microscopy (Eckert and Lazarides, 1978; Condeelis, 1981; Rubino et al., 1983) and by electron microscopy (Discussion) to be rich in actin filaments. Using conditions adapted from those developed for cultured vertebrate cells (Small and Celis, 1978; Höglund et al., 1980; Small, 1981), we have now been able to reveal the filamentous elements of Dictyostelium with more clarity. We show that the lamellipodia of spreading amoeba consist of complex cross-linked meshworks of actin filaments from which organized bundles form to produce filopodia at the cell periphery. Observations on the microtubule-organizing center of Dictyostelium are also reported.

MATERIALS AND METHODS

Cell cultures

Axenic cultures of Dictyostelium discoideum (A×2 strain) were grown in HL-5 medium containing glucose or maltose (Watts and Ashworth, 1970) at 22°C. Cells were used during the exponential phase of growth. For electron microscopy studies amoebae were attached to filmed (Parlodion or Collodion) silver hexagonal support grids (150 mesh, Teepe-Brandsma, Holland) pretreated with ultraviolet light for at least 60 min at 22°C. For indirect immunofluorescence the cells were attached to coverslips as described above or to slides as reported by Unger et al. (1979).

Preparation of cytoskeletons

Cytoskeletons were prepared using mixtures of 0.2% Triton-0.001% glutaraldehyde or 1% Triton-0.025% glutaraldehyde in a buffer containing 20 mM KCl, 2mM $MgSO_4$, 11 mM EGTA, 10 mM PIPES pH 7 (CB buffer) for 2 min. at 22°C and then postfixed in 1% glutaraldehyde in the same buffer for at least 10 min. In some cases, a passage through 2×10^{-5}M phalloidin was used before the fixation in 1% glutaraldehyde.

Negative staining

Negative staining with sodium silicotungstate was carried out at room temperature essentially as described elsewhere for tissue culture cells (Small, 1981).

Drying procedures for rotary shadowing

For freeze drying, grids (after extraction and fixation) were washed briefly in H_2O and 10% methanol and quickly dipped into liquid propane or nitrogen slush, mounted under liquid nitrogen and transferred into a Balzers freeze etching apparatus 301 onto a cold rotary stage and dried for at least 3 h at -95°C. After raising the temperature to -80°C for 2 h the rotary shadowing was done at -100°C using electron beam evaporation guns for coating the specimens with platinum/carbon (95:5% w/w) and a backing film of carbon.

Critical point drying was performed in a Polaron E 3001 apparatus using CO_2 after dehydration in ethanol and rotary shadowed as after freeze drying but at room temperature.

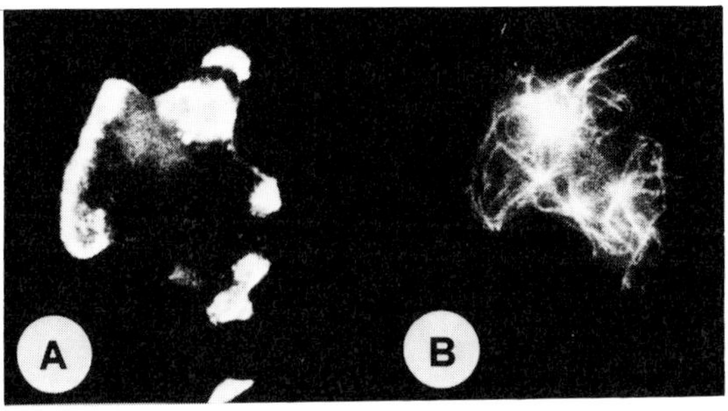

Figure 7.1. Double immunolabeling of actin with rhodamine conjugated phalloidin (A) and microtubules with monoclonal antiyeast tubulin (B). (3,000×)

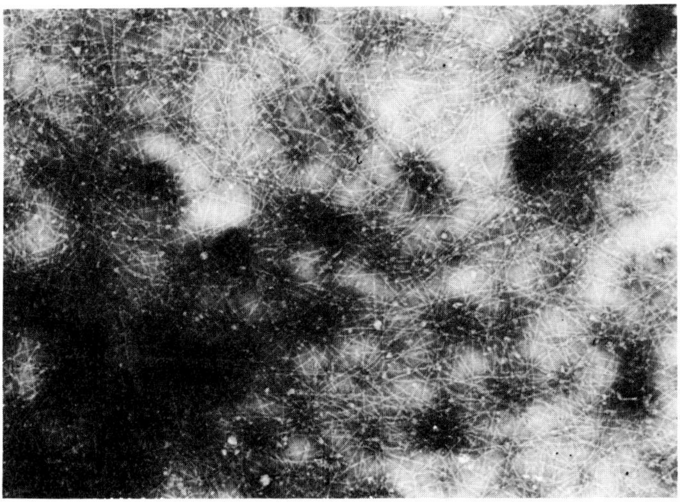

Fig. 7.2. Lamellipodium of Triton-glutaraldehyde cytoskeleton showing the complex network of actin filaments. (144,000×).

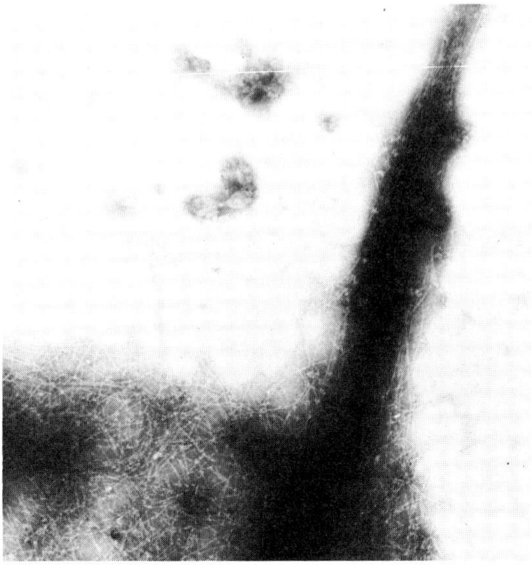

Fig. 7.3. Arrangement of filaments at the base of a filopodium. (46,000×).

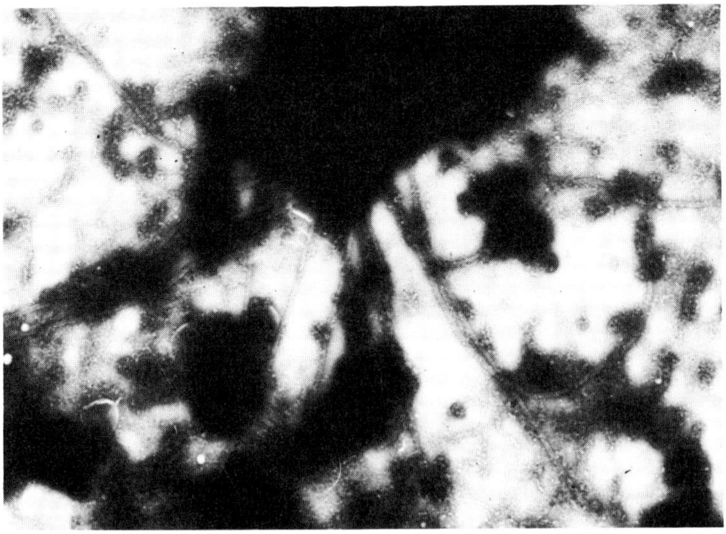

Fig. 7.4. Negative staining of a microtubule organizing center with microtubules originating. (34,000×).

Indirect immunofluorescence

Experiments were performed on either intact cells or cell cytoskeletons. For double labelling the cells were processed and fixed as described in the "Preparation of cytoskeleton." The Triton-glutaraldehyde cytoskeletons were treated with $NaHBO_4$ 2 times for 5 min. (Weber et al., 1978). In some case, the cells were fixed with a mixture of 3% formaldehyde-0.2% Triton X-100 in CB buffer for 10 min.

The cells were incubated for 60 min. at 37°C with a rat monoclonal antibody against yeast tubulin (a gift of Dr. J. Kilmartin, Cambridge, England, washed in buffer and then incubated for 60 min. at 37°C with a mixture of antirat antibodies conjugated with FITC (Dako) and rhodamine conjugated phalloidin (a gift of Prof. Th. Wieland, Heidelberg) at a final concentration of 2×10^{-5}M. After extensive washing the slides were mounted in Gelvatol (Monsato Corporation, N.Y.) containing n-propyl gallate (Sigma) to inhibit bleaching.

Fluorescence microscopy was carried out with a Leitz Dialux 20EB microscope equipped with epifluorescence optics and selective filters for fluorescein and rhodamine. Photographs were taken on Kodak TRI-X Pan film (Fig. 7.5). Microtubules

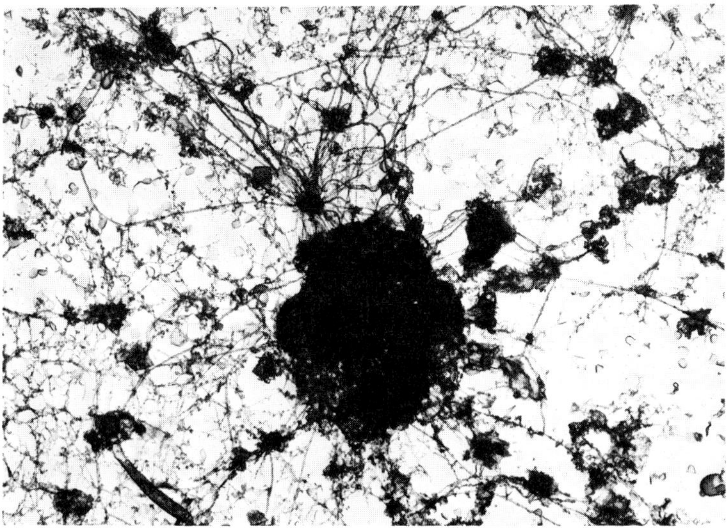

Fig. 7.5. Nucleus and microtubules organizing center in a freeze drying preparation. (22,410×).

radiating from the body were seen to penetrate into the actin meshworks and sometimes reached close to the cell periphery. No consistent association of microtubule ends with filopodia was apparent.

DISCUSSION

The cytoskeleton of Dictyostelium discoideum plays a primary role in such processes as cell motility, changes in cell shape, mitosis, salatory particle movement, endocytosis and phagocytosis (for review see Spudich and Spudich, 1982; Condeelis, 1981; Cappuccinelli et al., 1982). In this report the organization of cytoskeleton has been studied using an improved extraction method entailing the use of mixtures of glutaraldehyde and Triton X-100.

Microfilaments have been previously demonstrated in amoebae of Dictyostelium by electron microscopy in thin sections (Eckert et al., 1977; Maeda and Eguchi, 1977) or after negative staining of whole cells attached to a surface coated with polylysine (Clarke et al., 1975). However, in neither case were the delicate meshworks of actin in the subcortical cytoplasm preserved.

Our data obtained with indirect immunofluorescence or with negative staining for electron microscopy are consistent with a close association of actin with the cell membrane (Condeelis, 1981, 1979; Salisbury et al., 1983). Interestingly, the disposition of actin meshworks in the thin lamellopodia of well-spread Dictyostelium discoideum resembles, after negative staining, that seen in motile fibroblastic cells (Small, 1981; Small et al., 1982). In Dictyostelium, microtubules have been demonstrated by electron microscopy in mitosis and interphase (Moens, 1976; Roos, 1982; Eckert et al., 1977), by indirect immunofluorescence microscopy (Unger et al., 1979; Cappuccinelli et al., 1982; Rubino et al., 1983). Our present results confirm these data and suggest a possible interaction in the lamellopodia between microtubules and microfilaments. Microtubules in Dictyostelium discoideum have similar characteristics to those of mammalian cells; they are cold sensitive and are inhibited by microtubules poisons (Cappuccinelli and Ashworth, 1976; Rubino et al., 1982); they originate from a microtubule organizing center (MTOC) that corresponds to the nucleus associated organelle (Unger et al., 1979; Cappuccinelli et al., 1982) and radiate to the cell periphery. Microtubules seem to have a marginal role in cell locomotion in Dictyostelium as recently demonstrated by Rubino et al. (1983).

In conclusion, the glutaraldehyde-Triton fixation and negative staining techniques with Dictyostelium discoideum should be useful for the study of the interaction of actin with its associated proteins (Condeelis, 1981; Condeelis et al., 1983; Vahey et al., 1983), with microtubules and with the cell membrane.

ACKNOWLEDGMENTS

Many thanks are due to Prof. Th. Wieland for the gift of rhodamine conjugated phalloidin and to Dr. J. Kilmartin for the rat monoclonal antibodies against yeast tubulin. This work was supported by a grant from the Austrian Science Research Council (J.V.S.) and M.P.I.

REFERENCES

Cappuccinelli, P. & Ashworth, J. M. (1976) Effect of microtubule and microfilament function on the cellular slime mould Dictyostelium discoideum. Exp. Cell Res. 103, 387-93.

Cappuccinelli, P., Rubino, S., Fighetti, M. & Unger, E. (1982) Organization of microtubular system in Dictyostelium amoebae. In Microtubules in Microorganisms (edited by Cappuccinelli, P. & Morris, N.R.), pp. 71-97, New York and Basel: Marcel Dekker.

Clarke, M., Schatten, G., Maria, D., & Spudich, J. A. (1975) Visualization of actin fibers associated with the cell membrane in amoebae of Dictyostelium discoideum. Proc. Natl. Acad. Sci. USA 72, 1758-62.

Condeelis, J. (1979) Isolation of concanavalin A caps during various stages of formation and their association with actin and myosin. J. Cell Biol. 80, 751-58.

Condeelis, J. (1981) Microfilament-membrane interactions in cell shape and surface architecture. In International Cell Biology (edited by Schweiger, H. G.), pp. 306-320. Berlin and Heidelberg: Springer-Verlag.

Condeelis, J., Geosits, S. & Vahey, M. (1983) Isolation of a new actin-binding protein from Dictyostelium discoideum. Cell motility 2, 273-85.

Cooke, R., Clarke, M., von Wedel, R. J. & Spudich, J. A. (1976) Supramolecular forms of Dictyostelium actin. In Cell motility (edited by Goldman, R., Pollard, T. & Rosenbaum, J.), pp. 575-87. Cold Spring Harbor Laboratory, Cold Spring Harbor, New York.

Eckert, B. S. & Lazarides, E. (1978) Localization of actin in Dictyostelium amebas by immunofluorescence. J. Cell Biol. 77, 714-21.

Eckert, B. S., Warren, R. H. & Rubin, R. W. (1977) Structural and biochemical aspects of cell motility in amebas of Dictyostelium discoideum. J. Cell Biol. 72, 339-50.

Hoglund, A. S., Karlsson, R., Arro, E., Fredriksson, B. E. & Linberg, U. (1980) Visualization of the peripheral weave of microfilaments in glia cells. J. Muscle Res. Cell Motil. 1, 127-33.

Maeda, Y. & Eguchi, G. (1977) Polarized structures of cell in the aggregating cellular slime mould D. discoideum. An electron microscope study. Cell Struct. Funct. 2, 159-61.

Moens, P. B. (1976) Spindle and kinetochore morphology of Dictyostelium discoideum. J. Cell Biology 68, 113-22.

Roos, U. P. (1975) Fine structure of an organelle associated with the nucleus and cytoplasmic microtubules in the cellular slime mould Polysphondylium violaceum. J. Cell Sci. 18, 315-26.

Roos, U. P. (1982) Morphological and experimental studies on the cytocenter of cellular slime moulds. In microtubules in microorganisms (edited by Cappuccinelli, P. & Morris, N. R.) pp. 51-69. New York and Basel: Marcel Dekker.

Rubino, S., Unger, E., Fogu, G. & Cappuccinelli, P. (1982) Effect of microtubule inhibitors on the tubulin system of Dictyostelium discoideum. Z. Allg. Mikrobiol. 22, 127-31.

Rubino, S., Fighetti, M., Unger, E. & Cappuccinelli, P. (1983) Location of actin, myosin and microtubular structure during directed locomotion of Dictyostelium amoebae. J. Cell Biol. in press.

Salisbury, J. L., Condeelis, J. S. & Satir, P. (1983) Receptor-mediated endocytosis: machinery and regulation of the clalthrin-coated vesicle pathway. Int. Rev. Exp. Pathol. 24, 1-62.

Small, J. V. & Celis, J. E. (1978) Filament arrangements in negatively stained cultured cells: the organization of actin. Citobiologie 16, 308-25.

Small, J. V. (1981) Organization of actin in the leading edge of cultured cells: influence of osmium tetroxide and dehydration on the ultrastructure. J. Cell Biol. 91, 695-705.

Small, J. V., Rinnerthaler, G. & Hinssen, H. (1982) Organization of actin meshworks in cultured cells: the leading edge. Cold Spring Harbor Symposia on Quantitative Biology, vol. XLVI pp. 599-611, Cold Spring Harbor Laboratory.

Spudich, J. A. & Spudich, A. (1982) Cell motility. In The development of Dictyostelium discoideum (edited by Loomis W. F.) pp. 169-94, New York: Academic Press.

Unger, E., Rubino, S., Weinert, T. & Cappuccinelli, P. (1979) Immunofluorescence of the tubulin system in cellular slime moulds. FEMS Microbiol. Lett. 6, 317-20.

Vahey, M. Carboni, J., DeMey, J. & Condeelis, J. (1983) Properties of the 120,000 and 95,000 dalton actin binding proteins from Dictyostelium discoideum: 95,000 dalton protein is Dictyostelium α-actinin. Submitted.

Watts, D. J. & Ashworth, J. M. (1970) Growth of myxamoebae of the cellular slime mould Dictyostelium discoideum in axenic culture. Biochem. J. 119, 171-74.

Weber, K., Rathke, P. C. & Osborn, M. (1978) Cytoplasmic microtubular images in glutaraldehyde-fixed tissue culture cells by electron microscopy and by immunofluorescence microscopy. Proc. Natl. Acad. Sci. USA 75, 1820.

8 Purification and Physicochemical Properties of Brevin: A Calcium-dependent Actin Severing Protein

Zohra Soua, Françoise Porte,
Marie-Claude Kilhoffer, Dominique Gerard,
Jean-Paul Capony

Actin is a major constituent of many types of cells and tissues where it exists in a variety of structural forms. Numerous actin-binding proteins which regulate actin assembly have been described (Weeds, 1982), and three major classes have been identified: (1) actin cross-linking proteins; (2) severing and capping proteins; (3) G-actin stabilizing proteins. Brevin, also referred to as actin depolymerizing factor (ADF), isolated from serum and plasma (Norberg et al., 1979; Harris and Schwartz, 1981), belongs to the class of severing and capping proteins. Brevin severs actin filaments in a Ca^{2+} dependent manner, without net depolymerization.

We describe here the isolation of brevin from bovine serum and present some of its physical and chemical properties.

Table 8.1 summarizes a typical purification procedure of brevin. Thawed bovine serum was fractionated between 30 and 50% ammonium sulfate saturation. The resulting pellet was dissolved in 20 mM Tris-HCl (pH 7.8), .2 mM DTT, 2mM EGTA, 80 mM NaCl (Buffer A), dialyzed, and applied to a DEAE-Sepharose DL6B column equilibrated in Buffer A. Protein elution was performed with a linear gradient of 80-200 mM NaCl. Brevin eluted around 120 mM NaCl and was further fractionated on Sephacryl S-200 column equilibrated in 1M NaCl, .2 mM DTT, 1 mM EGTA, 20 mM Tris-HCl pH 7.3. Purification was achieved by using hydrophobic chromatography. Brevin was applied to phenyl-Sepharose in the presence of Ca^{2+}, at high ionic strength and eluted with EGTA in a low ionic strength buffer. Brevin obtained by this method was electrophoretically homogenous. The overall yield of the purification is approximately 60 mg protein per liter of serum.

Physical parameters of brevin are listed in Table 8.2. SDS polyacrylamide gel electrophoresis of purified brevin shows

Table 8.1. Purification of Brevin from Bovine Serum

Purification step	Total Protein mg	Total Activity units	Specific activity units/mg	Purification fold
Serum	42,640	36,670	0.86	1
30-50% $(NH_4)_2SO_4$	8,280	27,324	3.3	3.8
DEAE-sepharose CL6B	2,090	17,347	8.3	9.6
Sephacryl S-200	759	8,422	11.1	12.9
Phenyl sepharose	9.8	3,269	333.3	387.5

Note: Typical preparation of brevin starting from 250 ml of bovine serum. One unit of activity is defined as the amount of brevin, which reduces the final viscosity of a 0.5 mg/ml of actin solution by 50% from that obtained without inhibitor.

Table 8.2. Physical Parameters of Brevin

Parameter	Value	Determined by
Molecular weight	91,000	SDS-gel electrophoresis
Stokes radius	43.6 Å	Gel filtration
Sedimentation coefficient		
+ Calcium	4.44 S	Analytical ultra-centrifugation
- Calcium	4.78 S	
Isoelectric point (calcium free)	6.02-6.17	Isoelectric focusing gel
Partial specific volume	0.728 ml/g	
$E^{1\%}_{280nm}$	15.38	Ultraviolet spectrum

a single band corresponding to an Mr of 91,000 daltons. However, on two dimensional gel electrophoresis several spots corresponding to isoelectric points ranging between 6.02 and 6.17 can be observed. This may suggest the possible presence of prosthetic groups (glycosyl or phosphoryl) on brevin. Brevin contains a great amount of aromatic amino acids, tryptophan and tyrosin. This accounts for the high value of its extinction coefficient. The amino-acid composition of brevin shows a close similarity to that of gelsolin isolated from macrophages (Yin and Stossel, 1980) or spinal cord actin-fragmenting protein (Petrucci et al., 1983). However, up to now there is no evidence as to the identity between the different proteins.

Brevin, as well as other severing proteins, acts at substoichiometric concentrations. Indeed, 10 μg/ml brevin reduced the viscosity of 0.5 mg/ml actin by 50% (corresponding to a molar ratio actin to brevin of 1:100). The severing activity of brevin, followed by viscosimetric assays is strictly Ca^{2+}-dependent suggesting a possible Ca^{2+}-induced transconformation of the protein. Such a conformational change is also sustained by the difference in sedimentation coefficients and peptide patterns obtained by limited tryptic or chymotryptic proteolysis in the presence or absence of Ca^{2+}.

Absorption and fluorescence spectra of brevin are characteristic of a tryptophan containing protein. The maximum of fluorescence emission (for an excitation wavelength at 295 nm) is located at 324 nm suggesting that the tryptophan residues are buried within the protein molecule. Ca^{2+} does not affect the emission maximum. However, Ca^{2+} decreases both the

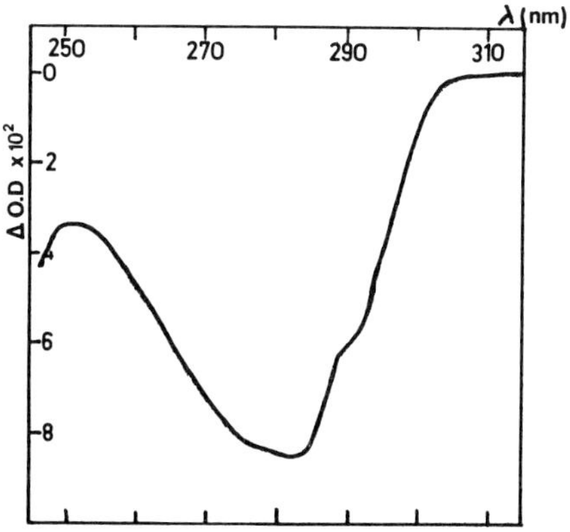

Fig. 8.1. Ultraviolet absorption differential spectrum of brevin (1×10^{-6} M) in 50 mM HEPES, 150 mM KCl, 10^{-4}M EGTA, 1.1×10^{-4}M $CaCl_2$ pH (i.e., 1.35×10^{-5}M free Ca^{2+}).

absorbance and the fluorescence intensity of brevin. Brevin ultraviolet differential spectrum is shown in Figure 8.1.

Preliminary results obtained after Ca^{2+} titration of brevin followed by the absorption decrease at 282 nm suggest the existence of two classes of Ca^{2+} binding sites. Class I shows high affinity for Ca^{2+} ($K_D \simeq 5 \times 10^{-8}$ M) whereas class II contains sites with lower affinity ($K_D \simeq 5 \times 10^{-6}$ M). However, the physiological relevance of these high affinity Ca^{2+} binding sites on brevin, as regards Ca^{2+} concentration in serum, is hitherto unknown.

Brevin, a high molecular weight calcium binding protein isolated from serum, is quite similar to gelsolin in its amino acid composition and activity toward actin filaments. However, the two proteins seem to differ notably in their Ca^{2+} binding properties and may be their localization. In addition, many questions remain unanswered concerning the physiological role of brevin in serum, and the existence of high affinity Ca^{2+} binding sites on an extracellular protein.

REFERENCES

Harris, D. and Schwartz, J. Characterization of brevin, a serum protein that shortens actin filaments. Proc. Natl. Acad. Sci. USA 78, 6798-6802 (1981).

Norberg, R., Thorstensson, R., Utter, G. and Fagraeus, A. F-actin depolymerizing activity of human serum. Eur. J. Biochem. 100, 575-83 (1979).

Petrucci, T., Thomas, C. and Bray, D. Isolation of a Ca^{2+}-dependent actin-fragmenting protein from brain, spinal cord and cultured neurones. J. Neurochem. 40, 1507-16 (1983).

Weeds, A. Actin-binding proteins-regulators of cell architecture and motility. Nature 296, 811-16 (1982).

Yin, H. and Stossel, T. Purification and structural properties of gelsolin, a Ca^{2+}-activated regulatory protein of macrophages. J. Biol. Chem. 255, 9490-93 (1980).

9 Monoclonal Antitubulin Antibody in the Study of Microtubules in Different Cell Types

P. Dráber, E. Vachoutová, V. Viklický

METHODS

Preparation of Immunogen and Immunization

Microtubule proteins (MTP) were prepared from porcine brains according to Shelanski et al. (1973). BALB/c mice were injected subcutaneously at three-week intervals with 100 μg of MTP emulsified in 0.1 ml of adjuvant Al-Span Oil (Sevac, Prague). On days 4, 3, and 2 before fusion they received another three intravenous injections of 100 μg of MTP in 0.1 ml saline.

Cell Hybridization and Cloning

Spleen cells from hyperimmune mice were fused with mouse myeloma cell line P3-X63-Ag8.653 as described by Galfré et al. (1977) using polyethylene glycol as a fusing agent. After fusion, the cell suspension was distributed in 96-well plates in H-MEMd medium containing HAT and 10% fetal calf serum. Supernatants were tested for antibody activity between 18th and 25th day after fusion. Cells from wells in which antibody secretion was demonstrated were cloned in soft agar with feeder cells.

Antibody Binding Assay

The supernatants from individual wells of the plate were tested by the enzyme immunoassay in Terasaki plates. Pig brain tubulin separated from the accessory proteins by phosphocellulose chromatography was used as an antigen. Bound antibodies were detected by affinity purified swine antimouse Ig antibody labelled with horseradish peroxidase.

Production of Ascitic Fluid and Antibody Purification

Large quantities of monoclonal antibody were prepared by growing hybridoma cells in mice, which had previously been injected with paraffin oil. The antibodies contained in the ascitic fluid were purified by ammonium sulfate precipitation followed by chromatography on DEAE cellulose. The subclass of the monoclonal antibody and the type of its light chain were determined using rabbit antibodies specific for the subclass and light chain type of murine immunoglobulins.

Gel Electrophoresis and Immunoblot Technique

SDS-polyacrylamide gel electrophoresis (Laemmli, 1970) was carried out on 1-mm thick slab gels. The α- and β-subunits of tubulin were effectively separated on 7.5% gels in the presence of SDS (Sigma). Whole cell lysates and pig brain protein extract were separated on 10% gels. After electrophoresis, proteins were electrophoretically transferred onto nitrocellulose sheets (Towbin et al., 1979). Confirmation of protein transfer was obtained by staining with amido black, and the other tracks were cut out and stained with purified monoclonal antibody using immunoperoxidase procedure with 4-chloro-1-naphthol as a chromogen.

Indirect Immunofluorescence

Fibroblasts were grown in 5% CO_2 in air in H-MEMd medium supplemented with 10% calf serum. MOLT 3 cells (human T-lymphoblastoid cell line), grown under the same conditions, were attached to the polybrene-coated coverslips. Cells were extracted with Triton X-100, fixed in glutaraldehyde and treated with borohydride (Schliwa et al., 1981). The coverslips were incubated with the monoclonal antibody and stained with FITC-labeled swine antimouse Ig antibody. Cells were mounted in glycerol and examined on a Leitz Orthoplan microscope. In control experiments mouse monoclonal antihuman transferrin antibody (IgG_1, κ) was used. To some of the cultures 10^{-4} M vinblastine (18 h), 10^{-5} M taxol (18 h), 10^{-4} M Colcemid (30 min) or 10^{-5} M cytochalasin B (30 min) were added.

RESULTS AND DISCUSSION

Of the 314 growing hybrid cells, 15 hybridomas were established. The hybridoma TU-01 was selected for further study. The monoclonal antibody produced by this hybridoma belongs to the IgG_1 subclass, kappa light chain. Ascitic fluid gave significant binding at dilutions up to 10^7-fold in the enzyme immunoassay. The purified monoclonal antibody was bound significantly at a concentration of 0.02 μg/ml. The specificity of the antibody was investigated by the immunoblot technique. As shown in Figure 9.1, only the α-subunit of purified porcine brain tubulin is strongly labelled. In the whole lysates of fibroblasts, pig brain protein extract, and neuroblasts, only the tubulin band was labelled by monoclonal antibody. Using mouse monoclonal antibody against human transferrin as a control, no binding of the second labelled antibody was observed.

The TU-01 antibody was further tested for its binding activity by indirect immunofluorescence. Abundant arrays of microtubules and mitotic figures, but no other structures were stained by monoclonal antibody in fibroblasts (Fig. 9.2a,c) and in the lymphoblastoid cell line (Fig. 9.2b). In controls, no cytoskeletal structures were detected. The conclusion that TU-01 antibody stained only microtubules in extracted cells was further confirmed by the following findings: no cytoskeletal structures were detectable after the pretreatment with Colcemid, microtubule arrangement was maintained in the presence of cytochalasin B, and the typical paracrystals following pretreatment with vinblastine were visible. The typical microtubular arrays were formed after pretreatment with taxol.

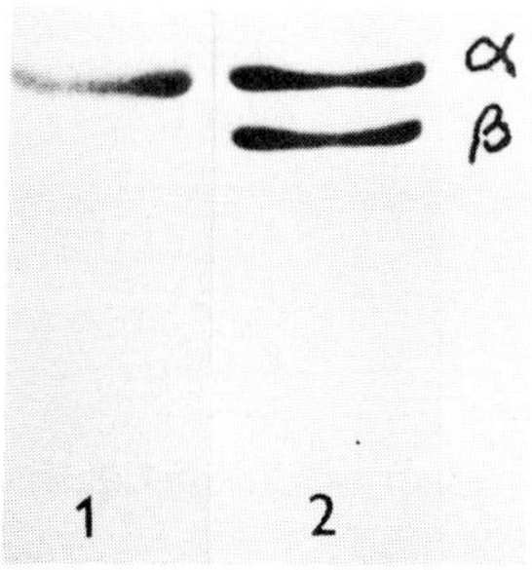

Fig. 9.1. Nitrocellulose protein blot transfer of purified pig brain tubulin separated into subunits on SDS-polyacrylamide gel. (1) Immunostaining by TU-01 antibody, (2) amido black staining.

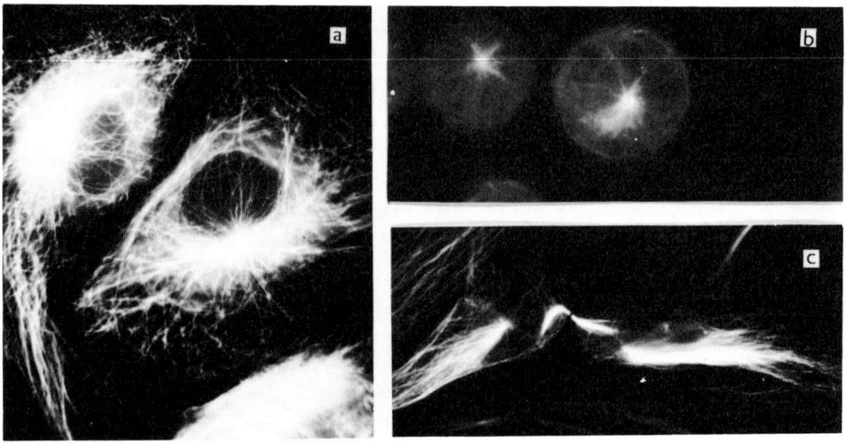

Fig. 9.2. Indirect immunofluorescent staining of different cell types by TU-01 antibody. (a) PTK_2 cells, (b) MOLT 3 cells, (c) LEP cells.

Table 9.1. Positive reactivity of TU-01 antibody with mammalian and bird cell microtubules in indirect immunofluorescence

Species	Cell type	Strain
Human	fibroblasts	LEP
Human	T-lymphoblastoid cell line	MOLT 3
Mouse	fibroblasts	L-A9
Mouse	Leydig's cells	primary culture
Mouse	embryonal carcinoma	F 9
Rat kangaroo	fibroblasts	PTK_2
Chicken	fibroblasts	primary culture

In the samples so far tested cross-reactivity or the presence of a target epitope on cytoskeletal structures other than microtubules was not found by the methods used. The TU-01 antibody, however, simultaneously binds to microtubules in a number of cell types of different species (Table 9.1). Although the target determinant of the TU-01 antibody has not yet been determined, one can conclude from the results described that it is different from the determinant recognized by the monoclonal antibody 1-6.1 (Asai et al., 1982).

REFERENCES

Asai, D. J., Brokaw, C. J., Thompson, W. C. and Wilson, L. (1982) Two different monoclonal antibodies to alpha-tubulin inhibit the bending of reactivated sea urchin spermatozoa. Cell Motility, 2, 599-614.

Galfre, G., Howe, S. C., Milstein, C., Butcher, G. W. and Howard, J. C. (1977) Antibodies to major histocompatibility antigens produced by hybrid cell lines. Nature (London), 266, 550-52.

Gozes, I. and Barnstable, C. (1982) Monoclonal antibodies that recognize discrete forms of tubulin. Proc. Natl. Acad. Sci. USA, 79, 2579-83.

Kilmartin, J. V., Wright, B. and Milstein, C. (1982) Rat monoclonal antitubulin antibodies derived by using a new nonsecreting rat cell line. J. Cell Biol., 93, 576-82.

Laemmli, U. K. (1970) Cleavage of structural proteins during the assembly of the head of bacteriophage T_4. Nature (London), 227, 680-85.

Schliwa, M., Euteneuer, U., Bulinski, J. C. and Izant, J. C. (1981) Calcium lability of cytoplasmic microtubules and its modulation by microtubule-associated proteins. Proc. Natl. Acad. Sci. USA, 78, 1037-41.

Shelanski, M. L., Gaskin, F. and Cantor, C. R. (1973) Microtubule assembly in the absence of added nucleotides. Proc. Natl. Acad. Sci. USA, 70, 765-68.

Towbin, H., Staehelin, T. and Gordon, J. (1979) Electrophoretic transfer of proteins from polyacrylamide gels to nitrocellulose sheets. Proc. Natl. Acad. Sci. USA, 76, 4350-54.

Viklický, V., Dráber, P., Hašek, J. and Bártek, J. (1982) Production and characterization of a monoclonal antitubulin antibody. Cell Biol. Int. Rep. 6, 725-31.

10 Preliminary Identification and Developmental Behavior of a Cytokeratinlike Set of Polypeptides in Drosophila

Margarita Cervera, Alberto Domingo,
Leandro Medrano, Roberto Marco

INTRODUCTION

Intermediate (8-10 mm in diameter) filaments (Lazarides, 1980) represent a distinct class of cytoskeletal structures different from actin filaments (6 mm), microtubules (25 mm), and myosin filaments (15 mm) which have not been reported to exist unambiguously in invertebrates, with the exception of the neurofilament protein from squid giant axons (Humeus & Davidson, 1970).

We have initiated a research program with the goal of identifying and characterizing these important cell constituents in Drosophila melanogaster to exploit the large background in genetics and developmental biology of this organism in the clarification of some of the unsolved questions posed about these cytoskeletal structures.

MATERIALS AND METHODS

Drosophila melanogaster Oregon R files were used. Intermediatelike filaments were prepared, washed, depolymerized, and repolymerized as described by Franke et al. (1981) with slight modifications.

Monoclonal antibodies against cytokeratins were prepared by L.M., immunizing BALB/c mice with weekly injections of human keratin during several months until consistent high titers in the mice sera were found. Monoclonal antibodies were produced after fusion with NS-1 myeloma lines and shown to react specifically with a broad spectrum of keratin bands from human epithelia by immunoprecipitation and immunoblotting. Immunofluorescence studies on frozen sections of human, mouse, and rat epidermal tissues, i.e., tongue, thymus, trachea, mammary gland, and different types of cultured cells, indicated an exclusive recognition of epithelial tissue. One of them, KDN-23, has been used in this study.

SDS-polyacrylamide gels were carried out with the method of Laemmli (1970) and two dimensional gels as described by O'Farrell (1975).

Immunoblotting was carried out as described by Towbi et al. (1979).

RESULTS

Characterization of a Set of Polypeptides with Cytokeratinlike Properties from *Drosophila* Adults

When the particulated fraction of *Drosophila melanogaster* adults homogenized in a buffer containing physiological concentrations of salt and 1% Triton was extracted with a series of buffers with different amounts of salt, a distinctive set of polypeptides remains consistently associated with the insoluble materials. This insoluble set of polypeptides can be solubilized using high concentrations of urea as reported by Fraser et al. (1972) and reconstituted as described by Franke et al. (1981) by dialysis against low salt buffers lacking urea. In Figure 10.1, the result of two depolymerization-polymerization cycles applied to the *Drosophila* insoluble material is shown. Poly-

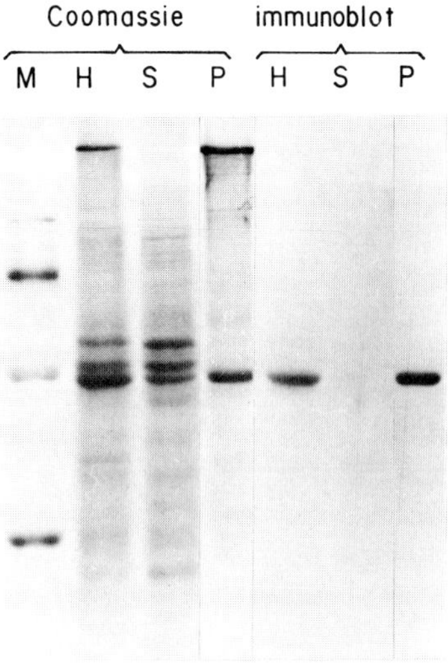

Fig. 10.1. Purification and identification of cytokeratinlike proteins. Adult flies were homogenized and the final pellet was prepared by extraction and two cycles of reconstitution. The 10% SDS-polyacrylamide gels and the immunoblotting were performed as described in Material and Methods. Lanes: M, molecular weight markers; H, whole *Drosophila* homogenate; S, soluble fraction; P, final pellet after reconstitution.

peptides of 200,000, 54,000, and 41,000 molecular weight appear enriched in the material obtained after these two cycles. In Figure 10.1, the reaction of different Drosophila fractions with an antihuman cytokeratin monoclonal antibody (see Materials and Methods) after immunoblotting of a parallel gel is also shown. The 41,000 band reacts conspicuously with this antibody. If the conditions of the reaction of the antibody are made stronger, some other insoluble polypeptides also react suggesting that there is a family of polypeptides of different molecular weights which are recognized by the antibody. Two-dimensional gels of the intermediate filament fraction purified by reconstitution show that the 41,000 band has several forms of different isoelectric point with pIs varying from 5.8 to 6.1.

Negative staining and electron microscopical observation of both purified intermediate filamentlike native fraction and a reconstituted fraction from purified 41,000 sent of polypeptides show 8 nm wide filaments.

Immunofluorescence distribution of the antihuman cytokeratin Drosophila components appears mainly localized in tissues of epithelial origin, ecto and endodermal.

Developmental Behavior of the Drosophila Cytokeratinlike Polypeptides

In Figure 10.2 the differences in the pattern of the intermediate filamentlike polypeptides during Drosophila development is shown. Two major points can be made from the figure. Although a similar set of polypeptides with solubility properties of intermediate filaments are present in all developmental stages, some changes in the pattern are clearly visible which allow to differentiate an embryonic from an adult pattern, which becomes progressively established during the larval stages. On the other hand, the antihuman cytokeratin monoclonal antibody reacting material, particularly the 41,000 reacting set of proteins is typical of the adult pattern, since it becomes clearly visible in extracts of 3d instar larvae (Fig. 10.2b), although if the antibody reaction conditions are forced, it can also be detected in the 2d instar and less readily in the 1st instar larvae extracts.

Extracts of Kc cells, in agreement with their embryonic origin, gave an embryonic pattern and no anticytokeratin cross-reacting material (Fig. 10.2b). Falkner et al. (1981) have presented evidence that a 46,000 and 40,000 protein present in Kc cells and which are modifiable by heat shock, are immunologically related to mammalian vimentin.

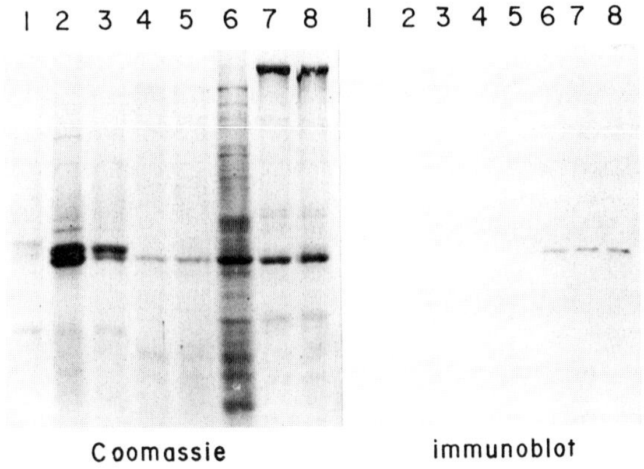

Fig. 10.2. Developmental behavior of the cytokeratinlike proteins. Equivalent amounts of adults, larvae, and embryos of different stages of development were homogenized in 10 mM Tris, 1.40 mM NaCl, 1% Triton pH 7.5. After low speed centrifugation, pellets were solubilized, 10% SDS-polyacrylamide gels and immunoblotting were carried out as described in Materials and Methods. Lanes: 1, Kc cells; 2, embryos of 9 hours of development; 3, embryos of 7 hours of development; 4, 1st instar larvae; 5, 2nd instar larvae; 6, 3rd instar larvae; 7, adult males and 8, adult females.

DISCUSSION

The study of intermediate filament has been until now limited to vertebrate. The implications of their diversity as a reliable test of cellular differentiation (Osborn & Weber 1982) make it even more interesting to find out the functional significance of these structures. In this respect, the absence of information about their presence in invertebrates was puzzling. Therefore, it is reassuring to find that when biochemical, immunological, and electron microscopical criteria for isolation and identification of intermediate filamentlike proteins are applied to Drosophila melanogaster extracts, a set of polypeptides are isolated which fulfill these criteria. The evidence argues quite strongly that the 41,000 set of polypeptides in the adult flies and possibly some of the additional anticytokeratin reacting material are the equivalent in Drosophila of the mammalian cytokeratins.

The identification of this class of cytoskeletal structures in Drosophila melanogaster opens the possibility of localizing the genes coding for them with all the advantages of the genetic, molecular, and developmental manipulations feasible in this organism. Interestingly, we report here a marked developmental regulation of the intermediate filamentlike polypeptides with the antihuman cytokeratin reacting material restricted to adult and larval tissues. This result suggest the possible existence in Drosophila of an embryonic set of intermediate filament (cytokeratin?)like proteins which is now under investigation.

ACKNOWLEDGMENTS

This work has been carried out with grants of the Fondo de Investigaciones Sanitarias and CAICYT to R.M. M.C. has a fellowship of the Spanish Ministry of Education and Sciences and A.D. from the Fondo de Investigaciones Sanitarias.

REFERENCES

Falkner, F. G., Saumweber, H. & H. Biessmann, "Two Drosophila melanogaster proteins related to intermediate filament proteins of vertebrate cells." J. Cell. Biol., 91, 175-83 (1981).

Franke, W. W., Winter, S., Ground, C., Schmid, E., Schiller, D. L. & E.-D. Jarasch, "Isolation and characterization of desmosome-associated tonofilaments from rat intestinal brush border." J. Cell. Biol., 90, 116-27 (1981).

Fraser, R. D. B., MacRae, T. P. & E. Suzuki. Keratins. Their Composition, Structures and Biosynthesis. Charles C. Thomas. Springfield, Il., 1972.

Huneeus, F. C. & P. F. Davison, "Fibrilar proteins from squid axons. I. Neurofilament protein." J. Mol. Biol., 52, 415-28 (1970).

Laemmli, U. K. "Cleavage of structural proteins during the assembly of the head of bacteriophage T_4." Nature, 227, 680-85 (1970).

Lazarides, E. "Intermediate filaments as mechanical integrators of cellular space." Nature, 283, 249-56 (1980).

O'Farrell, P. H. "High resolution two-dimensional electrophoresis of proteins." J. Biol. Chem., 250, 4007-4021 (1975).

Osborne, M. & K. Weber. "Intermediate filaments: Cell-type-specific markers in Differentiation and Pathology." Cell, 31, 303-06 (1982).

Towbin, H., Staehelin, T. & J. Gordon. "Electrophoretic transfer of proteins from polyacrylamide gels to nitrocellulose sheets: Procedure and some applications." Proc. Natl. Acad. Scil USA, 76, 4350-54 (1979).

11 Isoforms of Human Blood Platelet α-Actinin

Yannick Gache, Françoise Landon, Anna Olomucki

The polymorphism of muscle α-actinin has already been established. The α-actinins from red and white porcine muscle differ chemically and can be separated by polyacrylamide gel electrophoresis in nondissociating conditions (Suzuki et al., 1973). Similar studies also reveal the presence of a "hybrid type" of α-actinin (Robson & Zeece, 1973). α-Actinins from smooth and skeletal chicken muscle show chemical and immunological differences, as well (Bretscher et al., 1979).

In contrast, little is known about the polymorphism of nonmuscle α-actinin. We have previously shown that α-actinin purified from human blood platelets appears as a doublet when analyzed by polyacrylamide gel electrophoresis in dissociating conditions (Landon & Olomucki, 1983). The two bands of the doublet are polypeptide chains of about 100 kDa. Similarly, the presence of a doublet has been demonstrated in platelet α-actinin (Rosenberg et al., 1981) and in brain α-actinin (Duhaiman & Bamburg, 1982). Another study of Hela cells indicates the existence of two forms of α-actinin (Burridge & Feramisco, 1981).

The heterogeneity of the polypeptide chains suggests the presence of several dimeric isoforms. Here, the homodimeric and heterodimeric forms of platelet α-actinin have been analyzed.

Abbreviations: SDS-PAGE, polyacrylamide gel electrophoresis in the presence of sodium dodecyl sulfate; EGTA, ehtylene glycol bis (β-aminoethyl ether)-N,N'-tetraacetic acid.

METHODS

Purification of α-actinin: α-actinin was purified according to Landon & Olomucki, 1983. Human blood platelets were washed, frozen in liquid nitrogen, and stored at -80°C prior to extraction.

Viscosity was measured at 20°C as previously described (Landon & Olomucki, 1983).

Binding of α-actinin to F-actin: Rabbit skeletal muscle F-actin (0.4 mg/ml) and platelet α-actinin (40 μg/ml) were mixed in buffer (10 mM Tris-HCl, 0.1 mM ATP, 40 mM KCl, 1 mM 2-mercaptoethanol, pH 7.0) containing either 0.275 mM EGTA or 0.1 mM $CaCl_2$. After 30 min at room temperature, F-actin was sedimented by centrifugation for 40 min at 160,000 g. The distribution of αactinin in pellets and supernatants was examined by SDS-PAGE.

RESULTS

SDS-PAGE Analysis of Purified α-actinin

To improve the separation between the bands previously observed in purified α-actinin (Landon & Olomucki, 1983), our final α-actinin purification step involving hydroxyapatite chromatography was modified by performing the elution with a more gradual phosphate gradient (Fig. 11.1). Analysis by SDS-PAGE revealed three bands in the 100 kDa region, designated a, b, and c (Fig. 11.1). Four fractions were pooled from the major peak eluting at 0.1 M phosphate and the minor peak eluting at 0.18 M phosphate. Fraction I is an a-rich fraction with small quantity of b. Fraction II contains equal amounts of a and b. Fraction III is b-rich with a little of a and c. Fraction IV contains only c. (In our previous work, only the major peak was analyzed by SDS-PAGE and the a and b polypeptide chains were observed; the c polypeptide chain was just detectable at the trailing edge of the peak.) Fractions I, II, III, and IV represent 17-19%, 64-66%, 11-12%, and 5.3-5.4% of the total purified α-actinin, respectively (values from two purifications).

Interaction of the Different α-Actinin Fractions with F-Actin

α-Actinin composed of the a and b polypeptide chains is able to interact with F-actin in a calcium-sensitive manner

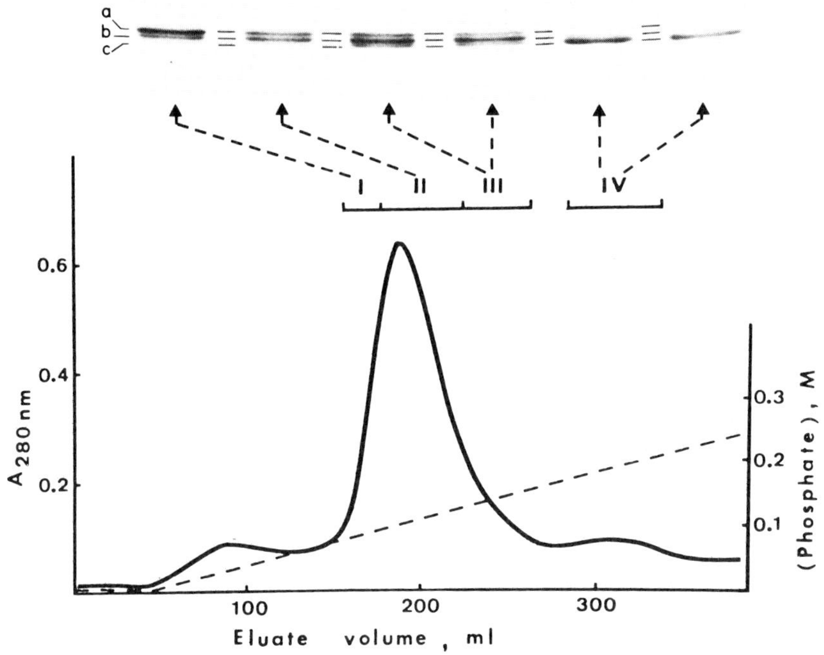

Fig. 11.1. Chromatography on hydroxyapatite. Partially purified platelet α-actinin was applied to a hydroxyapatite column which was equilibrated in 5 mM potassium phosphate, pH 6.8. The elution was performed using a linear gradient (5-240 mM) of potassium phosphate, pH 6.8. I, II, III, and IV represent the pooled fractions.

Upper part: SDS-PAGE of the fractions was performed according to Laemmli (1973) on a 11.25% acrylamide gel. Only the 100 kDa region of the gel stained with Coomassie blue is shown.

(Landon & Olomucki, 1983). The interactions of the b-rich fraction III and the c-rich fraction IV with F-actin were examined. In low free-calcium solutions (0.275 mM EGTA), both fractions increased the viscosity of F-actin to the same extent, and in high free-calcium solutions (10^{-4} M Ca^{2+}), the viscosity was unchanged even at molar ratios as high as 1:100 (α-actinin to actin).

The binding of α-actinin to F-actin was demonstrated by sedimentation experiments; in low free-calcium solutions, the a, b, and c polypeptide chains of fraction III sedimented with F-actin.

PAGE Analysis of Native α-Actinin and Subunit Composition

The PAGE pattern of fractions I, II, III, and IV is shown in Figure 11.2a. In nondissociating conditions, six forms were seen and numbered from 1 to 6 according to their anionic mobility. Forms 2 and 3 appear in the a-rich fraction I; forms 1, 2, and 3 in the (a+b)-rich fraction II; forms 1, 2, 3, 4, and 5 in fraction III, and form 6 in the c-rich fraction IV. The immunological cross-reactivity of the six forms with muscle α-actinin is shown in Figure 11.2b. (The rabbit antibody against porcine skeletal muscle α-actinin was a generous gift of Dr. B. Jackusch.)

In order to determine the polypeptide chain composition of these six forms, they were analyzed by SDS-PAGE (Fig. 11.2c). Forms 1, 3, and 6 result from the dimeric combination of identical polypeptide chains, bb, aa, and cc, respectively. However, forms 2, 4, and 5 are composed of different polypeptide chains, ab, bc, and ac, respectively.

One Dimensional Peptide Mapping

Limited proteolysis of a, b, and c polypeptide chains was performed according to Cleveland et al. (1977) to examine the similarity between the chains. All of the digested polypeptides yield some peptides with similar mobilities (i.e., 40, 30, 25, and 18 kDa). Nevertheless, only one peptide (25 kDa) is common to the three chains. On the basis of maps of a and b, at least one a-specific and b-specific peptide can be recognized, but most of the smaller peptides (<30 kDa) have similar or identical mobilities. In contrast, a comparison of the peptide maps of c and a revealed only two common peptides (Fig. 11.3).

DISCUSSION

The a, b, and c polypeptide chains observed in platelet α-actinin are subunits of the dimeric α-actinin (200,000 daltons). The three polypeptide chains have molecular weight close to 100,000. They cross-react immunologically with muscle α-actinin. Purified platelet α-actinin containing different ratios of the three polypeptide chains are able to induce gelation of F-actin and to bind to F-actin in a calcium-sensitive manner, similar to other nonmuscle α-actinins (Feramisco & Burridge, 1981).

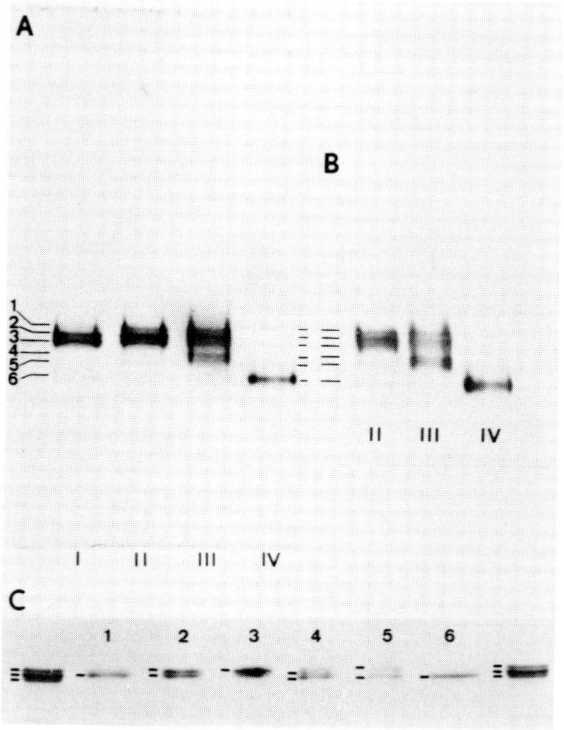

Fig. 11.2. Analysis of the multiple forms of α-actinin by PAGE in nondissociating conditions (A and B), and by SDS-PAGE (C). In A and B, the pooled fractions I, II, III, and IV were analyzed by PAGE in the conditions described by Neville (1971) except for the omission of SDS. In A, a 5% acrylamide gel was stained with Coomassie blue. In B, an immunoblot of this gel has been revealed by immunoperoxidase reaction (Towbin et al., 1979).

In C, each band visualized in A was excised, and then electrophoresced in dissociating conditions (see Fig. 11.1). Only the 100 kDa region of the gel is shown. As a reference, bands a, b, and c are shown in the first and last slots.

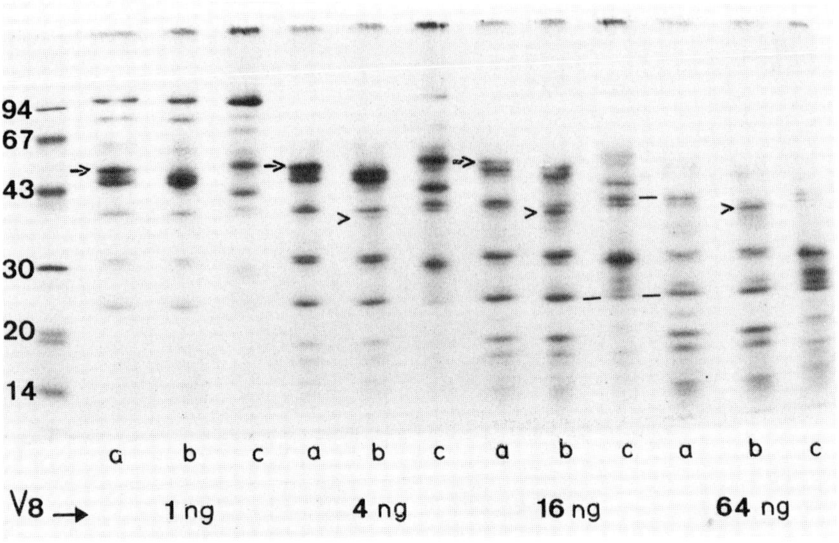

Fig. 11.3. One dimensional peptide mapping of the a, b, and c bands. The procedure of Cleveland et al., 1977, was used. a, b, and c bands, each containing about 2 μg of protein, were cut from SDS-gels and placed in sample wells of an SDS-gel. The gel slices were overlaid with different amounts of *Staphylococcus aureus* V8 protease. The 15% acrylamide resolving gel was stained with Coomassie blue. The first lane contains the molecular weight marker proteins (kDa). Peptides with similar mobilities are denoted by bars, and peptides characteristic of a (arrows) and of b (arrowheads) are indicated.

When fractions containing different ratios of the three subunits are analyzed in nondissociating conditions, six forms are observed. Analysis of the subunit composition of each form confirmed that the six dimers are: aa, ab, ac, bb, bc, and cc.

The a and b subunits appear very similar as determined by one dimensional peptide mapping. Since the b subunit is smaller than the a subunit, one cannot exclude the possibility that b is a degradation produce of a. A preliminary examination of platelet lysates prepared in the presence and absence of protease inhibitors has indicated that the b subunit is probably a proteolytic product of the a subunit. Although the c subunit has a smaller size than the a and b subunits, its distinct peptide map suggests that it is not a degradation product.

In platelets, there are probably only two actual subunits (a and c) and thus, three isoforms: two homodimers (aa, cc) and one heterodimer (ac).

ACKNOWLEDGMENT

Hélène Gerbail is greatly acknowledged for her expert technical assistance in the preparation of α-actinin and actin.

REFERENCES

Bretscher, A., Vandekerckhove, J., & Weber, K., α-Actinins from chicken skeletal muscle and smooth muscle show considerable chemical and immunological differences, Eur. J. Biochem. 100, 237-43 (1979).

Burridge, K., & Feramisco, J. R., Non-muscle α-actinins are calcium-sensitive actin-binding proteins, Nature 294, 565-67 (1981).

Cleveland, D. W., Fischer, S. G., Kirschner, M. W., & Laemmli, U. K., Peptide mapping by limited proteolysis in sodium dodecyl sulfate and analysis by gel electrophoresis, J. Biol. Chem. 252, 1102-1106 (1977).

Duhaiman, A. S., & Bamburg, J. R., Comparative studies on muscle and brain α-actinin, J. Cell Biol. 95, 286a (1982).

Landon, F., & Olomucki, A., Isolation and physico-chemical properties of blood platelet α-actinin, Biochim. Biophys. Acta 742, 129-34 (1983).

Laemmli, U. K., & Favre, M., Maturation of the head of bacteriophage T4, J. Mol. Biol. 80, 575-599 (1973).

Neville, D. M., Molecular weight determination of protein-dodecyl sulfate complexes by gel electrophoresis in a discontinuous buffer system, J. Biol. Chem. 246, 6328-6334 (1971).

Robson, R. M., & Zeece, M. G., Comparative studies of α-actinin from porcine cardiac and skeletal muscle, Biochim. Biophys. Acta 295, 208-224 (1973).

Rosenberg, S., Stracher, A., & Burridge, K., Isolation and characterization of a calcium-sensitive α-actinin-like protein from human platelet cytoskeletons, J. Biol. Chem. 256, 12986-12991 (1981).

Suzuki, A., Goll, D. E., Stromer, M. H., Singh, I., & Temple, J., α-Actinin from red and white porcine muscle, Biochim. Biophys. Acta 295, 188-207 (1973).

Towbin, H., Staehelin, T., & Gordon, J., Electrophoretic transfer of proteins from polyacrylamide gels to nitrocellulose sheets: Procedure and some applications, Proc. Natl. Acad. Sci. USA 76, 4350-4354 (1979).

12 Alpha-Actinin in Filaments of the Z Band Lattice

M. A. Goldstein, D. B. Zimmer,
J. P. Schroeter, R. L. Sass

The Z band is a dynamic complex protein lattice. A diagram of a longitudinal section of several Z band subunits is shown in Figure 12.1. The simplest form of Z band is shown at the top. The orientation is what we call the 24 nm orientation. The perpendicular distance between two adjacent axial filaments from the same sarcomere is 24 nm. The zig-zag appearance is due to the arrangement of the cross-connecting filaments. Fast skeletal muscle such as rabbit psoas, has one complete subunit with two layers of cross-connecting filaments separated by an axial distance of 38 nm (shown in the middle of Fig. 12.1). Cardiac and slow skeletal muscle most often have three subunits (as shown at the bottom of Fig. 12.1). The basic subunit has the dimensions of 24 × 24 × 38 nm.

Cross sections of the Z band show that the ends of the thin filaments interdigitate to form a centered square array of crosscut axial filaments. The two longitudinal planes of section that give maximum reinforcement of the axial filaments, as expected from tetragonal symmetry are at 24 nm and at 17 nm, the half-diagonal of the 24 nm square. We have carried out three-plane analyses of Z bands in cardiac and skeletal muscle using electron microscopy, optical diffraction (Goldstein et al., 1977), optical reconstruction (Goldstein et al., 1979; Goldstein et al., 1982), computer modeling (Schroeter et al., 1983), and most recently immunoelectron microscopy (Zimmer, 1983).

Optically filtered reconstructions of both cardiac and skeletal muscle suggest that the axial filaments are the ends of the thin filaments from neighboring I bands (Goldstein et al., 1979). Electron micrographs of logitudinal sections show a thickened region of the thin filament in the last 38 nm interval just before the Z band, suggesting some additional protein is added here.

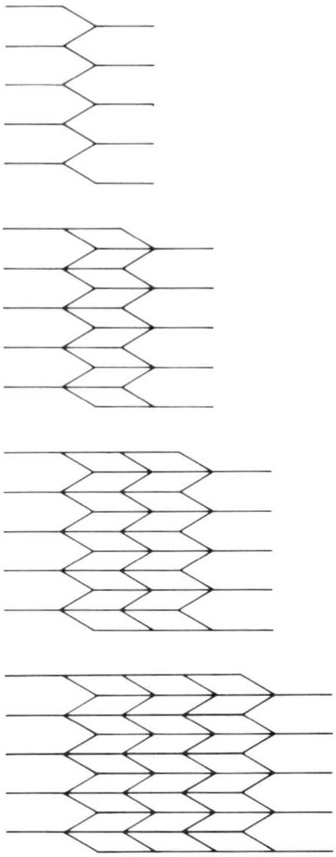

Fig. 12.1. A diagram of a longitudinal section of several Z band subunits in the 24 nm orientation. The simplest form of Z band (shown at the top) consists of axial filaments and one set of cross-connecting filaments. A complete subunit has two sets of cross-connecting filaments separated by an axial distance of 38 nm.

The Z band lattice in cross section can have two different arrays of cross-connecting filaments, called small square and basket weave (Fig. 12.2). Optical reconstructions of electron micrographs of Z bands in cross section show both arrays in the same myofibril (Goldstein et al., 1979, 1982). Since the interaxial spacing (i.e., 24 nm) is the same, these data suggest that the cross-connecting filaments can move independently of the spacing of the axial filaments. The projected profiles of the cross-connecting filaments give the appearance of small squares or of pin wheels. Each axial filament has four cross-connecting filaments emanating. Each cross-connecting filament is attached at one end to the axial filament of one sarcomere and the other end is attached to the axial filament from the adjacent sarcomere. We have modeled in cross-sectional projections how the cross-connecting filaments can move (Goldstein et al., 1982). By changing the apparent diameter of cross-connecting filaments and the curvature of these filaments one can generate images that resemble electron micrographs of Z bands and in particular the optically filtered images of these Z bands.

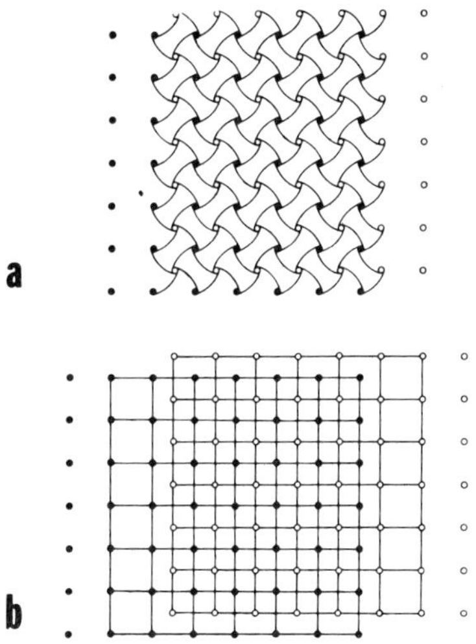

Fig. 12.2. Diagram of two different arrays of cross-connecting filaments in cross sections of the Z band lattice, called basket weave and small square.

How we have gone to computer modeling to help display a unified three dimensional model. After changing various parameters such as diameter, curvature, and angle of the cross-connecting filaments, we can display the model at various orientations (Fig. 12.3). Computer generated arrays of mutiple subunits can then be photographically reduced and compared to an optical reconstruction. For example, using an angle of 30-35° we can explain the change in curvature and length of the cross-connecting filaments, that is the change from small square pattern to basket weave pattern, by a 60° rotation of the axial filaments such that the ends of the cross-connecting filaments wrap around the axial filaments. The data obtained so far from all these methods suggest that the Z band is a complex filamentous lattice.

The most well-characterized protein specifically localized to the Z band is alpha-actinin. When we began our experiments to localize alpha-actinin, most people still thought that alpha-actinin was the amorphous component. Our structural studies, however, suggested to us that alpha-actinin was probably more of interest as an actin-binding protein since we believed that a major component of the axial filaments was actin. Initially we predicted a distribution of binding only at the points were the cross-connecting filaments link to the axial filaments, that is, in longitudinal sections at every 38 nm in the Z lattice. Alpha-actinin antibody coupled to fluorescein showed the characteristic distribution in the light microscope at the Z bands. Direct binding of the Fab fragment of the alpha-actinin antibody

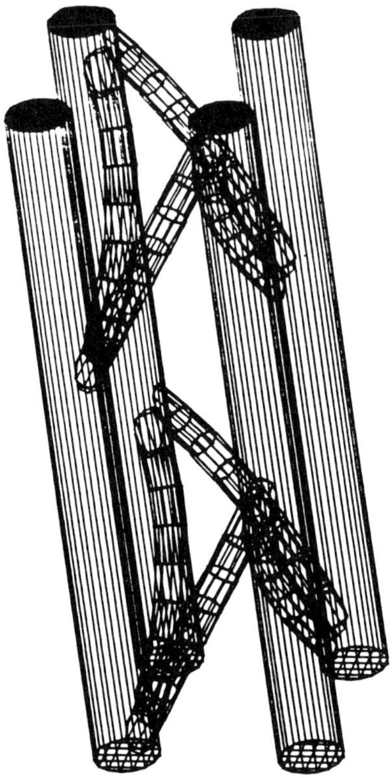

Fig. 12.3. Computer generated display of model subunit in three dimensions shown in perspective from a near 17 nm orientation.

showed a uniform distribution within the Z lattice as confirmed by optical diffraction. The increase in electron density in the Z lattice in antibody treated specimens versus control was quantitated using a scanning laser densitometer. A comparison of average densities for the four groups—antialpha-actinin, buffer control, antialpha-actinin + alpha-actinin, and alpha-actinin—showed no significant difference between the three controls and a significant difference ($p < 0.001$) between the treated and the controls. The axial filaments and the cross-connecting filaments of the Z lattice in antialpha-actinin labelled Z bands were significantly larger in diameter than those of control samples. The Fab antialpha-actinin antibody was confined to the Z lattice and did not extend into the I band, even along the thickened region of 38 nm at the edge of the Z band.

In summary, our Z band model differs significantly from previous models. The thin filaments do not subdivide at the edge of the Z but interdigitate. They individually traverse to the opposite side of the Z lattice maintaining polarity. The cross-connecting filaments appear to be uniform but can move to form two different arrays even at the same interaxial filament spacing. The model accounts for all of the major features observed so far but does not preclude additional proteins in the Z lattice. We have presented data that indicate that alpha-actinin is a major component of the cross-connecting filaments and that alpha-actinin is distributed to some extent along the axial filaments. The evidence so far indicates that actin and alpha-actinin are the primary components of the Z band lattice. The intriguing question is why and how do the cross-connecting filaments move.

REFERENCES

Goldstein, M. A., Schroeter, J. P., and Sass, R. L. 1977. Optical diffraction of the Z lattice in canine cardiac muscle. J. Cell Biol. 75:818-36.

Goldstein, M. A., Schroeter, J. P., and Sass, R. L. 1979. The lattice in canine cardiac muscle. J. Cell Biol. 83:187-204.

Goldstein, M. A., Schroeter, J. P., and Sass, R. L. 1982. The Z band lattice in a slow skeletal muscle. J. Muscle Res. Cell Motility 3:333-48.

Schroeter, J. P., Sass, R. L., and Goldstein, M. A. 1983. Modeling filamentous structures using DRAW-3D. Proceedings of Prophet Users Annual Colloquium, Airlie, Virginia, June 6-9, 1983.

Zimmer, D. B. 1983. Alpha-actinin is a component of the cross-connecting and axial filaments of the Z band lattice. Biophys. J. 41(2): 99a abstract.

13 Spatial Organization of Cytoskeletons and Myofibrils in Embryonic Chick Skeletal Muscle Cells in Vitro

Yuji Isobe and Yutaka Shimada

Many eucaryotic cells are known to contain a highly interconnected intracellular filamentous network usually known as the cytoskeleton. The organization of the cytoskeletal network varies in different cell types and also in different parts of a single cell (Temmink and Spiele, 1980; Schliwa and van Blerkom, 1981). In developing skeletal muscle cells, myofibrils are formed within the cytoskeletal lattice as a result of the regular assembly of contractile and regulatory proteins. Since many of these muscle proteins are ubiquitous in the cytoskeletons of nonmuscle cells, the myofibrillar system can be regarded as a specialized form of the cytoskeleton. Thus, studies relating cytoskeletal elements to the assembly of myofibrils appear to be important in understanding the phylogeny of the motile systems. The present paper gives an ultrastructural description of the spatial organization of the cytoskeleton and myofibrils in embryonic skeletal muscle cells in vitro by scanning (SEM) and transmission electron microscopy (TEM), using detergent-extraction with or without cryofracturing.

CYTOSKELETONS IN MYOGENIC CELLS AS SEEN BY SEM

Embryonic skeletal muscle cells cultured on gold-coated coverslips were extracted with Triton X-100 (Triton) and prepared for SEM by critical point drying followed by metal coating (isobe and Shimada, 1983). By this procedure, most membranous material was solubilized, leaving behind filament networks.

Cytoskeletons of a young myotube appeared to form a loose network composed of filaments that generally aligned along the

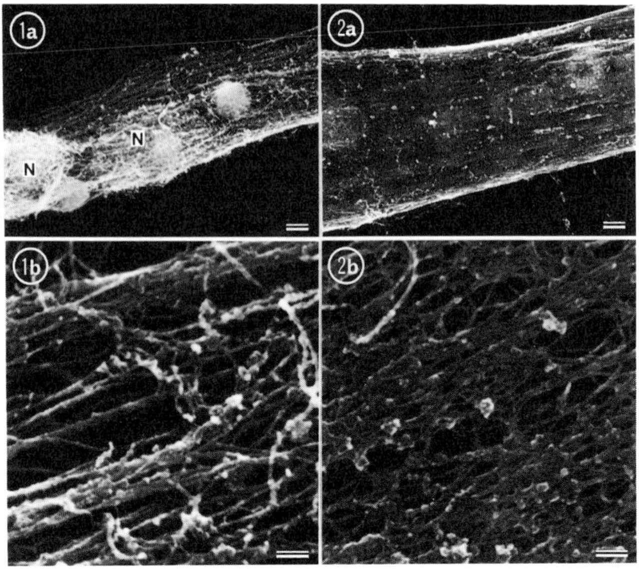

Figs. 13.1 and 13.2. Scanning electron micrographs of myogenic cells cultured on gold-coated coverslips. Cell membranes and soluble cytoplasm have been extracted with Triton X-100 (Triton), exposing cytoskeletal lattices.

Fig. 13.1a. Cytoskeletons of a young myotube within 1-day-old culture appear to form a loose network of filamentous elements. Nucleus, N. ×4,800. Scale = 1 μm.

Fig. 13.1b. An enlargement of Fig. 13.1a. Cytoskeletal elements align generally along the long axis of the cell. ×72,000. Scale - 0.1 μm.

Fig. 13.2a. A myotube at a more advanced stage (6-day-old culture) possesses a dense sheetlike cytoskeletal network that positions just underneath the cell membrane. Internal structures such as myofibrils that locate deep in the cell are obscured by this sheetlike structure. ×4,800. Scale = 1 μm.

Fig. 13.2b. A higher magnification view of Figure 13.2a. A dense subsarcolemmal cytoskeletal network appears to be composed of longitudinal filamentous elements interconnected with those aligned randomly. ×72,000. Scale = 0.1 μm.

cell axis (Fig. 13.1). At a more advanced state of myogenesis, cytoskeletons of a myotube were different from those at the early stages: the subsarcolemmal cytoskeleton formed a dense sheetlike structure, thus obscuring the cytoskeletal network and myofibrils both located depp in the cell (Fig. 13.2a). At a higher magnification the surface cytoskeleton was found to be composed of filamentous elements aligned logitudinally and interconnected with those arrayed randomly (Fig. 13.2b). The difference in the appearance of the Triton-extracted myogenic cells at these two stages appeared to show the different developmental stages of the cytoskeletal organization under the sarcolemma.

CYTOSKELETONS AND MYOFIBRILS REVEALED BY TEM OF PLATINUM REPLICAS OF CRYOFRACTURED SAMPLES

In order to observe the spatial relation between the cytoskeletal network and myofibrils located depp in the cell, we examined platinum replicas of Triton-extracted and cryofractured muscle cells grown on coverslips by TEM (Figs. 13.3 and 13.4). With cryofracturing and higher resolving power of TEM, two major cytoskeletal domains were distinguished by distinctly different arrangements of various filaments: (a) a dense network of highly branched and anastomosed filaments just underneath the cell membrane, and (b) a loose filament network filling the endoplasmic space. The transition between these two domains was gradual. The former structure corresponds to the sheetlike architecture observed by SEM. In both cytoskeletal domains, all the filaments were categorized in three different size classes: 2-4 nm (3 nm filaments), 8-11 nm (microfilaments), and 12-14 nm (intermediate filaments) in diameter. Microtubules with a diameter of 22-24 nm were easily distinguished from these filaments. Larger than usual diameter of each filamentous structure seemed to be caused by the platinum replication. Nascent myofibrils were clearly observed as long filament bundles within the loose endoplasmic network. A, I, and Z bands were discernible in these myofibrils.

By deep-etching of the cryofractured samples, the relation between the cytoskeletal elements and myofibrils was clarified. All the filament types and microtubules were found to be associated either at their lateral surfaces with one another in a crossover pattern or at their ends with the surfaces of each other, resulting in a triangular pattern. 3 nm filaments were seen to connect these cytoskeletal elements with each other and also to

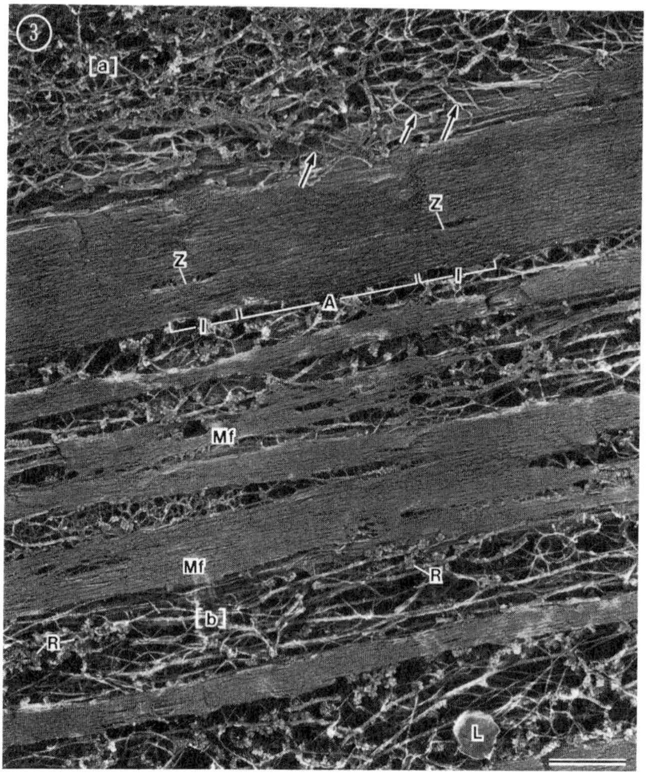

Figs. 13.3 and 13.4. Transmission electron micrographs of platinum replicas from Triton-extracted, cryofractured and freeze-dried cytoskeletons in myotubes cultured for 6 days.

Fig. 13.3. Low magnification view of a cryofractured cytoskeleton of a myorube. Two major cytoskeletal domains are demonstrated: (a) dense network of filaments just beneath the cell membrane, and (b) loose cytoskeletal network located deep in the cell. Myofibrils (Mf) are discernible in the deep area as filament bundles possessing A (A), I (I) and Z (Z) bands. The circumferences of myofibrils are intimately associated with cytoskeletal filaments (arrows). 12-14 filaments are continuous with the subsarcolemmal network composed mainly of finer filaments (8-11 nm). Lipid droplet, L; ribosomes, R ×35,000. Scale - 0.5 μm.

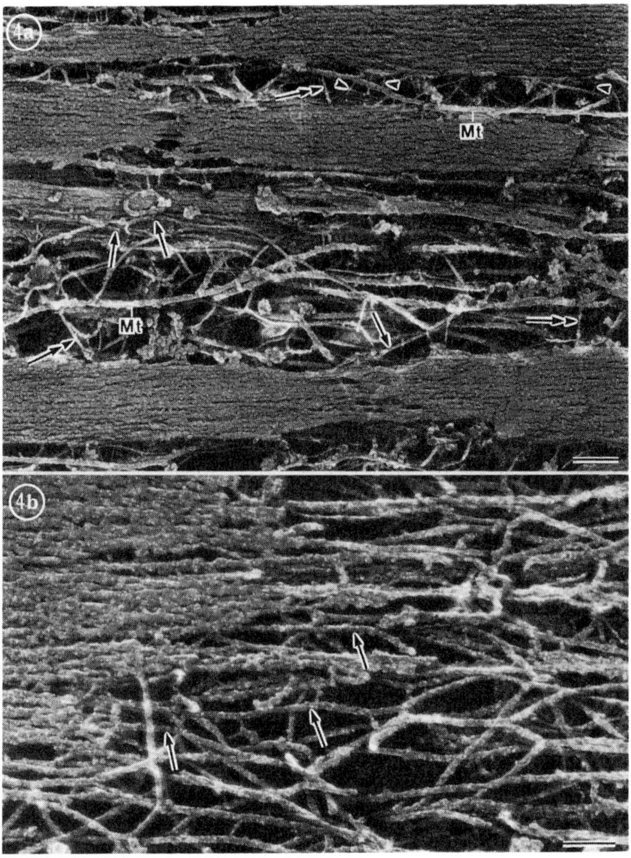

Fig. 13.4a. Close-up view of cytoskeletons and myofibrils. The circumferences of myofibrils are connected with the lateral surfaces of intermediate filaments at the Z band (single arrows), and also with the ends of those filaments at the A band (double arrows). Short 3 nm filaments (arrowheads) are occasionally seen to interlink microtubules (Mt) with myofibrils. ×59,000. Scale - 0.2 μm.

Fig. 13.4b. High magnification view of an obliquely fractured myofibril. Many intermediate filaments appear to fan out from the myofibril (arrows). ×120,000. Scale = 0.1 μm.

link the other types of cytoskeletal elements with myofibrils (Fig. 13.4a). In addition to 3 nm filaments, the circumferences of myofibrils were intimately related with the ends or lateral surfaces of intermediate filaments (Figs. 13.3 and 13.4a). Some of such filaments associated with myofibrils on the one hand were seen to be continuous with the subsarcolemmal filament network on the other (Fig. 13.3a). At the places where myofibrils were fractured obliquely, these cytoskeletal filaments were seen to fan out from the myofibrils (Fig. 13.4b).

REMARKS

In this study, we showed that incipient myofibrils were connected with the cytoskeletal network. Such connection has also been recognized by some workers in their whole-mount studies of muscle cells (Porter and Anderson, 1980; Peng et al., 1981). In adult skeletal muscles, Lazarides (1980) has shown that myofibrils are interconnected with each other by intermediate filaments in the form of a lattice. Recently, Wang and Ramirez-Mitchell (1983) demonstrated the existence of an extensive network of longitudinal intrafibrillar bridges of intermediate filaments that connected adjacent sarcomeric structures of the same myofibrils and transverse interfibrillar bridges that linked neighboring myofibrils in KI-extracted adult muscle. They suggested that these filamentous bridges would be capable of transmitting tension generated by myofibrils in both directions of the cell. Therefore, the close association of cytoskeletal filaments with myofibrils and sarcolemma in developing muscle cells observed in this study may represent the early stages of the apparatus of anchoring and tension-transmission of myofibrils to the neighboring myofibrils and to the cell membranes.

REFERENCES

Isobe, Y. and Shimada, Y. Myofibrillogenesis _in vitro_ as seen with the scanning electron microscope. Cell Tissue Res. _231_, 481-94 (1983).

Lazarides, E. Intermediate filaments as mechanical integrators of cellular space. Nature _283_, 249-56 (1980).

Peng, H. B., Wolosewick, J. J. and Cheng, P.-C. The development of myofibrils in cultured muscle cells: a whole-mount and thin-section electron microscopic study. Develop. Biol. 88, 121-36 (1981).

Porter, K. R. and Anderson, K. The morphogenesis of myofibrils from the microtrabecular lattice as observed in cultured myoblasts. In Muscle Contraction: Its Regulatory Mechanisms. (S. Ebashi et al., eds.), pp. 527-40. Japan Sci. Soc. Press, Tokyo/Springer-Verlag, Berlin (1980).

Schliwa, M. and van Blerkom, J. Structural interaction of cytoskeletal components. J. Cell Biol. 90, 228-35 (1981).

Temmink, J. H. M. and Spiele, H. Different cytoskeletal domains in murine fibroblasts. J. Cell Sci. 41, 19-32 (1980).

Wang, K. and Ramirez-Mitchell, R. A network of transverse and longitudinal intermediate filaments is associated with sarcomeres of adult vertebrate skeletal muscle. J. Cell Biol. 96, 562-70 (1983).

14 Localization of Myosinlike ATPase Activity in Non-muscle Cells: An Histochemical Study

G. Giordano-Lanza, V. Cimini, S. Montagnani

INTRODUCTION

Actin and myosin may provide the force required for all cytoplasmic and cell movements in a variety of nonmuscle cells, but while in muscle myosin is the major protein, it is a minor protein in nonmuscle cells and may have different sizes and shapes. However, the essential features of myosin are its ability to bind reversibly to actin and its ATPase activity.

In both muscle and nonmuscle cells myosin heads contain the catalytic site for ATP hydrolysis and the actin-binding site. With the calcium-cobalt method it is possible to visualize the produce of ATP hydrolysis. This reaction is widely used to classify various muscle fiber types on the basis of the differences due to the acid or alkali stability of the myofibrillar ATPase reaction.

We tried to use a slightly modified calcium-cobalt method to localize the product of ATP hydrolysis in some nonmuscle cells containing contractile proteins and to characterize the type of ATPase activity using some specific inhibitors. Our purpose was to provide some information about the localization and the activity of myosin in the intact nonmuscle cell.

METHODS

The cell strains used were NIH 3T3 transformed with Balb MSV, NIH 3T3 transformed with LK2V, FRT of epithelial origin, thyreoglobulin producing TL 5 and thyroid 1 5G from a thiouracil induced tumor.

Cells in their plastic dishes were fixed for 5 min with 3% paraformaldehyde in PBS (pH 7.4) at room temperature, then washed in PBS and stained for myofibrillar ATPase after acid (Brooke and Kaiser, 1970) or alkaline preincubation (Guth and Samaha, 1969); the histochemical method used for the localization of ATPase activity was that of Padykula and Herman.

Cells were incubated for 30 min at 37°C in the presence of 18 mM Ca and 4 mM ATP at pH 9.4; the buffering agent was sodium barbital. After incubation, cells were washed twice in 2% calcium chloride for 3 min, transferred to 2% cobaltous chloride for 3 min, washed several times in distilled water, immersed in 1% yellow ammonium sulfide for 1 min, washed under the tap, and mounted with a drop of glycerol. Permeabilized cells were treated with 0.1% Triton X-100 in PBS for 15 min at room temperature after fixation.

All inhibitors were incorporated into the incubating medium and thus exerted their effects in the presence of the substrate, except for FITC that was added from a freshly made dimethylformamide solution to an incubation medium containing 0.2 M sucrose, 25 mM Tris/Cl, 25 mM sodium glycine, and 0.1 mM $CaCl_2$ (Pick, 1981). After 10 min at 23°C cells were washed and then submitted to the histochemical reaction as before. Control dishes incubated in the medium without ATP were prepared for each experiment. All reagents were from Sigma Chemical Co., St. Louis 13, Missouri.

RESULTS

Reaction precipitates are visible in the cytoplasm of the cells, where the reaction is known to be limited to the myofibrils (Brooke and Kaiser, 1969); therefore, the histochemical reaction appears altered in the pattern of distribution for some cells examined, depending upon the conditions of preincubation. It is well known that, in the muscle, two myosin ATPase systems may be characterized histochemically by their lability at various pH values.

The specificity of the reaction was checked by using some specific inhibitors (Table 14.1); among them, NEM and PCMB prevent the reaction; ouabaine has no effect on the formation of the precipitate; EGTA in concentration that has been reported to inhibit myosin ATPase prevents the reaction almost totally (Shibata, 1972).

FITC has been reported to compete with ATP (Pick, 1982) in a 1:1 proportion for the active binding-site in sarcoplasmic ATPase; if it is an analogue of ATP, it has to bind also to the ATP reacting site of myosin, and it should inhibit the activity

Table 14.1

Inhibitors used	Inhibiting effect
N-etilmaleimide	+++
FITC	+++
EGTA	++
ouabaine	———
PCMB	+++

of myosin ATPase. In fact, as shown, FITC enters into the cell and completely inhibits the reaction.

Treatment of the cells with the nonionic detergent Triton X-100 (Marchisio, 1982) has been also reported to inhibit the membrane Na-K-ATPase (Clarke, 1975) without influencing the myosin ATPase; in fact, in the cells where there is a strong evidence of the reaction the formation of the precipitates in the inner regions of the cells does not appear to be affected.

DISCUSSION

Recently, contractile proteins similar to actin, myosin, and associated proteins of muscle cells have been described in several nonmuscle cells. While in nonmuscle cells actin is seen ultrastructurally as a class of cytoplasmic fibers known as microfilaments (Goldman, 1979), nonmuscle myosin is rarely seen in situ also by electron microscopy (Pollard, 1982), but it can be isolated as a complex with actin.

The localization of myosin in nonmuscle cells is essential for elucidating its possible functions: since myosin molecules cannot be seen directly, we have attempted to localize their activity. Myosin molecules from both muscle and nonmuscle cells are morphologically indistinguishable (Elliot, 1976); they have two globular heads joined to a tail whose length does not differ by more than 10 nm from species to species.

Myosin isolated from all sources have certain common properties: they all bind to actin and this binding enhances their ATPase activity. We tried to localize the product of ATP hydrolysis by myosin with a histochemical method to obtain some information on the site and the type of the ATPase activity of myosin in some strains of nonmuscle cells.

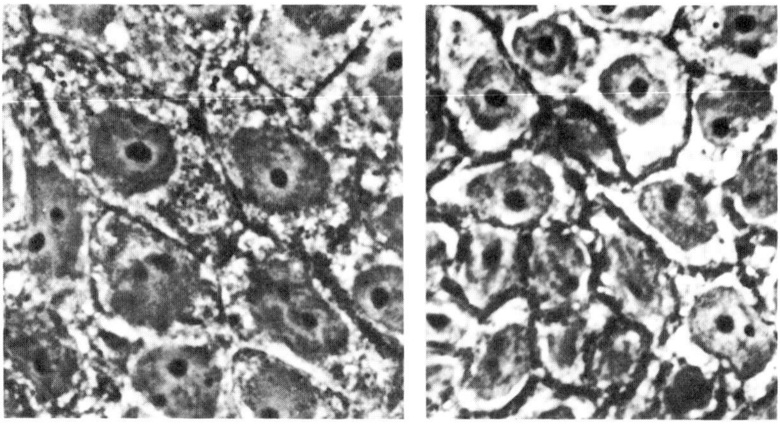

Fig. 14.1. FRT epithelial cells: (a) ATPase localization at pH 9.4; (b) ATPase localization after preincubation at pH 4.3. Phase contrast, 1,060×.

The first evidence is the fact that some nonmuscle cells contain myosin ATPase in a quantity that is conspicuous enough to permit the formation of visible precipitates; this finding is in accord with the rather high quantity of myosin isolated from several nonmuscle cells (Pollard, 1975).

Another evidence is the possibility of a different localization of ATPase activity at different pH values for thyroid epithelial cells tested for the myosin ATPase reaction (see Fig. 14.1a and b). It was found (Kobayashi, 1977) that myosin isolated from the thyroid gland is slightly different from other tissue myosin for its pH curves. The cellular line we have tested, however, has TSH receptors but does not produce tyreoglobulin. It has the typical feature of an epithelial cell line with desmosomes and characteristic disposition of microfilaments around the cytoplasmic membrane. Other microfilaments are present in the cytoplasm. Similar localization of ATPase activities at different pH values is also observed in 1 5G tumor cells from thyroid. Strong acid-stable ATPase activity appears to be associated with membrane protrusions toward neighboring cells, while the alkali-stable activity is mainly in the cytoplasm. The fact that in some cells, normal or transformed but both of epithelial thyroid origin, the acid-stable ATPase activity appears to be localized near the cytoplasmic membrane and the alkali-stable in the cytoplasm seems to indicate that both activities can be present at the same time and in the same cells. Further investigation is needed to clarify the physiological significance of this finding.

REFERENCES

Brooke MH, Kaiser KK: Some comments on the histochemical characterization of muscle ATPase. J. Histochem. Cytochem. 17:431-32 (1969).

Elliott A, Offer G, Burridge K: Electron microscopy of myosin molecules from muscle and non-muscle sources. Proc. R. Soc. Lond. B 193:45-53 (1976).

Goldman R, Milsted A, Schloss J, Starger J: Cytoplasmic fibers in mammalian cells: cytoskeletal and contractile elements. Ann. Rev. Physiol. 41:702-22 (1979).

Kobayashi R, Goldman R, Hartshorne D, Field J: Purification and characterization of myosin from bovine thyroid. The J. Biol. Chem. vol 252, N. 22, 8285-91 (1977).

Padykula H and Herman E: The specificity of the histochemical method for adenosine triphosphatase. J. Histochem. Cytochem. 3:161 (1955).

Pick U: Interaction of Fluorescein Isothiocyanate with nucleotide binding sites of the Ca-ATPase from sarcoplasmic reticulum. Eur. J. Biochem. 121, 187-95 (1981).

Shibata N, Tatsumi N, Tanaka K, Okamura Y: A contractile protein from leucocytes: its extraction. Bioch. Biophys. Acta 256, 565 (1972).

15 Immunochemical and Immunocytochemical Localization of S-100 Protein in the Cilia of Cell Types of Different Species

D. Cocchia, F. Michetti, S. Raffioni, R. Donato

INTRODUCTION

S-100 is an acidic Ca^{2+}-binding protein (Moore, 1965; Calissano et al., 1969) structurally but not immunologically related to calmodulin and other Ca^{2+}-binding proteins (Isobe and Okuyama, 1978). Formerly considered to be specific to the nervous system, S-100 was subsequently shown to be also present in a variety of nonnervous organs in well-defined cell types (Møller et al., 1978; Cocchia and Miani, 1980; Nakajima et al., 1980; Cocchia and Michetti, 1981; Cocchia et al., 1981; Takahashi et al., 1981; Stefansson et al., 1982; Michetti et al., 1983).

We present here data showing that S-100 is localized, inter alia, at the level of the axonemes of the cilia of cell types of different species.

EXPERIMENTAL PROCEDURES

S-100 was purified from ox brain by the method of Moore (1965) slightly modified as reported (Donato, 1978).

The anti-S-100 antiserum was obtained and characterized as described by Zuckerman et al. (1970).

Rat ependyma, planarians, and protozoa were fixed in 4% paraformaldehyde and treated with anti-S-100 antiserum diluted 1:500 using the PAP method as described elsewhere (Cocchia and Michetti, 1981).

RESULTS AND DISCUSSION

By immunocytochemistry, S-100 is localized at the level of the axonemes of the cilia of ependymal cells in the rat brain

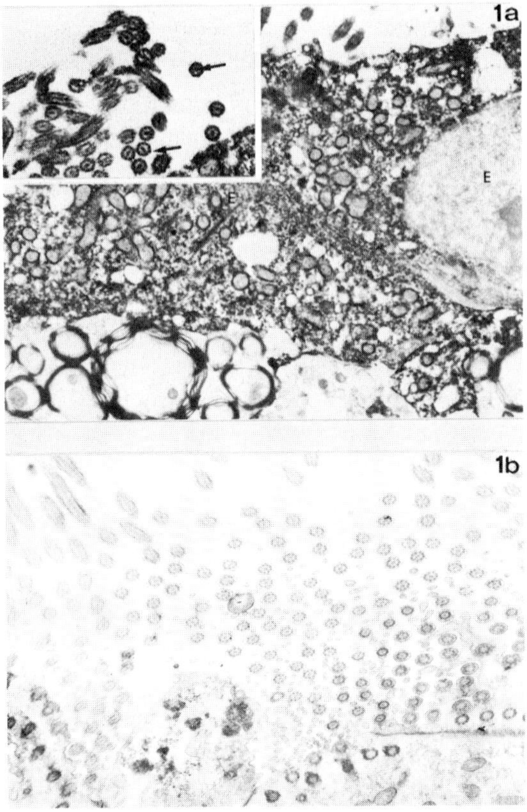

Fig. 15.1. a. Rat ependymal cells treated with anti-S-100 antiserum show reaction product in the cytoplasm as well as in the cilia. ×12,000. Inset. Arrows indicate immunoreaction product at the level of the axonemes. ×12,000.

b. Epithelial ciliated cells of rat trachea treated with anti-S-100 antiserum do not show immunoreaction product. ×12,500.

(Fig. 15.1a), of the epidermal cells of the planarian Dugesia gonocephala (Fig. 15.2a), and of the cilia of the marine protozoan Euplotes crassus (Fig. 15.3a). The preimmune serum and the anti-S-100 antiserum previously absorbed with S-100 fail to stain the cilia (Figs. 15.2b and 15.3b). No immunoreactivity is detected in the epithelial ciliated cells of the rat trachea (Fig. 15.1b), confirming the specificity of the reaction product, where present.

These data indicate that S-100 is structurally related to microtubules (MTs) in different but definite cell types in a wide

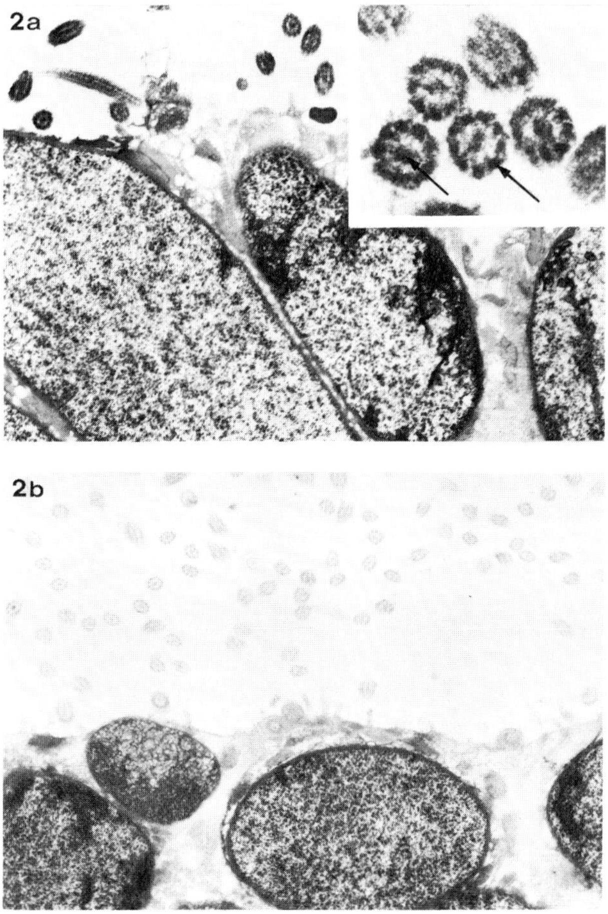

Fig. 15.2a. Planarian treated with anti-S-100 antiserum. Reaction product is present in the cilia (arrows) of the epidermal cells. The cytoplasm and the cytoplasmic rhabdites appear devoid of reaction product. ×20,000. Inset. Arrows indicate S-100 immunoreactivity at the level of the axonemes. ×50,000.

b. Planarian treated with anti-S-100 antiserum previously absorbed with S-100. No reaction product is detectable. ×20,000.

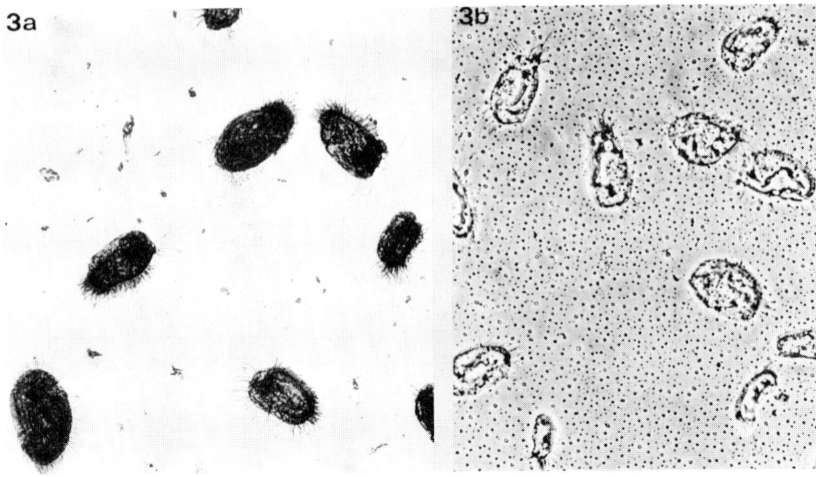

Fig. 15.3a. Euplotes crassus treated with anti-S-100 antiserum. Immunoreaction product is detectable at the level of the cilia. ×75.

b. Euplotes crassus treated with anti-S-100 antiserum previously absorbed with S-100. No immunoreaction deposit is present. ×75.

variety of species. In this respect, it is noteworthy that ciliated cells of different embryonic origin such as tracheal cells of rat are not S-100 immunoreactive.

Consistent with these observations is the recent finding that S-100 inhibits the assembly of brain MT proteins in a dose- and Ca^{2+}-dependent way (Donato, 1983) and potentiates the disassembling effect of Ca^{2+} (Baudier et al., 1982; Donato, 1983), in vitro.

S-100 is regarded to consist of at least two isomers (Isobe et al., 1977) with similar antigenic sites. In this regard, we are not able at present to state whether the molecule(s) responsible for the immunoreactivity in the planarian and in Euplotes crassus corresponds to one or more of the S-100 molecules identified in vertebrates, or is a different molecule bearing the S-100 antigenic determinants. In any case, ox S-100 and the S-100-like immunoreactive material in planarian have been shown not to be immunologically identical (Michetti and Cocchia, 1982). Data are lacking in this respect, concerning the S-100 immunoreactivity in Euplotes crassus.

REFERENCES

Baudier J., Briving C., Deinum J., Haglid K., Sörskog L. and Wallin M. Effect of S-100 proteins and calmodulin on Ca^{2+}-induced disassembly of brain microtubule proteins in vitro. FEBS Lett. 147:165-67 (1982).

Calissano P., Moore B. W. and Friesen A. Effect of calcium ion on S-100, a protein of the nervous system. Biochemistry 8:4318-26 (1969).

Cocchia D. and Miani N. Immunocytochemical localization of the brain-specific S-100 protein in the pituitary gland of adult rat. J. Neurocytol. 9:771-82 (1980).

Cocchia D. and Michetti F. S-100 antigen in satellite cells of the adrenal medulla and the superior cervical ganglion of the rat. An immunochemical and immunocytochemical study. Cell Tissue Res. 215:103-12 (1981).

Cocchia D., Michetti F. and Donato R. Immunochemical and immunocytochemical localization of S-100 antigen in normal humal skin. Nature 294:85-87.

Donato R. The specific interaction of S-100 protein with synaptosomal particulate fractions. Site-site interactions among S-100 binding sites. J. Neurochem. 30:1105-11 (1978).

Donato R. Effect of S-100 protein on assembly of brain microtubule proteins in vitro. FEBS Lett., in press.

Isobe T., Nakajima T. and Okuyama T. Reinvestigation of extremely acidic proteins in bovine brain. Biochim. Biophys. Acta 494:222-32 (1977)

Isobe T. and Okuyama T. The amino-acid sequence of S-100 protein (PAP I-b protein) and its relationship to the calcium-binding proteins. Eur. J. Biochem. 89:379-88 (1978).

Michetti F. and Cocchia D. S-100-like immunoreactivity in a planarian. An immunochemical and immunocytochemical study. Cell Tissue Res. 223:575-82 (1982).

Michetti F., Dell'Anna E., Tiberio G. and Cocchia D. Immunochemical and immunocytochemical study of S-100 protein in rat adipocytes. Brain Res. 262:352-56 (1983).

Moore B. W. A soluble protein characteristic of the nervous system. Biochem. Biophys. Res. Commun. 19:739-44 (1965).

Møller M., Ingild A. and Bock E. Immunohistochemical demonstration of S-100 protein and GFA protein in interstitial cells of the rat pineal gland. Brain Res. 140:1-13 (1978).

Nakajima T., Yamaguchi H. and Takahashi K. S-100 protein in folliculostellate cells of the rat pituitary anterior lobe. Brain Res. 191:523-31 (1980).

Stefansson K., Wollmann R. L., Moore B. W. and Arnason B. G. W. S-100 protein in human chondrocytes. Nature 295:63-64 (1982).

Takahashi K., Yamaguchi H., Ishizeki J., Nakajima T. and Nakazato Y. Immunohistochemical and immunoelectron microscopic localization of S-100 protein in the interdigitating reticulum cells in the human lymph node. Virchow Arch. (Cell Pathol.) 37:125-35 (1981).

Zuckerman J. E., Herschman H. R. and Levine L. Appearance of a brain specific antigen (S-100 protein) during foetal development. J. Neurochem. 17:247-51 (1970).

16 The Fine Structure of Z-discs and Z-rods in Cryosections

L-E Thornell and S. C. Watkins

INTRODUCTION

The ultrastructure of the myofibrillar Z-disc has been extensively investigated in a variety of animals, c.f., Fish (Franzini-Armstrong, 1973); Amphibian (Knappeis and Carlsen, 1962; Franzini-Armstrong and Porter, 1964; Kelly, 1967); Rodent (Reedy, 1964; Landon, 1970; Ullrick et al., 1977; Rowe, 1971); Canine (Goldstein et al., 1977); and Human muscle (Fardeau, 1969; Macdonald and Engel, 1971). These studies have resulted in several different models to explain Z-disc structure (for reviews see Squire, 1981; Landon, 1982). A common observation is that the Z-disc is highly sensitive to the processing technique employed, and quite dramatic differences are seen when for example osmium tetroxide and glutaraldehyde are interchanged as the primary fixatives.

The steps involved in standard processing for electron microscopy involve fixation of the tissue, dehydration, and embedding the tissue in plastic. This process inevitably leads to shrinkage and distortion of the tissue fine structure (pepe and Drucker, 1972). To avoid these effects of standard electron microscopical processing we have employed cryoultramicrotomy which involves only a light glutaraldehyde fixation, no dehydration or embedding. Because of the paucity of studies of the structure of the Z-disc in human skeletal muscle we have studied the ultrastructure of Z-discs and nemaline rods in normal and myopathic human muscle to gain new insights into the structure of these myofibrillar components.

MATERIALS AND METHODS

Biopsies have been taken for diagnostic purposes from different limb muscles of over 100 patients suffering from muscle disorders, six of which had a nemaline myopathy. Biopsies have also been taken from volunteers with no symptoms or history of myopathic disorders. The specimens were lightly fixed for 1 hour in 2.5% glutaraldehyde in Tyrodes buffer at 4°C. The biopsies were then cut into small pieces (1 mm × 0.5 mm × 0.5 mm) and passed through a series of glycerol concentrations (10%, 20%, 30%, 30 mins each), mounted on copper stubs, and frozen in Freon 22, chilled in liquid nitrogen.

The blocks were mounted in an LKB cryokit and ultrathin sections cut using glass knives with a trough solution of 50% dimethyl sulphoxide. The grids were mounted on coated grids, negatively strained with a 2% solution of ammonium molybdate in distilled water, air-dried, and examined using a goniometer stage in a Philips EM300. Portions of all biopsies were processed for standard electron microscopy and viewed in the same way.

RESULTS

In plastic section the Z-disc is generally seen as a dense central core with fuzzy material adhering (Fig. 16.1). The width of the Z-disc varies slightly between different fibers. In favorable areas further details may be resolved (see insets). It is clear that the I-band filaments are continuous with interdigitating Z-disc filaments. There is also a marked change in electron density of the filaments just outside the Z-disc. Furthermore, it is possible that oblique filaments interlink opposing filaments within the Z-disc.

In negatively stained cryosections (Fig. 16.2) the overall morphology of muscle fibers is very different to that seen in plastic sections. Periodicities exist throughout the myofibril but with a variety of spacings. The number of periodicities within the M-band and Z-disc reflect the "fiber type" of the parent fiber; both the M-band and Z-disc are wider and have more periodicities in the Type I fibers than in the Type II. The distance between the periodicities within the Z-disc is about 18 nm. A similar periodicity is seen in sections of nemaline rods of the same thickness. When micrographs of very thin cryosections are viewed obliquely the same periodicity is seen traversing both Z-discs and nemaline rods (Fig. 16.3), however. When the micrographs are viewed from directly above, further distinct microstructural features become apparent. The Z-disc and Z-rod filaments are interdigitating. A further notable feature is that these axial filaments are thinner (about 5 nm) than the I-band filaments either side (about 7 nm) (Fig. 16.2,

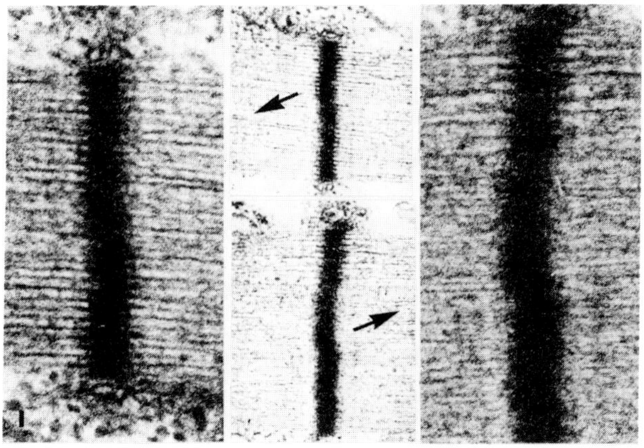

Fig. 16.1. The Z-disc as seen in plastic section. The two central figures are of different fiber types (Type I below, Type II above) (45,000×) flanked by higher power images of the same discs (110,000×). The Z-disc region is slightly wider in the Type I fibers than in the Type II. The higher power micrographs show that filaments within the disc are continuous with the I-band filaments though more electron dense. The lateral margins of the disc are obscured by fuzzy material bordering the central core.

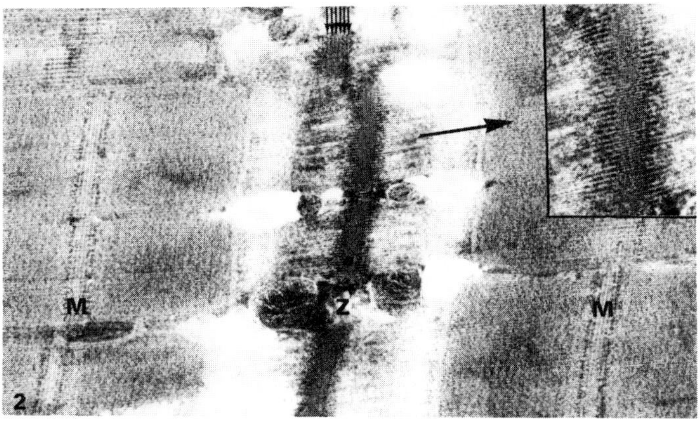

Fig. 16.2. The Z-disc as seen in cryosection (45,000×). The inset shows the same disc at a higher power (90,000×). The I-band filaments can be seen to enter the Z-disc region in an interdigitated fashion. Within the disc a periodicity is apparent (arrows). The lateral margins of the disc are clearly delimited. In the M-band a clear periodicity of five lines is apparent.

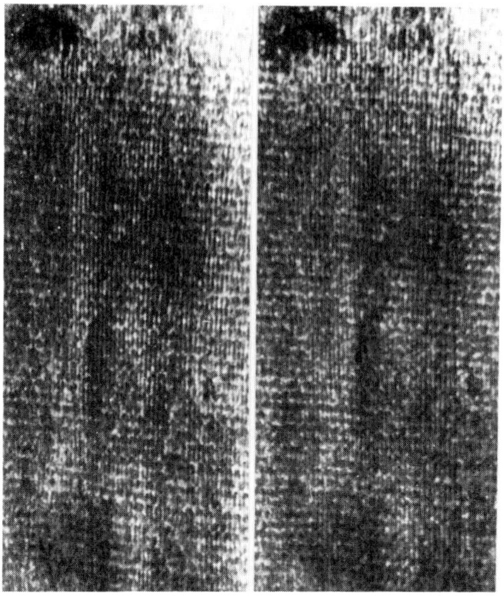

Fig. 16.3. A stereo pair of a Z-rod. The longitudinal axial filaments, oblique interlinking filaments, and less distinct intermediate periodicity are present. When the two micrographs are compared a transformation of the image is apparent when the section is tilted. (125,000×)

inset). The axial filaments of the Z-disc and Z-rod are interlinked at 38 nm intervals by oblique thin (about 3 nm) filaments. Intermediate between this repeat is a less distinct transverse periodicity. The resolution of the structures is affected by the angle at which the sections are viewed as tilting of the section obscures the pattern in some regions and reveals the same pattern in others. Therefore the use of stereo pairs (Fig. 16.3) is of benefit.

DISCUSSION

The staining properties of negative stain are very different from normal electron microscopical stains. Rather than binding to structures within the section, the stain collects in irregularities in the surface and at points where there is a large change in mass density. This, coupled with the absence of severe fixation, dehydration, and embedment in cryosections, is the reason for their different morphology. Thus cryosections allow new insights into the fine structure of Z-discs and nemaline rods.

From the evidence presented here it appears that the Z-disc consists of interdigitating I-band filaments interlinked to filaments of opposite polarity by oblique filaments. This model is similar to that presented by Goldstein and her colleagues (1982). However, further than this it appears that other Z-disc proteins may be present along the whole of the axial Z filaments especially in the intermediate zone between the oblique Z-filaments. Two factors make this conclusion possible:

(1) Positive stain binds to the axial filaments making them more electron dense and somewhat thicker than the I-band filaments.

(2) In the cryosections the negative stain penetrates into the Z-disc giving the longitudinal filaments a smaller diameter than the I-band filaments and an intermediate periodicity not related to the oblique Z filaments.

ACKNOWLEDGMENTS

This project was supported by the Swedish Medial Research Council (12X393). The Muscular Dystrophy Association, U.S.A., The University of Umeå, Umeå, Sweden, and the Royal Society of Great Britain.

REFERENCES

Fardeau M, 1969 Ultrastructure des fibres musculaire squelettiques. Presse medicale 77:1341.

Franzini-Armstrong C, 1973 The structure of a simple Z-line. J Cell Biol 58:630.

Franzini-Armstrong C, Porter KR, 1964 The Z-disc of skeletal muscle fibrils. Zeitschrift fur Zellforschung und Mikroscopiche Anatomie 61:661.

Goldstein MA, Schroeter JP, Sass RL, 1977 Optical difraction of the Z lattice in canine cardiac muscle. J Cell Biol 75:818.

Goldstein MA, Schroeter JP, Sass RL, 1982 The Z-band lattice in slow skeletal muscle. J Muscle Res and Cell Motil 3:333.

Kelly DE, 1967 Models of muscle Z-band fine structure based on a looping filament configuration. J. Cell Biol 34:827.

Knappeis GG, Carlsen F, 1962 The ultrastructure of the Z-disc in skeletal muscle. J Cell Biol 13:323.

Landon DN, 1969 The influence of fixation on the fine structure of the Z-disk or rat striated muscle. J Cell Sci 6:257.

Landon DN, 1982 Skeletal muscle-normal morphology, development and innervation. In Skeletal Muscle Pathology. pp. 1-88. Eds Mastaglia FL, Walton JN, Pub: Churchill Livingstone.

MacDonald RD, Engel AG, 1971 Observations in the organisation of Z-disk components and on rod bodies of Z-disk origin. J Cell Biol 48:431.

Pepe FA, Drucker B, 1972 The myosin filament. IV. Observations of the internal structural arrangement. J Cell Biol 52:255.

Reedy MK, 1964 The structure of actin filaments and the origin of the axial periodicity in the I substance of vertebrate striated muscle. Proc Roy Soc B 160:458.

Rowe RW, 1971 Ultrastructure of the Z-line of skeletal muscle fibers. J Cell Biol 51:674.

Squire J, 1982 In: The structural basis for muscle contraction. Pub: Plenum.

Ullrick WC, Toselli PA, Saide JD, Phear WPC, 1977 Fine structure of the vertebrate Z-disc. J Molec Biol 115:61.

17 Properties of a 90-kDa Protein-Actin Complex Isolated from Human Blood Platelets

Martine Coué, Claude Huc,
Francine Lefebure, Anna Olomucki

INTRODUCTION

In response to stimuli, blood platelets exhibit dramatic structural transformations. Upon activation, the discoidal cells round up, extend pseudopodia, aggregate, and undergo an internal reorganization (White & Gerrard, 1979). It has been suggested that these events result from the activation and interaction of cytoskeletal proteins. Actin, one of the most abundant proteins in platelets, seems to be particularly involved in the shape changes and the internal movements during activation (Nachmias, 1980).

In fact, platelet activation is accompanied by the polymerization of the major part of actin which is maintained in nonpolymerizable state in intact platelets (Carlsson et al., 1979; Pribluda & Rotman, 1982). Moreover, it was also shown that calcium mediates the regulation of actin assembly in human platelet extracts (Carlsson et al., 1979).

These phenomena raise the problem of the mechanism of a calcium sensitive regulation of the ultrastructure of actin.

Several groups have found that platelets contain a 90-kDa protein capable of modulating actin assembly in a calcium sensitive manner (Wang & Bryan, 1981; Coué et al., 1982; Markey et al., 1982; Lind et al., 1982).

We have isolated from human blood platelets the 90-kDa protein as a stable complex with actin and we tried to gain an insight into the mechanism of its interaction with actin, in the presence and in the absence of free calcium.

METHODS

DNAase I-Sepharose column was prepared by the method of Bretscher and Weber (1980). Actin was isolated from acetone powder of whole platelets, as described by Coué et al. (1982).

The interaction of 90-kDa-actin complex with actin was followed by viscosity and ΔA at 232 nm measurements and by electron microscopy.

RESULTS AND DISCUSSION

We isolated a 90-kDa-actin complex starting from acetone powder of whole platelets. The low ionic strength buffer extract was sieved on Sephadex G-150 column and the filtrate was purified by two methods.

The first consists in chromatography on DEAE-cellulose column and elution by 0.25 M KCl, followed by the precipitation at 35 to 45% saturation of ammonium sulfate, and finally filtration on Sephacryl S-300 in a buffer containing 2 mM EGTA and 0.15 M NaCl.

In the second method, the filtrate was applied to Sepharose-DNase I column and the complex eluted by a buffer containing 5 mM EGTA. The final step involves filtration on Sephacryl S-300 under the same conditions as for method 1.

The first method is much longer but gives a three-times higher yield.

Analysis by gel electrophoresis in SDS of the final product shows (Fig. 17.1) principally two bands of polypeptides, one of them at 90-kDa and the other one, corresponding to actin, at 42-kDa. Densitometry of the Coomassie blue stained gels shows that the ponderal ratio of the two proteins is about 1:1, corresponding to 1 mole of 90-kDa protein to 2 ± 0.2 moles of actin. This ratio was confirmed by the estimation of the molecular weight of the complex from gel filtration behavior.

The complex is not dissociable by EGTA. Under high ionic strength such as 0.8 M KCl, the ratio of actin decreases to about 1.6 actins for one 90-kDa protein. Only a treatment by 8 M urea allows to dissociate the complex.

Finally we confirmed that antibodies against macrophage gelsolin cross-react with the 90-kDa component of the complex. After isolation of the complex, we undertook a study of its effect on actin assembly.

To study the polymerization of actin in the presence of Ca^{2+}, we followed the changes in absorbance at 232 nm and in the

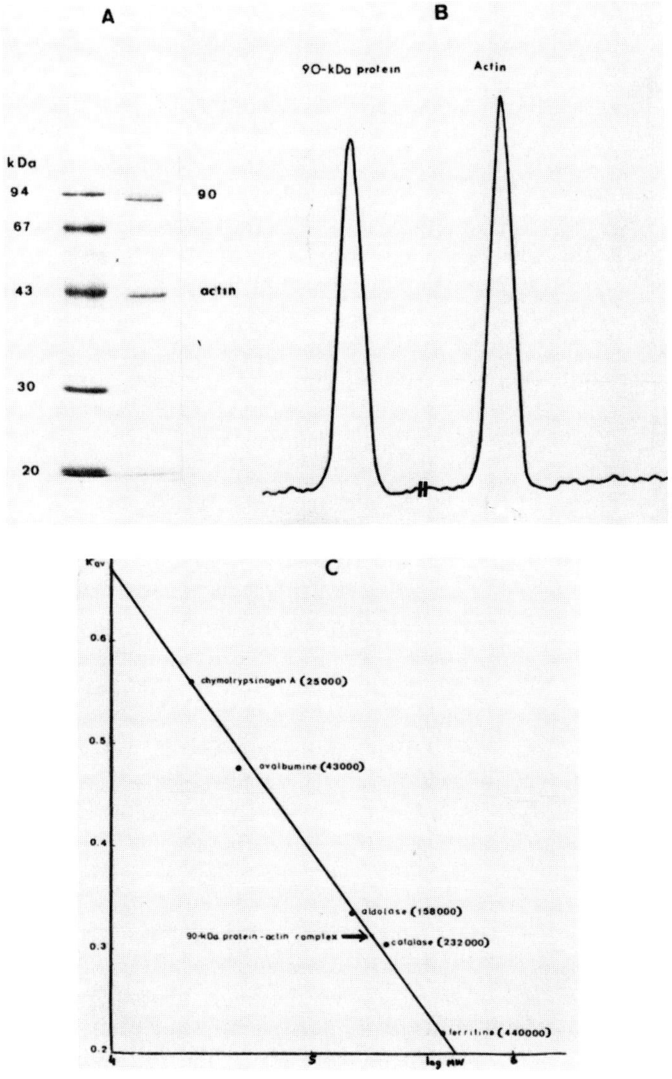

Fig. 17.1. Analysis of the isolated 90-kDa protein-actin complex. (a) 10% polyacrylamide gel electrophoresis in SDS Coomassie blue staining. (b) Densitometry of a. (c) Determination of molecular weight on Sephacryl S-300 column in 2 mM EGTA.

specific viscosity. The polymerization was initiated by 0.76 mM $MgCl_2$ under conditions which favor the sigmoid kinetics of the control actin (Fig. 17.2a). In the presence of the complex the lag time is abolished, the initial rate of polymerization increases, and the final level of viscosity decreases. At a molar complex to actin ratio 1 to 165 the final viscosity is reduced by 70%. Under conditions of slower polymerization (such as in 0.5 mM $MgCl_2$), the inhibition is even higher and ataains 84%. A similar effect of the complex on actin polymerization can be observed by following the changes of absorbance at 232 nm (not shown).

The above results indicate that the complex acts as an actin nucleation factor inducing the formation of more but shorter polymers. This is confirmed by the electron microscope pictures of solutions of actin polymerized in the presence of the complex.

In the absence of free calcium (Fig. 17.2b), i.e., in 2 mM EGTA, the final viscosity is also reduced, but this effect is smaller as compared to the one appearing in the presence of Ca^{2+}.

There is also a different effect on the initial rate of polymerization which decreases with addition of increasing amounts of the complex. To explain these results it is necessary to keep in mind that in the presence of EGTA a spontaneous nucleation takes place so that the additional nucleation by the complex seems canceled. The decrease of the initial rate of polymerization and of final viscosity shows that even in the absence of free Ca^{2+} the complex is still capable of interacting with actin, probably by capping the fast growing ends of nuclei and thus slowing down the elongation.

With a low complex to actin ratio (1:830) both the initial rate and the final viscosity are only weakly reduced, the majority of spontaneously generated nuclei probably grow fast, and only few of them can be capped by the complex. For a higher (1:165) ratio the initial rate and final viscosity are both strongly reduced although always less than in the presence of Ca^{2+}.

The effect of the complex on preformed actin filaments (Fig. 17.3) was studied in the presence and in the absence of free Ca^{2+}. In both cases we observed an immediate drop of viscosity when the complex is added, followed by a slow equilibration. This effect is much weaker than the one obtained when the same amount of complex is added to G-actin before the initiation of the polymerization.

Anyway, all the effects on preformed actin filaments are somewhat lower in the presence of EGTA.

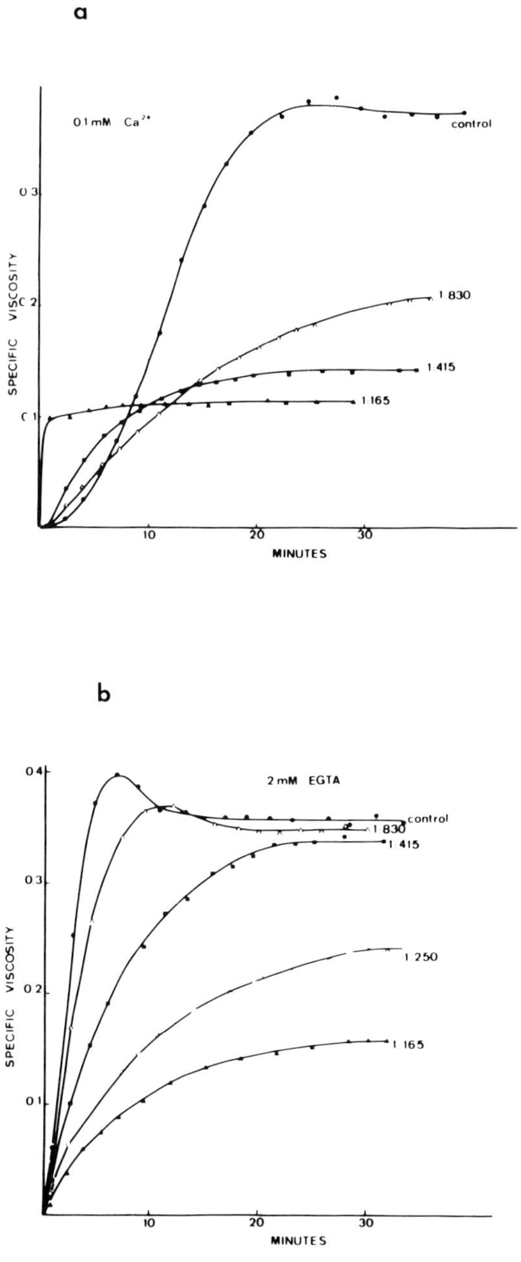

Fig. 17.2. Effect of complex on the time course of 9.5 μM actin polymerization in the presence of 0.1 mM $CaCl_2$ or 2 mM EGTA. The polymerization was initiated by the addition of 0.76 mM $MgCl_2$. Complex-actin ratio as indicated. Measurements were performed at 25°.

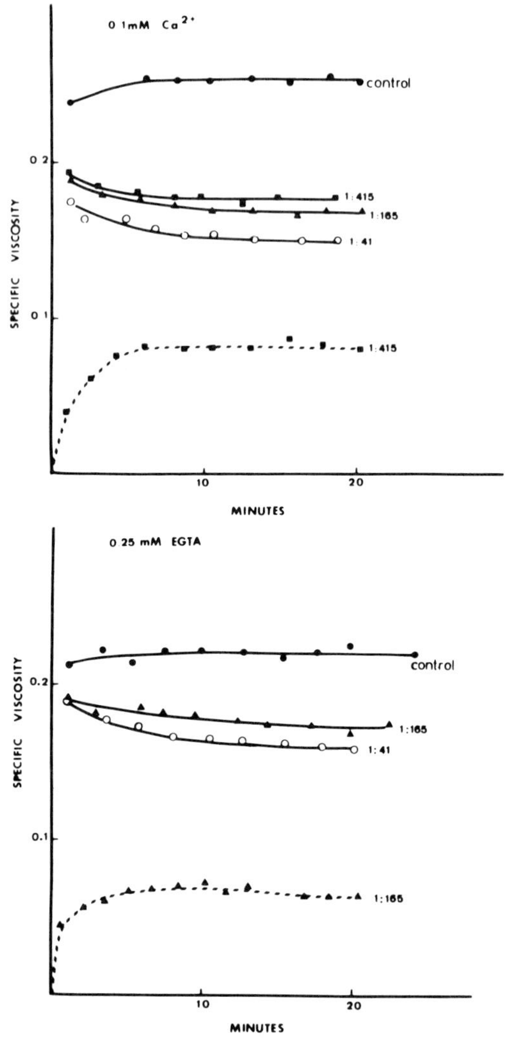

Fig. 17.3. Effect of 90-kDa protein-actin complex on the viscosity of preformed actin filaments in the presence of 0.1 mM $CaCl_2$ or 0.25 mM EGTA. The complex was added in the indicated molar ratios to 4.8 μM F-actin (prepared in 100 mM KCl and 2 mM $MgCl_2$). Dotted lines: effect of the complex on the polymerization of G-actin under the same conditions. Measurements at 25°.

CONCLUSION

The 90-kDa-actin complex isolated from platelet is an inhibitor of actin polymerization; it accelerates the nucleation step; and it reduces the extent of actin polymerization at the steady state. The decrease of the final viscosity is primarily due to the formation of more numerous but shorter filaments. The reduction of the steady state absorbance at 232 nm signifies that the total amount of F-actin is diminished.

The complex thus resembles some other actin-modulating proteins such as intestinal villin (Glenney et al., 1981), macrophage gelsolin (Yin et al., 1981), Physarum fragmin (Hasegawa et al., 1980) and serum ADF (Harris and Weeds, 1983).

By some of its other properties the complex resembles the fragmin heterodimer (Hinssen, 1981): little ability to depolymerize F-actin and low calcium sensitivity of its effects on actin assembly. Fragmin and 90-kDa protein form stable, non-EGTA-dissociable complexes with actin.

The 90-kDa protein from platelets seems to belong to the class of actin end-blocking proteins.

ACKNOWLEDGMENTS

The authors are indebted to Dr. A. Fattoum for performing the immunological analysis and to P. Desbois for her expert technical assistance.

REFERENCES

Bretscher, A., & Weber, K., Villin is a major protein of the microvillus cytoskeleton which binds to G and F actin in a calcium-dependent manner. Cell 20, 839-47 (1980).

Carlsson, L, Markey, F., Blikstad, I., Persson, T., & Lindberg, U., Reorganization of actin in platelets stimulated by thrombin as measured by the DNase I inhibition assay. Proc. Natl. Acad. Sci. USA 76, 6376-80 (1979).

Coué, M., Landon, F., & Olomucki, A., Comparison of the properties of two kinds of preparations of human blood platelet actin with sarcomeric actin. Biochimie 64, 219-26 (1982).

Coué, M., Lefébure, F. & Olomucki, A., Studies of two protein factors from blood platelets which regulate the polymerization of actin. Biology of the Cell 45, 605 (1982).

Glenney, J. R. Jr., Kaulfus, P., & Weber, K., F actin assembly modulated by villin: Ca^{2+}-dependent nucleation and capping of the barbed end. Cell 24, 471-80 (1981).

Harris, H. E., & Weeds, A. G., Plasma actin depolymerizing factor has both calcium-dependent and calcium-independent effects on actin. Biochemistry 22, 2728-41 (1983).

Hasegawa, T., Takahashi, S., Hayashi, H., & Hatano, S., Fragmin: a calcium ion sensitive regulatory factor on the formation of actin filaments. Biochemistry 19, 2677-83 (1980).

Hinssen, H., An actin-modulating protein from Physarum polycephalum. II. Ca^{2+}-dependence and other properties. Europ. J. Cell Biol. 23, 234-40 (1981).

Lind, S. E., Yin, H. L., & Stossel, T. P., Human platelets contain gelsolin, a regulator of actin filament length. J. Clin. Invest. 69, 1382-87 (1982).

Markey, F., Persson, T., & Lindberg, U., A 90,000-dalton actin-binding protein from platelets. Comparison with villin and plasma brevin. Biochim. Biophys. Acta 709, 122-33 (1982).

Nachmias, V. T., Cytoskeleton of human platelets at rest and after spreading. J. Cell Biol. 86, 795-802 (1980).

Pribluda, V. & Rotman, A., Dynamics of membrane-cytoskeleton interactions in activated blood platelets. Biochemistry 21, 2825-32 (1982).

Wang, L. L. & Bryan, J., Isolation of calcium-dependent platelet proteins that interact with actin. Cell 25, 637-49 (1981).

White, J. G. & Gerrard, J. M., Interactions of microtubules and microfilaments in platelet contractile physiology. Meth. Achiev. Exp. Pathol. 9, 1-39 (1979).

Yin, H. L., Harting, J. H., Maruyama, K., & Stossel, T. P., Ca^{2+} control of actin filament length. Effects of macrophage gelsolin on actin polymerization. J. Biol. Chem. 256 9693-97 (1981).

Yin, H. L., Harting, J. H., Maruyama, K., & Stossel, T. P., Ca^{2+} control of actin filament length. Effects of macrophage gelsolin on actin polymerization. J. Biol. Chem. 256 9693-97 (1981).

18 Mechanism of the Interaction of Serum Vitamin D Binding Protein with Actin

A. Olomucki, M. Coué, M. Viau, J. Constans

INTRODUCTION

The human serum vitamin D binding protein (DBP) is an α-globulin of about 56,000 d (Coue et al., 1983). It was initially known as a "group specific component" Gc acting as an anti-rachitic sterol carrier (Hirschfeld et al., 1960).

The discovery of the capacity of DBP to give a complex with cytoplasmic actin raises the question of the physiological role of this protein having two entirely different functions. Van Baelen et al. (1980) have first observed that DBP is capable of depolymerizing filaments of actin and they proposed for this protein the role of an actin filaments scavenger in circulating blood. Some other groups confirmed this interaction of DBP with actin (Constans et al., 1981; Haddad, 1982; Vandekerckhove et al., 1982). We now tried to elucidate more precisely the mechanism of this interaction in vitro in order to approach the understanding of the physiological significance of the binding of DBP to actin.

METHODS

DBP was isolated from human serum and purified using immunoaffinity chromatography as described by Coué et al. (1983).

The interaction of DBP with actin was followed by viscosity and ΔA at 232 nm measurements and by electron microscopy. Determination of isoelectric points of different forms of DBP and their complexes was performed by electrophoresis in two dimensions according to Constans et al. (1980).

RESULTS

We obtained 16 mg of pure DBP from 100 ml of human serum. The DBP corresponding to two phenotypes was purified separately. The protein produced by the Gc_2 allele (DBP_2) is characterized by the presence of a single band after isoelectrofocusing (Fig. 18.1a), while the Gc_1 protein (DBP_1) is revealed as a double band, the more acidic protein, Gc_{1a} containing a sialic acid residue.

The isoelectric points of purified DBP_1 (Gc_{1a} and Gc_{1c} proteins) and DBP_2 were estimated to be 4.85, 4.95, and 5.1, respectively.

The purified DBP is still able to bind 25-hydroxy-D_3 vitamin D derivative. The isoelectric points of the three holoforms are equal to 4.70, 4.80, and 4.90, respectively.

Determination of the isoelectric points of DBP-actin complexes was possible only after isoelectrofocusing in 0.5 M urea. Under these conditions the complexes are sharply focused.

The isoelectric points of complexes of DBP with muscle and platelet actins have significantly different values (Fig. 18.1a and b).

The isoelectric points of all DBP-actin complexes are lower than those of the corresponding DBP forms. Moreover, DBP complexes with platelet actin are less acidic than DBP complexes with muscular actin. This is due to the charge differences between two isoforms of actin and signifies that in the interaction with DBP the N-terminal part of actin polypeptide is apparently not involved.

Effect of DBP on the Polymerization of Muscle Skeletal Actin

As shown in Figure 18.1a and b the initial rate of polymerization of G-actin decreases in the presence of DBP. The final extent of actin assembly is also strongly diminished by addition of DBP. This decrease of steady state absorption at 232 nm is proportional to the amount of DBP added (Fig. 18.1a, inset). For a ratio of DBP to actin of 0.3, the final specific viscosity is diminished by about 50% (Fig. 18.1b).

In order to demonstrate the effect of DBP on elongation independently of its effect on the nucleation step we added an excess of actin nuclei to the system. Figure 18.2a shows that DBP also inhibits actin filament elongation.

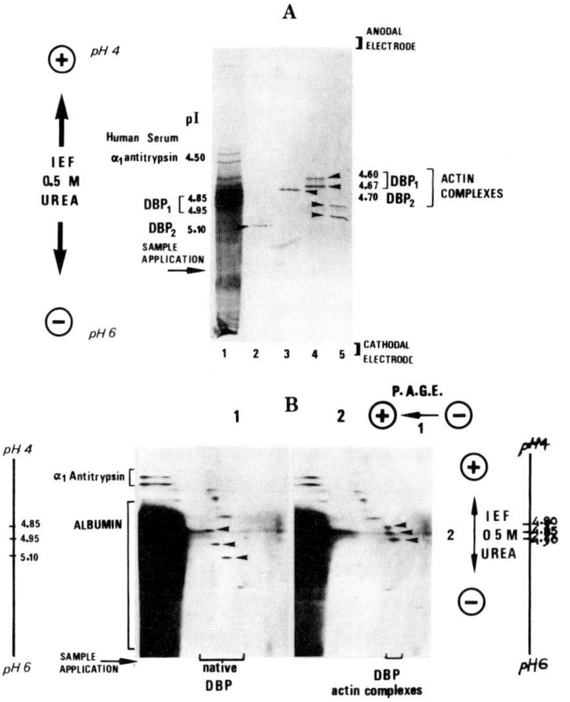

Fig. 18.1. a: Isoelectric focusing of DBP and of its complexes with muscle actin. 1, Untreated human plasma; 2, purified DBP_2 isoform; 3, purified DBP_2-actin complex; 4, purified DBP-actin complex; 5, purified DBP_1 isoform. b: Electrophoretic pattern obtained after migration in two dimensions: first, polyacrylamide gel electrophoresis in SDS; second, isoelectrofocalization in 0.5 M urea. 1, Untreated human plasma (native DBP); 2, 10 μM platelet actin added to 3.5 μM native DBP. Gel prepared according to Constans et al. (1980).

The evaluation of the effect of DBP on the apparent critical concentration for actin polymerization was performed in 100 mM KCl (Fig. 18.2b). As can be observed, the increase of the concentration of nonpolymerized actin depends on the concentration of DBP. Above this apparent critical concentration the reduced viscosity (slope of specific viscosity versus actin concentration) is the same as for the control (actin alone) Fig. 18.2b). Since the reduced viscosity depends on filament lengths this result is in favor of the hypothesis that DBP has no effect on the polymer size distribution in the steady state. The final

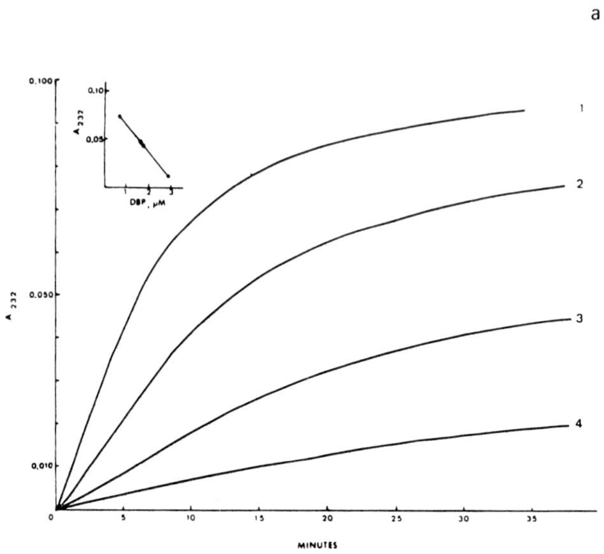
a
0,100
0,050
0,010
A 232
0
5
10
15
20
25
30
35
MINUTES
1
2
3
4
0,10
0,05
A 232
DBP, μM

Fig. 18.2.

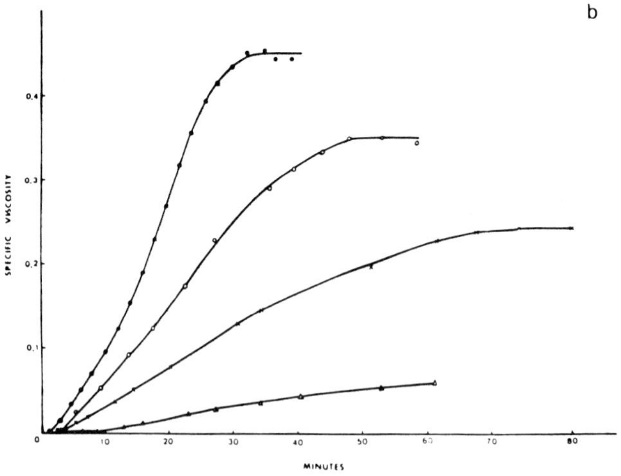
b
0,4
0,3
0,2
0,1
SPECIFIC VISCOSITY
0
10
20
30
40
50
70
80
MINUTES

viscosity is lower in the presence of DBP because the amount of nonpolymerized actin is higher.

This hypothesis is confirmed by electron microscopic analysis of samples of actin polymerized in the presence of DBP (not shown). It appears that the overall quantity of filaments decreases but the relative length of remaining filaments seems to be unchanged as compared to the control.

Effect of DBP on Preformed Actin Filaments

When DBP is added to F-actin, the actin slowly depolymerizes as shown by the decrease of its specific viscosity (Fig. 18.3). The final viscosity was the same as the one reached when the same amount of DBP was added to the G-actin in polymerizing buffer. Analogous results were obtained following the changes of absorbence at 232 nm (not shown).

DISCUSSION

The aim of our work was to try to elucidate more precisely the effect of human serum vitamin D binding protein on actin polymerization.

Our results allow us to propose the following characteristics of the mechanism of this reaction. When added to G-actin in polymerizing buffers, DBP inhibits the nucleation and elongation steps and when added to preformed actin filaments it depolymerizes them. The two isoforms DBP_1 and DBP_2 have similar

Fig. 18.2. Kinetics of actin polymerization in the presence of DBP monitored: in a, by the relative increase in absorbence at 232 nm; in b, by the increase of viscosity. a: To 6.1 μM G-actin in buffer G containing 0.1 mM ATP was added $MgCl_2$ to a concentration of 2 mM and DBP at a final molar ratio actin to DBP equal to 8.8 for curve 2; to 3.6 for curve 3; and to 2.1 for curve 4. Control (curve 1) is actin without DBP. Inset: plot of ΔA_{232} at 35 min vs DBP concentration. DBP_2 (•) and DBP_1 (○). b: G-actin (9.5 μM) in buffer G was mixed with 0 μM (•-•); 1.4 μM (○-○); 2.75 μM (×-×) or 6 μM (Δ-Δ) CBP and incubated 7 min at 25°C (room temperature). The polymerization was then initiated by addition of KCl to a final concentration of 100 mM and followed by the increase in viscosity at 25°C.

a

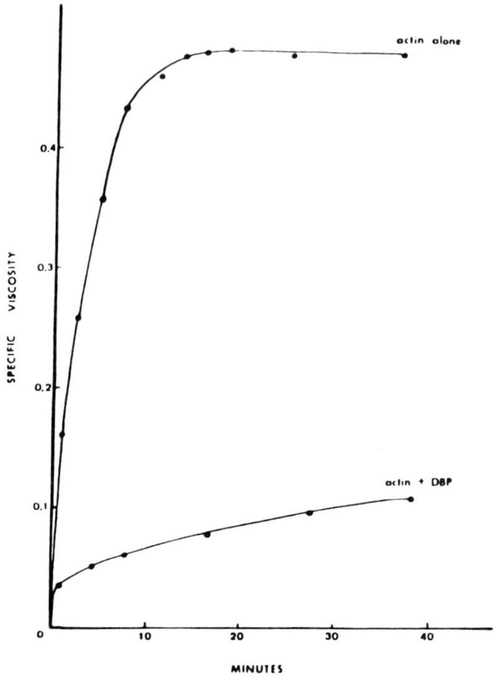
actin alone
actin + DBP
SPECIFIC VISCOSITY
0.4
0.3
0.2
0.1
0
10
20
30
40
MINUTES

Fig. 18.3

b

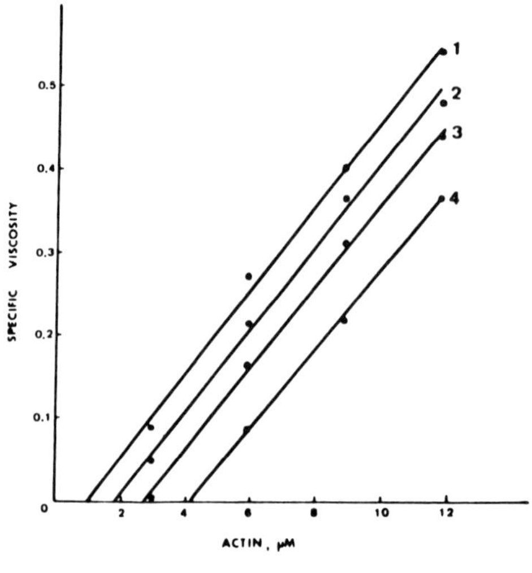
1
2
3
4
SPECIFIC VISCOSITY
0.5
0.4
0.3
0.2
0.1
0
2
4
6
8
10
12
ACTIN, μM

efficiency. All the effects of DBP on the polymerization state of actin are entirely calcium independent.

DBP seems to belong to a class of actin modulating proteins such as profilin and DNAase I (Korn, 1982) which decrease the concentration of polymerized actin without interacting directly on the filaments. In the presence of such a protein the critical concentration of actin monomer C_0 (steady state concentration of free actin monomers in equilibrium with F-actin) is unaffected, while the total amount of nonpolymerized actin (C_0 + actin-DBP complex) increases, and hence the concentration of polymerized actin is decreased.

This hypothesis is favored by the following observations: (i) similar results are obtained when the effect of DBP on actin polymerization is followed by absorbence measurements at 232 nm and by the changes in viscosity; (ii) electron microscope images of the mixtures of DBP and actin in different ratios show the absence of filaments or a decrease of their concentration but never the change of the polymer size distribution in the steady state. In addition, the very strong (K_D = 0.06 - 0.12 μM; Haddad, 1982; Vandekerckhove, 1982) 1:1 complex which DBP forms with actin certainly decreases the concentration of G-actin and displaces G $\rightleftarrows$ F equilibrium to the left.

In the light of this new information concerning the mechanism of DBP binding to actin, we try to understand the physiological significance of this reaction. Why has the seric protein, having a function of a sterol-carrier, the capacity to produce a very strong, stoichiometric complex with actin, thus inhibiting its polymerization?

When damaged cells release actin into the blood stream, the capture of this actin could be partially performed by DBP as previously suggested by Van Baelen (1980). However, this

Fig. 18.3. a: Effect of DBP on the time course of nucleated actin polymerization. G-actin (9.5 μM) in buffer G was mixed with 6.1 μM DBP and incubated 7 min at 25°C. The polymerization was initiated by addition of 0.73 μM F-actin nuclei and by 100 mM KCl. The increase of viscosity was measured at 25°C.
b: Dependence of the extent of actin polymerization on the concentration of actin and DBP. Actin alone (curve 1) or in presence of DBP: 0.96 μM (cirve 2), 1.79 μM (curve 3), and 2.89 μM (curve 4) was polymerized at 25°C for 15 h in buffer G plus 100 mM KCl. Critical concentration of actin: 1, C_0 = 0.899 μM; 2, C_0 = 1.8 μM; 3, C_0 = 2.7 μM; 4, C_0 = 4.2 μM.

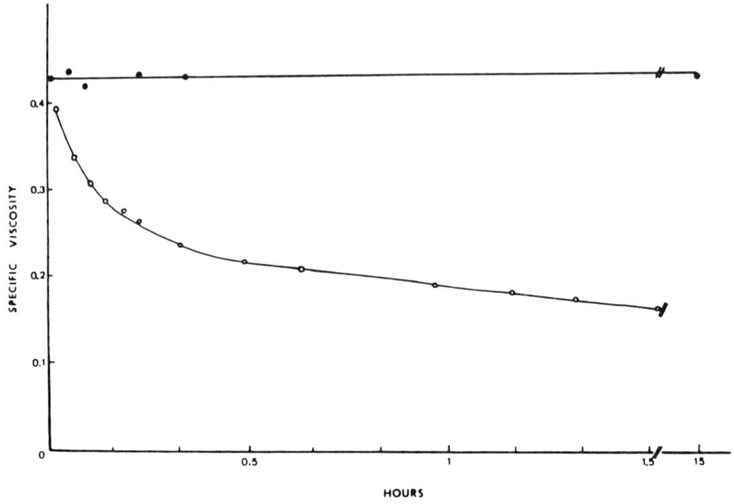

Fig. 18.4. Effect of DBP on the viscosity of steady state actin filaments. 9.5 μM G-actin in buffer G was polymerized 15 h at 25°C in the presence of 100 mM KCl. At zero time 0.07 ml of DBP in the same buffer was added to 0.5 ml of F-actin to give a final concentration: DBP, 9.6 μM, and actin, 8.4 μM. (●-●) actin alone; (o-o) F-actin plus DBP.

is probably the role of ADF (Harris and Schwartz, 1981; Harris and Weeds, 1983) equally abundant in the serum which in substoichiometric amounts is able to produce a faster effect on F-actin.

In the present stage of this work, we cannot yet propose a clear explanation of the role of DBP-actin interaction.

ACKNOWLEDGMENTS

The authors are indebted to F. Lefébure for performing the electron microscopy in S.C. No. 18 of INSERM, Laboratoire de Pathologie Cellulaire, with the collaboration of G. Geraud and J. Bureau, and to C. Dubord for differential spectroscopy measurements.

REFERENCES

Constans, J., Viau, M., and Buisson, C., Affinity differences for the 25-OH-D_3 associated with the genetic heterogeneity of the vitamin D-binding protein. FEBS Lett. 111, 107-111 (1980).

Constans, J., Oksman, F., and Viau, M., Binding of the apo- and holoforms of the serum vitamin D-binding protein to human lymphocyte cytoplasm and membrane by indirect immunofluorescence, Immunol. Lett. 3, 159-62 (1981).

Coué, M., Constans, J., Viau, M., and Olomucki, A., The effect of serum vitamin D-binding protein on polymerization and depolymerization of actin is similar to the effect of profilin on actin. Biochim. Biophys. Acta 759, in press (1983).

Haddad, J. G., Human serum binding protein for vitamin D and its metabolites (DBP); evidence that actin is the DBP binding component in human skeletal muscle. Arch. Biochem. Biophys. 213, 538-44 (1982).

Harris, D. A., and Schwartz, J. H., Characterization of brevin, a serum protein that shortens actin filaments. Proc. Natl. Acad. Sci. USA 78, 6798-6802 (1981).

Harris, H. E., and Weeds, A. G., Plasma actin depolymerizing factor has both calcium-dependent and calcium-independent effects on actin. Biochemistry 22, 2728-41 (1983).

Hirschfeld, J., Jonsson, B., and Rasmuson, M., Inheritance of a new group-specific system demonstrated in normal human sera by means of an immunoelectrophoretic technique. Nature 185, 931-32 (1960).

Korn, E. D., Actin polymerization and its regulation by proteins from non-muscle cells. Physiol. Rev. 62, 672-737 (1982).

Van Baelen, H., Bouillon, R., and Demoor, P., Vitamin D-binding protein (Gc-globulin) binds actin. J. Biol. Chem. 255, 2270-72 (1980).

Vandekerckhov, J. S., and Sandoval, I. V., Purification and characterization of a new mammalian serum protein with the ability to inhibit actin polymerization and promote depolymerization of actin filaments. Biochemistry 21, 3983-3991 (1982).

19 Heterogeneity of Tropomyosin in Earthworm Body Wall Muscle

Angela Ditgens and Jochen D'Haese

INTRODUCTION

The body wall muscle of the earthworm Lumbricus terrestris is dual regulated, that is Ca^{++}-regulatory proteins are located on the myosin as well as on the actin filaments. Within our efforts to characterize the actin-linked regulatory system, we isolated the regulatory protein complex, which showed properties similar to the tropomyosin-troponin of vertebrate skeletal muscle. From earthworm troponin we characterized a troponin-I- and a troponin-C-like subunit. Earthworm tropomyosin was so far described as a heterodimer with a 34,000 and a 37,000-Da subunit (D'Haese & Ditgens, 1980; Ditgens et al., 1980; Ditgens et al., 1982; Ditgens, 1983). Besides showing considerable similarities to vertebrate striated muscle tropomyosin this protein showed a remarkably higher inhibitory effect on the acto-HMM ATPase activity. The results presented here deal with the heterogeneity of tropomyosin in this obliquely striated annelid muscle.

DETECTION AND PURIFICATION OF THE TROPOMYOSIN ISOFORMS

The SDS-polyacrylamide gel electrophoresis (SDS-PAGE) of crude earthworm actomyosin (control slot in Fig. 19.1) shows two major bands with a slightly faster mobility than actin, representing the tropomyosin dimer we already described (Ditgens et al., 1982). When the actomyosin was separated by two-dimensional electrophoresis (Fig. 19.1) four points of nearly equal protein content appeared in the area which is typical for the position of tropomyosin. Under the same condi-

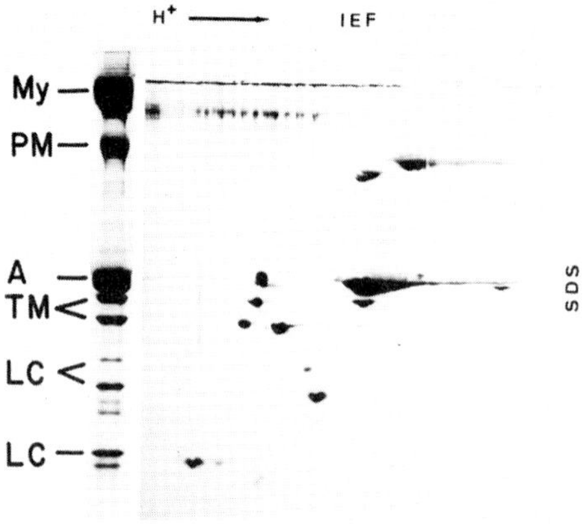

Fig. 19.1. Two-dimensional electrophoresis of earthworm actomyosin. The slot on the left shows the components of acto-actomyosin separated by SDS-PAGE. Four tropomyosin spots are visible only in the two-dimensional system. A, actin; LC, light chains of myosin; My, heavy chains of myosin; PM, paramyosin; TM, tropomyosin. Two-dimensional electrophoresis was performed according to O'Farrell (1975), SDS-PAGE was performed as previously described (Ditgens et al., 1982).

tions rabbit skeletal muscle tropomyosin is separated as two spots, the α- and β-chains, being present in the proportion of 4:1 (Cummins & Perry, 1973). From this comparison it was assumed that four tropomyosin subunits exist in the earthworm body wall muscle. Five different polypeptides of tropomyosin were recently detected in cultured cells. It was found that especially the tropomyosin pattern changed after transformation by DNA virus (Matsumura et al., 1983). As the presence of four different tropomyosin subunits was not described so far for muscular tissues, we made efforts to characterize the multiple forms of tropomyosin from the earthworm body wall muscle.

From their purification behavior it follows, that the four subunits form two heterodimeric tropomyosins. When crude tropomyosin (Fig. 19.2a) was chromatographed on DEAE-sepharose Cl-6B (comp. Ditgens et al., 1982) it was separated into two main fractions, the last of which—already identified as tropomyosin—was called tropomyosin$_1$ (Fig. 19.2b). The other main fraction, named tropomyosin$_2$, contained two bands with molecular weights of 33,000 and 42,000 (Fig. 19.2c). Further separation of the subunits of these dimeric proteins

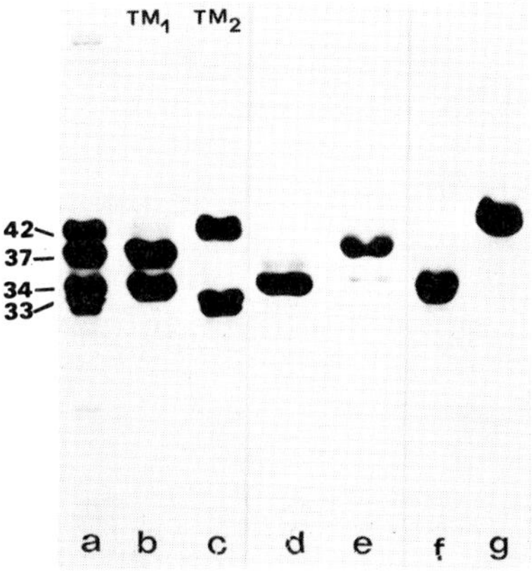

Fig. 19.2. SDS-PAGE of earthworm tropomyosin fractions. (a) crude tropomyosin; (b) $tropomyosin_1$ (TM_1); (c) $tropomyosin_2$ (TM_2); (d) and (e) subunits of $tropomyosin_1$ with molecular weights of 34,000 and 37,000, separated by CM-cellulose chromatography in the presence of 8 M urea; (f) and (g) subunits of $tropomyosin_2$ with molecular weights of 33,000 and 42,000, separated by DEAE-sepharose chromatography in the presence of 7 M urea.

was possible only in the presence of urea. $Tropomyosin_1$ was separated by CM-cellulose chromatography in the presence of 8 M urea into its two subunits (Fig. 19.2d and e). The isolation of the two subunits of $tropomyosin_2$ was performed by DEAE-sepharose chromatography in the presence of 7 M urea (Fig. 19.2f and g).

DEMONSTRATION OF DIVERSITY

We wanted to assure that both tropomyosin isoforms isolated from the earthworm body wall muscle were distinct proteins. Therefore, a proteolytic digestion with Staphylococcus aureus protease was performed. Figure 19.3 shows the SDS-PAGE of rabbit tropomyosin and the two earthworm tropomyosins before (Fig. 19.3a-c) and after (Fig. 19.3d-f) digestion with the protease (Fig. 19.3g). The digestion products of all three tropomyosins show considerable differences. Slight similarities were rather found between rabbit tropomyosin and earthworm $tropomyosin_1$ than among the earthworm proteins.

Another proof for the diversity of both earthworm tropomyosins is the lack of any immunological cross-reactivity.

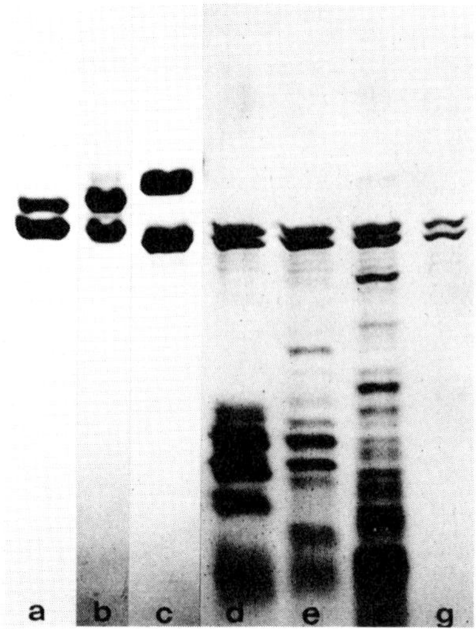

Fig. 19.3. SDS-PAGE of rabbit and earthworm tropomyosin before (a-c) and after (d-f) incubation with Staphylococcus aureus protease; (a) and (d) rabbit tropomyosin; (b) and (e) earthworm tropomyosin$_1$; (c) and (f) earthworm tropomyosin$_2$; (g) Staphylococcus aureus protease. Note that the proteolytic digestion products of all three tropomyosins show considerable differences in their band patterns; similarities exist between rabbit and earthworm tropomyosin$_1$.

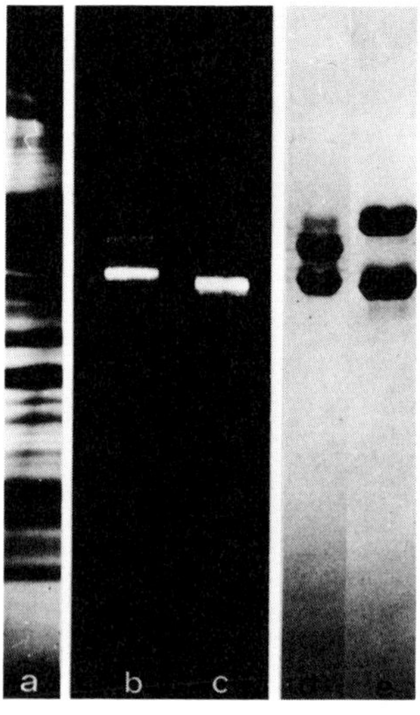

Fig. 19.4. Demonstration of immunological diversity of both earthworm tropomyosins. Crude earthworm actomyosin, separated by SDS-PAGE (a), was electrophoretically transferred on nitrocellulose sheets. The blots were incubated with antiearthworm antibodies and with ^{125}J-labelled protein-A as indicator before exposure to X-ray film. Blot treated with whole tropomyosin$_1$ (b), Tropomyosin$_{2(33)}$ subunit (c). For orientation, SDS-gels of purified tropomyosin$_1$ (d) and tropomyosin$_2$ (e) are shown. There was no cross-reactivity between both tropomyosin isoforms and the antibodies. Immunoblotting was performed according to Towbin et al. (1979).

Nitrocellulose blots of crude earthworm actomyosin (Fig. 19.4a) were incubated with antibodies against either earthworm tropomyosin$_1$ or the 33,000-Da subunit of tropomyosin$_2$. The immunoblots (Fig. 19.4b and c) show that the antibody reaction was specific for only the particular tropomyosin isoform.

FUNCTIONAL PROPERTIES

With the following assays the functional properties of the tropomyosin$_2$ type are demonstrated. It is known that rabbit skeletal muscle tropomyosin binds to actin and is able to inhibit the acto-HMM ATPase activity independently of the Ca^{++} concentration (Eaton et al., 1975). In combination with the troponin complex it is essential for the suppression of the actin myosin interaction in the absence of Ca^{++} (Ebashi & Endo, 1968).

That tropomyosin$_2$ binds to rabbit F-actin was established by the ultracentrifugation of a mixture of pure rabbit actin and the earthworm tropomyosin$_2$ with subsequent examination of the pellet by two-dimensional electrophoresis (Fig. 19.5). Whereas in one dimensional SDS-PAGE the upper band of the tropomyosin comigrates with actin (control slot in Fig. 19.5), the two-dimensional system reveals that both polypeptides had bound to actin in rather equal amounts.

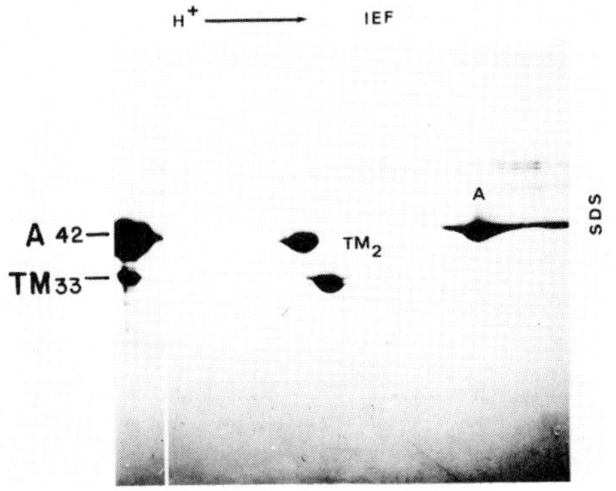

Fig. 19.5. Binding of earthworm tropomyosin$_1$ to rabbit skeletal muxcle F-actin. 1 mg/ml actin and 0.5 mg/ml tropomyosin$_2$ (TM$_2$) were mixed and the actin was sedimented. Two-dimensional electrophoresis reveals that both tropomyosin subunits had bound to actin. Slot on the left shows the actin pellet separated by SDS-PAGE. A, actin; TM, tropomyosin.

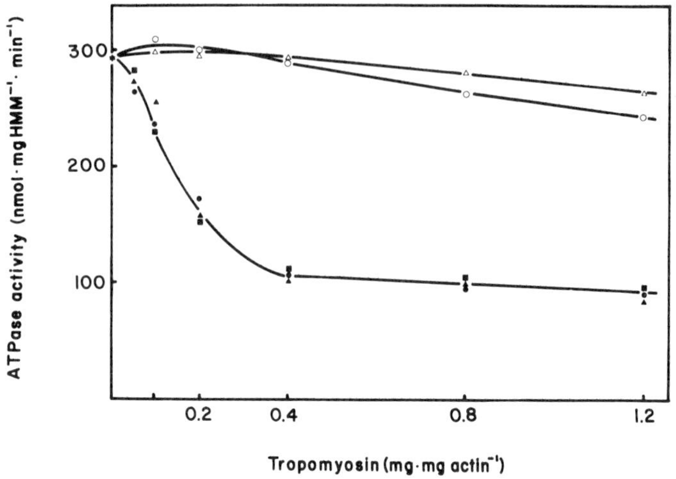

Fig. 19.6. Influence of intact earthworm tropomyosin and its isolated subunits on the rabbit acto-HMM ATPase activity. Tropomyosin fraction added to acto-HMM: tropomyosin$_1$ (■), tropomyosin$_2$ (▲), isolated subunit tropomyosin$_{2(33)}$ (○), isolated subunit tropomyosin$_{2(42)}$ (△), recombined subunits of tropomyosin$_2$ (●). Note that only the unfractionated isoforms are able to inhibit the acto-HMM ATPase activity up to 70%. The isolated subunits have almost no effect. The concentrations of actin and HMM were 0.25 and 0.5 mg/ml respectively. ATPase assays were performed as previously described (Ditgens et al., 1982).

The strong inhibitory effect of earthworm tropomyosin$_1$ on the rabbit acto-HMM ATPase activity was already demonstrated. At a constant ratio (w/w) of actin:HMM of 1:2 a 70% inhibition of the ATPase activity took place (Ditgens et al., 1982). Figure 19.6 shows that tropomyosin$_2$ has the very same effect on the ATPase activity. At physiological actin:tropomyosin ratios the maximum inhibition is reached.

A quite different result was obtained when the isolated subunits tropomyosin$_{2(33)}$ and tropomyosin$_{2(42)}$ were added to acto-HMM. Nearly no inhibitory effect was measurable. An inhibition to the same extent as found for both unfractionated tropomyosins (TM$_1$ and TM$_2$) could be achieved when both subunits were recombined in a 1:1 ratio (Fig. 19.6). In contrast to this reaction Cummins & Perry (1973) found that the isolated components of rabbit skeletal muscle tropomyosin, while forming homodimers, possess the same biological activity as the unfractionated $\alpha\beta$-tropomyosin. Our result led to the conclusion that the isolated subunits of the worm tropomyosin are not able to form homodimeric tropomyosin molecules.

There is strong evidence for the identification of the second isoform as tropomyosin given its cooperation with rabbit troponin within a hybrid tropomyosin-troponin complex. Figure 19.7 demonstrates that both earthworm tropomyosins are able to replace tropomyosin in the Ca^{++}-sensitive acto-HMM system of rabbit skeletal muscle. Only in the presence of Ca^{++} is the strong inhibition of the acto-HMM ATPase activity exhibited by both earthworm tropomyosins neutralized.

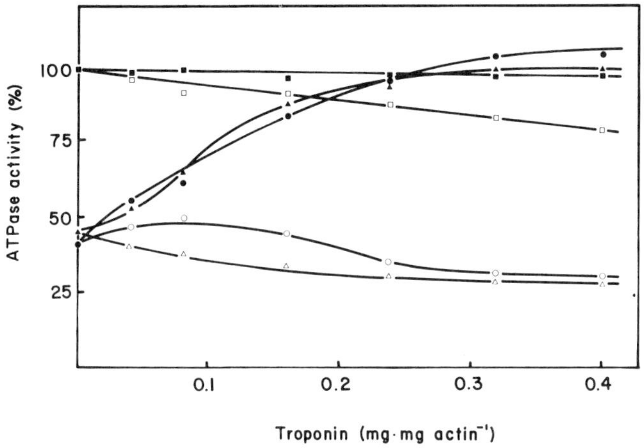

Fig. 19.7. Interaction of earthworm tropomyosin and skeletal muscle troponin in the rabbit skeletal acto-HMM system. Increasing amounts of rabbit troponin were added to acto-HMM in the absence (squares) and presence of either $tropomyosin_1$ (triangles) or $tropomyosin_2$ (circles). Activities in the presence (0.1 mM $CaCl_2$) or absence (2 mM EGTA) of Ca^{++} are marked by closed and open symbols, respectively. Note that both earthworm tropomyosins are able to replace rabbit tropomyosin in the Ca^{++}-sensitive acto-HMM system. Protein concentrations: actin, 0.25 mg/ml; HMM, 0.5 mg/ml; tropomyosin, 0.08 mg/ml. The maximum ATPase activity, square on the ordinate, was 300 nmol Pi $\times$ mg HMM^{-1} $\times$ min^{-1}.

STRUCTURAL PROPERTIES

A structural property of tropomyosin is its formation of paracrystalline aggregates in the presence of divalent cations. Figure 19.8 shows paracrystals of the earthworm tropomyosin isoforms compared to tactoids of rabbit tropomyosin. All three tropomyosins differ in their band patterns (Fig. 19.8d-f) whereas the period length is identical with 40 nm (Fig. 19.8a). In addition to needle-shaped aggregates produced by both earthworm tropomyosins (Fig. 19.8g and h), the tropomyosin_1 formed square and hexagonal nets (Fig. 19.8b and c) with a 40 nm internode spacing. From this structural analysis it follows that both earthworm tropomyosins have the same molecular length as other tropomyosins from muscular sources (Caspar et al., 1969; Millward & Woods, 1970).

CONCLUDING REMARKS

The results presented here demonstrate that in the earthworm body wall muscle four tropomyosin polypeptides exist in rather equal amounts. It was shown that they constitute two heterodimeric tropomyosin isoforms and no homodimers. They both show features characteristic for tropomyosin under functional as well as structural aspects.

An important question is whether there is a special distribution of the isoforms within the worm muscle. For rabbit skeletal muscle it was shown that neither subunit is strictly specific for either fast or slow muscles (Bronson & Schachat, 1982) contrary to the earlier proposal of Dhoot & Perry (1979). First immunological staining of the earthworm body wall and isolated cells with antibodies against both tropomyosin isoforms revealed that they are homogeneously distributed within the two muscle layers and within the single cells. Work is now in progress to investigate the tropomyosin distribution on the molecular level to find some insights in the functional significance of two tropomyosin isoforms in the dual regulated actomyosin.

ACKNOWLEDGMENT

This work was supported by a grant from the Deutsche Forschungsgemeinschaft (DFG).

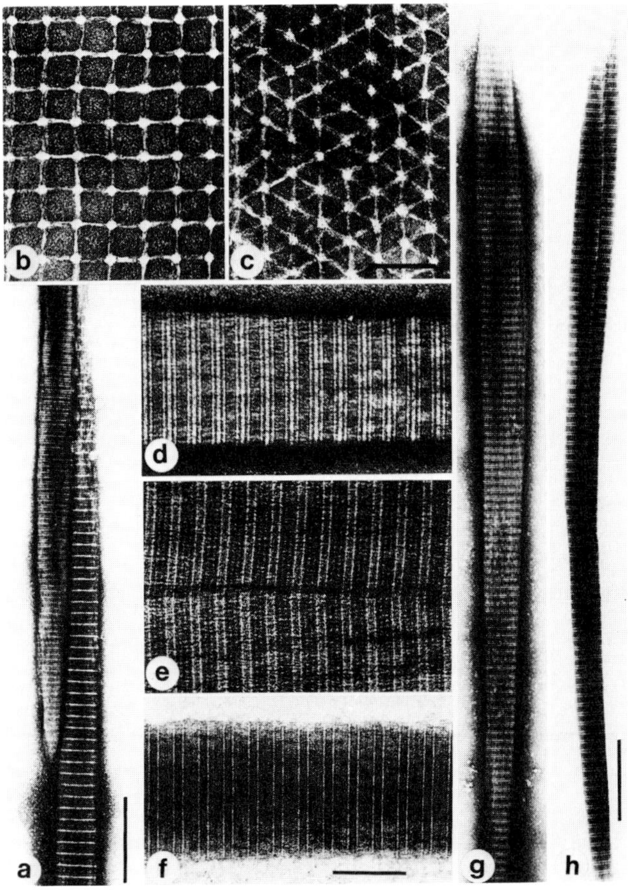

Fig. 19.8. Electron micrographs of tropomyosin paracrystals and nets. The crystallization medium contained 60 mM KCl, 50 mM $CaCl_2$, and 30 mM Tris-HCl (pH 7.6). (a) mixture of rabbit tropomyosin and earthworm $tropomyosin_1$, showing the identical periodicity (rabbit tropomyosin left, earthworm tropomyosin right); (b) and (c) square and hexagonal nets of earthworm $tropomyosin_1$; (d-f) detailed banding pattern of tactoids from (d) earthworm $tropomyosin_1$, (e) earthworm $tropomyosin_2$, (f) rabbit tropomyosin; (g) paracrystal from earthworm $tropomyosin_1$; (h) paracrystal from earthworm $tropomyosin_2$. The bars represent (a) 300 nm, (b and c) 80 nm, (d-f) 160 nm, (g and h) 400 nm.

REFERENCES

Bronson, D. D. & Schachat, F. H. (1982) Heterogeneity of contractile proteins. J. Biol. Chem. 257, 3937-44.

Caspar, D. L. D., Cohen, S. & Longley, W. (1969) Tropomyosin: crystal structure, polymorphism and molecular interactions. J. Mol. Biol. 41 87-107.

Cummins, P. & Perry, S. V. (1973) The subunits and biological activity of polymorphic forms of tropomyosin. Biochem. J. 433, 765-77.

D'Haese, J. & Ditgens, A. (1980) Double regulation in the obliquely striated muscle of Lumbricus terrestris. J. Muscle Res. Cell Motil. 1, 208.

Dhoot, G. K. & Perry, S. V. (1979) Distribution of polymorphic forms of troponin components and tropomyosin in skeletal muscle. Nature 278, 714-18.

Ditgens, A. (1983) Strukturelle und biochemische Untersuchungen an der schräggestreiften Muskulatur des Regenwurms Lumbricus terrestris. Dissertation an der Universität Düsseldorf.

Ditgens, A., D'Haese, J. & Sobieszek, A. (1980) Tropomyosin and the actin-linked Ca-regulation of the obliquely striated body wall muscle of the earthworm Lumbricus terrestris. J. Muscle Res. Cell Motil. 1, 467.

Ditgens, A., D'Haese, J., Small, J. V. & Sobieszek, A. (1982) Properties of tropomyosin from the dual regulated obliquely striated body wall muscle of the earthworm (Lumbricus terrestris L.). J. Muscle Res. Cell Motil. 3, 57-74.

Eaton, B. L., Kominz, D. L. & Eisenberg, E. (1975) Correlation between the inhibition of the acto-heavy meromyosin ATPase and the binding of tropomyosin to F-actin: effects of Mg, KCl, troponin I, and troponin C. Biochemistry 14, 2718-25.

Ebashi, S. & Endo, M. (1968) Calcium ion and muscle contraction. Progr. Biophys. Mol. Biol. 18, 123-83.

Matsumura, F., Yamashiro-Matsumura, S. & Lin, J. J.-C. (1983) Isolation and characterization of tropomyosin-containing microfilaments from cultured cells. J. Biol. Chem. 258, 6636-44.

Millward, G. R. & Woods, E. F. (1970) Crystals of tropomyosin from various sources. J. Mol. Biol. 52, 585-88.

O'Farrell, P. H. (1975) High resolution two-dimensional electrophoresis of proteins. J. Biol. Chem. 250, 4007-21.

Towbin, H., Staehelin, T. & Gordon, J. (1979) Electrophoretic transfer of proteins from polyacrylamide gels to nitrocellulose sheets: procedure and some applications. Proc. Natl. Acad. Sci. U.S.A. 76, 4350-54.

20 Myosin Isoforms in Earthworm Body Wall Muscle

Jochen D'Haese and Dietrich Carloff

INTRODUCTION

In the last few years myosin-linked regulatory systems have been found to play an important role in the conversion of the Ca^{++}-signal to the contractile apparatus in muscular systems and nonmuscle cells. In addition, it turned out that the majority of contractile systems is dual regulated, though the regulatory proteins of the actin- and myosin-linked systems are not always of equal importance. As it will be shown later, both regulatory systems of the earthworm actomyosin are able to almost completely inhibit the actomyosin ATPase activity in the same range of Ca^{++} concentration. Therefore, the annelid muscle appeared to be a suitable object to study possible interactions between the actin- and myosin-linked regulation. The Ca^{++}-sensitivity of the myosin component was found to be, like in some molluscan muscles (Szent-Györgyi et al., 1973; Kendrick-Jones et al., 1976; Nishita et al., 1979; Ashiba et al., 1980; Konno et al., 1980) dependent on the presence of a regulatory light chain, which can be reversibly removed by treatment with EDTA (D'Haese, 1980). The desensitized myosin could be resensitized by readdition of the extracted light chain fraction. A similar effect was achieved by the addition of P-light chains from chicken gizzard and rabbit skeletal muscle myosin. Earthworm myosin is probably regulated by direct binding of Ca^{++}, because it binds two mols Ca^{++} with high affinity per mol of myosin. As a heterogeneity was now found with respect to the myosin light chains it was necessary to get more information about the functional significance of these isoforms.

MYOSIN-LINKED REGULATION

The myosin-linked regulation can be clearly demonstrated when isolated myosin from earthworm body wall (Fig. 20.2b) is mixed with increasing amounts of pure rabbit skeletal muscle actin (Fig. 20.1a). The actomyosin ATPase activity is inhibited up to 90% in the absence of Ca^{++} by the myosin-linked regulatory system.

For further characterization of the actin- and myosin-linked regulation, we tested for differences between the two regulatory systems in their response to changes of the Ca^{++} concentration. As differences of the regulatory properties have been described for fast decapod and vertebrate skeletal muscle myosin (Lehman, 1977, 1978; Pulliam et al., 1983), the assays were performed at 20 mM (closed circles) and 80 mM KCl (open circles in Fig. 20.1b and c). The Ca^{++} dependence of hybridized actomyosin consisting of earthworm thin filaments and rabbit skeletal muscle myosin (Fig. 20.1b) and the reaction of actomyosin containing a mixture of earthworm myosin with pure rabbit actin (Fig. 20.1c) are shown for comparison. Obviously the range of Ca^{++} concentration in which the regulatory systems influence the actomyosin-ATPase activity is nearly the same for both systems. When the ionic strength is increased the activities of both hybridized actomyosins are diminished and the Ca^{++} sensitivity is reduced. The open triangles in Figure 20.1c represent a different myosin preparation. The reason why some myosin preparations are insensitive to changes in the KCl concentration is uncertain at this time.

Fig. 20.1. Functional properties of the myosin- and actin-linked regulatory systems. (a) Ca^{++}-dependent activation of earthworm myosin (0.3 mg/ml) by pure rabbit skeletal muscle actin. Closed and open circles correspond to the activities in the presence (0.1 mM $CaCl_2$) and absence (2 mM EGTA) of Ca^{++}. Assay conditions: 20 mM KCl, 30 mM Tris-maleate buffer, pH 7.0, 2 mM $MgCl_2$, 1 mM ATP, 0.1 mM $CaCl_2$ or 2 mM EGTA. Ca^{++} dependence of hybridized actomyosins containing (b) earthworm thin filaments (1 mg/ml) and rabbit skeletal muscle myosin (0.5 mg/ml) and (c) earthworm myosin (0.5 mg/ml) and rabbit skeletal muscle actin (1 mg/ml). Closed and open circles represent measurements at 20 mM and 80 mM KCl. Triangles and circles are representative curves of different myosin preparations. Assay conditions as in (a) using a 2 mM Ca^{++}/EGTA buffer according to Weber (1969).

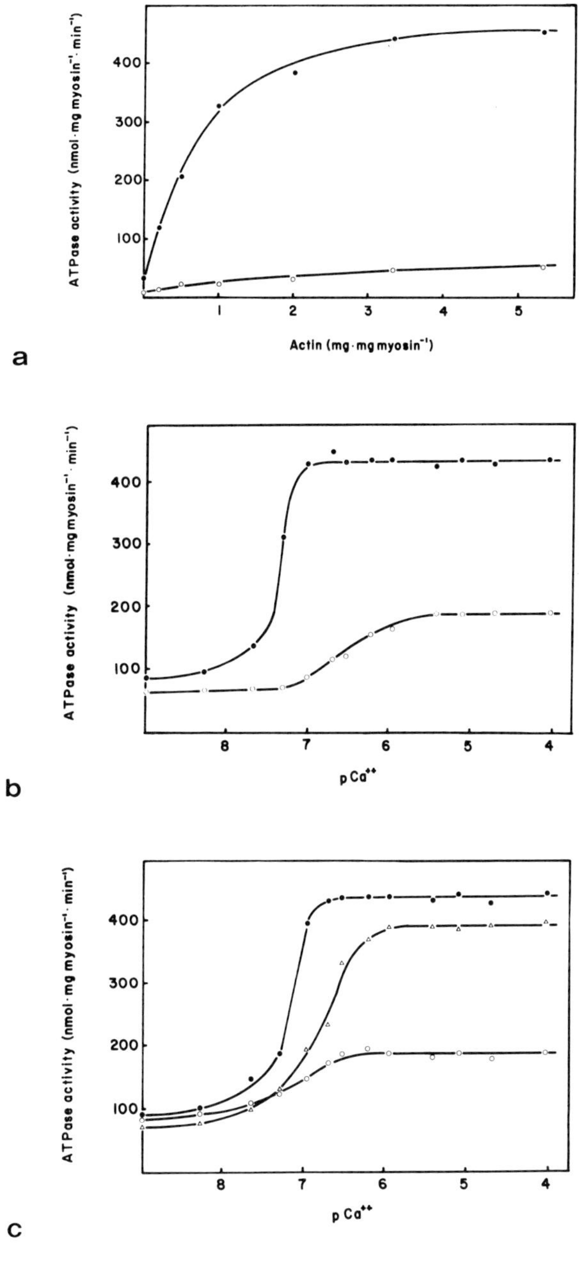
ATPase activity (nmol · mg myosin⁻¹ · min⁻¹)
400
300
200
100
1
2
3
4
5
Actin (mg · mg myosin⁻¹)
a
ATPase activity (nmol · mg myosin⁻¹ · min⁻¹)
400
300
200
100
8
7
6
5
4
pCa⁺⁺
b
ATPase activity (nmol · mg myosin⁻¹ · min⁻¹)
400
300
200
100
8
7
6
5
4
pCa⁺⁺
c

ISOLATION OF MYOSIN LIGHT CHAINS

Earthworm myosin (Fig. 20.2b) isolated from actomyosin extracts (Fig. 20.2a) shows two main light chain bands with molecular weights of 25 k and 18 k. The 25 k light chain, the regulatory one, is removed when the actomyosin (Fig. 20.2h) is desensitized by treatment with 10 mM EDTA at 30° (Fig. 20.2i). Myosin (Fig. 20.2k) prepared from desensitized actomyosin contains three main low molecular weight bands, and it is assumed that the additional band represents another light chain with a molecular weight of 28 k, being enriched by this procedure. The same band–less pronounced–was found in all myosin preparations. To clarify the possible presence of two myosin isoforms with different light chains we isolated the 25 k and 28 k proteins, both of which are extracted when the myosin is treated with the thiol reagent DTNB (5,5'-dithiobis-(2 nitrobenzoic acid) (Fig. 20.2g). The final isolation procedure includes Affi-gel blue and DEAE-sepharose C1-6B chromatography of total guanidine-HCl extracts of isolated myosin. Under the conditions used (comp. Toste & Cooke, 1979) both DTNB light chains were bound to the Affi-gel column (Fig. 20.2d) and were eluted in one step with high salt. The 18 k light chain was not bound (Fig. 20.2e). The DTNB light chains were separated by ionic exchange chromatography using a linear phosphate gradient from 100 to 300 mM according to the method of Holt and Lowey (1975) into three peaks–one containing the 25 k light chain (Fig. 20.2e), and two peaks containing the 28 k light chain (Fig. 20.2f).

REGULATORY PROPERTIES OF THE ISOLATED LIGHT CHAINS

All three light chain fractions obtained after DEAE-chromatography were tested with respect to their ability to resensitize the myosin-linked regulation. The results are summarized in Figure 20.3. Fully desensitized actomyosin (0.5 mg/ml), which had lost the myosin- as well as the actin-linked regulatory properties, was mixed with 0.02 mg of the particular light chain fraction. It is obvious that all three fractions restore the sensitivity to nearly the same extend. From this result it is evident that both 28 k molecular weight fractions can function as regulatory light chains. In view of the following results it is assumed that the separation of two 28 k light chain peaks is caused by slight modifications of the same light chain.

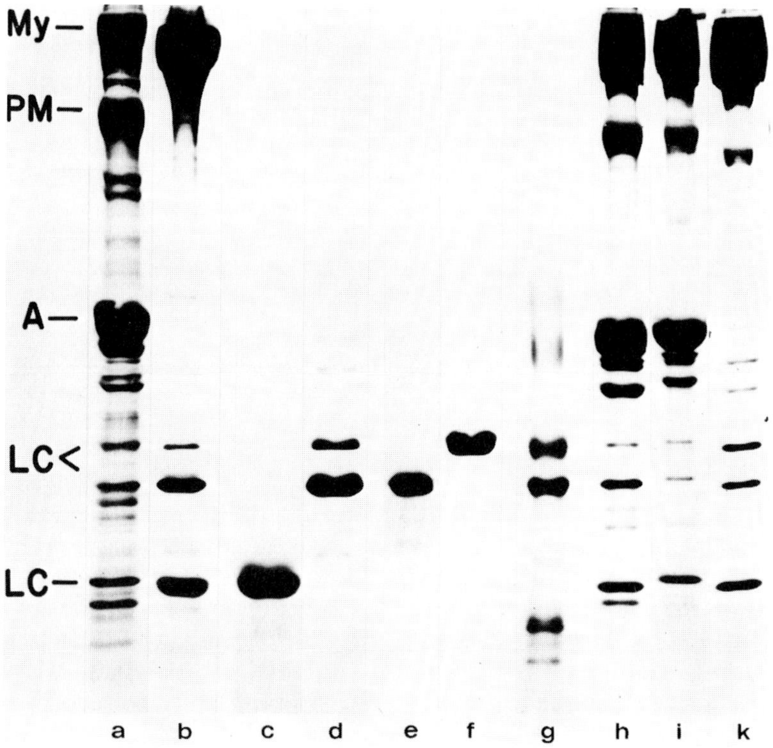

Fig. 20.2. SDS-Polyacrylamide gel electrophoresis of earthworm protein fractions to illustrate the preparation of actomyosin (a), myosin (b), and myosin light chains (c-g), as well as of desensitized actomyosin (h,i), and of myosin from desensitized actomyosin (k). Whereas a DTNB-extract (g) contains only the 28 k and 25 k myosin light chains (LC), all three light chains are extracted by Gdn-HCl and are separated by Affi-gel blue chromatography into an 18 k LC (c) and 25/28 k LC-fraction (d). DEAE-chromatography leads to the elution of one peak with the 25 k and two peaks with the 28 k LC. Note that myosin prepared from desensitized actomyosin (i) contains equal amounts of 25 and 28 k LC. The bands of the myosin heavy chains (My), paramyosin (PM), actin (A), and the myosin light chains (LC) are marked.

	Mg-ATPase activity Ca	EGTA	Ca-sensitivity (%)
Actomyosin (0.5 mg/ml) (desensitized)	277	255	8
Actomyosin (0.5 mg/ml) (desensitized)			
+ LC-25 (0.02 mg/ml) 1st DEAE peak	287	125	56
+ LC-28 (0.02 mg/ml) 2nd DEAE peak	289	125	57
+ LC-28 (0.02 mg/ml) 3rd DEAE peak	284	104	63

Fig. 20.3. Resensitization with different light chain fractions. Treatment of earthworm actomyosin with 10 mM EDTA according to Chantler et al. (1980) led to a fully desensitized actomyosin having lost the actin- as well as the myosin-linked regulatory properties. A resensitization of such actomyosin was possible with all three light chain fractions obtained by DEAE-chromatography.

IMMUNOLOGICAL LOCALIZATION OF LIGHT CHAINS

Antibodies against the 25 k and the 28 k (first DEAE-peak) light chains were raised in rabbits by low dose immunization. The specificities of the antisera and IgG fractions were tested by enzyme linked immunosorbent assay (ELISA), radial immunodiffusion and blotting techniques. The antibody against the 28 k light chain could not discriminate between both 28 k DEAE-peak fractions. The isolated IgG fractions showed low cross-reactivity with other proteins from crude earthworm actomyosin and the other light chains, and were used to get some information about the light chain distribution in the earthworm body wall muscle.

Cryostat cross sections of the body wall incubated with antibodies against the 25 k and 28 k light chains (Fig. 20.4a and b) clearly reveal a distinct distribution of both light chains. The 25 k light chain is almost exclusively located at the periphery of each muscle layer whereas the 25 k light chain, which is by far the prominent light chain of the body wall muscle, is found

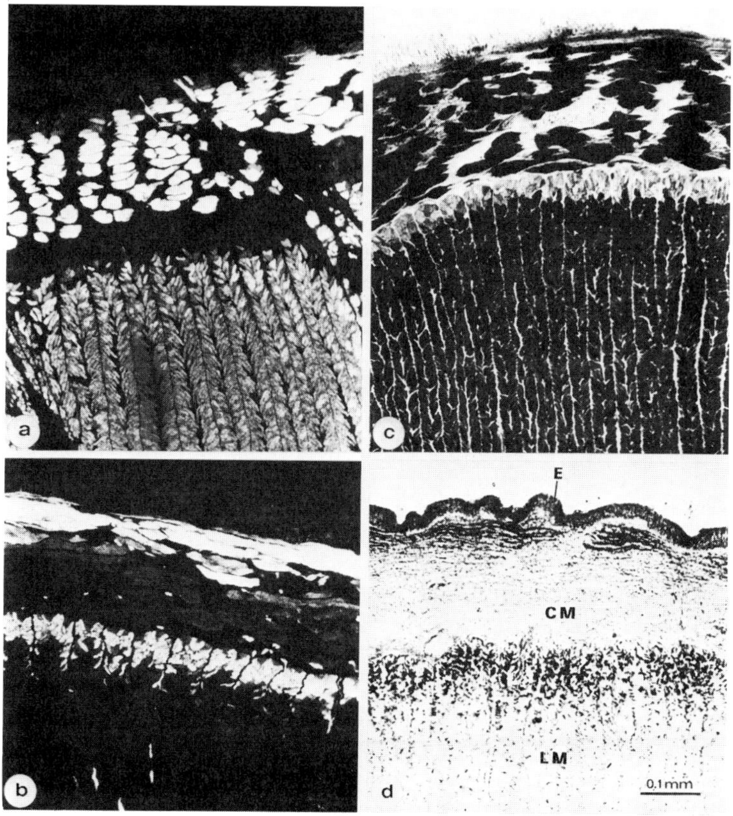

Fig. 20.4. Cryostat cross sections of the earthworm body wall to demonstrate the light chain distribution (a and b) and the activities of myosin ATPase and succinic dehydrogenase (SDH) (c and d). The sections (a and b) were incubated with antibodies against the 25 k (a) and the 28 k (b) myosin light chains. FITC-labelled antirabbit IgG was used as second antibody. The sections (c and d) were treated histochemically to demonstrate Ca^{++}-activated myosin ATPase at pH 9.4 (incubation time 5 min., 20°C) (c) and SDH (incubation time 1 h, 37°C) (d). The distribution of light chains corresponds very well with the enzyme activities. E, epidermis; CM, circular muscle layer; LM, longitudinal muscle layer.

in the inner proximal portions of the circular as well as the longitudinal muscle layer. This special location and the ratio of light chains is apparently not dependent on earthworm development. Young worms still in the cocoon exhibit the same distribution.

CORRELATION OF LIGHT CHAIN DISTRIBUTION AND ENZYME ACTIVITY

To see if there is some functional significance of this light chain distribution, enzyme-histochemical tests were made. Two of them are shown in Figure 20.4c and d. The frozen cross section in Figure 20.4c was stained for Ca^{++}-activated myosin ATPase activity (according to Khan et al., 1972) and that shown in Figure 20.4d for succinic dehydrogenase activity (according to Nachlas et al., 1957). Interestingly, both stainings exactly correspond to the described light chain distribution. The staining demonstrates that the myosin containing the 25 k light chain has a higher ATPase activity compared to that with the 28 k light chain. This correlates with the finding that the activity of succinic dehydrogenase, a typical enzyme of an oxidative metabolism, is concentrated in the periphery of the muscle layers where the less active myosin was found.

CONCLUDING REMARKS

The earthworm body wall muscle contains two different DTNB-light chains. Both light chains are equally suited to resensitize desensitized actomyosin. The stainings by antibodies against these 25 k and 28 k light chains complete each other so that both stainings together cover the total body wall muscle. The observed differences in the activities of myosin-ATPase and succinic dehydrogenase can be directly correlated to the distribution of the 25 k and 28 k light chains. These findings demonstrate that the different DTNB-light chains define the two homodimeric myosin isoenzymes of the body wall muscle. Thus both muscle layers apparently contain "fast" muscle cells and at the periphery "slow" fibers with an oxidative metabolism to fulfill the physiological requirements.

Quite different results were obtained so far for the nematode *Caenorhabditis elegans*, the body wall of which contains two myosin isoforms with distinguishable myosin heavy chains being present within the same muscle cell (Mackenzie et al., 1978).

As the myosin fraction the properties of which we determined contained a mixture of both isoforms, we are now trying to separate the isoforms for the characterization of their individual regulatory properties.

ACKNOWLEDGMENT

This work was supported by a grant from the Deutsche Forschungsgemeinschaft (DFG).

REFERENCES

Ashiba, G., Asada, T. & Watanabe, S. Calcium regulation in clam foot muscle. J. Biochem. 88, 837-46 (1980).

Chantler, P. D. & Szent-Györgyi, A. G. Regulatory light chains and scallop myosin. J. Mol. Biol. 138, 473-92 (1980).

D'Haese, J. Regulatory light chains of myosin from the obliquely-striated body wall muscle of Lumbricus terrestris. FEBS Lett. 121, 243-45 (1980).

Holt, J. C. & Lowey, S. An immunological approach to the role of the low molecular weight subunits in myosin. I. Physical-chemical and immunological characterization of the light chains. Biochemistry 14, 4600-4620 (1975).

Kendrick-Jones, J., Szentkiralyi, E. M. & Szent-Györgyi, A. G. Regulatory light chains in myosins. J. Mol. Biol. 104, 747-75 (1976).

Khan, M. A., Papadimitriou, P. G., Holt, P. G. & Kakulas, B. A. A calcium-citro-phosphate technique for the histochemical localization of myosin ATPase. Stain Technol. 47, 277-81 (1972).

Konno, K., Arai, K.-I. & Watanabe, S. Myosin-linked calcium regulation in squid muscle. In: Muscle Contraction: Its Regulatory Mechanisms, eds. Ebashi, S. et al., Japan Sci. Soc. Press, Tokyo/Springer Verlag Berlin, pp. 391-99 (1980).

Lehman, W. Calcium ion-dependent myosin from decapod-crustacean muscles. Biochem. J. 163, 291-96 (1977).

Lehman, W. Thick-filament-linked calcium regulation in vertebrate striated muscle. Nature 274, 80-81 (1978).

Mackenzie, J. M., Schachat, F. & Epstein, H. F. Immunocytochemical localization of two myosins within the same muscle cell in Caenorhabditis elegans. Cell 15, 413-19 (1978).

Nachlas, M. M., Tsou, K. C., de Souza, K., Cheng, C.-S. & Seligman, A. M. Cytochemical demonstration of succinic dehydrogenase by the use of a new p-nitro-phenyl substituted ditetrazole. J. Histochem. Cytochem. 5, 420-37 (1957).

Nishita, K., Ojima, T. & Watanabe, S. Myosin from striated adductor muscle of Chlamys nipponensis akazara. J. Biochem. 86, 663-73 (1979).

Pulliam, D. L., Sawyna, V. & Levine, R. J. C. Calcium sensitivity of vertebrate skeletal muscle myosin. Biochemistry 22, 2324-31 (1983).

Szent-Györgyi, A. G., Szentkiralyi, E. M. & Kendrick-Jones, J. The light chains of scallop myosin as regulatory subunits. J. Mol. Biol. 74, 179-203 (1973).

Toste, A. P. & Cooke, R. Interactions of contractile proteins with free and immobilized Cibacron Blue F3GA. Anal. Biochem. 95, 317-328 (1979).

Weber, A. Parallel response of myofibrillar contraction and relaxation to four different nucleoside triphosphates. J. Gen. Physiol. 53, 781-91 (1969).

PART II

FUNCTION OF CONTRACTILE AND CYTOSKELETAL PROTEINS

21 Time-resolved X-ray Diffraction Studies of Muscle Contraction

H. E. Huxley

The physical agents which generate the sliding force between actin and myosin filaments in muscle are the so-called crossbridges, structural elements which protrude from the backbone of the myosin filaments and can interact with the actin filaments alongside (Fig. 21.1). These represent the enzymatically active part of the myosin molecules, which can split ATP, thereby yielding the energy for contraction, and can combine with actin, thereby, in principle, allowing a force to be exerted on the actin filaments by the myosin. There are some interesting points about how the actin and myosin molecules are all arranged in the structure to have the appropriate structural polarity to interact in a very specific way, but the central question I wish to consider here concerns exactly where and how the sliding force is developed.

The distance through which a given actin and myosin filament can slide past each other during muscle contraction may be several thousands of Angstrom units (depending on the extent of shortening), so it is clear from the dimensions of the crossbridges (the Head units are about 150 Å in length) and their axial separations (143 Å intervals along the thick filaments) that in an efficient system they must act repetitively and cyclically many times during the whole period of a contraction. The model which has been generally accepted is one in which, during activity, a crossbridge first attaches to a specific site on an actin filament; a structural change then takes place within the actin-crossbridge complex so that the actin filament is pulled along a distance of about 120 Å; and the crossbridge then detaches, ready to reattach again at a new site further along the actin and bring about a further unit of shortening. This cycle is repeated many times at all the crossbridges during the time the muscle is actively contracting.

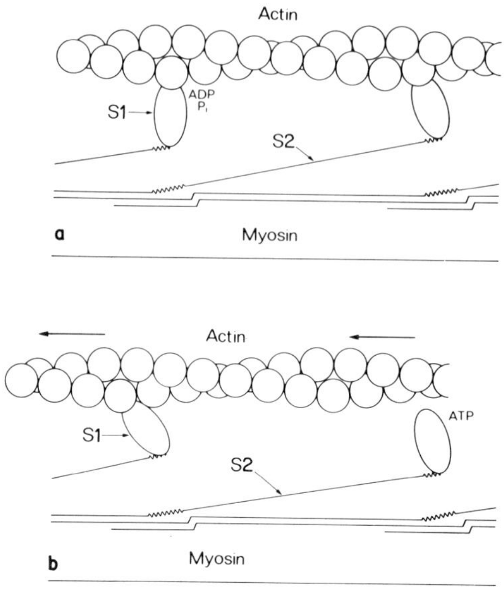

Fig. 21.1. Active change in angle of attachment of crossbridges (S_1 subunits) to actin filaments could produce relative sliding movement between filaments maintained at constant lateral separation (for small changes in muscle length) by long range force balance. Bridges can act asynchronously since subunit and helical periodicities differ in the actin and myosin filaments. (a) Left hand bridge has just attached; other bridge is already partly tilted. (b) Left hand bridge has just come to end of its working stroke; other bridge has already detached, and will probably not be able to attach to this actin filament again until further sliding brings helically arranged sites on actin into favorable orientation. A similar mechanism could be based on a shape change in the attached S_1 subunits.

At the time this general model became accepted—at least as a good working hypothesis—the basis for this acceptance was much more a matter of what seemed to be common sense rather than what was experimentally verifiable or theoretically proved. That is, the overall nature of the overlapping filament structure, the existence of the crossbridges and their identity as the active ends of the myosin molecules were not in any serious doubt, but there was essentially no direct experimental evidence about their actual behaviour, i.e., whether they really did undergo structural changes during contraction which would enable them to exert a sliding force on actin. Nevertheless, the most common reaction which I found when I pointed out that the moving crossbridge model was only an hypothesis which still needed a lot of experimental verification was "yes, but what other way could the system possibly work?" I can think of several ways—and so can other people—and the important thing was to devise experiments which begin to probe the actual structural behavior of the crossbridges during contraction.

What I would like to do here is to summarize as objectively as I can the results of various kinds of experiments directed toward this end, so that you can see the present status of the problem, and the unanswered questions. My own work has primarily involved X-ray diffraction studies on contracting muscle. These were originally very laborious, because the patterns given are relatively weak ones andused to take several hours to record. However, in the last few years we have been able to use synchrotron radiation from electron storage rings as a very intense X-ray source, giving a gain in recording speed by a factor of about 1,000 times at present over the best rotating anode X-ray tube. This, together with the use of electronic X-ray counters instead of film giving a further gain in speed and convenience, has enabled us to explore the changing patterns during different types of contraction with a time resolution of 1 millisecond or less on the stronger reflections.

The results give us information about two main aspects of the structural nature and arrangements of the crossbridges. The equatorial parts of the X-ray diffraction diagram give information about the lateral distribution of density and consist of a series of reflections from the hexagonal lattice of actin and myosin filaments, of which the inner two reflections, the [10] and [11] are particularly prominent. The spacing of these reflections changes very little during isometric contraction (i.e., when the muscle develops full tension but is kept at constant length), but their intensities change by a large amount (the [10] falling to about 0.6 of its resting intensity, the [11] increasing to about double its resting intensity). This change shows that there has been a large transfer of material from around the thick filaments to positions much nearer the axis of the thin filaments (5,4). I think there can be no doubt that it corresponds to the attachment of a considerable proportion of the crossbridges to the actin filaments during tension development. If we look at the time course of this change during the course of a short contraction, we see that to a first approximation the structural change corresponds in its time course quite closely with the onset of tension development following stimulation and its decay during relaxation. More precisely, the structural change leads tension development by 5-10 msecs., which may indicate that crossbridge attachment has to be followed by another step with a finite rate constant before that particular cross-bridge generates tension. Somewhat surprisingly, at moderate velocities of shortening, when the tension generated may be (for example) half its maximum isometric value, the X-ray pattern indicates that the same number of crossbridges are attached (12), suggesting to the model-builders that the rate-limiting step is not attachment but a subsequent one. In very rapid shortening, however, attachment does seem to become rate-limiting (7).

These results provide very strong support for one aspect of the moving crossbridge model, namely the attachment of myosin heads to actin in order to develop tension, and a cycling of that attachment during shortening and sliding, so that there are always a substantial proportion of crossbridges attached.

Another experiment one can do is to apply a small but very rapid length decrease to a contracting muscle, corresponding, for example, to about 100 Å of filament sliding in about 1 msec. This should bring about a large simultaneous change in the configuration of all the attached crossbridges, of the same kind that happens asynchronously during their normal operation. Surprisingly, only very small changes in the intensities of the equatorial reflections are produced when this is done. However, calculations show that with the known crossbridge head mass distribution (from 3D reconstruction studies of "decorated" actin (1)) where much of the density lies close to actin, the result neither eliminates nor places strong restrictions on the possible movement or shape change of the part of the attached myosin head more distal to actin.

Clearly, having established cross-bridge attachment, what we need next is information about changes in the longitudinal position and structure of the crossbridges during contraction, and this is provided by the axial part of the X-ray diagram. In resting muscles this shows a rich pattern of layer line reflections from the helical arrangements (with quite different symmetries) of both myosin and actin molecules in their respective filaments. During isometric contraction the actin reflections remain almost unchanged, but the off-meridional myosin reflections all become very much weaker, and no new reflections or intensification of the actin pattern is seen (8,10). We would expect the myosin helical arrangement to become disrupted on attachment to actin. The absence of any new pattern indicates that the myosin heads are attached to actin in a variety of configurations, which is what we would expect if the myosin heads underwent a substantial structural change during their working strokes (which will be asynchronous). These changes in pattern are also closely linked to the onset and decay of tension, and, like the equatorial changes, run slightly ahead of initial tension development (10).

Thus in general terms the X-ray results and particularly the time-resolved ones obtained with synchrotron radiation provide good experimental evidence for the moving, cycling, cross-bridge model, of a kind that previously was lacking. However, what I have described so far only represents the very beginning of a picture of what kind of structural change is taking place in the crossbridge to produce the force and movement. I will spend the remainder of my time trying to say a little more on this crucial subject.

There are, basically, three ways in which a cross-bridge mechanism might work, and it is useful to look at each of them and see whether our experimental evidence can either disprove or provide support for any of them.

In the first mechanism most or all of the myosin S_1 head would behave as a rigid body and alter its angle of attachment to actin, either by local changes in the binding site or sites, or as a result of a change in the actin monomer or monomers to which it is attached. This is one of the simplest models we can think of, and it is also one of the simplest to show in a diagram, so it is widely used by muscle people as a kind of shorthand for a cross-bridge structural change, even when they have in mind something a lot more complicated. However, the evidence available at present indicates that this model cannot be correct. The evidence I have in mind includes the electron spin resonance studies (2) which show that these S_1 heads of myosin which appear to be attached to actin in a contracting muscle have the spin labels which have been specifically attached to them all oriented at a very specific angle. That is, the portion of the head where the label is attached does not change its angular orientation during the working stroke of the cross-bridge. A somewhat similar result has been found by measurements of the orientation of an optical probe (15). So we can, provisionally, cross off this model, unless the full movement takes place immediately a crossbridge develops tension, which seems less likely for other reasons.

The second fairly simple model is one in which the S_1 head of myosin attaches to actin in only one orientation, and does not alter its shape, and where the movement is produced by active shortening of the S_2 region of the molecule (3). This is an interesting and attractive model in some ways, but there are three kinds of experimental evidence, and one somewhat abstract argument, which indicate that it is not correct. First, there is now evidence from model systems—actin filaments attached to surfaces with a preferred orientation which will give rise to movement either with myosin subfragment 1 [S_1] alone in solution (16), or with S_1 joined to S_2 with the latter attached to an inert surface (13)—which appears to indicate that S_1 on its own without the active help of S_2 can support movement. Secondly, there is the X-ray finding that a contracting muscle shows no sign of the characteristic axial pattern given by a muscle in rigor, in which many of the actin reflections are intensified. This shows that the attached myosin heads have a large part of their mass disordered with respect to the symmetry of the actin helix. This would be consistent with them being attached over a range of configurations, but not with them being all in the same structural state as which would be expected if all the movement were in S_2. There is a further piece of X-ray evidence which indicates that the length of S_2 remains approximately constant during the working stroke. Despite the fact that the X-ray diagram from contracting muscle shows that very little if any of the myosin head structure is ordered with respect to either the actin or myosin filament helical repeat, there is still a strong meridional reflection at 143 Å, showing that a significant amount of the structure is ordered with respect to the myosin filament axial repeat (7). Since there is good evidence (9,10) that this reflection arises mainly from the S_1 subunits rather than from the backbone of the myosin filaments, this indicates very strongly that the required change of about 120 Å in the effective axial length of the crossbridges during the working stroke cannot be occurring in S_2. If it were, the reflection would become very weak during contraction since the axial position of the whole of S_1 would be variable. Instead, the S_2 structure must be approximately constant in length, and by acting as a hinge must allow sufficient radial and azimuthal movement for attachment to actin to occur—and for the helical reflections to be greatly weakened. On this model, we would have to assume that sufficient of the S_1 near to the S_1S_2 junction remains close to the 143 Å axial repeat to still give a strong reflection, even though the more distal parts undergo axial displacement during the working stroke. (In a resting muscle the reflection may be weakened by disorder of the unattached bridges.) The theoretical argument is simply that since the enzyme site which interacts with actin and splits ATP is in the S_1 portion, one might expect the structural change to occur there rather than at the end of a two chain α-helical structure some 400 Å away.

The third type of mechanism is one in which S_2 remains approximately constant in length, but where a major structural rearrangement takes place within the myosin head during the working stroke, but one or more regions within the head maintains an approximately constant orientation (Fig. 21.2). One such region could be the part involved in the binding site to actin, or it could be the region near the S_1S_2 junction—or even both! A restricted region fixed on actin would have only a small effect on the X-ray diagram but might account for the constant orientation of the labels seen in the ESR and optical experiments. Alternatively, such labels might be located on a fixed site near S_2.

Finally, there is one very positive piece of recent experimental evidence showing a change in configuration of attached crossbridges during their operation (9,11). This comes from observations of the behavior of the 143 Å meridional reflection during small rapid releases of an otherwise isometrically contracting muscle, i.e., ones that produce a movement of about 100 Å between adjacent actin and myosin filaments. Such a release brings about a very large decrease in intensity of the reflection, taking place within about 1 millisecond, but delayed by about half a millisecond behind the length change. What we think may be happening is the following. Before the release, the actin is fairly uniformly labelled with attached crossbridges which are at various stages of their working stroke. Because of the constancy of the S_2 length, the ends of the crossbridges distal to actin still maintain a 143 Å axial repeat. After the quick-release, the crossbridges have all undergone variable amounts of structural change, depending on how far they were from the end of their working strokes at the moment of release. Thus their ends distal to actin will have moved by variable amounts and the 143 Å repeat will no longer be maintained. Put another way, the S_2 region will be compressed by variable amounts—perhaps this is the function of its more flexible portion!

Altogether then, I think the balance of evidence favors a model in which the active structural change takes place within the myosin head subunit. However, there are still a number of puzzling features in the various experimental results, and it would be rash to rule out the possibility of structural changes elsewhere in the crossbridge-actin structure—say in the S_2 link.

I think I have told you now enough to show that we still have an interesting problem on our hands, but one where

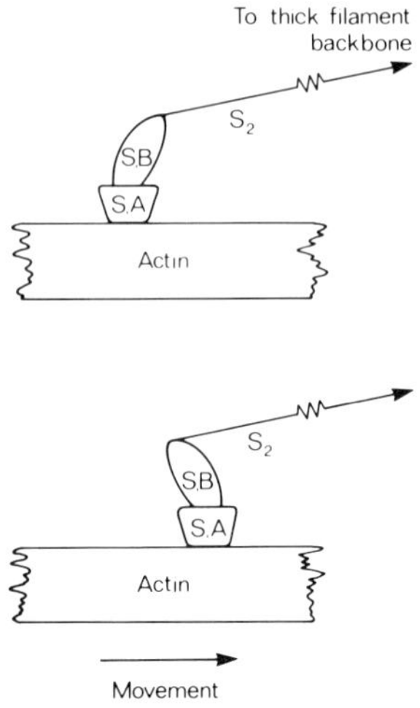

Fig. 21.2. Another possible model of cross-bridge action, in which a change in the relative sliding positions of different structural domains in the head subunit brings about movement of the actin filament to which it is attached. One domain could remain fixed and thereby account for the constant orientation of various electron-spin-resonance and optical probes seen in contracting muscle, if they all happen to attach to this domain. At present, our knowledge of the detailed structure of the myosin head is insufficient to assess the likelihood of this mechanism.

experiments are still making progress. However, we are a very long way from describing the system in molecular terms. We need more detailed analysis and modelling of the X-ray diagrams from muscles during contraction, under different conditions, and these need to be supplemented by electron microscope observations of the corresponding structures arrested in some way in different functional states. These together should give us a realistic overall picture of the structural changes during contraction, which we can then begin to try to understand in terms of the detailed atomic structure of actin and myosin—which may soon begin to emerge from X-ray crystallographic studies of these proteins.

REFERENCES

1. Amos, L. A., Huxley, H. E., Holmes, K. C., Goody, R. S. & Taylor, K. A. (1982) Nature (London) 299, 467-69.
2. Cooke, R., Crowder, M., Barnett, V. A. & Thomas, D. D. (1982) Biophys. J. 37, 117a.
3. Harrington, W. F. (1979) Proc. Natl. Acad. Sci. USA 76, 5066-70.
4. Haselgrove, J. C. & Huxley, H. E. (1973) J. Mol Biol. 77, 549-68.
5. Huxley, H. E. (1968) J. Mol. Biol. 37, 507-520.
6. Huxley, H. E. (1975) Acta. Anat. Nippon. 50, 310-25.
7. Huxley, H. E. (1979) In 'Cross-bridge Mechanism in Muscle Contraction' (Sugi & Pollack, eds.) pp. 391-401, University of Tokyo Press, Tokyo.
8. Huxley, H. E. & Brown, W. (1967) J. Mol. Biol. 30, 383-434.
9. Huxley, H. E., Simmons, R. M., Faruqi, A. R., Kress, M., Bordas, J. & Koch, M. H. J. (1981) Proc. Natl. Acad. Sci. USA 78, 2297-2301.
10. Huxley, H. E., Faruqi, A. R., Kress, M., Bordas, J. & Koch, M. H. J. (1982) J. Mol. Biol. 158, 637-84.
11. Huxley, H. E., Simmons, R. M., Faruqi, A. R., Kress, M., Bordas, J. & Koch, M. H. J. (1983) J. Mol. Biol. 169, 469-506.
12. Podolsky, R. J., St. Onge, R., Yu, L. & Lymn, R. W. (1976) Proc. Natl. Acad. Sci. USA 73, 813-17.
13. Sheetz, M. P. & Spudich, J. A. (1983) Nature 303, 31-35.
14. Taylor, K. A., Amos, L. A. (1981) J. Mol. Biol. 147, 297-324.
15. Yanagida, T. (1981) J. Mol. Biol. 146, 539-60.
16. Yano, M., Yamamoto, Y. & Shimizu, H. (1982) Nature 299, 557-58.

22 Movement of Myosin Molecules in Vitro: A Quantitative Assay

James A. Spudich, Michael P. Sheetz, Gary Peltz,
Paula Flicker, Linda M. Griffith, Peter Parham

The foundation of the work described here resulted from a very enjoyable collaboration with Dr. Michael Sheetz which began in September of 1983 when he joined my laboratory on sabbatical leave from the University of Connecticut. Important contributions to this work have come from a number of colleagues in the Cell Biology Department. In collaboration with Dr. Peter Parham, Dr. Gary Peltz prepared monoclonal antibodies against Dictyostelium myosin and characterized them with respect to their effects on myosin function. Dr. Paula Flicker localized these antibodies on the myosin molecule by rotary shadowing. Dr. Linda Griffith is purifying myosin light chain kinase from Dictyostelium and is exploring the role of light chain phosphorylation on the functions of the myosin. The experiments mentioned below concerning smooth muscle myosin were carried out in collaboration with Dr. Jim Sellers (National Heart and Lung Institute, Bethesda, Maryland).

As discussed in this symposium in the elegant preceding talk by Dr. Hugh Huxley there is little doubt that muscle contraction as well as many forms of nonmuscle motility derive from the repetitive interaction of myosin heads with actin filaments. The actin filaments are polar, and in the muscle sarcomere they are attached to the Z lines at their so-called barbed ends. Thus as the sarcomere contracts, myosin molecules are believed to move along the actin filaments from their "pointed" ends toward their barbed ends.

There are many fundamental aspects of the interaction of myosin with actin which need to be elucidated. For example, what parts of the myosin molecule are critical for the production of motive force? In what ways can one alter the molecule such that the rate at which it moves and the force involved are affected? These and other critical questions have been difficult to answer due to the lack of a quantitative in vitro assay for myosin movement. There are two requirements for establishing an in vitro assay. First, an oriented polar array of actin filaments must be obtained to act as tracks along which the myosin can move. Second, one must devise a means to monitor the position of the myosin molecule as it moves along these tracks. The second requirement is easily satisfied by covalently coating otherwise inert beads (1 to 100 μm diameter) with myosin and observing these myosin-beads in the light microscope. The first requirement is more difficult to fulfill than the second, which probably explains why we and others failed in earlier attempts to develop an in vitro assay for myosin movement.

Success was obtained in establishing an assay for myosin movement when Dr. Sheetz and I took advantage of a naturally occurring parallel array of polar actin filaments from the alga Nitella (Sheetz and Spudich, 1983a). Cells of Nitella are very large and cylindrical (about 4 cm long and about 1 mm wide). Because of their large size they stream their cytoplasm actively up one edge of the cell and down the other. This streaming presumably serves to mix the cytoplasm for more efficient diffusion of nutrients. Nitella has been extensively studied as a model system for cytoplasmic streaming by Drs. Kamiya, Kuroda, Higashi-Fujime, Kersey, Wessells, and others (for review, see Kamiya, 1981). The cytoplasmic face of the cell membrane has rows of chloroplasts which are fixed in place. They do not move. Attached to these rows of chloroplasts are rows of actin cables which are oriented with uniform polarity over long distances (Kersey and Wessells, 1976).

Our approach to using this array of polar filaments was to cut open a cell by a longitudinal incision along one side. The cell was then laid out flat and pinned such that the cytoplasmic face of the cell membrane was exposed. In our first experiments we added myosin-beads that were labeled with fluorescein. In the fluorescence microscope the chloroplasts appear red due to fluorescence of the chlorophyll and the myosin-beads appear green. When we deposit the myosin-beads on the actin cables of the Nitella, we see Brownian motion until the beads reach the actin cables. At this time the beads stop their Brownian motion and then move in a directed fashion. In Figure 22.1 notice that many myosin-beads have not yet settled down onto the surface. In contrast, two linear bead aggregates in the left portion of the photograph have linked to actin cables and are moving along the horizontal chloroplast rows at a rate greater than 1 μm/sec.

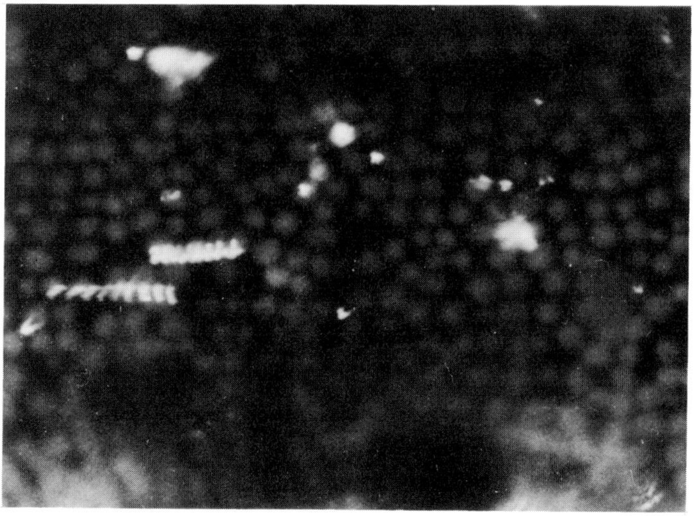

Fig. 22.1. Light micrograph showing movement of myosin-beads along chloroplast rows. The myosin used was from rabbit skeletal muscle. The photograph was produced by a series of 1 sec exposures (taken every few seconds; total exposure ~ 20 sec). Conditions and techniques for this experiment are described in Sheetz and Spudich (1983a). ×900.

An interesting feature of Nitella is that there are two sets of actin filaments with opposite polarities. This arrangement is required since streaming goes up one side of the cell, reverses at the cell terminus, and comes back down the other side. Indeed Kersey and Wessells (1976) observed the switch in polarity of the filaments as one crosses over the so-called indifferent zone that separates these two sets of filaments. As we described earlier (Sheetz and Spudich, 1983a), the myosin-beads all move in one direction (the direction of cytoplasmic streaming) on one side of the indifferent zone and in the opposite direction on the other side. The movement is totally dependent on ATP and furthermore is inhibited by reacting the sulfhydryl groups of the myosin heads with N-ethyl maleimide, a well characterized poison of motive force production in muscle. These observations demonstrate the in vitro movement of myosin molecules on actin cables. The rest of this paper will be devoted to additional work which establishes that our in vitro assay for myosin movement is indeed a quantitative one.

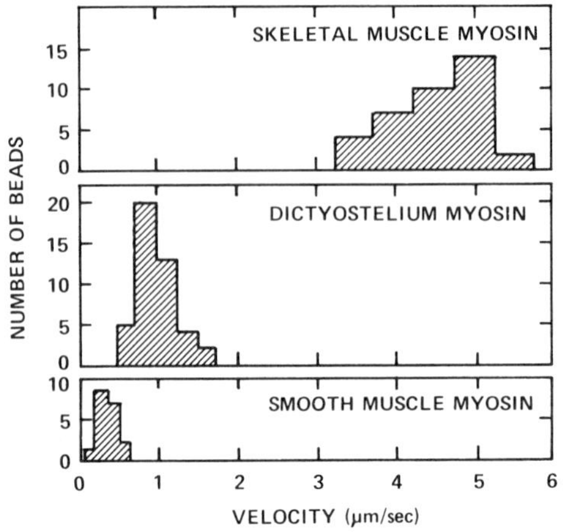

Fig. 22.2. Dependence of rates of myosin-bead movement on the myosin source. Myosins from rabbit skeletal muscle, Dictyostelium, and turkey gizzard smooth muscle are compared. The histograms relate the number of moving myosin beads analyzed and rate of movement. Rabbit skeletal muscle myosin was prepared by a modification of Margossian and Lowey (1982) and Kielley and Harrington (1960). Dictyostelium myosin (with partially phosphorylated LC_{18}) was purified by a modification of Mockrin and Spudich (1976). Gizzard smooth muscle myosin was prepared and phosphorylated by Dr. Jim Sellers according to Sellers et al. (1981).

First, we have compared the rates of movement of myosin purified from three different cells. Figure 22.2 shows histograms of the number of myosin-beads or bead aggregates that move at a particular rate. It is striking that the average rate of movement depends on the source of myosin. Skeletal muscle myosin moves with a velocity of about 5 μm/sec, which is close to the rate at which myosin heads move past actin filaments during the contraction of skeletal muscle (Squire, 1981). Smooth muscle myosin on the other hand moves only one-tenth of that rate, in keeping with the slower rate of contraction of smooth muscle. Dictyostelium myosin moves at its own characteristic rate which is between that of the smooth muscle and the skeletal muscle. It appears that an isolated purified myosin molecule has all of the information built into it to determine the rate at which contraction occurs in vivo.

Does phosphorylation of myosin affect its rate of movement? As we have previously shown (Kuczmarski and Spudich, 1980), the Dictyostelium myosin light chain (LC_{18}) is phosphorylated in vivo during growth of the amoebae. Dr. Linda Griffith has been able to remove the Dictyostelium light chain phosphate with a Dictyostelium phosphatase and rephosphorylate the light chain using Dictyostelium light chain kinase. Interestingly, the dephosphorylated myosin-beads do not move. The myosin on these beads can be rephosphorylated by incubation of the bead suspension with the light chain kinase and ATP, and these rephosphorylated myosin-beads move at about 1 μm/sec. We have carried out similar experiments in collaboration with Dr. Jim Sellers using smooth muscle myosin phosphorylated to varying levels. There is very little or no movement if the smooth muscle myosin is dephosphorylated, whereas the myosin moves at about 0.4 μm/sec when the light chain is fully phosphorylated. A good correlation exists between the rate of movement and the fraction of the molecules on the bead that are phosphorylated. These Dictyostelium and smooth muscle myosin phosphorylation experiments serve to emphasize that we do indeed have a quantitative assay for myosin movement.

Our third approach to study the myosin molecule used Dictyostelium myosin and monoclonal antibodies. The object of these experiments was to localize the position of binding of these monoclonal antibodies on the myosin molecule and then to examine the effects of that binding on the ability of the myosin to move. Antibody binding sites were determined from images of myosin molecules with bound antibody contrasted by rotary shadowing. The top row of Figure 22.3 shows five examples of Dictyostelium myosin without any antibody bound. The middle row shows four examples of myosin with one of the monoclonal antibodies, known as My10, in its IgG form. In the bottom row are five examples of myosin with the Fab' fragment from My10. These examples clearly show that My10 binds specifically approximately two-thirds of the way down the tail from the head-tail junction. Analysis of fifty examples such as these gives the position of binding more accurately as shown in Figure 22.4. It is apparent that out of fifty examples all of the binding was localized approximately two-thirds of the way down the tail from the head-tail junction and no examples were found with the My10 monoclonal antibody bound elsewhere on the molecule, for example to the myosin heads. We found the Dictyostelium myosin tail to be about 1850 Å. This value, which confirms the data of Claviez et al. (1982), is somewhat

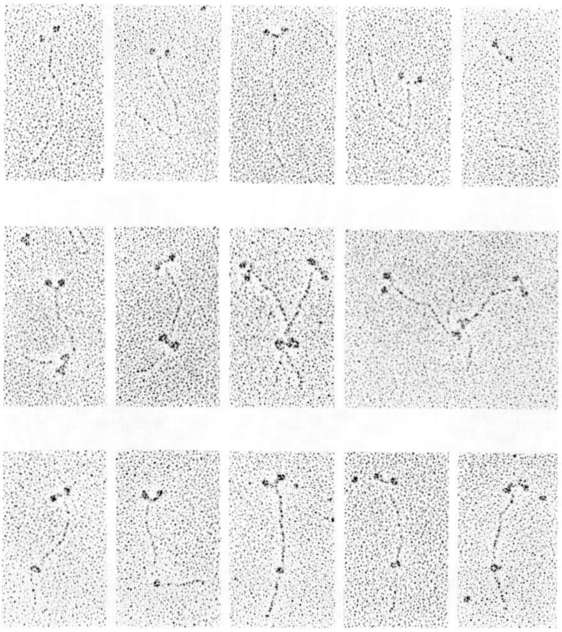

Fig. 22.3. Electron micrographs of rotary shadowed Dictyostelium myosin molecules decorated with My10 monoclonal antibody. Rotary shadowed replicas were prepared as described by Flicker et al. (1983). Upper panel: myosin alone. Middle panel: myosin with attached My10 IgG. Lower panel: myosin with attached Fab' fragment of My10. ×125,000.

larger than that of muscle myosin (about 1560 Å; Elliot and Offer, 1978). The positions of monoclonal antibody binding sites and the effects of these antibodies on motility are summarized in Figure 22.5. Shown below the schematic drawing are the velocities at which the myosin moved when those sites were blocked by the respective antibodies. In the absence of antibody the myosin moves at about 1 μm/sec. As expected, antibodies that bind to the myosin head region inhibit the myosin movement. These include My6 which binds to the heavy chain near the head region and inhibits the rate of movement by 50%, and My8 which binds to the 18,000 dalton light chain and inhibits movement completely. Antibodies which bind to various positions along the length of the tail and have no effect on the rate of movement of the myosin include My2, My3, and My4. Surprisingly, three other monoclonal antibodies, My1, My5, and My10,

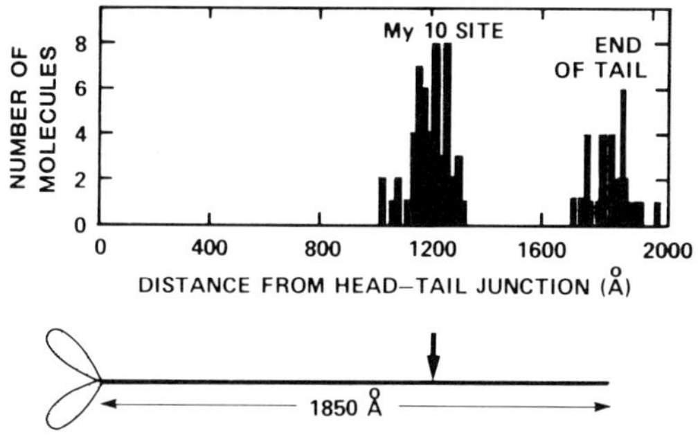

Fig. 22.4. Analysis of location of My10 binding to the *Dictyostelium* myosin molecule. Tail lengths and the position of binding of My10 monoclonal antibody were measured for individual myosin molecules from electron micrographs of rotary shadowed specimens. Distances from the head-tail junction are shown.

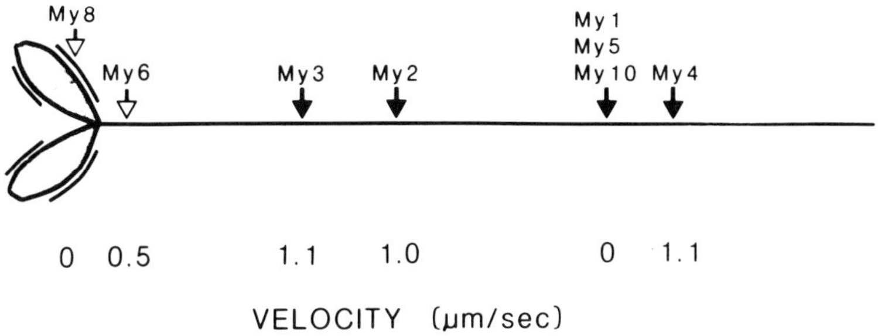

Fig. 22.5. Schematic drawing relating positions of eight monoclonal antibodies against *Dictyostelium* myosin and their effects on the motility of myosin in vitro. Control myosin in the absence of antibody moved at about 1 μm/sec. The positions of all of the antibodies with the exception of My6 and My8 were determined by measurements from electron micrographs of rotary shadowed specimens. The locations shown for My6 and My8 are inferred from the knowledge that My6 binds to the heavy chain portion of the myosin head region while My8 binds to the 18,000 dalton light chain (Peltz, Spudich and Parham, in preparation).

which are specific for the tail, all inhibit myosin movement completely. These antibodies bind distal to the heads approximately two-thirds of the distance from the head-tail junction to the end of the tail. The Fab' fragments of My1 and My10 have also been used to test their effects on myosin movement. They both block movement completely, indicating that the inhibition is not due to cross-linking of myosin molecules. There are a variety of possible explanations for this inhibition. An exciting possible explanation is that this portion of the tail may be directly involved in motive force production, in keeping with the model put forth by Harrington and his colleagues (1979). Another category of explanation would be that this particular region of the tail can for some reason fold back and physically interact with the head region of the molecule. In the case of smooth muscle myosin as well as thymus myosin, such a folded back structure has been described (Trybus et al. (1982); Suzuki et al. (1978); Craig et al. (1983)). We are currently examining the Dictyostelium myosin in more detail by electron microscopy to determine whether such a structure exists in this case.

DISCUSSION

There are several implications of the work presented here. We have demonstrated for the first time that myosin molecules are capable of walking along actin filaments in vitro, and more importantly we have established that this in vitro assay is reliable and quantitative. Although we have begun to investigate aspects of myosin involved in its ability to move, a great deal more needs to be done and the future looks extremely exciting. For example, much higher resolutzion information should be obtainable about the structure-function relationships of the myosin molecule. Molecular genetics techniques can be used to alter the molecule by site specific mutagenesis. The altered myosins can then be examined for their ability to move. Such information coupled with high resolution structural data from X-ray crystallography will provide critical insights into this molecule and how the energy derived from the hydrolysis of ATP is converted into mechanical movement. Another implication of our work is that structures are capable of moving along polar actin cables if those structures have myosin molecules bound to them (Sheetz and Spudich, 1983b). It seems likely that vesicular elements within cells move by a similar mechanism in vivo, and many forms of movement may not require bipolar thick filaments interacting with oppositely oriented actin filaments. Determination of the arrangement of myosin molecules on the beads may provide insight into possible mechanisms for such movements. It will be important to isolate vesicular components from a variety of types of cells, explore their ability to move on the Nitella actin substratum, and determine if and how myosin is associated with these structures.

ACKNOWLEDGMENTS

This work was supported by grants from the NIH to J.A.S. (GM 25240 and GM 33289), and postdoctoral fellowships from the NIH to G.P. (CA-09302) and L.M.G. (CA-09151). M.P.S. is an American Heart Association Established Investigator and was on sabbatical leave from the Department of Physiology, University of Connecticut Health Center, Farmington, Connecticut 06032, USA.

REFERENCES

Claviez, M., Pagh, K., Maruta, H., Baltes, W., Fisher, P., and Gerisch, G. (1982) Electron microscopic mapping of monoclonal antibodies on the tail region of Dictyostelium myosin. EMBO J. 1:1017-22.

Craig, R., Smith, R., and Kendrick-Jones, J. (1983). Light-chain phosphorylation controls the conformation of vertebrate non-muscle and smooth muscle myosin molecules. Nature 302:436-38.

Elliot, A. and Offer, G. (1978). Shape and flexibility of the myosin molecule. J. Mol. Biol. 123:505-19.

Flicker, P. F., Wallimann, T., and Vibert, P. (1983). Electron microscopy of scallop myosin: Location of regulatory light chains. J. Mol. Biol. 169:723-41.

Harrington, W. F. (1979) Proc. Natl. Acad. Sci. USA 76:5066-70.

Kamiya, N. (1981) Physical and chemical basis of cytoplasmic streaming. Annu. Rev. Plant. Physiol. 32:205-36.

Kersey, Y. M. and Wessells, N. K. (1976) Localization of actin filaments in internodal cells of characean algae. J. Cell Biol. 68:264-75.

Kielley, W. W. and Harrington, W. F. (1960) A model for the myosin molecule. Biochim. Biophys. Acta 41:401-21.

Kuczmarski, E. R. and Spudich, J. A. (1980) Regulation of myosin selfassembly: Phosphorylation of Dictyostelium heavy chain inhibits thick filament formation. Proc. Natl. Acad. Sci. USA 77:7292-96.

Margossian, S. S. and Lowey, S. (1982) Preparation of myosin and its subfragments from skeletal muscle. Meth. Enz. 85B: 55-71.

Mockrin, S. C. and Spudich, J. A. (1976) Calcium control of actin-activated myosin ATPase from Dictyostelium discoideum. Proc. Natl. Acad. Sci. USA 73:2321-25.

Sellers, J. R., Pato, M. D., and Adelstein, R. S. (1981) Reversible phosphorylation of smooth muscle myosin, heavy meromyosin, and platelet myosin. J. Biol. Chem. 256:13137-42.

Sheetz, M. P. and Spudich, J. A. (1983a) Movement of myosin-coated fluorescent beads on actin cables in vitro. Nature 303:31-35.

Sheetz, M. P. and Spudich, J. A. (1983b) Movement of myosin-coated structures on actin cables. Cell Motility, in press.

Squire, J. The Structural Basis of Muscular Contraction. Plenum Press, New York, 1981.

Suzuki, H., Onishi, H., Takahashi, K., and Watanabe, S. (1978) Structure and function of chicken gizzard myosin. J. Biochem. (Tokyo) 84:1529-42.

Trybus, K. M., Huiatt, T. W., and Lowey, S. (1982) A bent monomeric conformation of myosin from smooth muscle. Proc. Natl. Acad. Sci. USA 79:6151-55.

23 On the Presence of a 100 K Protein in Rabbit Skeletal Muscle: The Brush Border of Small Intestine, Culture Fibroblasts and Red Blood Cells

E. E. Alia and N. Arena

INTRODUCTION

In a previous paper (Alia et al. 1968) we suggested that the morphologic disorganization of the ischaemic striated muscle could be related to a disruption of sarcomeres, which in turn may be caused by the rupture of the Z-lines.

Z-lines have been shown to contain a large amount of alpha-actinin, a protein that seems to sustain most of the cytoskeletal function of the Z-line. In order to study the behavior of the alpha-actinin during the disruption of the Z-line, induced in the striated muscle by ischaemia, we have prepared antibodies against purified alpha-actinin and have utilized them for immunohistochemical demonstration of the alpha-actinin in various cells and tissues, under various experimental conditions. In addition to the ischaemic striated muscle, we have explored the microvilli of rabbit small intestine, culture rabbit fibroblasts and, finally, human and rabbit red blood cells.

MATERIALS AND METHODS

Extraction of Alpha-Actinin

We have followed the procedure suggested by Arakawa et al. (1970), except that the ionic exchange resin has been substituted by DEAE-Sephacel and the elution has been performed on a continuous gradient of KCl from 0 to 0.5 M. The electrophoretic isolation of our protein has been obtained as suggested by Laemli, on 10% PAA-gel in SDS. We have obtained 44 mg of isolated protein of 100 K M.W. from 200 g of rabbit striated muscle.

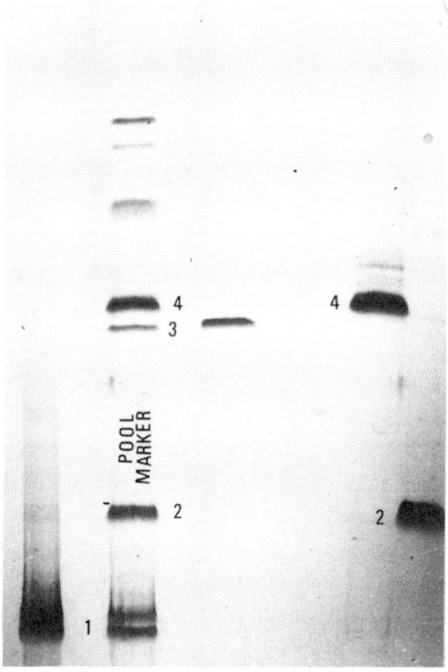

Fig. 23.1. PAA-Gel Electrophoresis of different proteins purified in our laboratory. 1: myosin—2: alpha-actinin—3: vimentin—4: desmin.

Antibody Production

Anti-alpha-actinin antibodies have been obtained from sheep, by injecting 2.5 ml of a protein solution (2.5 mg/ml final concentration) plus 2.5 ml of Freund complete antigen. Such treatment has been performed each ten days for a total of 3 times. One week after the last injection, a single intra-arterial dose of 1 ml of protein was given, and then, after 24 hours an adequate amount of blood was taken, from which total serum was extracted. Gel immunodiffusion of such a serum showed a strong reaction if tested for alpha-actinin (Fig. 23.2).

Ischaemic Muscle

Ischaemia was induced in rabbit skeletal muscles by ligature of the femoral artery and stenosis of homolateral ilyac artery, for a maximal period of ten hours. Biopsies have been taken 1, 2, 3, 5, and 8 hours after the ischaemia was established, both from ischaemic leg and from contralateral leg, which was used as a control. All samples were fixed in a picric acid-formaldehyde fixative (Zamboni et al., 1967) and processed according to the usual techniques for embedding in paraffin or in Epon resin. Thick sections (1 μm) from paraffin embedded samples have been prepared for immunofluorescence with anti-sheep rabbit serum labelled with fluorescine (FITC). After this procedure we have been able to obtain fluorescent images only from control muscles or from muscles after one hour of ischaemia. Samples taken after longer periods did not show noticeable fluorescence. For these samples we have used 1 μm thick sections from Epon embedded specimens, which were indirectly stained for Peroxidase-antiperoxidase, according to Sternberger et al. 1970.

Small Intestine Sampling

Samples taken from rabbit small intestine were subjected to the same fixation, embedding, and immunohistochemical reactions as described above for striated muscle.

Culture Fibroblasts

Culture rabbit fibroblasts were established from the kidney of a one week old rabbit. Kidneys were finely chopped and treated with trypsin; after a very gentle centrifugation, the supernatant was incubated with Dulbecco's Medium in 100 mm Petri dishes in a 5% CO_2 at 37°C. Part of the dishes were subjected to several changes of the medium to obtain more

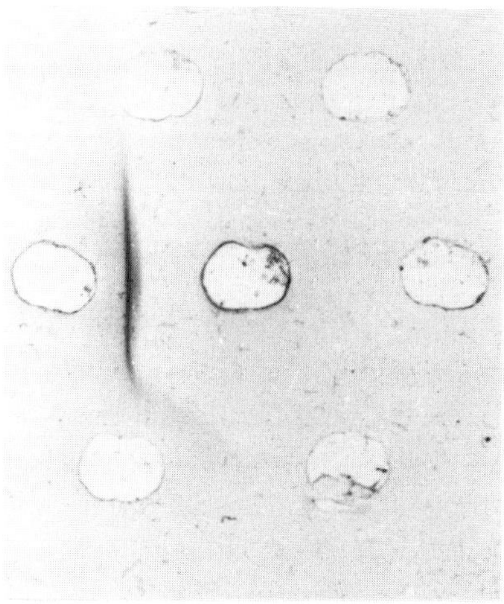

Fig. 23.2. Immunodiffusion Plate for the Ag-Ab Tests between alpha-actinin and anti-alpha-actinin.

numerous mitosis and younger cells, while the remaining dishes had only a few changes in order to obtain more differentiated and/or slightly swollen fibroblasts.

Direct and indirect immunofluorescence and indirect immunoperoxidase reaction were performed using the same anti-alpha-actinin antibodies described above. After a short fixation in 2.5% glutaraldehyde and several washings with PBS, cells were treated with Triton X-100 to permeabilize the plasmamembrane; part of these cells have been utilized for scanning electron microscopy and part for fluorescence by DAB according to Grube (1970), in those samples which were positive for peroxidase reaction. All the fluorescence observations have been carried out with a Dialux 20 EB Leitz optical microscope, equipped for fluorescence with a filter for K 460-490.

Blood Red Cell and Ghost Preparation

We have prepared ghosts from rabbit blood cells according to Steck et al. (1970). Whole blood was washed several times with saline and fixed by the Karnovski's fixative (4% formaldehyde + 0.7% glutaraldehyde in cacodylate buffer). Using the same fixed cells we have prepared smears for the indirect peroxidase reaction; the endogenous peroxidase activity was inhibited by a previous treatment with H_2O_2. Observations and micrographs were carried out by phase contrast, interferential contrast and fluorescence optical microscope, according to Grube (1970).

The other part of fixed blood cells was processed through all steps for the embedding in Epon resin. Thick sections (1 μm) from such preparation were treated with Na-methylate to partially remove the Epon resin, with H_2O_2 to neutralize the endogenous peroxidase and, finally, with labelled antibody. A double control was performed either by using a neutralized serum (serum + an excess of alpha-actinin), or by omitting in the procedure the treatment with the antibody itself. Thin sections (60-70 nm) were used for an immunoreaction which was evidenced by the complex Protein A (Staphylococcus aureus)++colloidal gold particles, according to Roth et al. (1978).

RESULTS AND DISCUSSION

Time-Course of the Ischaemic Treatment (Fig. 23.4)

The PAP reaction on ischaemic striated muscle was seen at different times from the starting of the ligature. After one

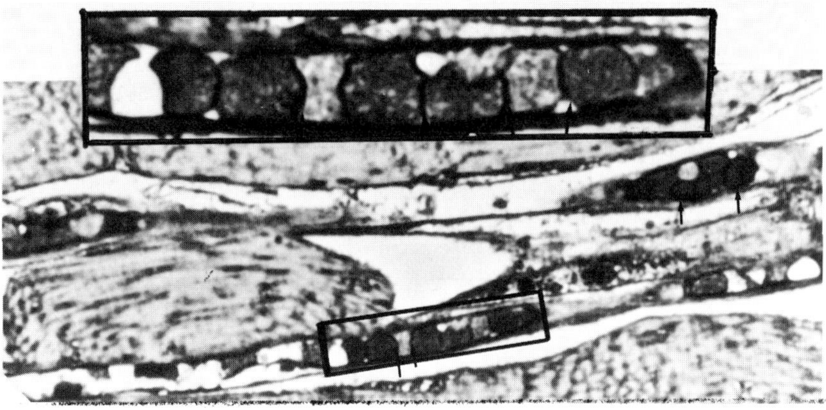

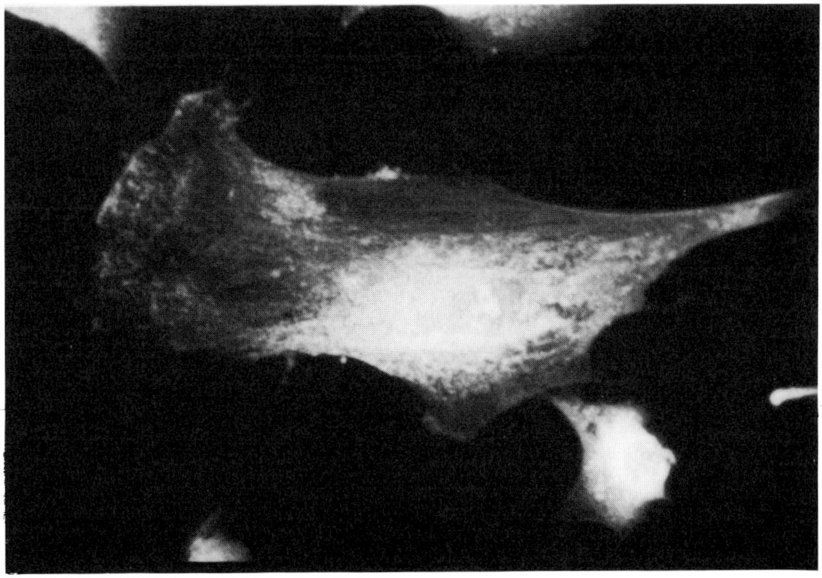

Fig. 23.4. Diffused fluorescence shown on the cytoplasm and the cell membrane of a stretched fibroblast, while in young fibroblast, Fluorescence is shown only in the perinuclear regions. Direct immunofluorescence with TRITC. 2,800×.

and two hrs. of ischaemia the PAP reaction pattern indicated an increasing displacement of the alpha-actinin from the classical sites, i.e., Z-lines. After three hrs. the PAP reaction started to decrease, leaving empty holes, including in those regions in which in the previous times the alpha-actinin was diffusely present, and especially at sarcolemmal level. After 5-8 hrs. from the ligature, only rare and weak zones of reaction persisted, despite very irregular Z-lines were still recognizable. These time-course observations suggest that alpha-actinin of the Z-lines is first dislocated and then possibly denatured; further, that this event may represent an early step in the disorganization of the sarcomere of the ischaemic striated muscle (Alia et al., 1978).

Culture Fibroblasts

Direct immunofluorescence on adult fibroblasts (Fig. 23.5) showed that alpha-actinin is localized not only in the perinuclear

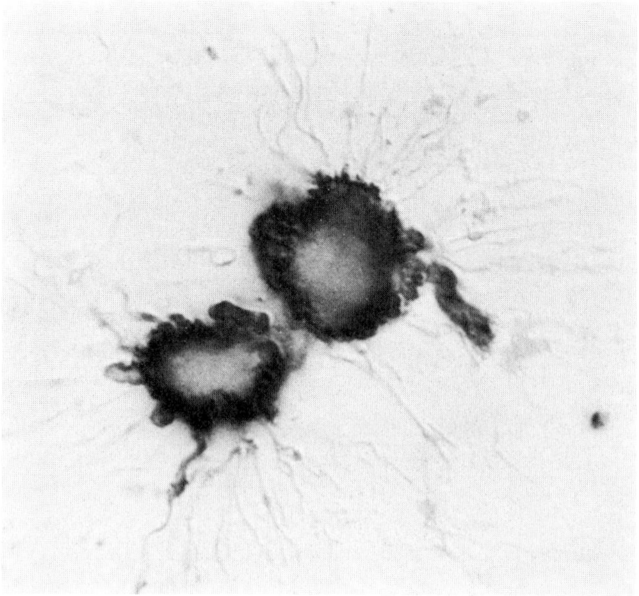

Fig. 23.5. Mitotic fibroblasts subjected to indirect immunoperoxidase reaction observed by Nomarsky Optical Microscope. The dense ring around the cell membrane indicates the presence of alpha-actinin. 2,800×.

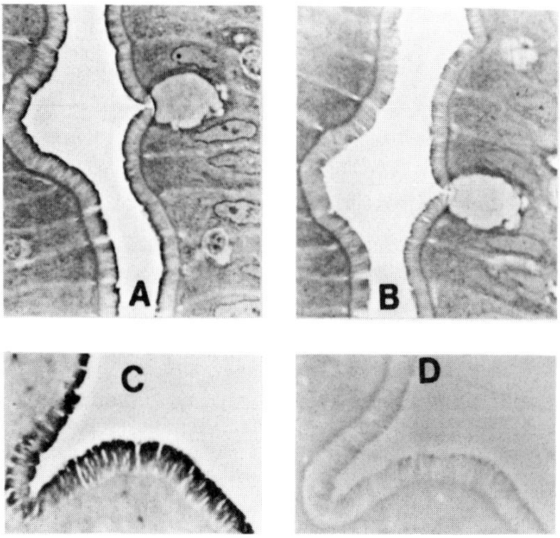

Fig. 23.6. Semithick section of rabbit small intestine treated for immunoperoxidase reaction to evidentiate alpha-actinin. A–PAP positive reaction; B–Control reaction in which the antibody was omitted; C–PAP-positive reaction; D–Control reaction in which a mixture of antigen + primary antibody was used.

regions, but also is present in plasmamembrane and stress fibers. Such a picture appears to be different from that described by other authors (Schollmayer et al., 1976; Lazarides et al., 1975), mainly because the localization around the nucleus (Alia et al., 1981). In young mitotic fibroblasts the indirect immunoperoxidase brown reaction was present only at the periphery of the cell, suggesting a protective cytoskeletal ring for the progressing mitosis. In addition "philopodia," blebs, and other cytoplasmic evaginations showed also a positive reaction, indicating the alpha-actinin as an important component of the adherence plaques, according to that shown by others (Alia et al., 1982; Comoglio et al., 1982).

Small Intestine

Thick sections of embedded rabbit small intestine which were treated for indirect PAP reaction showed that alpha-actinin is localized at the apex of the brush border microvilli, according to the hypothesis suggested by Mooseker and Tilney (1975).

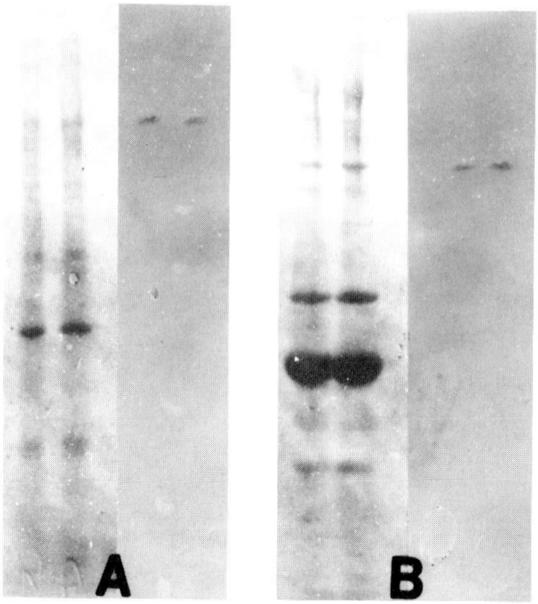

Fig. 23.7. Immunoblottings. (A) Electrophoresis and immunoblotting on the total protein extract from rabbit small intestine brush border. (B) Electrophoresis and immunoblotting on the total protein extract from rabbit striated skeletal muscle. In both cases the reaction is present at level of the 100 K protein, i.e., alpha-actinin.

To further test the specificity of the anti-alpha-actinin antibody prepared in our laboratory and also to verify its affinity for the brush border alpha-actinin, we performed immunoblottings on microelectrophoresis strips of integral (whole) proteins extracted from purified muscle myofibrils, or of whole (integral proteins extracted from brush border preparation (Mooseker and Tilney, 1975). Figure 23.8 shows that our antibody identifies (reacts with) the same 100 K protein in electrophoresis strip, both from muscle fibril and brush border integral extracts; and further suggests that 100 K protein from brush border is also alpha-actinin.

Red Blood Cells and Ghosts

Occasional observations during ischaemic muscle studies of RBC which displayed a strong PAP reaction (Fig. 23.4), have

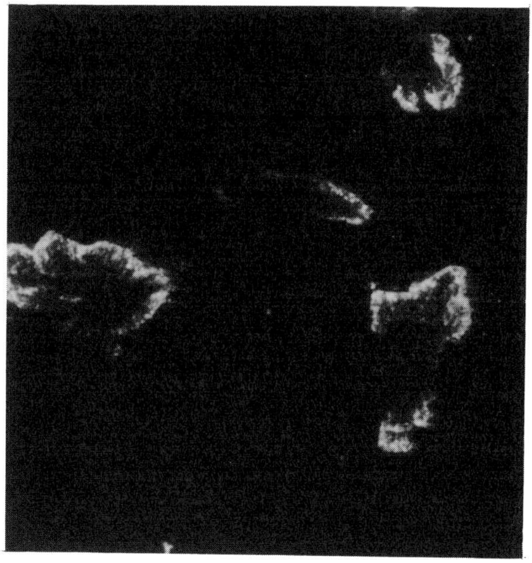

Fig. 23.8. Human erythrocyte ghosts treated for direct immunofluorescence with TRICT for the 100 K protein. A positive reaction is evident.

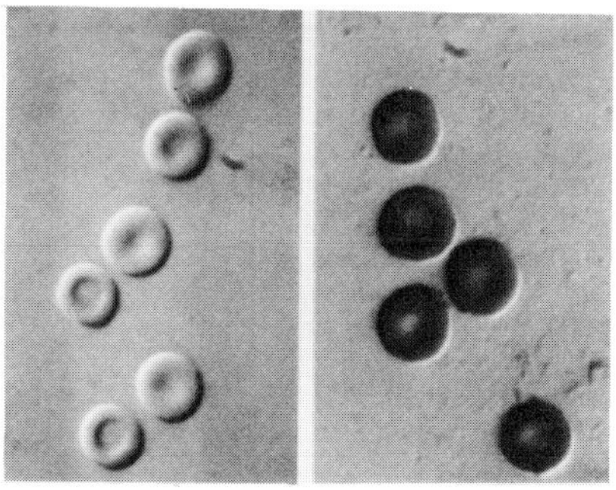

Fig. 23.9. Smear of blood on which PAP-indirect reaction was performed, observed by Nomarsky Microscope. A: Control– B: Positive PAP-reaction. 4,500×.

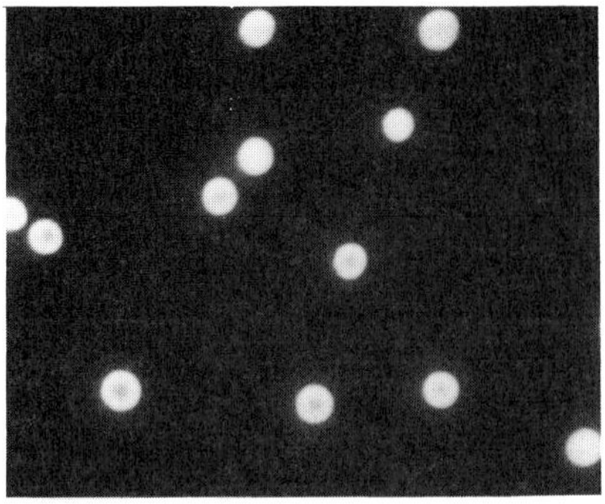

Fig. 23.10. Smear of Erythrocytes on which PAP-reaction was performed, observed with a Grube-Fluorescence Microscope. A strong fluorescence is present all over the cell surface. 2,500×.

suggested an inquiry into the presence of the alpha-actinin in the isolated RBC and their ghosts. In Figure 23.9 ghosts from human erythrocytes show an evident positive reaction when observed at direct immunofluorescence. This feature is confirmed by the positive PAP reaction evidenced in intact RBC by interferential contrast optical microscopy (Fig. 23.10), by fluorescence according to Grube (1970) (Fig. 23.11) and, finally, by contrast optical microscopy in semithick sections (Fig. 23.12). In the last case the reaction appears to be localized specifically along the plasmamembrane, and this is confirmed by the immunoreaction with protein A-colloidal gold particles complex performed on thin sections of Epon embedded rabbit RBC. Figure 23.13a shows that gold particles

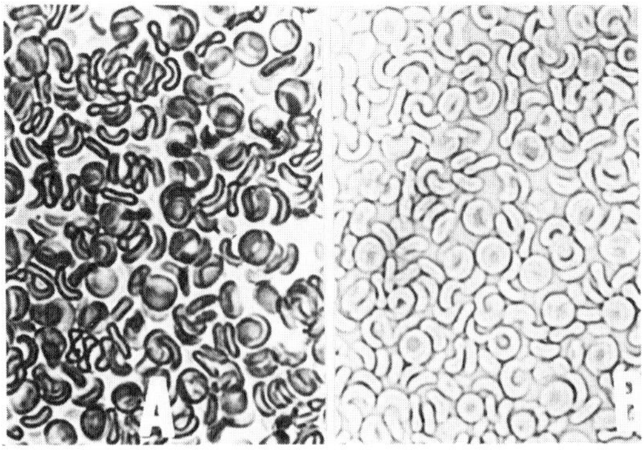

Fig. 23.11. Semithin section of blood embedded in Epon, on which PAP-reaction was carried out. An evident reaction is present along the plasmamembrane. Contrast-Phase Microscopy observation. 2,500×. A: Positive PAP-reaction. B: Control.

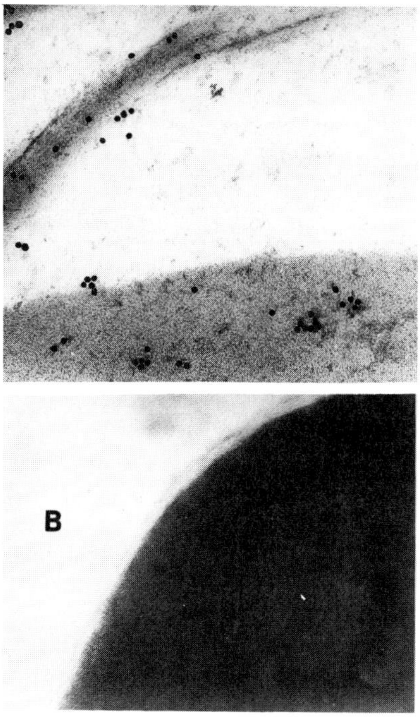

Fig. 23.12. Ultrathin sections of erythrocytes embedded in Epon and subjected to the protein A-Gold Complex Reaction. A: Complexes are visible mainly inside the erythrocytes and on the internal side of the plasmamembrane. B: Control. T.E.M. observation. (100,000×)

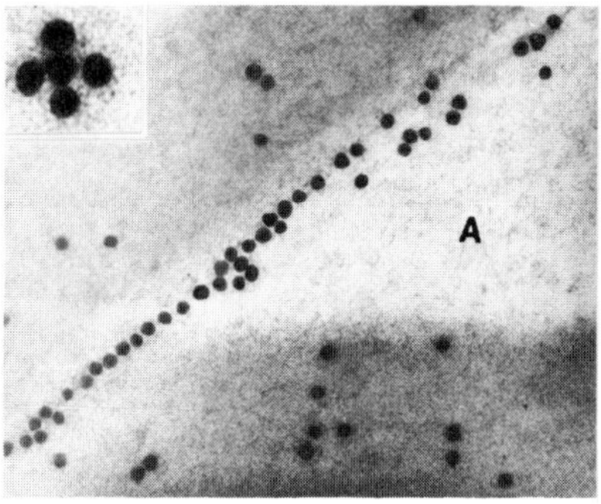

Fig. 23.13. Ultrathin sections of erythrocytes embedded in Epon subjected to the protein A-Gold Complex Reaction. A: Evident reaction on the cell membrane. (100,000×). B: High magnification of protein A-Gold Complex. (200,000×).

are mainly localized close or at the internal part of the plasma-membrane. However, we have been unable to have positive immunoblotting with total protein extract from RBC and the identification of the protein which is responsible for all the positive immune reactions in RBCs is still uncertain.

CONCLUSIONS

All the observations presented here on ischaemic striated muscle, on culture fibroblasts, on small intestine and RBC, suggest that the 100 K protein that have been isolated in our laboratory may be alpha-actinin; the 100 K MW has been confirmed by others (Bretscher et al., 1979; Geiger et al., 1979; Feramisco et al., 1980; Goll et al., 1982; Suzuki et al., 1976; Pinter et al., 1980). However, other authors have shown a 95 K MW (Podlubnaja et al., 1975) or a 94 K MW (Langer et al., 1980). Such discrepancy may be due to tissue or animal origin of the alpha-actinin and/or to the procedure for its extraction and purification.

Our protein shows an electrophoretic speed identical to that of a 100 K protein marker used as a control, and, further, it reacts at the immunoblotting in the identical migratory band when total protein extracts are used either from total striated muscle, either from purified myofibrils or brush border. We suggest that the lack of reaction in the immunoblotting with 100 K protein extracted from RBC may be due to an artifactual alteration of the molecule related to the procedure for ghost preparation or for the protein extraction.

In addition, considering that alpha-actinin displays a partial specie-specificity, we tested our protein by immunoblotting with human, pig, guinea pig and rat alpha-actinins. Human alpha-actinin had a strong reaction similar to that seen with rabbit alpha-actinin; pig alpha-actinin showed a moderate reaction, while with guinea pig and rat alpha-actinins the reaction was completely absent. No linkage (cross-reaction) was present with myosin, actin, spectrin, vimentin and desmin.

A recent paper (Goll et al., 1983) have shown that alpha-actinin after enzymatic digestion, is composed of different subunits. We are planning to test such different fractions against our antibody to identify the specific antigen(s) which in turn could be used as a more specific and powerful immunogen. One or more of these fractions could be identified with some monomers of the band #3 of the total protein extract from RC membrane (Marchesi et al., 1972; Marchesi et al., 1976; Branton et al.,

1981; Fairbanks et al., 1971). Finally we suggest that RBC alpha-actinin may be more abundant in younger cells or RC precursors, especially reticulocytes and erythroblasts.

REFERENCES

Alia E.E. (1978): "Sulle alterazioni dei miofilamenti di actina in corso di ischemia arteriosclerotica nell'uomo. (Ricerche ultrastrutturali)." Boll. Soc. It. Biol. Sper. LIV (19): 1777-83.

Alia E.E. and Arena N. (1981): "Local izzazione dell'alfa-actinina in fibroblasti di coniglio mediante l'immunofluorescenza diretta ed indiretta con TRIC e FITC." Atti del Conv. Naz. Soc. Ital. di Anatomia di Pisa. In press.

Alia E.E. and Arena N. (1982): "Red blood cell alpha-actinin: a likely relationship to cell membrane and co-capping with surface receptors." Letter to Editor. Basic and Appl. Histoch. 26(2):131-32.

Alia E.E. and Arena N. (1982): "Immunoreattività della membrane eritrocitaria umana con anticorpi anti-alfa-actinina di coniglio." Atti del Conv. della Soc. Ital. di Anatomia. Ed. Intern. C.I.C. Roma:33-34.

Arakawa N., Robson R.M., Goll D.E. (1970): "An improved method for the preparation of alpha-actinin from rabbit skeletal muscle." Bioph. Bioch. Acta, 200:284-95.

Arena N., Vergani G., Madeddu R., Satta G., Deiana L., Alia E.E. (1982): "Immunohistochemical observations on the tubulin and alpha-actinin during mitosis of rabbit fibroblasts." Boll. Soc. It. Biol. Sperim. LVIII(23):1552-57.

Arena N., Vergani G. und Alia E.E. (1983): "Beobachtungen über alpha-actinin, dargestellt in den Microvilli des Kaninchen-Dündarmes durch indirekte Immunoperoxidase Reaktionen." Verh. Anat. Ges. 77(S):565-66.

Branton D., Cohen D.M. and J. (1981): "Interaction of Cyto-skeletal Proteins on the human Erythrocyte Membrane." Cell 24:24-32.

Bretscher A. and Weber K. (1978): "Localization of actin and microfilament Associated proteins in the microvilli and the Terminal web of the intestinal brush border by immunofluorescence microscopy." J. Cell Biol. 79:839-45.

Bretscher A., Vandekerkhove J. and Weber K. (1979): "Alpha-Actinins from chicken skeletal muscle and smooth muscle show considerable chemical and immunological differences." Eur. J. Bioch. 100:237-43.

Comoglio P., Tarone G., Prat M. (1982): "Il ruolo della membrana plasmatica nella riproduzione cellulare." Atti del Conv. Soc. It. di Anatomia. Roma. Ed. Intern. C.I.C.:11-12.

Fairbanks G., Steck T.L. and Wallach D.F.H. (1971): "Electrophoretic Analysis of the Major Polypeptides of the Human Erythrocyte Membrane." Biochem. 10(13):2606-17.

Feramisco J.R. and Burridge K. (1980): "A rapid Purification of Alpha-Actinin, Filamin and a 130,000-dalton Protein from smooth muscle." The J. of Biolog. Chem. 255(3):1194-99.

Geiger B. and Singer J.S. (1979): "The participation of Alpha-Actinin in the capping of Cell Membrane components." Cell 16:213-22.

Geiger B., Tokuyasu K.T. and Singer S.J. (1979): "The immunomechanical localization of Alpha-Actinin in intestinal epithelial Cells." Proc. Nat. Acad. Sci. U.S.A. 76:2833-37.

Goll D.E., Suzuki A., Temple J. and Holmes G.R. (1972): "Studies on purified Alpha-Actinin." J. Molec. Biol. 67: 469-88.

Goll D.E., Muguruma M. (1983): "Interaction of Alpha-Actinin with Actin." Acta of Int. Sympos. on Contr. Prot. Alia and Arena Eds. Praeger Scient. Publ. New York.

Grube D. (1980): "Immunoperoxidase Methods: Increased Efficiency using fluorescence microscopy for 3,3'-Diaminobenzidine (DAB) stained semithin sections." Histoch. 70:19-22.

Laemmli U.K. (1970): "Cleavage of structural proteins during the assembly of the head of bacteriophage T 4." Nature (London), 227: 680-85.

Langer B.G. and Pepe F.A. (1980): "New, rapid methods for purifying Alpha-Actinin from chicken gizzard and chicken pectoral muscle." The J. of Biol. Chem. 255(11):5429-34.

Lazarides E. and Burridge K. (1975): "Alpha-Actinin: Immunofluorescent localization of a muscle structural Protein in Non-Muscle Cells." Cell 6:289-98.

Marchesi V.T. and Furthmayr H. (1976): "The red cell membrane." Annual Rev. of Bioch. Snell E.E. & C. Eds. Palo Alto, Ca. 45:667-98.

Marchesi V.T., Tillack T.W., Jackson R.L., Segrest J.P. and Scott R.E. (1972): "Chemical characterization and surface orientation of the major glicoprotein of the human erythrocyte membrane." Proc. Nat. Acad. Sci. U.S.A. 69(6):1445-49.

Mooseker M.S. and Tilney L.G. (1975): "Organization of an actin filament-membrane complex." J. Cell Biol. 67:725-43.

Ouctherlomy O. (1962): "Progress in Allergy." Kallos P. and Wolksman B.H. Edts. Basel (Karger).

Pintér K., Jancso A., Birò E.N.A. (1980): "A Simple Procedure for the preparation of Electrophoretically homogeneous Alpha-Actinin from rabbit muscle." Acta Bioch. et Bioph. Acad. Hung. 15(3):217-22.

Podlubnaya Z.A., Tskhovrebova S., Zaalishvili M.M., Stefanenko G.A. (1975): "Electron microscopic studies of Alpha-Actinin." Letter to the Editor. J. Mol. Biol. 92:357-59.

Roth J., Bendyan M. and Orci L. (1978): "Ultrastructural localization of intracellular antigens by the use of Protein A-Gold Complex." The J. of Histoch. & Citoch. 26:1074-81.

Schollmeyer J.E., Furcht L.T., Goll D.E., Robson R.M. and Stromer M.H. (1976): "Localization of Contractile Proteins in smooth muscle cells and in normal and transformed fibroblasts." Cell Motility, Book A: Motility, muscle and non-muscle cells. Goldman R. & C. Edts. Cold Spring Harbor, New York. 3:361-88.

Steck T.L., Weinstein R.S., Strauss J.H., Wallach D.F.H. (1970): "Inside-out red cell membrane vesicles: Preparation and Purification." Science 168:255-57.

Stenberger A., Hardy P.H., Coculis J.J., Meyer H.G. (1970): "The unlabelled antibody enzyme method of immunohistochemistry. Preparation and properties of soluble Ag-Ab complex (horseradish peroxidase-antihorseradish peroxidase) and its use in identification of spirochetes." J. Histoch. & Citoch., 18:315-33.

Suzuki A., Goll D.E., Singh I., Allen R.E., Robson R.M. and Stromer M.H.: "Some properties of purified skeletal muscle alpha-actinin." J. Biol. Chem. 251:6860-70.

Zamboni L. and De Martino C. (1967): "Buffered Picric Acid-Formaldehyde: A New, Rapid Fixative for Electron Microscopy." J. Cell Biol. 35(2):148 A.

24 Histochemical and Contractile Characteristics of Aging Skeletal Muscle in Humans

Lars Larsson

INTRODUCTION

Senile muscle atrophy is one of the most conspicuous alterations in the senile motor system and is, as well, by far the most commonly encountered kind of muscle atrophy in man. In spite of this, senile atrophy, particularly the influence of atrophy on muscle function, has so far received relatively little scientific attention. The aims of the present investigations were to study basic characteristics, morphological and functional, of skeletal muscle as well as the influence of physical exercise on these characteristics in man at different ages.

MATERIALS AND METHODS

Subjects

Fifty-five healthy male subjects, 22-65 years of age, volunteered for the studies. They were all white-collar workers and were employees of the same insurance company. The volunteers had a low physical activity level and all could be classified in groups I-II according to Saltin and Grimby's (1968) classification of occupational and spare-time physical activity. On physical examination they were found to be without locomotory defects. An oral consent was obtained from each individual after having explained the aims and possible discomforts of the tests. The study was approved by the Ethical Committee at the Karolinska Institute.

Height, body weight, skinfold thickness, and skeletal width were measured for each subject to estimate fat free soft tissue weight (FFS). The circumference of the thigh was taken in a horizontal plane just under the gluteal furrow.

Muscle Sampling

Muscle samples were taken from the middle portion of the resting quadricep femoris muscle (m. vastus lateralis) using the needle biopsy technique (Bergström 1962). The vastus lateralis muscle is a major contributor to the force generated during knee extensions and is located at a site convenient for muscle biopsy sampling. The choice of the quadricep muscle had an additional advantage because it has been shown that aging processes, such as a decline in muscle strength and volume, occur relatively early in this muscle (for ref. see Larsson 1978, 1982a).

Histochemistry

Each sample was trimmed, mounted, frozen in 3-methylbutane cooled by liquid nitrogen, and then stored at -80°C until further processing. Serial transverse sections (10 μm) were cut with a cryotome at -20°C, and stained for myofibrillar ATPase after alkaline or acid preincubation (Brooke and Kaiser 1970, Dubowitz and Brooke 1973, Padykula and Herman 1955) or for NADH tetrazolium reductase (Novikoff et al. 1961). Fibers were classified into types I (slow twitch) and II A, B, C (fast twitch) according to their myofibrillar ATPase staining patterns (muscle fibers were classified into type I and II in all 55 subjects while type II fiber subgroups were classified in 41 subjects). Fiber diameters were measured from the sections stained for NADH tetrazolium reductase using the "lesser fiber diameter" method (see Dubowitz and Brooke 1973). For comparative purposes, fiber areas were calculated assuming a circular cross section of the fibers.

Mechanical Measurements

Maximum isometric and dynamic strengths were measured in the left knee-extensor muscles using an isokinetic dynamometer (Cybex II, Lumex Inc., New York). The subjects were seated in an adjustable chair with support for the back, shoulders, and hips. The hip angle was fixed at 90° and the lower leg moved the lever of the dynamometer. The level was kept at a constant length and attached to the tibia. The center of the dynamometer's axis of rotation was aligned with the anatomical axis of rotation, i.e., the knee joint. The angular velocity was controlled at a preset level by an internal resistance, which accommodates to the muscular force applied. The angular movement of the knee joint was from 100 to 0°, i.e., full knee extension. Isometric strength was measured at selected knee angles (30, 60 and 90°), whereas the dynamic strength was recorded over the whole range of motion at different angular velocities. The velocities studied were 30, 60, 120 and 180° × s -1. Two attempts were allowed at each velocity or knee angle, and the highest value was noted. The produced torque was calculated as the force times the length of the lever, and unit given is Nm. The measurements were made in sequence from slow to fast speeds with 30 s recovery between each contraction. High precision and accuracy for the torque registrations have earlier been reported (for ref. see Thorstensson 1976).

Physical Exercise Program

A subsample of the volunteers described above, eighteen male subjects between 22 and 65 years of age, were involved in a strength training program twice a week for 15 wk. The training was supervised by two experienced physical education teachers. After a 10-min warm-up period a circuit strength training program with 10 different stations was begun. The knee extensor muscles were involved in every second station (knee extensions with weighted shoes, squatting, skipping, jumping over a bench, and stepping up on a bench) and the remaining stations mainly involved the abdominal, back, and upper arm muscles. The resistance was kept relatively low (body weight or light weights), and a relatively large number of repetitions (gradually increased to 20-30) was performed at each station. The training was considered intense and lasted 60-80 min. Muscle morphological and strength measurements were performed in an identical way after the training as before.

Statistics

Arithmetic means, standard errors of the means (S.E.), and linear regression coefficients (r) were calculated from individual values. Intraindividual differences were tested using the paired student's t-test. Where the data was considered to be affected by more than one independent variable, the statistical analysis was based on multiple regression and partial correlation was calculated. These calculations are part of the SPSS

Table 24.1. Anthropometric characteristics, fiber type distribution, fiber areas, and distribution of fiber types within the type II fiber polulation. Values are means ± SE

Age group (yr)	n	Mean age (yr)	Height (cm)	Body-weight (kg)	Type I %	Type I area ($100 \times \mu m^2$)	Type II %	Type II area ($100 \times \mu m^2$)	Percent of type II fibers IIA	IIB	IIC
20-29	11	26±1	182±2	76±3	40±4	29.5±2.5	60±4	36.6±2.2	59±2	35±3	6±1
30-39	12	35±1	180±2	77±3	36±2	29.2±1.2	64±2	36.7±2.9	61±4	34±3	5±2
40-49	10	43±1	181±1	77±2	48±4	31.9±1.9	52±4	32.6±2.4	70±7	27±7	3±1
50-59	12	54±1	180±1	80±2	52±3	28.8±1.6	48±3	28.0±1.2	60±6	37±5	3+1
60-65	10	62±1	175±1	74±4	55+4	22.6±2.4	45±4	21.2±1.7	62±7	32±8	6±2

statistical package (Nie et al. 1975) calculated on an IBM 370/165 computer.

RESULTS

Anthropometric Characteristics

The anthropometric data in the different age groups are presented in Table 24.1. No significant difference in these characteristics was seen between the groups. Height and body weight in the 20-29, 30-39, . . . to the 60-65-year-old group were compared with the corresponding age groups of a larger (n = 749) Swedish male population examined in these respects (Levin, personal communication). Except for the 50-59-year-old group, where height was found to be higher ($p < 0.05$) in the present subjects, no significant differences were seen in the corresponding age groups.

Histochemical Characteristics

Fiber type distribution data displayed a linear decrease ($r = -0.46$; $p < 0.001$) in the proportion of type II fibers with increasing age (Table 24.1). The distribution of subtypes within the type II fibre population did not, on the other hand, change with age (Table 24.1).

The average area of type II fibers declined with age ($r = -0.60$; $p < 0.001$) as did type II/I area ratio ($r = -0.47$; $p < 0.001$), whereas no statistically significant linear change was seen in type I fiber area (Table 24.1). The percentage distribution of muscle fiber diameters are shown in Figure 24.1, demonstrating a displacement to smaller diameters of type II fibers with age with only a minor change in the distribution pattern of type I fiber diameters.

Muscle Function and Its Relation to Muscle Morphology

Maximum strength reached its peak in the 20-29-year-old group, remained almost constant to the 40-49-year-old group and then declined with age ($p < 0.05$-0.001). (Table 24.2). This was true either if strength was given in absolute values or corrected for body weight or FFS weight. The relation

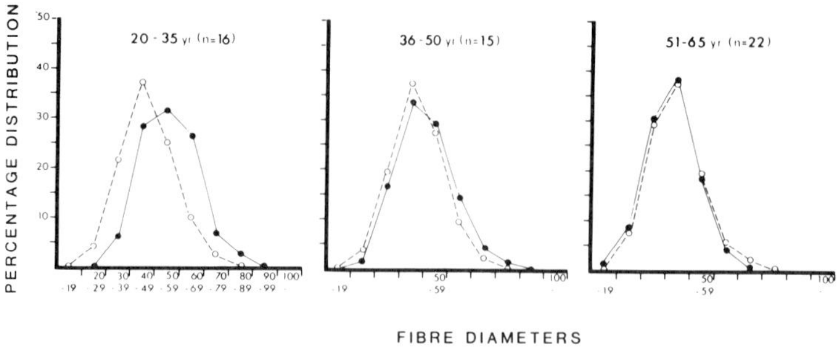

Fig. 24.1. Percentage distribution of muscle fiber diameters of types I (unfilled dots) and II (filled dots) fibers. The number of subjects in each age group is given within brackets. Type II/I fiber area ratio is 1.33, 1.12, and 0.99 in subjects of 20-35, 36-50, and 51-65 years. Arbitrary units are used for muscle fiber diameters..

between maximum strength and age was similar for isometric and dynamic strength at all angular velocities studied. High correlations were also found between isometric and dynamic strengths ($r = 0.86-0.89$) as well as between dynamic strengths at the different angular velocities ($r = 0.86-0.95$).

Significant linear correlations ($r = 0.44-0.54$; $p < 0.01-0.001$) were found between strength, isometric as well as dynamic, and type II fiber area. Weaker, although statistically significant, correlations were seen between strength and type I fiber area ($r = 0.28-0.39$; $p < 0.05-0.01$).

In stepwise multiple regression analysis it was found that chronological age, FFS weight, and type II fiber area were the only variables that had a significant independent influence on strength, isometric and dynamic. When the influence of type II fiber area was eliminated by multifactor analysis age still had a statistically significant negative correlation with maximum strength. This indicates that other factors than the fiber atrophy were responsible for the age-related strength decline.

Physical Exercise Effects

The anthropometric variables and distribution of fiber types were not affected by the training program. Before the training period the histochemical and the functional muscle characteristics correlated with age in a similar way in the

Table 24.2. Isometric (0 deg. xs^{-1}) and dynamic strengths (30, 60, 120, and 180 deg. xs^{-1} in the different age groups. Values are means ± SE

Age group (yr)	Isometric and dynamic strengths (Nm)				
	0 deg. xs^{-1}	30 deg. xs^{-1}	60 deg. xs^{-1}	120 deg. xs^{-1}	180 deg. xs^{-1}
20-29	221 ± 12	211 ± 7	201 ± 9	169 ± 7	139 ± 8
30-39	212 ± 8	207 ± 11	199 ± 9	163 ± 8	137 ± 6
40-49	223 ± 11	214 ± 8	199 ± 5	169 ± 3	139 ± 4
50-59	197 ± 10	191 ± 10	182 ± 8	155 ± 7	127 ± 4
60-65	163 ± 9	145 ± 9	137 ± 7	115 ± 5	91 ± 5

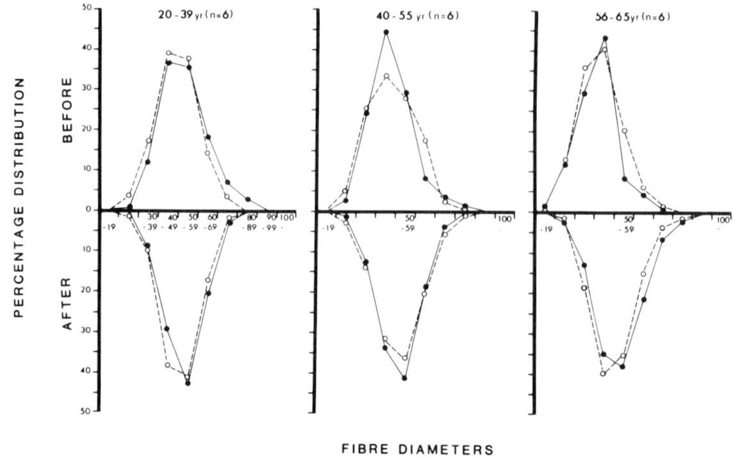

Fig. 24.2. Percentage distribution of fiber diameters, type I (unfilled dots) and II (filled dots), before and after the physical exercise program in the 20-39-, 40-55, and 56-65-year-old groups, respectively. The number of subjects in each age group is given within brackets.

subsample (n = 18) that was involved in the exercise program as was seen in the original population (n = 55) (Larsson 1982b). After the training period, on the other hand, no change could be seen in fiber size with age. This was due to an increase ($p < 0.05$-0.01) in fiber areas, type II in particular, in the oldest age groups. The distribution pattern of fiber diameters was almost identical in the different age groups after the training. (Fig. 24.2). The anthropometric variables and the distribution of fiber types were not affected by the training program. Maximum strength, irrespective of speed of movement, tended to increase in all age groups, but the average increase was slightly higher in the older age groups as demonstrated by a 2.9, 3.5, and 7.5 percent change in the 20-39-, 40-55-, and 56-65-year-old groups, respectively. This is in agreement with the more marked increase in mean fiber areas in the older subjects. However, despite the slight increase in strength in the older subjects, strength still decreased ($r = -0.53$ to -0.71; $p < 0.05$-0.001) with age after the training period.

DISCUSSION

Subjects

In the study of aging, a large number of methodological problems arise that may bias the data for the investigator. This appears to be the most plausible explanation for the divergent and even conflicting results obtained in different aging studies. The most commonly used method to study aging in man is the cross-sectional approach (comparing subjects of different ages). The reliability of this method rests largely on the use of as homogenous populations as possible in all respects except for age. In cross-sectional aging studies of skeletal muscle characteristics, the environmental influence of physical activity has a strong influence on muscle volume, metabolism and function and should therefore be held as constant as possible. The presence of disuse in old age secondary to diseases of cardiovascular or locomotor systems may distort a material completely and thereby obscure "primary" aging mechanisms. In the present study, the subjects were carefully selected to have equal occupational and leisure physical activity levels (i.e., male white collar workers between 22 and 65 years of age with a low physical activity level and with no prior history of cardiovascular disease or locomotor deficiency).

Body-size characteristics of the present subjects conformed well with a representative age-matched Swedish male population and the histochemical muscle characteristics in the youngest age groups were in accordance with previously reported values in young sedentary persons (Larsson 1978; 1982a; 1983; Larsson et al. 1978; Örlander 1980).

Histochemical Characteristics

The main finding from the histochemical studies was the altered relative occurence of fiber types, showing a decline in the percentage of type II fibers, together with the decline in fiber size during aging, preferentially affecting type II fibers. The altered proportion of fiber types with age is in accordance with animal studies of heterogenous or fast skeletal muscles (for ref. see Larsson 1982a) and has recently been confirmed in human studies with subjects of similar as well as higher ages (Larsson and Örlander 1983; Scelsi et al. 1980; Sjöström et al. 1980) (Fig. 24.3).

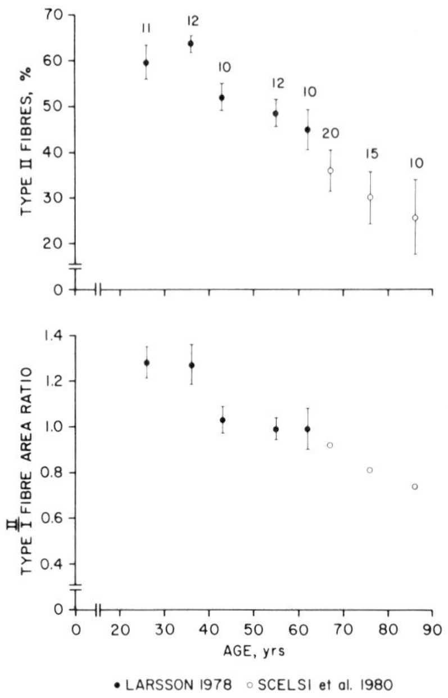

Fig. 24.3. Fiber type distribution and type II/I fiber area ratio in the quadriceps muscle of one hundred 22- to 80-year-old sedentary subjects. The filled and unfilled dots represent data from Larsson (1978) and Scelsi et al. (1980). Mean values ± SE and number of subjects in each group are presented. (The figure is published with permission from Acta Physiol Scand).

The mechanisms underlying these changes are not fully understood but a dominant influence of neuropathic factors is suggested from extensive electrophysiological and histological studies. The decreased number of motor units as well as the increased size of the remaining motor units imply a denervation process with reinnervation via collateral sprouting (for ref. see Larsson 1982a).

The pathogenesis and site of this partial denervation process in aged muscle are controversial and many different and even conflicting hypotheses have been presented. The neuromuscular junction has been proposed to be the site of the denervation process. This is supported by the profound ultrastructural changes seen in the neuromuscular junction during aging (for ref. see Larsson 1982a). Recent electrophysiological studies have, however, shown that these morphological changes are either not rate-limiting or are well compensated for (Robbins and Kelly 1981). Wohlfart (1939, 1957) suggested that the age-related denervation process in humans was caused by a loss of large efferent nerve fibers and possibly also anterior horn cells. This hypothesis is supported by a large number of studies showing a decline of myelinated nerve fibers in spinal roots and peripheral nerves (for ref. see Larsson 1982a). Further, Caccia et al. (1979) could show that the loss of motor units with age was accompanied by a similar loss of axons in the innervating nerve trunks. A shift to smaller diameters in the caliber spectrum of dorsal and ventral root fibers has been observed at advanced ages in man and was interpreted to be due to a selective atrophy and degeneration of the largest and fastest conducting nerve fibers (Rexed 1944). In peripheral nerves a similar loss of the largest myelinated fibers has been shown in older persons (Cottrell 1940; Swallow 1966). This is of specific interest in this context since large and fast conducting nerve fibers mainly innervate type II muscle fibers (see Burke and Edgerton 1975). It is suggested that the altered fiber type proportions with age giving a higher proportion of type I fibers is an early sign of the denervation reinnervation process preceding the senile fiber type grouping with clusters mainly containing type I fibers.

The preferential atrophy of type II fibers during aging is in conformity with other studies in man as well as in rat (for ref. see Larsson 1982a). The present results show, however, that this fiber atrophy can be counteracted by an increased physical activity. It is further suggested that the age-related fiber atrophy, at least up to 65 years of age, is secondary to a decreased nerve impulse activity related to a reduced physical activity level. This probably takes place as a slow adaptation to a more sedentary way of living with increasing age, i.e., a older person's preferring to use an elevator instead of climbing stairs etc., in spite of the fact that the present subjects were selected to have of almost equal occupational and leisure time physical activity levels.

Functional Characteristics

Isometric strength performance at different ages has been the subject of scientific interest since Quetelet's (1836) pioneering study. Recent studies have shown an improved maintenance of strength in the elderly compared to the work of Quetelet, showing peak isometric strength in the third decade, a slow or imperceptible decrease to the fifth decade, and thereafter an accelerated decline (for ref. see Larsson 1982a). From the present results it is evident that dynamic muscular strength follows a similar relation with age as previously reported for isometric strength (Larsson et al. 1979).

Since muscular strength is proportional to the cross-sectional area and contractility of the active muscle, a decrease in muscle fiber size and/or fiber number, and/or impaired excitation-contraction coupling appears to be the most plausible cause(s) for the lower strength in old age.

It has been suggested that the decrement in fiber size is the dominant cause for the impaired strength since fiber area and strength decline to the same extent during aging (for ref. see Larsson 1982b). Cursory inspection of the present data support this view. However, these figures may be deceptive because, first, the muscle fiber atrophy mainly affected type II fibers that are known to be contained in motor units with high activation thresholds. There is reason to believe that there is an inability to recruit all motor units under voluntary conditions and it seems reasonable that some of the high threshold motor units were never recruited in the type of maximal voluntary effort studied. Accordingly, they had no affect on the measured age-related strength decline. Second, in multiple regression analysis it was found that when the independent influence of type II fiber area on strength was eliminated, age still had a significant negative partial correlation with strength, suggesting that other factors also contributed to the decline in strength (Larsson 1978, Larsson et al. 1979). Third, when the age-related fiber atrophy was diminished after the four month physical exercise program, strength still declined with age (Larsson 1982b).

Despite the lack of direct measurements of total muscle fiber number in man during aging, there is reason to believe that it follows a similar decrease as has been reported in other mammals (for ref. see Larsson 1982a) and may thus play an important role for the age-related strength decline.

The effects of aging on the excitation-contraction coupling has, so far, received sporadic scientific interest and those published are conflicting and rather preliminary. On the one hand, it has been claimed that the decrement in strength during aging is solely due to a loss of contractile elements via a decrease in fiber size and/or number (Ermini 1976; McCarter 1978).

On the other hand, others have suggested an impaired excitation-contraction coupling during aging (Gutmann and Hanzlikova 1972; Gutmann et al. 1971; Matsuki et al. 1966).

At present, a study is undertaken where the contractile characteristics of fast and slow single motor units and of whole fast and slow muscles are measured in young, middle-aged, and old rat skeletal muscles. (Larsson and Edström, unpublished). The contractile characteristics of the single motor units are measured and related to the histochemical type, cross-sectional area and number of muscle fibers in the unit (Edström and Kugelberg 1968; Kugelberg and Edström 1968). The aim of the study is to shed some light on the relative importance of the excitation-contraction coupling versus the quantity of contractile material on the altered contractile characteristics of aging mammalian skeletal muscle.

ACKNOWLEDGMENTS

The study was supported by grants from the Swedish Medical Research Council, Karolinska Institutets Fonder, the Swedish Society of Medical Sciences, and the Swedish Sports Federation.

REFERENCES

Bergström, J. 1962. Muscle electrolytes in man. Scand. J. Clin. Lab. Invest. Suppl. 68.

Brooke, M. H., and Kaiser, K. K. 1970. Three "myosin ATPase" systems: the nature of their pH lability and sulfhydryl dependence. J. Histochem. Cytochem. 18:670-72.

Burke, R. E., and Edgerton, V. R. 1975. Motor unit properties and selective involvement in movement. In Exercise and Sport Science Reviews, 3:31-82. New York: Academic Press.

Caccia, M. C., Harris, J. B., and Johnson, M. A. 1979. Morphology and physiology of skeletal muscle in aging rodent. Muscle Nerve 2:202-12.

Cottrell, L. 1940. Histologic variations with age in apparently normal peripheral nerve trunks. Arch. Neurol. Psychiat. 43:1138-50.

Cowdry, E. V. 1952. Aging of individual cells. In Cowdry's Problems of Aging, ed. A. I. Lansing, pp. 50-88. Baltimore: Williams and Wilkins.

Dubowitz, W., and Brooke, M. H. 1973. Muscle biopsy: a modern approach. In: The Series Major Problems in Neurology. Philadelphia, PA: Saunders, vol. 2.

Edström, L., and Kugelberg, E. 1968. Histochemical composition, distribution of fibers and fatiguability of single motor units. J. Neurol. Neurosurg. Psychiat. 31:424-33.

Ermini, M. 1976. Aging changes in mammalian skeletal muscle. Biochemical studies. Gerontology 22:301-16.

Gutmann, E., and Hanzlikova, V. 1972. Age Changes in the Neuromuscular System. Bristol: Scientechnica.

Gutmann, E., Hanzlikova, V., and Vyskocil, F. 1971. Age Changes in cross striated muscle of the rat. J. Physio. 219:331-43.

Kugelberg, E., and Edström, L. 1968. Differential histochemical effects of muscle contractions on phosporylase and glucogen in various types of fibers: relation to fatigue. J. Neurol. Neurosurg. Psychiat. 31:415-23.

Larsson, L. 1978. Morphological and functional characteristics of the aging skeletal muscle in man. A cross-sectional study. Acta Physiol Scand. Suppl. 457.

Larsson, L. 1982a. Aging in mammalian skeletal muscle. In: The aging motor system (ed. F. J. Pirozzolo and G. J. Maletta). pp. 60-98. Praeger, New York.

Larsson, L. 1982b. Physical training effects on muscle morphology in sedentary males at different ages. Med. Sci. Sports Exercise 14:203-06.

Larsson, L. 1983. Histochemical characteristics of human skeletal muscle during aging. Acta. Physiol. Scand. 117: 469-71.

Larsson, L., and Örlander, J. 1983. Muscle morphology, enzyme activities and strength in relation to smoking. A study of smoking discordant monozygous twins. Acta Physiol. Scand. In press.

Larsson, L., Grimby, G., and Karlsson, J. 1979. Muscle strength and speed of movement in relation to age and muscle morphology. J. Appl. Physiol. 46:451-56.

Larsson, L., Sjödin, B., and Karlsson, J. 1978. Histochemical and biochemical changes in human skeletal muscle with age in sedentary males, age 22-65 years. Acta Physiol. Scand. 103:31-39.

Matsuki, H., Takeda, Y., and Tonomura, Y. 1966. Changes in biochemical properties of isolated human skeletal myofibrils with age and in Myasthenia gravis. J. Biochem. 59:122-25.

McCarter, R. 1978. Effects of age on contraction of mammalian skeletal muscle. In Aging, vol. 6. eds. G. Kaldor and W. J. D. Battista, pp. 1-21. New York: Raven Press.

Nie, N. H., Hull, C. H., Jenkins, J. G., Steinbrenner, K., and Bent, D. H. 1975. Statistical Package for the Social Sciences. (SPSS) (2nd ed). New York: McGraw.

Novikoff, A. B., Shin, W. Y., and Drucker, J. 1961. Mitochondrial localization of oxidation enzymes: staining results with two tetrazolium salts. J. Biophys. Biochem. Cytol. 9:47-61.

Örlander, J. 1980. Skeletal muscle metabolism in sedentary men in relation to age, low intensity training and smoking. Act Universitatis Upsaliensis. 550 (Thesis).

Padykula, H. A., and Herman, E. 1955. The specificity of the histochemical method of adenosine triphosphatase. J. Histochem. Cytochem. 3:170-95.

Quetelet, A. 1936. Sur l'homme et le development de ses facultes. Bruxelles: L. Hauman and Cie.

Rexed, B. 1944. Contributions to the knowledge of the postnatal development of the peripheral nervous system in man. Acta Psychiat. Neurol. Suppl. 33:121-93.

Robbins, N., and Kelly, S. S. 1981. Evoked transmitter release in young and old mouse neuromuscular junction. Neurosci. Abstr. Vol. 7.

Rubinstein, L. J. 1961. Aging changes in muscle. In Structural Aspects of Aging, ed. G. H. Bourne, pp. 209-26. London: Pitman Medical.

Saltin, B., and Grimby, G. 1968. Physiological analysis of middle-aged and old former athletes. Circulation 38: 1104-15.

Scelsi, R., Marchetti, C., and Poggi, P. 1980. Histochemical and ultrastructural aspects of m. vastus lateralis in sedentary old people (age 65-89 years). Acta Neuropathol. (Berl) 51: 99-105.

Sjöström, M., Ängqvist, K.-A. and Rais, O. 1980. Intermittent claudication and muscle fiber fine structure: correlation between clinical and morphological data. Ultrastruct. Pathol. 1:309-326.

Swallow, M. 1966. Fiber size and content of anterior tibial nerve of the foot. J. Neurol. Neurosurg. Psychiat. 29: 205-13.

Thorstensson, A. 1976. Muscle strength, fiber types and enzyme activities in man. Acta Physiol. Scand. Suppl. 443.

Wilcox, H. H. 1959. Structural Changes in the Nervous System Related to the Process of Aging, eds. J. E. Birren, II. A Imus, and W. F. Windle, pp. 16-23. Springfield, Ill.: Thomas.

Wohlfart, G. 1939. Histo-pathological studies on muscular atrophy. III 3rd Congr. Neurol. Intern. p. 465.

Wohlfart, G. 1957. Collateral regeneration from residual motor nerve fibers in Amyotrophic Lateral Sclerosis. Neurology (Minneap.) 7:124-34.

25 The Rotation Model as a Basis for Nitella Filament-Dynamics

Ilse Foissner

INTRODUCTION

Actin filament bundles which create the motive force for cytoplasmic streaming in characean internodes (Nagai and Rebhun 1966; Palevitz and Hepler 1975) may detach from the chloroplast files to which they normally adhere. Besides particle transport the filament bundles then perform a sliding motion in the opposite direction, may assemble into rings, show wave propagation and dislocation of branchings. The fibrils can be studied in isolated cytoplasm (e.g., Jarosch 1976), in in vitro preparations (Higashi-Fujime 1980) but also in intact cells (Kamitsubo 1966) which indicates that these movements are not artificial. The knowledge of the bundle behavior allows conclusions about the properties of actin filaments and is therefore important for the understanding of the mechanism of cytoplasmic streaming.

MATERIAL AND METHODS

Nitella flexilis was collected from the Hellbrunner Bach in Salzburg. Cytoplasmic droplets were prepared by pressing excised internodes on a slide (Jarosch 1956). Thin preparations were viewed with Reichert Anoptral optics.

RESULTS

A small cytoplasmic droplet (d_1, Fig. 25.1) was found to perform a circular motion. Close observation revealed that this droplet was connected with a large one (d_2) via two filament bundles. One filament bundle was straight, the other bent and wound around the straight one (b_1 and b_2 in the wire model of Fig. 25.2). Both filament bundles disintegrated into fine fibrils in the large droplet where they showed remarkable network dynamics (compare Foissner and Jarosch 1981). In the small droplet the filament bundles terminated abruptly and one formed a distinct protrusion of the plasma membrane. The motion of the small droplet was due to the rotation of the bent fibril. The clockwise movement, as viewed from the larger droplet, could be followed from the different positions of the bend (white arrow in Fig. 25.1). The rotary velocity amounted to 5 s per revolution but was irregular and sometimes slowed down to 1 min per revolution. The fibrils did not translocate particles along their surface. After 10 min from the beginning of the observation the two droplets fused and the fibrils dissolved into fine motile components. The straight filament bundle seemed to rotate only passively.

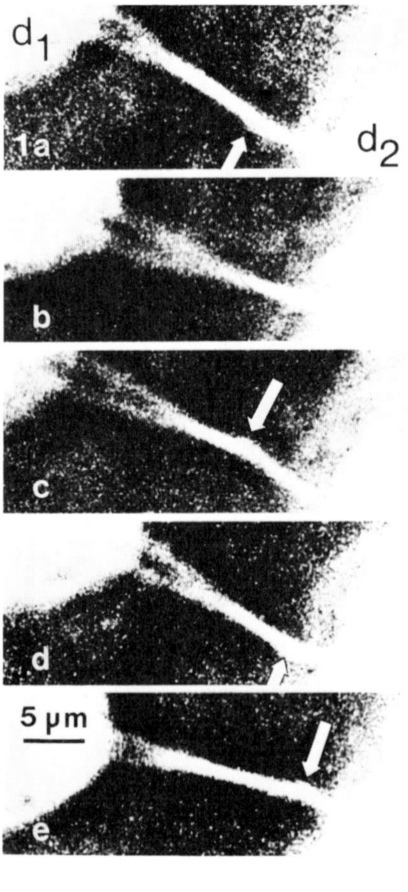

Fig. 25.1a-d. Two cytoplasmic droplets (d_1, d_2) are joined by two filament bundles. At least one filament bundle rotates as indicated by the different positions of a bend (arrow). Time interval 1-10 s.

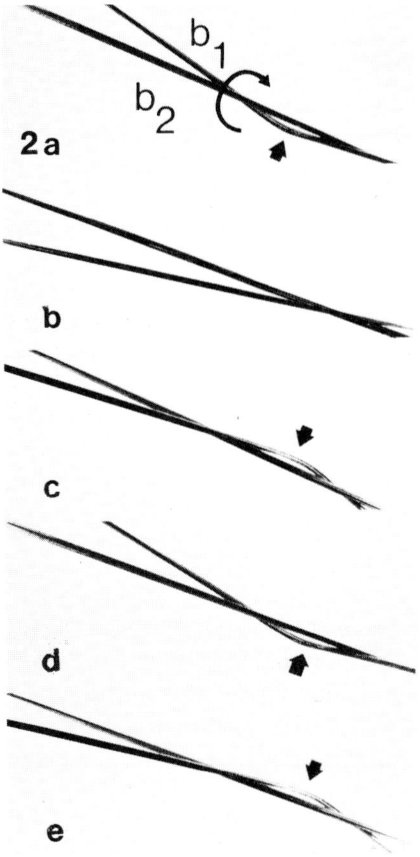

Fig. 25.2a-d. Wire model of the filament bundles in Fig. 25.1. Bundle b_1 is bent and wound around the straight bundle b_2. The bend (arrow) changes its position according to a rotary motion (curved arrow).

DISCUSSION

The rotation of Nitella fibrils around their main axis is not exceptional because similar motions were already mentioned by another author (Jarosch 1976). The rotation seemed to be active and not the mechanical consequence of particle transport which was not observed along the fibrils. This type of motion is especially interesting because it is also shown by the actin filament bundle in the acrosomal process of Limulus sperm (Tilney 1975; DeRosier et al. 1980, 1982). Rotation of that process is brought about by changes in the supercoiling of the bundle. Supercoiling arises from rotational motions of the tightly connected filaments which are caused by a change in the actin filament twist. This behavior is very similar to the mechanism proposed for cytoplasmic streaming where actin filaments are thought to rotate and shift the cytoplasm by the hereby occurring helical waves (Jarosch 1964, 1979; Foissner and Jarosch 1981; Jarosch and Foissner 1982, 1983). The rotational energy, however, is supposed to be provided by associated proteins and not by the actin filaments themselves.

In the supercoiled state the Limulus filament bundle is not smoothly coiled but bent (Tilney 1975). A similar kink was observed in one of the Nitella fibrils studied here. Bends are very common in fibrillar rings of Nitella which were therefore described as polygons (Jarosch 1958). Thus it is reasonable to conclude that the bends in Nitella fibrils are also due to supercoiling. That the fibril is not smoothly curved points to a certain stiffness of the superhelix which allows only bending at "weak" sites. Supervoiling can occur when the free rotation of individual filaments is hindered by firm links as in Limulus sperm. Another reason could be extreme viscous surroundings which would explain the enhanced appearance of polygons in elder cytoplasmic drops. Neither of these possibilities is realized in Nitella fibrils which move in artificial (i.e., diluted!) medium and only occasionally show bends (Higashi-Fujime 1980). As expected, these smooth rings consist of parallel nonconnected filaments and show no supercoiling. Unfortunately, polygons in cytoplasmic droplets have been examined insufficiently up to now with regard to their ultrastructure. The bend in the fibril described here seemed to be rather stiff probably because it was stabilized by the straight fibril. Bends in Nitella fibrils normally behave like flexible joints with varying angles (Jarosch 1976) and thus are again very similar to the bends in the acrosomal process (DeRosier et al. 1982) not only in form but also in function.

Although this motion of Nitella actin filament bundles is not the normal one and usually does not occur in intact cells, it allows important conclusions regarding the mechanism of cytoplasmic streaming. The similarity of the Nitella bundle with the acrosomal process suggests a common mechanism of motility. Since the actin filaments of the process are known to rotate and characteristic features of Nitella filament bundles such as particle transport and wave propagation can be imitated in detail by rotating helices (Jarosch 1976; Foissner and Jarosch 1981; Jarosch and Foissner 1983), the filament rotation model for cytoplasmic streaming is strongly supported.

ACKNOWLEDGMENT

This work was supported by the Österreichische Nationalbank, Projekt Nr. 1927.

REFERENCES

DeRosier, D., L. Tilney, and P. Flicker (1980): A Change in the Twist of the Actin-containing Filaments Occurs during the Extension of the Acrosomal Process in Limulus Sperm. J. Mol. Biol. 137. 375-89.

DeRosier, D. J., L. G. Tilney, E. M. Bonder, and P. Frankl (1982): A Change in Twist of Actin Provides the Force for the Extension of the Acrosomal Process in Limulus Sperm: The False-discharge Reaction. J. Cell Biol. 93, 324-37.

Foissner, I. and R. Jarosch (1981): The Motion Mechanics of Nitella Filaments (Cytoplasmic Streaming): Their Imitation in Detail by Screw-Mechanical Models. Cell Motility 1, 371-85.

Higashi-Fujime, S. (1980): Active Movement In Vitro of Bundles of Microfilaments Isolated from Nitella Cell. J. Cell Biol. 87, 569-78.

Jarosch, R. (1956): Plasmaströmung und Chloroplastenrotation bei Characeen. Phyton (Arg.) 6, 87-107.

Jarosch, R. (1958): Die Protoplasmafibrillen der Characeen. Protoplasma 50, 93-108.

Jarosch, R. (1964): Screw-mechanical basis of protoplasmic movement. In: "Primitive Motile Systems in Cell Biology" (eds. R. D. Allen and N. Kamiya), 599-622, Academic Press, New York and London.

Jarosch R. (1976): Dynamisches Verhalten der Aktinfibrillen von Nitella auf Grund schneller Filament-Rotation. Biochem. Physiol. Pflanz. 170, 111-31.

Jarosch, R. (1979): The torsional movement of tropomyosin and the molecular mechanism of the thin filament motion. In: Cell Motility: Molecules and Organization (eds. S. Hatano, H. Ishikawa, and H. Sato), 291-319, University of Tokyo Press.

Jarosch, R. and I. Foissner (1982): A rotation model for microtubule and filament sliding. Eur. J. Cell Biol. 26, 295-302.

Jarosch, R. and I. Foissner (1983): The rotation model for filament sliding as applied to the cytoplasmic streaming.

In: The Application of Laser Light Scattering to the Study of Biological Motion (eds. J. C. Earnshaw and M. W. Steer), 545-57, Plenum Publishing Corporation.

Kamitsubo, E. (1966): Motile Protoplasmic Fibrils in Cells of Characeae. I. Movement of Fibrillar Loops. Proc. Japan Acad. 42, 507-511.

Nagai, R. and L. I. Rebhun (1966): Cytoplasmic Microfilaments in Streaming Nitella Cells. J. Ultrastruct. Res. 14, 571-89.

Palevitz, B. A. and P. K. Hepler (1975): Identification of actin in situ at the ectoplasm-endoplasm interface of Nitella. J. Cell Biol. 65, 29-38.

Tilney, L. G. (1975): Actin filaments in the acrosomal reaction of Limulus sperm. J. Cell Biol. 65, 289-310.

26 The Rotation Model Applied to Melano(Erythro)phore Motility

Ilse Foissner and Robert Jarosch

The pigment migration in the dermal chromatophores of teleosts and amphibia leading to color changes and adaptation is especially suited for the analysis of a cellular motility mechanism because pigment aggregation and dispersion differ in many respects (reviewed by Porter 1976; Dustin 1978). These differences must be demonstrated by a good model. According to the rotation model particle transport along cytoskeletal elements means winding up and unwinding of associated helical filaments (Jarosch and Foissner 1982). In principle, there exist three possibilities: (1) aggregation is mainly combined with winding (Fig. 26.1), dispersion with unwinding, (2) aggregation is mainly combined with unwinding (Fig. 26.3), dispersion with winding, (3) during aggregation and dispersion winding and unwinding exist simultaneously side by side (Fig. 26.5). In the last case the outer rods in Fig. 26.5 correspond to the rod in Figure 26.1 and the central rod corresponds to the rod in Fig. 26.3. Accordingly the helices of the central rod would wind up along the outer rods during aggregation. During aggregation most microtubules of the erythrophores of *Holocentrus ascensionis* obviously behave like the model in Figure 26.1, i.e., they become wrapped by helices which previously existed in the cytoplasm. This is illustrated in the model of a whole erythrophore sector (Fig. 26.4) where pieces of paper are shifted by the helical waves. Since these "particles" adhere to the helices they return to the same location after winding and unwinding as do the real pigment particles (Porter and McNiven 1982). This behavior is in accordance with the displacement of the filamentous network toward the cell center (see the helices in Fig. 26.4b), the shortening of the "microtrabeculae" (Luby and Porter 1980; Porter and McNiven 1982), and the displacement of the majority of the microtubules into

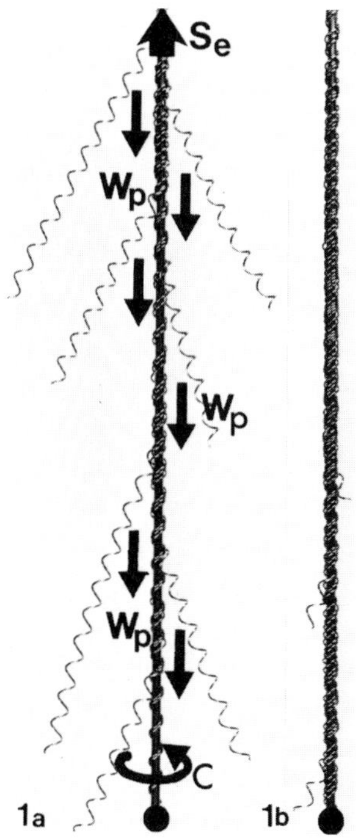

Fig. 26.1. Displacement (arrows W_p) by winding of helices on a rod before (a) and after (b) rotation (arrow C). The arrow S_e shows the direction of a possible screwing motion.

the "cortices on the cytoplasmic surface of the plasma membrane" (Byers and Porter 1977) during aggregation (screwing motion S_e in Fig. 26.1a).

In consequence of winding the microtubules must show a denser wrapping of filamentous substance in the aggregated state than in the dispersed state and the surrounding cytoplasm will therefore appear empty (Luby and Porter 1980). A decrease in the number of microtubules is not obvious here (with Holocentrus) and also not with Fundulus heteroclitus (Murphy and Tilney 1974) but very prominent in Pterophyllum scalare where they are often replaced by dense regions (Schliwa and Bereiter-Hahn 1973; Schliwa and Euteneuer 1978; Schliwa et al. 1979). This disappearance of microtubules can be interpreted according to the model in Figure 26.5b: complete unwinding of tubulin-binding MAPs will cause the depolymerization of certain microtubules. Associated helices which wind up from the MTOC (microtubule-organizing center) (compare Fig. 26.1a) will appear densely wrapped round the microtubules (Fig. 26.1b). Therefore they are only prominent in the dispersed state (see Fig. 26.6a) and correspond to the increased number of "electron dense aggregates" in the central apparatus (Schliwa 1978;

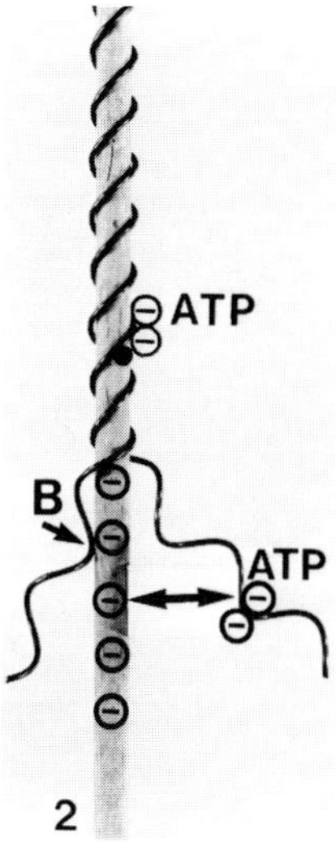

Fig. 26.2. Model to explain promotion of microtubule rotation by ATP ("electroscope effect"). The "shearing brake" (arrow B) of the lateral branch at the left is loosened by electrostatis ATP-repulsion (double arrow at the right).

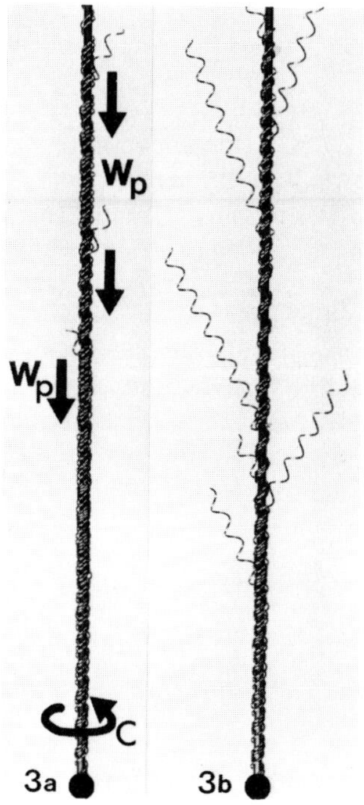

Fig. 26.3. Displacement (arrows W_p) by unwinding of helices from a rod before (a) and after (b) rotation (arrow C).

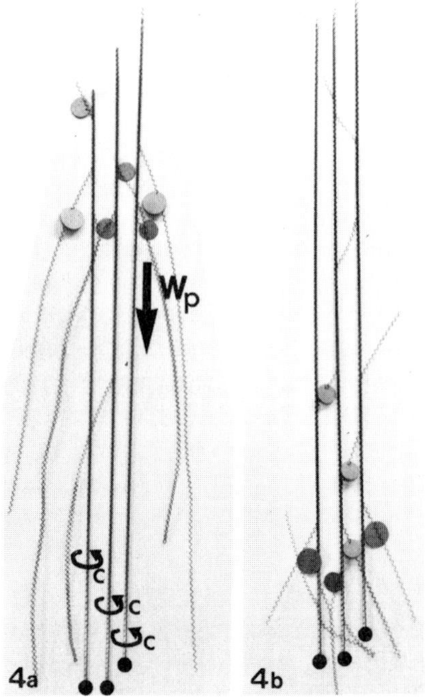

Fig. 26.4. Displacement (arrows W_p) of paper particles by winding of long double helices. Some helices have unwound at the top of Fig. 26.4b.

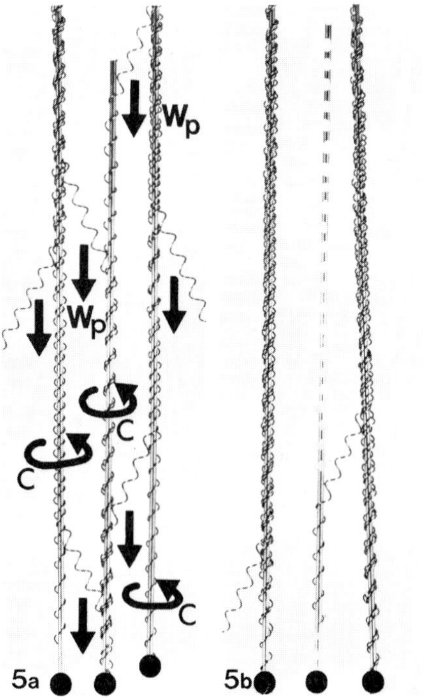

Fig. 26.5. Displacement (arrows W_p) by winding <u>and</u> unwinding. Unwinding causes "depolymerization" of the middle "microtubule" (punctuated).

Schliwa et al. 1979; McNiven and Porter 1981), or the appearance of dense aggregates of 10 nm filaments in the place of Mts in the dispersed state of frog dermal melanocytes (Moellmann and McGuire 1975).

According to the experiments of Schliwa and Bereiter-Hahn (1975) the melanophores disperse their pigment proportional to a mechanic force applied to the aggregated cell. This seems to be similar to the "force enhancement" by stretching of plasmodia and activated muscles: stretching of the cell surface will lead to an extension of the microtubule-associated filaments (microtrabeculae) in the cortex which are coextensive with the plasma membrane (Porter and McNiven 1982) and to a passive unwinding from the microtubules (=dispersion). This implies a reversible increase of the torsional force—similar to the pulling down of a roller blind. The roller blind shows the least torsional force when totally hoisted. In our model this corresponds to the aggregated state, i.e., the state of maximal entropy, also shown by dying cells (Porter and McNiven 1982). EDTA (EGTA) treatment distinctly reduces the counterforce to this dispersion (Schliwa and Bereiter-Hahn 1975) and thus supports the view that the state of winding depends on the concentration of Ca^{2+}. The "pulsating" particle movements (see Porter and McNiven 1982, Stearns and Ochs 1982) imply a periodic change in the Ca^{2+} concentration as in plasmodia (Yoshimoto et al. 1981, compare Jarosch and Foissner 1983).

THE CHEMISTRY OF UNWINDING: ION EFFECTS, CONSEQUENCES OF LINKS BETWEEN MAPs AND MICROTUBULES, AND THE ATP HYDROLYSIS

According to the model three consequences of chemical effects can be distinguished: (1) Changes in the charge state of the MAP side chains triggering torsional rotations. This concerns primarily the rise in the Ca^{2+} concentration necessary for aggregation (Luby-Phelps and Porter 1982). (2) Influences on possible links or electrostatic repulsions between MAPs (see below). (3) Influences on bonds between MAPs and microtubules. In the latter case the model most easily allows detailed statements. The MAPs have been characterized as "polycations" because of their high content of strongly basic polypeptides (Erickson 1976; Bärmann et al. 1982) and the tubulins possess strongly negatively charged regions at the C-end (Ponstingl et al. 1983). Therefore, winding will proceed more easily and more quickly than unwinding, which must overcome the electro-

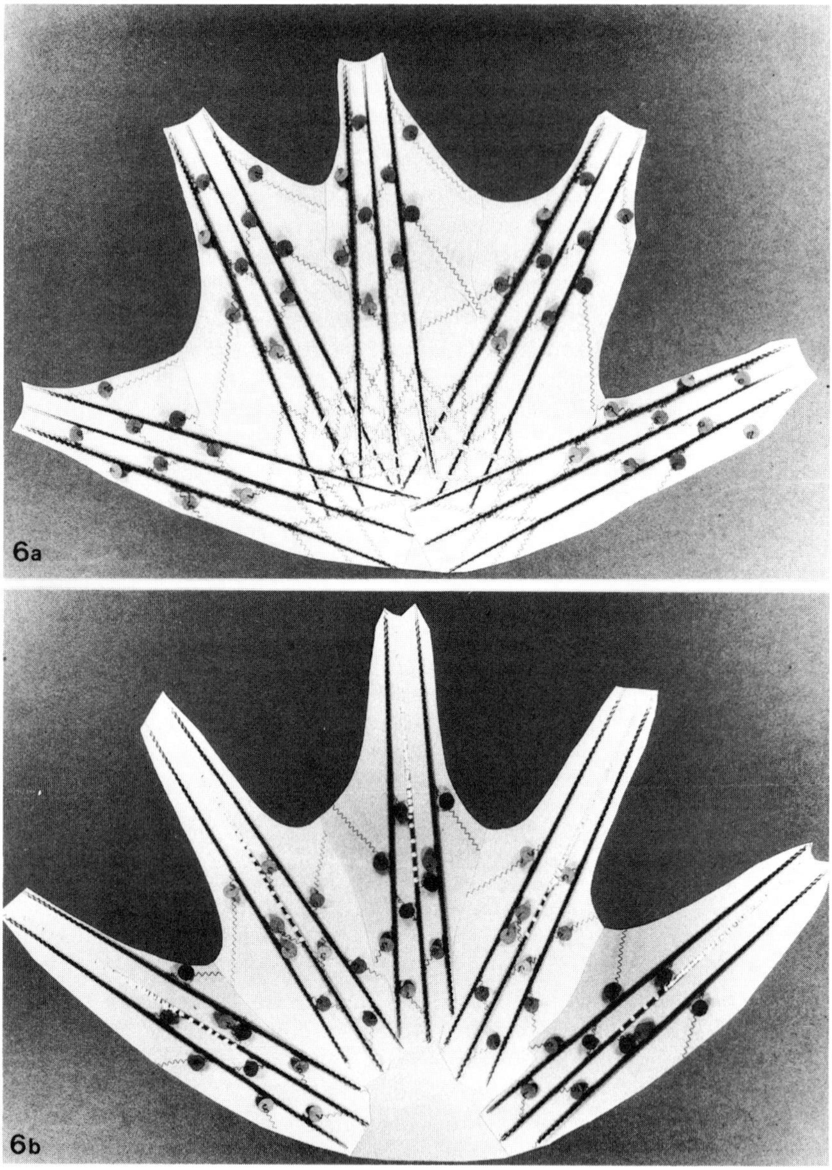

Fig. 26.6. Microtubule model of a part of a chromatophore, consisting of five Fig. 26.5-models, in the "dispersed" (26.6a) and in the more "aggregated" (26.6b) state.

static attraction. In fact, dispersion proceeds irregularly and takes 2-3 times longer than aggregation (Green 1968; Porter 1976; Byers and Porter 1977).

Besides direct bonds between microtubules and MAPs also links by protein-independent bridge molecules may be expected which can then be described as substrate molecules or co-substrates, because winding means controlled bringing together and therefore facilitated binding of ligands ("proximity effect"). Unwinding may then lead to rupture of bridges and formation of radicals. $Mg^{2+}ATP^{4-}$ complexes could in this way link MAPs and MAPs or MAPs and microtubules with their negatively charged phosphate chains and their adenine heads respectively. These bridges would hinder or totally inhibit unwinding (hindrance effect of ATP!). If the rotational force overcomes the totality of the binding forces these ATP bridges must rupture at the sites of unwinding: ATP hydrolysis in the case of ATPases, phosphorylation of associated proteins, its ligands or of tubulin in the case of kinases. According to this hypothesis the accelerating effect of cAMP on dispersion only (Novales and Fujii 1970; Schliwa and Bereiter-Hahn 1974) may be interpreted as a lack of ATP bridges that hinder unwinding by the blockade of adenine binding sites by cAMP.

AN ATP EFFECT AS "ELECTROSCOPE EFFECT"?

The fact that only dispersion, which according to our model means mainly unwinding, is favored by ATP (Junqueira et al. 1974; Luby and Porter 1980) can be explained not only by secondary effects like a decrease in the concentration of free Ca^{2+} or formation of cAMP by adenyl cyclase (Luby and Porter 1980) but also by a possible "electroscope effect" of ATP as shown in Figure 26.2. The greatest friction that hinders rotation like a brake at the last attachment site of the associated helix (arrow B) is overcome by an increased inclination angle of the lateral branch. This could be achieved by electric repulsion as in the case of ATP addition (double arrow) and could then be similar to the plasticizing effect in muscle. It should be noted that the possibility of ATP action proposed here is only indirectly connected with ATP hydrolysis and that ATP is supposed to be polyfunctional.

ACKNOWLEDGMENT

This work was supported by the Österreichische Nationalbank, Jubilaumsfondsprojekt Nr. 1927.

REFERENCES

Bärmann, M., K. Mann, and H. Fasold (1982): Strongly basic polypeptides among microtubule associated proteins. Biochem. Biophys. Res. Comm. 105, 653-58.

Beckerle, M. C. and K. R. Porter (1983): Analysis of the role of microtubules and actin in erythrophore intracellular motility. J. Cell. Biol. 96, 354-62.

Bikle, D., L. G. Tilney, and K. R. Porter (1966): Microtubules and pigment migration in the melanophores of Fundulus heteroclitus L. Protoplasma 61, 322-45.

Byers, H. R. and K. R. Porter (1977): Transformations in the structure of the cytoplasmic ground substance in erythrophores during pigment aggregation and dispersion. J. Cell Biol. 75, 541-58.

Dustin, P. (1978): Microtubules. Springer, Berlin.

Erickson, H. P. (1976): Facilitation of microtubule assembly by polycations. In: Cell Motility (eds. R. Goldman, T. Pollard, and J. Rosenbaum), Book C, 1069-80, Cold Spring Harbor Laboratory.

Green, L. (1968): Mechanism of movement of granules in melanocytes of Fundulus heteroclitus (L.). Proc. Nat. Acad. Sci. USA 59, 1179-86.

Jarosch, R. and I. Foissner (1982): A rotation model for microtubule and filament sliding. Eur. J. Cell Biol. 26, 295-302.

Jarosch, R. and I. Foissner (1983): The rotation model for filament sliding as applied to the cytoplasmic streaming. In: The application of laser light scattering to the study of biological motion (eds. J. C. Earnshaw and M. W. Steer), 545-58, Plenum Publ. Corp.

Junqueira, L. C., E. Raker, and K. R. Porter (1980): The control of pigment migration in the melanophores of the teleost Fundulus heteroclitus (L.). Arch. Hist. Jap. 36, 339-66.

Luby, K. J. and K. R. Porter (1980): The control of pigment migration in isolated erythrophores of Holocentrus ascensionis (Osbeck). I. Energy requirements. Cell 21, 13-23.

Luby-Phelps, K. and K. R. Porter (1982): The control of pigment migration in isolated erythrophores of Holocentrus ascensionis (Osbeck). II. The role of Ca^{2+}. Cell 29, 441-50.

McNiven, M. A. and K. R. Porter (1981): The microtubule organizing center in erythrophores. J. Cell Biol. 91, 334a.

Moellmann, G. and J. McGuire (1975): Correlation of cytoplasmic microtubules and 10-nm filaments with the movement of pigment granules in cutaneous melanocytes of Rana pipiens. Ann. N.Y. Acad. Sci. 253, 711-22.

Murphy, D. G. and L. G. Tilney (1974): The role of microtubules in the movement of pigment granules in teleost melanophores. J. Cell Biol. 61, 757-99.

Novales, R. R. and R. Fujii (1970): A melanin-dispersing effect of cyclic adenosine monophosphate on Fundulus melanophores. J. Cell. Physiol. 75, 133-35.

Ponstingl, H., E. Kraus, and M. Sittle (1983): Tubulin amino acid sequence and consequences. J. Submicros. Cytol. 15, 359-62.

Porter, K. R. (1976): Introduction: Motility in cells. In: Cell Motility (eds. G. Goldman, T. Pollard, and J. Rosenbaum), Book A, 1-28. Cold Spring Harbor Laboratory.

Porter, K. R. and M. A. McNiven (1982): The cytoplast: A unit structure in chromatophores. Cell 29, 23-32.

Schliwa, M. (1978): Microtubular apparatus of melanophores. Three-dimensional organization. J. Cell Biol. 76, 605-614.

Schliwa, M. and J. Bereiter-Hahn (1973): Pigment movements in fish melanophores: Morphological and physiological studies II. Cell shape and microtubules. Z. Zellforsch. 147, 107-25.

Schliwa, M. and J. Bereiter-Hahn (1974): Pigment movements in fish melanophores: Morphological and physiological studies IV. The effect of cyclic adenosine monophosphate on normal and vinblastine treated melanophores. Cell Tissue Res. 151, 423-32.

Schliwa, M. and J. Bereiter-Hahn (1975): Pigment movements in fish melanophores: Morphological and physiological studies V. Evidence for a microtubule=independent contractile system. Cell Tiss. Res. 158, 61-73.

Schliwa, M. and U. Euteneuer (1978): Quantitative analysis of the microtubule system in isolated fish melanophores. J. Supramol. Struct. 8, 177-90.

Schliwa, M., U. Euteneuer, W. Herzog, and K. Weber (1979): Evidence for rapid structural and functional change of the melanophore microtubule-organizing center upon pigment movements. J. Cell Biol. 83, 623-32.

Stearns, M. E. and R. L. Ochs (1982): A functional in vitro model for studies of intracellular motility in digitonin-permeabilized erythrophores. J. Cell Biol. 94, 727-39.

Yoshimoto, Y. T. Sakai, and N. Kamiya (1981): ATP oscillation in Physarum plasmodium. Protoplasma 109, 159-68.

27 Filament Winding and Unwinding as a Basis for Microtubule-behavior: A Theoretical Analysis with Helix-models

Robert Jarosch and Ilse Foissner

In contrast to the polymerization of MTs from solutions of high tubulin concentration in vitro, the growth of MTs out from a MTOC is directed and is also possible at a relatively low tubulin concentration if MAPs are present (Sloboda et al. 1976; Weber and Osborn 1979; Brown et al. 1982). MT assembly could then occur by winding up of tubulin carrying MAPs to a double or manifold helix in the MTOC leading to controlled subunit addition (Jarosch and Foissner 1982). In the model of Figure 27.1a winding of the two helices (MAPs, small arrows) causes the helix rod complex to rotate (arrow c) and to screw upwards (arrow S_e). This corresponds to a sliding process plus growth. The part of the rod not yet wrapped is punctuated. Contrary to the enhanced subunit addition at the free MT ends in in vitro experiments, the tubulin supply here is from the basis (the plus end), in the MTOC, as already discussed by Weber and Osborn (1979) and determines the MT length (Brinkley et al. 1981). The growth ceases when the supply is exhausted (Fig. 27.1b). An anchorage of the MT in the MTOC may nonetheless persist (compare Fig. 27.1c, 27.1d). The MT assembled in this way may perform further rotations which cause "sliding" or particle transport respectively without considerable polymerization or depolymerization at its ends ("stable MT"). The anchor filaments (compare Fig. 27.1c, 27.1d), however, would have to change their twist. It should be noted that the particle transport by the helical waves (arrow W_p in Fig. 27.1c) always proceeds in the direction opposite to the screwing motion of the MTs (arrow S_e in Fig. 27.1a). Therefore the MTs in isolated erythrophores tend to "screw out" during pigment aggregation (see Fig. 27.5b in Beckerle and Porter 1983). Another impressive example is the poleward transport in spindles and asters (see Bajer and Mole-Bajer 1982 for review), corresponding to

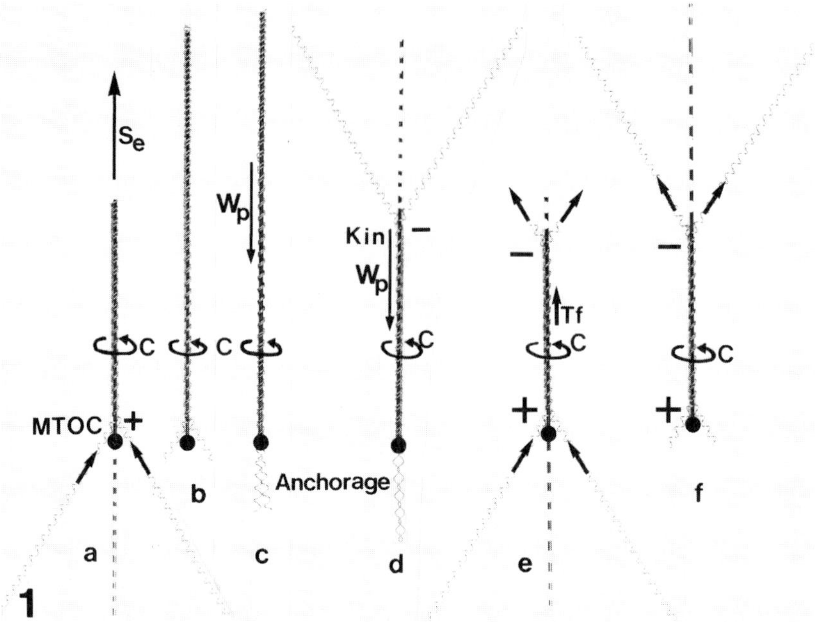

Fig. 27.1. Rod-helix model showing winding and unwinding to explain MT polymerization (a, b), particle transport (c, d), depolymerization (d) and tubulin flux (e, f). Arrow C means clockwise rotation as seen from the MTOC. For further explanation see the text.

arrow W_p in Figure 27.1c and the "aster migration" which occurs only in asymmetric asters in the direction (arrow S_e in Fig. 27.1a) of the longer MTs. In the cytopharynx of many ciliates and in suctorian tentacles pieces of the prey cytoplasm are shifted into the cell along MT rows. This corresponds to the translocation of paper particles (arrow W_e) in the model of Figure 27.5 where each helix winds and unwinds at four different rods. A large model was used to illustrate the shift of prey cytoplasm in suctorian tentacles (Fig. 27.6). If the MTOC in suctorian tentacles is distally situated, we have to assume a change in the rotational direction of the MTs (arrow c → arrow cc) after their completed growth in the direction S_e (Fig. 27.6a, compare Fig. 27.1a). Starting from the resting state (Fig. 27.6a) the MT rows of the tentacle first shift distally in the direction of the anchorage ("knob-eversion," arrows S_p in Fig. 27.6b) comparable to the opening of an umbrella (Bardele 1974; Tucker 1974; Tucker and Mackie 1975). Only then does the displacement of the prey cytoplasm, represented by a piece of paper, start in the direction of the arrow W_e (Fig. 27.6b, 27.6c).

When the function of the MTs is accomplished their depolymerization may occur by unwinding of tubulin binding MAPs. A change in the direction of rotation would cause unwinding to start at the MTOC. Whether this possibility is realized in vivo must remain open (see Weber and Osborn 1979). On the other hand, depolymerization by unwinding from the free MT end toward the MTOC is obviously realized with the kinetochore MTs during the anaphase A movement of the chromosomes (Fig. 27.1d). In this case, the same particle translocating waves occur (arrow W_p) as in Figure 27.1c. The direction of rotation (arrow c) is the same as during polymerization. The fine filaments produced at kinetochores (compare Schibler and Pickett-Heaps 1980) could combine with and unwind the tubulin bound MAPs and in this way the spindle MT becomes a kinetochore MT. In accordance with the conception of Grudzdev (1972) and the results of Forer (1976) the kinetochore MTs disassemble at the kinetochore end during anaphase A. The funnel-shaped end of the kinetochore MTs described from electromicrographs (Bajer and Mole-Bajer 1972) is also evident in a model with unwinding helices (Fig. 27.2). Another possible way in which MAPs could wind or unwind at the MT ends of the kinetochores or MTOC is depicted in the structure model of Figure 27.3.

How is it possible to explain MT depolymerization at low temperature or high hydrostatic pressure (compare e.g., Murphy

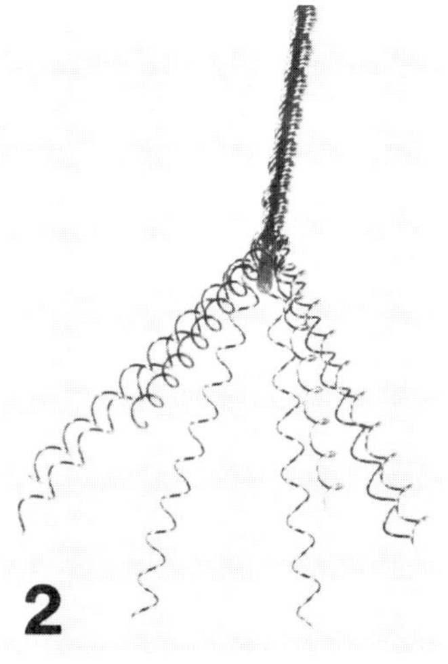

Fig. 27.2. Funnel-shaped unwinding of the helices from the rod as an imitation of the MT end at a kinetochore.

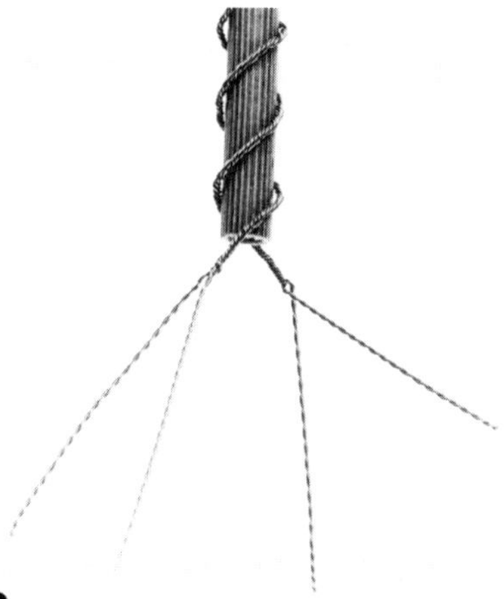

3

Fig. 27.3. Proposal for MAP winding or unwinding at the end of an MT.

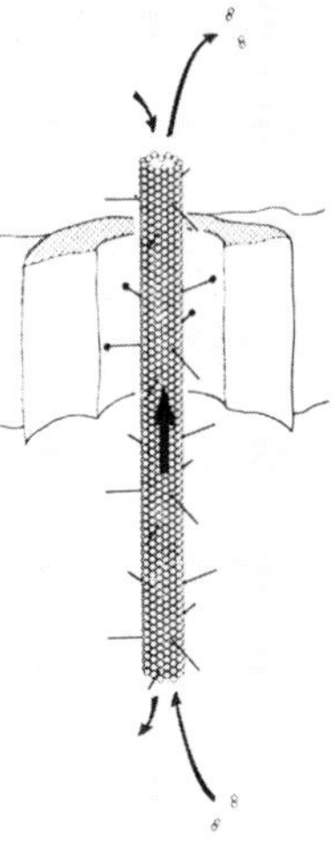

4

Fig. 27.4. Tubulin flux in the "treadmilling model" (from Margolis and Wilson 1981 with permission, compare Fig. 27.1e).

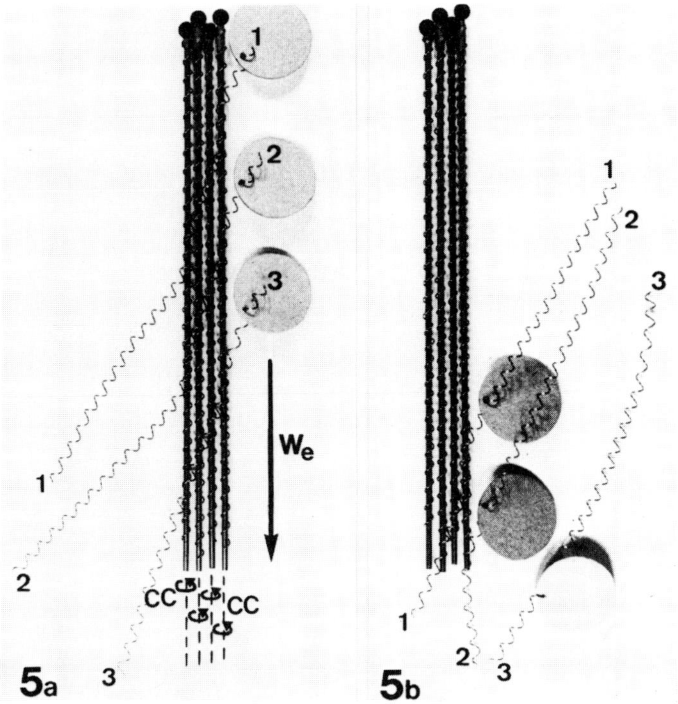

Fig. 27.5. Rod-helix model to explain particle transport (arrow W_e) along an MT ribbon, arrows cc mean counterclockwise rotation as seen from the MTOC. For further explanation see the text.

and Tilney 1974) by means of the rotation model? According to the model the MT rotates because of the rotational forces transmitted by winding or unwinding of associated proteins. Chilling or increase in hydrostatic pressure enhances the viscosity of the MT surroundings and therefore the friction resistance to rotation. Because of that the torsional stress increases which induces the appearance of "wavy" and "twisted" MTs (compare Toyohara et al. 1978). Just as a screw can be removed by torsion when the resistance is too high, the rotational forces of the associated proteins may overcome the bindings between the tubulin subunits. The subunits will then unwind with the MAPs. Fragmentation of MTs after a rise in the concentration of Ca^{2+} (Weber and Osborn 1979) could be interpreted as a similar effect of enhanced torsional stress. MT dissociating agents such as vinblastine could lower the cohesion between the MT subunits which would also facilitate unwinding (see Schliwa and Bereiter-Hahn 1975 and p. 000 in this volume).

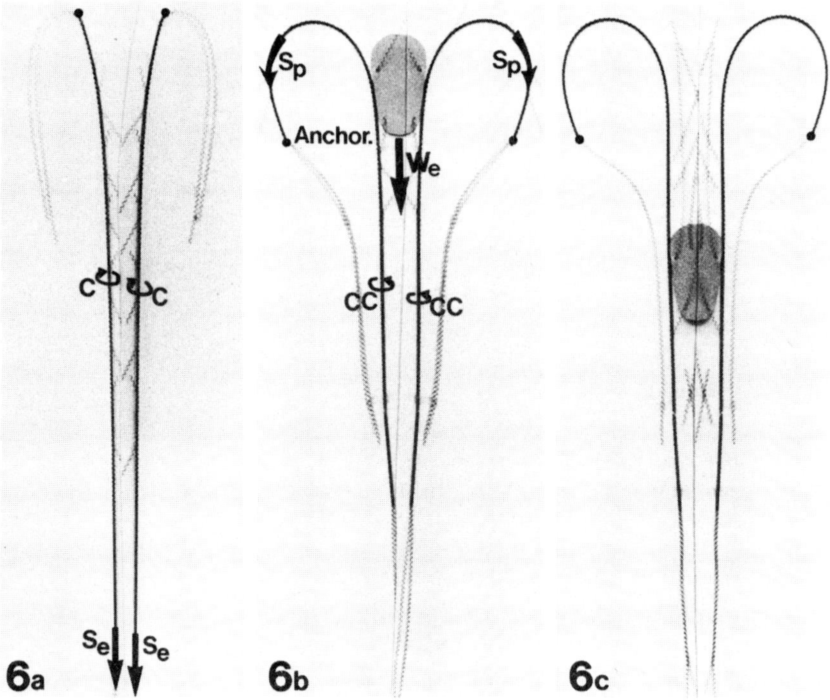

Fig. 27.6. Model of a suctorian tentacle during polymerization of MTs (a) "knobeversion" and feeding (b, c). For further explanation see the text.

During polymerization of spindle MTs or depolymerization of kinetochore MTs respectively, winding and unwinding proceed successively in time. These processes may occur simultaneously too as described in the model of Figure 27.1e, f. The direction of helix winding and unwinding is indicated by small arrows and the rod displacement corresponds to the tubulin flux (arrow T_f in Fig. 27.1e) as also postulated in the "treadmilling model" of Margolis and Wilson (1981) (Fig. 27.4). In a real treadmill the treadling person must have an appropriate weight in order to overcome the resistance of the mill. The addition of only thermodynamically supplied tubulin to the MT ends may certainly elongate the MT but hardly move it. In our model tubulin addition and MT displacement are due to the winding and screwing motion of MAPs which are provided with torsional energy. When tubulin supply ceases at the plus end (Fig. 27.1f) and subunits are further lost at the minus end, the fixed MT may shorten at the minus end with a concomittant dislocation of waves basally (arrow W_p) as already shown in Figure 27.1d. The other possibility—without anchorage—would be a screwing motion upwards whereby the plus end, now without supply, is translocated upwards.

ACKNOWLEDGMENT

This work was supported by the Österreichische National-bank, Jubiläumsfondsprojekt Nr. 1927.

REFERENCES

Bajer, A. S. and J. Molè-Bajer (1972): Spindle dynamics and chromosome movements. Academic Press, New York and London.

Bajer, A. S. and J. Molè-Bajer (1982): Asters, poles, and transport properties within spindlelike microtubule arrays. In: Organization of the cytoplasm. Cold spring harbor symposia on quantitative biology, Vol. 46, 263-83. Cold Spring Harbor Laboratory.

Bardele, C. F. (1974): Transport of materials in suctorian tentacle. Symp. Soc. Exp. Biol. 27, 191-208.

Beckerle, M. C. and K. R. Porter (1983): Analysis of the role of microtubules and actin in erythrophore intracellular motility. J. Cell Biol. 96, 354-62.

Brinkley, B. R., S. M. Cox, D. A. Pepper, L. Wible, S. L. Brenner, and R. L. Pardue (1981): Tubulin assembly sites and the organization of cytoplasmic microtubules in cultured mammalian cells. J. Cell Biol. 90, 554-62.

Brown, D. L., M. E. Stearns, and T. H. Macrae (1982): Microtubule organizing centers. In: The cytoskeleton in plant growth and development (ed. C. W. Lloyd), 55-83, Academic Press, London and New York.

Forer, A. (1976): Actin filaments and birefringent spindle fibers during chromosome movements. In: Cell motility (eds. R. Goldman, T. Pollard, and J. Rosenbaum), Book C, 1273-93, Cold Spring Harbor Laboratory.

Gruzdev, A. D. (1972): Tsitologiya 14, 141-49 (in Russian, English translation in: NRC Technical Translation 1758, National Research Council of Canada, Ottawa).

Jarosch, R. and I. Foissner (1982): A rotation model for microtubule and filament sliding. Eur. J. Cell Biol. 26, 295-302.

Margolis, R. L. and L. Wilson (1981): Microtubule treadmills—possible molecular machinery. Nature (London) 293, 705-11.

Murphy, D. B. and L. G. Tilney (1974): The role of microtubules in the movement of pigment granules in teleost melanophores. J. Cell Biol. 61, 757-79.

Schibler, M. J. and J. D. Pickett-Heaps (1980): Mitosis in Oedogonium: spindle microfilaments and the origin of the kinetochore fiber. Eur. J. Cell Biol. 22, 687-98.

Schliwa, M. and J. Bereiter-Hahn (1975): Pigment movements in fish melanophores: Morphological and physiological studies V. Evidence for a microtubule-independent contractile system. Cell Tiss. Res. 158, 61-73.

Sloboda, R. D., W. L. Dentler, R. A. Bloodgood, B. R. Telzer, and S. G. L. Rosenbaum (1976): Microtubule-associated proteins (MAPs) and the assembly of microtubules in vitro. In: Cell Motility (eds. R. Goldman, T. Pollard, and J. Rosenbaum), Book C, 1171-1212, Cold Spring Harbor Laboratory.

Toyohara, A., Y. Shigenaka, and H. Mohri (1978): Microtubules in protozoan cells. II. Ultrastructural changes during disintegration and reformation of heliozoan microtubules. J. Cell Sci. 32, 87-98.

Tucker, J. B. (1974): Microtubule arms and cytoplasmic streaming and microtubule bending and stretching of intertubule links in the feeding tentacle of the suctorian ciliate Tokophrya. J. Cell Biol. 62, 424-37.

Tucker, J. B. and J. B. Mackie (1975): Configurational changes in helical microtubule frameworks in feeding tentacles of the suctorian ciliate Tokophrya. Tissue & Cell 7, 602-12.

Weber, K. and M. Osborn (1979): Intracellular display of microtubular structures revealed by indirect immunofluorescence microscopy. In: Microtubules (eds. K. Roberts and J. S. Hyams), 279-313, Academic Press, London, New York.

28 The Filament Rotation Model: Molecular and Screw-mechanical Details

Robert Jarosch

A model should agree with all known facts. Since the life motions are confined to proteins the models for life motions should consider all possibilities of the protein structure and should not comprise unfounded statements. The ability for active rotation is the natural consequence of charge modifications at the side chains of α-helical coiled-coils. This is more precisely described in what follows.

Table 28.1 shows the amino acid sequence of α-tropomyosin (Stone et al. 1975) arranged in clusters of seven (a, b, c, d, e, f, g). The strongly charged side chains are underlined. The seven positions are seen in the cross-sectioned coiled-coil after McLachlan and Stewart (1975) (Fig. 28.1a). Figure 28.1b shows the charge distribution along one α-helix rolled down on a plane. 136 of the 284 side chains are strongly charged with a distinct surplus of negative charges (81 negative, 55 positive). The 82 positions a and d are located at the inner concave site of the α-helix (compare Fig. 28.1a and Fig. 28.3a). Their side chains are mainly hydrophobic. A quantitative evaluation of the interactions that occur here is possible by summarizing the attractions and repulsions between neighboring side chains with the aid of the interaction coefficients (IC) described by Kuntz et al. (1976). This is only a rough method because the interactions of one side chain with its six neighbors are not equivalent. The interhelical attraction (-) and repulsion (+) of the side chains a-a and d-d yield a ratio of -638 to +100. This shows the high stability of the coiled-coil which is necessary

---

*Dedicated to the pioneer of molecular biological thinking, Prof. Hans Linser, on the occasion of his 76th birthday.

	a	b	c	d	e	f	g
1	Met-	Asp-	Ala-	Ile-	Lys-	Lys-	Lys-
	Met-	Gln-	Met-	Leu-	Lys-	Leu-	Asp-
15	Lys-	Glu-	Asn-	Ala-	Leu-	Asp-	Arg-
	Ala-	Glu-	Glu-	Ala-	Glu-	Ala-	Asp-
29	Lys-	Lys-	Ala-	Ala-	Glu-	Asp-	Arg-
	Ser-	Lys-	Gln-	Lau-	Glu-	Asp-	Glu-
43	Leu-	Val-	Ser-	Leu-	Gln-	Lys-	Lys-
	Leu-	Lys-	Gly-	Thr-	Glu-	Asp-	Glu-
57	Leu-	Asp-	Lys-	Tyr-	Ser-	Glu-	Ala-
	Leu-	Lys-	Asp-	Ala-	Gln-	Glu-	Lys-
71	Leu-	Glu-	Leu-	Ala-	Glu-	Lys-	Lys-
	Ala-	Thr	Asp-	Ala-	Glu-	Ala-	Asp-
85	Val-	Ala-	Ser-	Leu-	Asn-	Arg-	Arg-
	Ile-	Gln-	Leu-	Val-	Glu-	Glu-	Glu-
99	Leu-	Asp-	Arg-	Ala-	Gln-	Glu-	Arg-
	Leu-	Ala-	Thr-	Ala-	Lau-	Gln-	Lys-
113	Leu-	Glu-	Glu-	Ala-	Glu-	Lys-	Ala-
	Ala-	Asp-	Glu-	Ser-	Glu-	Arg-	Gly-
127	Met-	Lys-	Val-	Ile-	Glu-	Ser-	Arg-
	Ala-	Gln-	Lys-	Asp-	Glu-	Glu-	Lys-
141	Met-	Glu-	Ile-	Gln-	Glu-	Ile-	Gln-
	Leu-	Lys-	Glu-	Ala-	Lys-	His-	Ile-
155	Ala-	Glu-	Asp-	Ala-	Asp-	Arg-	Lys-
	Tyr-	Glu-	Glu-	Val-	Ala-	Arg-	Lys-
169	Leu-	Val-	Ile-	Ile-	Glu-	Ser-	Asp-
	Leu-	Glu-	Arg-	Ala-	Glu-	Glu-	Arg-
183	Ala-	Glu-	Leu-	Ser-	Glu-	Gly-	Lys-
	Cys-	Ala-	Glu-	Leu-	Glu-	Glu-	Glu-
197	Leu-	Lys-	Thr-	Val-	Thr-	Asn-	Asn-
	Leu-	Lys-	Ser-	Leu-	Glu-	Ala-	Gln-
211	Ala-	Glu-	Lys-	Tyr-	Ser-	Gln-	Lys-
	Glu-	Asp-	Lys-	Tyr-	Glu-	Glu-	Glu-
225	Ile-	Lys-	Val-	Leu-	Ser-	Asp-	Lys-
	Leu-	Lys-	Glu-	Ala-	Glu-	Thr-	Arg-
239	Ala-	Glu-	Phe-	Ala-	Glu-	Arg-	Ser-
	Val-	Thr-	Lys-	Leu-	Glu-	Lys-	Ser-
253	Ile-	Asp-	Asp-	Leu-	Glu-	Asp-	Glu-
	Leu-	Tyr-	Ala-	Gln-	Lys-	Leu-	Lys-
267	Tyr-	Lys-	Ala-	Ile-	Ser-	Glu-	Glu-
	Leu-	Asp-	His-	Ala-	Leu-	Asn-	Asp-
281	Met-	Thr-	Ser-	Ile			

Table 1

Fig. 28.1. Amino acid sequence of α-tropomyosin (Stone et al. 1975) grouped in clusters of seven. The side chains that are strongly charged at pH 7 are underlined.

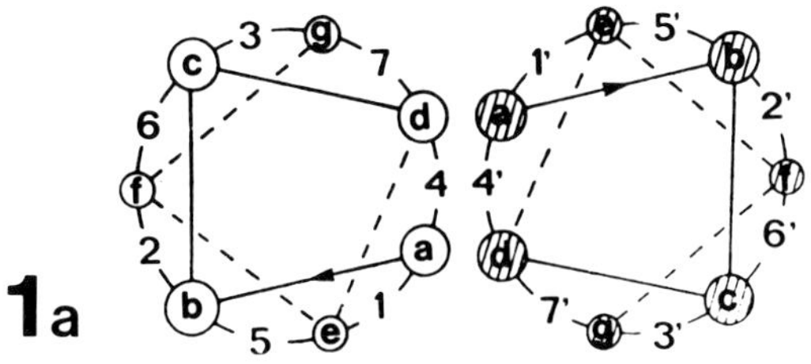

Fig. 28.1a. The residue positions in the cross-sectioned coiled-coil (according to McLachlan and Stewart 1975). The numbers indicate the positions of the hydrogen bonds.

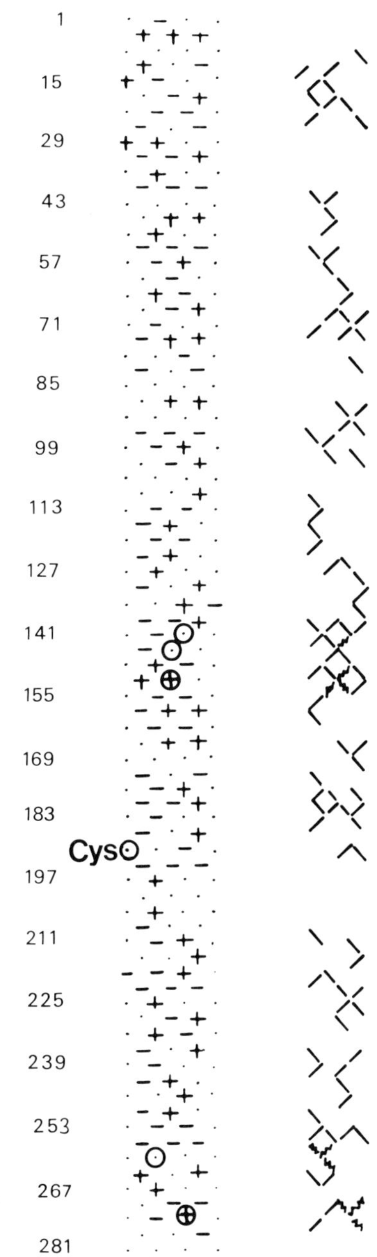

Fig. 28.1b (left) . The charge pattern of one tropomyosin α-helix projected on a plane.

Fig. 28.2 (right). Sites of strong repulsion (lines) which stretch the hydrogen bonds. Only four positions show attraction (wavy lines) that shorten the hydrogen bonds.

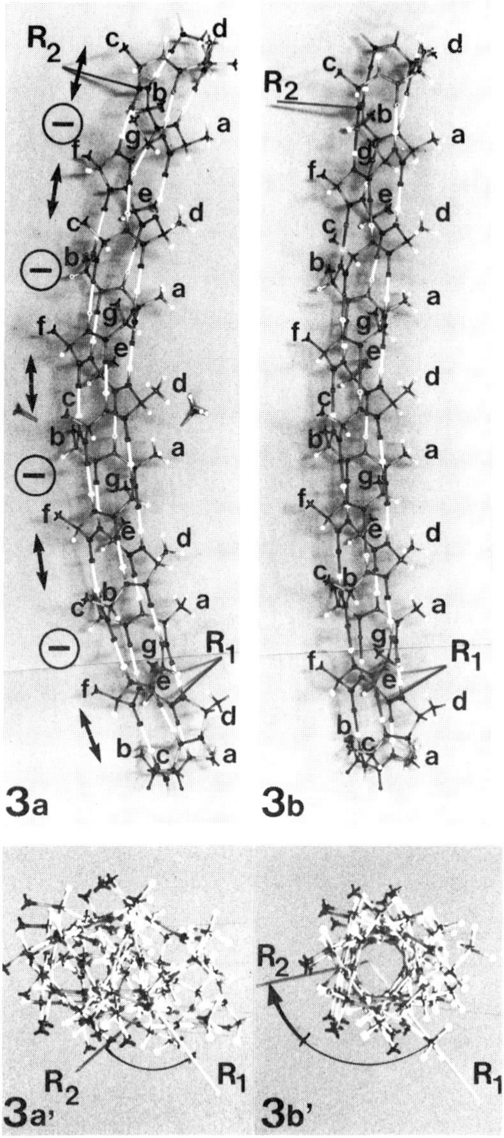
R2
R1
a
b
c
d
e
f
g
3a
3b
3a'
3b'

for life function and which is caused by hydrophobic attraction (the cystein interaction and the salt bridges between e and g are not yet included here). 189 interactions are possible between the strongly charged neighboring residues of one of the two α-helices. According to Pauling and Corey (1953) they influence the length of the α-helical hydrogen bonds. This is, however, not true for the strong interactions between those side chains which are neighbored perpendicularly to the helical axis (e.g., b-c and e-f) and also not for oppositely charged side chains (IC = -10) which cannot attract each other in the direction of the hydrogen bonds owing to the larger rotational freedom of the long positively charged side chains (compare Jarosch 1979). Only 98 strongly repelling (of those 59 with IC +25, 39 with IC +10 or +15) and 10 strongly attracting interactions (IC -10 or -15) remain then. Their position is indicated in Figure 28.2: the sums of repulsion and attraction are in the ratio of about +1,900 to -120. If the weak interactions (IC -2 and -5) which influence the hydrogen bonds are also considered, the ratio is +2047 to -498. In any case, the outside of the α-helices shows a strong intrahelical repulsion which stretches the external hydrogen bonds (Fig. 28.3a) and in this way causes the typical coiled-coil conformation. Even a small charge-decrease at the surface, e.g., the binding of an ion, must influence the hydrogen bonds. This causes an increase in the α-helical twist as indicated by the angle between the rods R_1 and R_2 in Figure 28.3b' (for details see Jarosch 1979). The fundamental significance of this torsion for life motions has been overlooked or ignored up to now (e.g., Phillips et al. 1980).

Figure 28.4 summarizes the described effects on a coiled-coil of about 41 nm length, the tropomyosin dimer, which is assumed to be fixed at the lower end. After ion addition (by raising the Ca^{2+}-concentration) a complete torsional rotation of the upper end (arrow C) is shown in Figure 28.4, whereby the two α-helices unwind somewhat and the pitch enlarges. Ion removal (by lowering the Ca^{2+}-concentration) means a rotation

Fig. 28.3 (left). A molecule model of a coiled-coil α-helix shows impressively the conformational change, which occurs after a decrease in the electrostatis repulsion of the side chains (double arrows in Fig. 28.3a). Shortening of the outer hydrogen bonds 2, 3, 5, 6 by about 0.25 Å leads to a decrease of the helical bend (Fig. 28.3b) but to an increase of the helical twist (compare the angle between the rods R_1 and R_2).

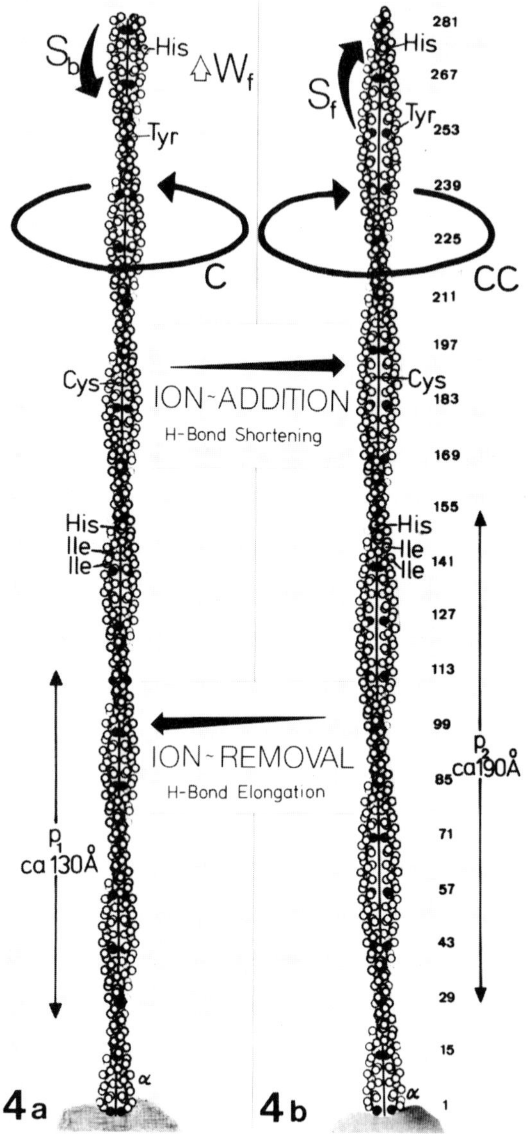

Fig. 28.4. Summarizing description of ion (Ca^{2+})-dependent conformational change in a coiled-coil (tropomyosin), assumed to be anchored at the bottom. Arrow C: clockwise rotation, arrow CC: anticlockwise rotation, as seen from bottom. A figure from Fraser and McRae (1973) according to Crick, was used. For further details see the text.

in the opposite sense (arrow CC). The state in Figure 28.4b is probably more flexible and less stable because of the weaker interactions between the side chains and the salt bridges. The importance of additional proteins (e.g., troponin) for these processes are not considered here. The following screw-mechanical effects are expected during the boiled-coil rotation: a helix which rotates because of the decrease of its pitch always shows waves moving toward the anchorage (Jarosch 1964). Shortening of the hydrogen bonds in the α-helix would thus cause the waves of the coiled-coil, which twists in the opposite way, to move in the direction of the arrow W_f (Fig. 28.4a). But the high resistance in the surroundings of a molecular helix does not permit a translocation of waves! Instead of that the helix end can screw in the direction of the arrow S_b ("sliding back") like a corkscrew in the cork and compress the coiled-coil. On the contrary, extension of the hydrogen bonds means

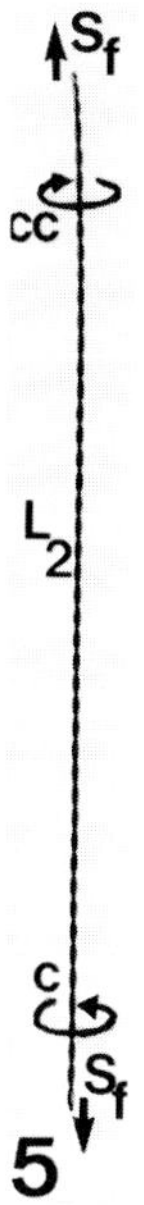

Fig. 28.5. Ion removal causes the rotating helix ends to screw distally (arrows S_f) leading to stretching of the filament.

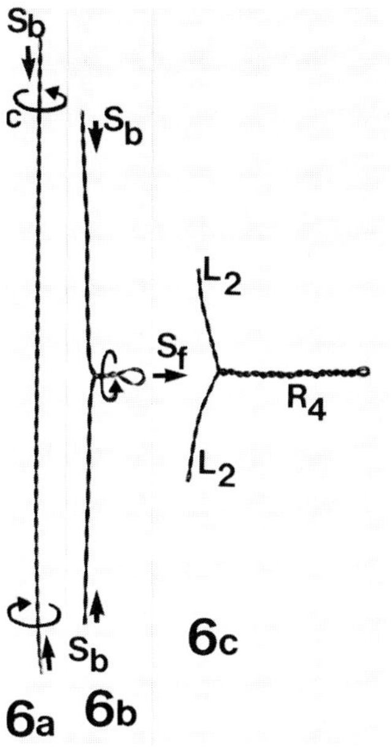

Fig. 28.6. Ion addition causes the helix ends to screw toward the center (arrows S_b) and thus promotes self-intertwining to a right-handed helix (R_4).

"sliding forth" (arrow S_f in Fig. 28.4b) and stretching of the coiled-coil. Only "sliding back" (arrows S_b in Fig. 28.6a, b) will favor self-intertwining which causes the left-handed coiled-coil (L_2) to wind up as a fourfold right-handed helix (R_4 in Fig. 28.6c). This helix rotates in such a way that it screws forward (arrow S_f). The torsional self-intertwining is better demonstrated by a rubber string which has been previously twisted in the direction of the arrow T (Fig. 28.7a). Self-intertwining often occurs in fibrillar cell structures (compare e.g., Jarosch 1972) and is especially prominent as a primitive contraction in spasmoneme-like filaments after Ca^{2+} addition (Weis-Fogh and Amos 1972; Amos 1975) and is "much like the wound rubber band engine of a model airplane" in ciliary roots (Salisbury 1983). Figure 28.8 shows this transformation to glomerate structures with a strongly twisted, irregular rubber string. $MgATP^{2-}$ brings about relaxation of the spasmonemes

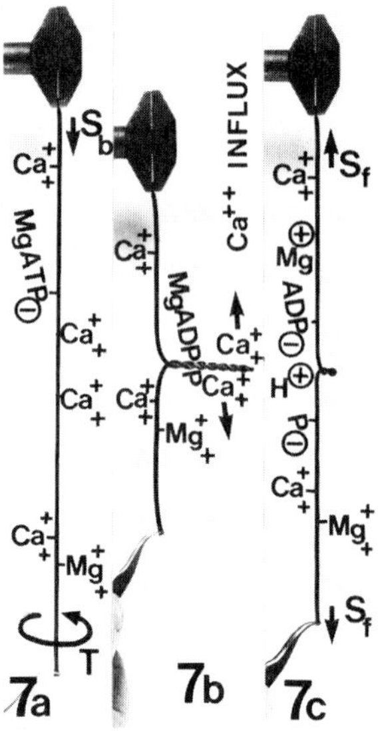

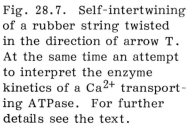

Fig. 28.7. Self-intertwining of a rubber string twisted in the direction of arrow T. At the same time an attempt to interpret the enzyme kinetics of a Ca^{2+} transporting ATPase. For further details see the text.

Fig. 28.8. Strong torsion in an irregularly built rubber-string mimics the behavior of Ca^{2+} activated filaments in flagellar roots.

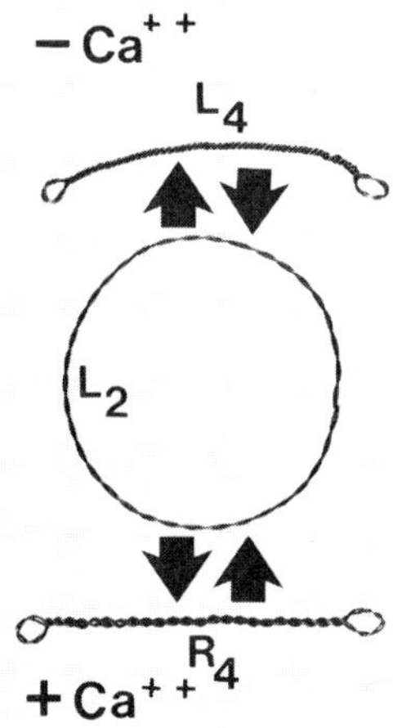

9

Fig. 28.9. Formation of a right-handed (R_4) or left-handed (L_4) helix by self-intertwining of a circularly arranged coiled-coil in dependence of Ca^{2+} concentration.

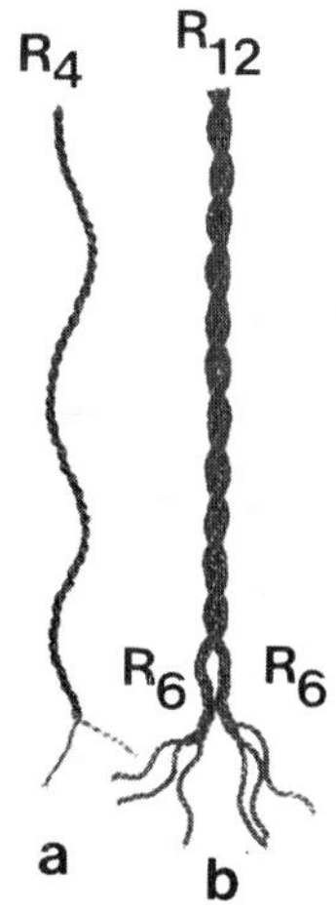

10

Fig. 28.10. (a) Model of a right-handed 3rd order super-helix (R_4) made by winding up of two left-handed coiled-coils. (b) Possible model of an "intermediate filament."

and produces cyclic contractions and relaxations (Hoffmann-Berling 1958)—an effect which disappears upon the addition of a Ca-EGTA buffer (Amos 1975). An interpretation of the kinetics of these processes and at the same time a certainly oversimplified mechanism of a Ca^{2+}-transporting ATPase is proposed in the model in Figure 28.7: addition of Ca^{2+} and $MgATP^{2-}$ causes torsion (Fig. 28.7a). The resulting self-intertwining leads to the displacement of a part of the previously attached Ca^{2+} (ion pump-effect, Ca^{2+} influx), and MgATP (written as MgADP-P, see Inesi 1981) is bound to both filaments (Fig. 28.7b). Ion displacement leads again to unwinding during which ADP-P is ruptured and the PO_3^{2-}-radical is hydrolyzed (Fig. 28.7c). The resulting charge separation where four ions (Mg^{2+}, ADP^{3-}, H^+, P_i^{2-}) arise from one $MgATP^{2-}$ complex is important for energy transformation: this local disturbance of the ionic balance could promote further torsional oscillations which will soon cease without further ATP supply. However, ATP-resynthesis may occur during winding up if the ionic conditions (no Ca^{2+} during one phase) allow the binding of orthophosphate. Figure 28.9 demonstrates the possible formation of a left- or right-handed fourfold helix according to ion removal or ion addition. The interpretation of certain membrane transports probably requires more complex winding of the membrane proteins than those in Figures 28.7 and 28.9. Data indicate that the enzymes ("gyrases") which coil and uncoil circular DNA (compare Fig. 28.9) consist of coiled-coils that function by torsional changes (see e.g., Abdel-Monem et al. 1977; Bauer et al. 1980). The "mot proteins" of bacterial cell membranes which turn the flagella (see Macnab 1983) may work similarly.

The winding of oligomers or polymers of the coiled-coil type could produce superhelices of a higher order (compare Fig. 28.10a) which are seen on electron micrographs in association with membranes (e.g. in Nagai and Hayama 1979). They may also exist freely in the cytoplasm as "microtrabeculae" (Porter and Tucker 1981). The intermediate filaments, e.g. neurofilaments (compare Fig. 28.10b) are mainly composed of coiled-coil helices of higher order (see Krishnan et al. 1979, Lazarides 1981, Franke et al. 1982, Geisler et al. 1982, McLachlan

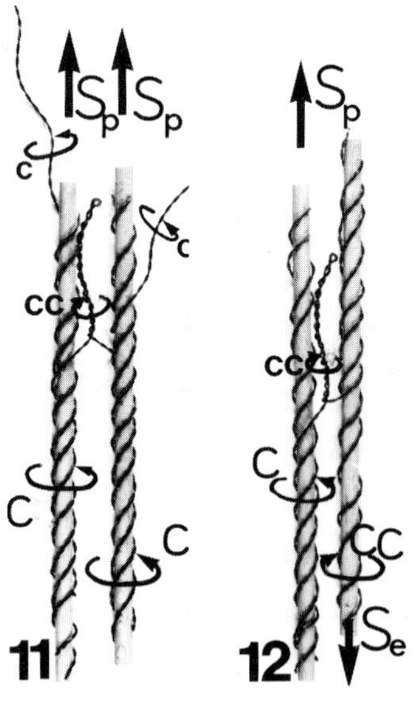

Figs. 28.11 and 28.12. Models of two neighboring microtubules during "parallel" (Fig. 28.11) and "antiparallel" (Fig. 28.12) sliding which are caused by the unwinding of coiled-coils.

and Stewart 1982). Furthermore, coiled-coils such as tropomyosin, actin-binding protein and myosin wind about actin filaments. Certain MAPs and dynein coil around microtubules. Thereby these coarse cytoskeletal elements get a distinct helical surface so that they can "slide" when they are rotated by winding and unwinding of the associated filaments. The "parallel sliding" (arrows S_p, S_p) and "antiparallel sliding" (arrows S_p, S_e) in the models of Figures 28.11 and 28.12 are shown in the state of helix unwinding. For example, the thick myosin filaments of Taenia coli smooth muscle cells may disappear during contraction (Ohashi and Nonomura 1979) because the myosin subunits are wound around the actin filaments. During one oscillation period of a plasmodium the fine myosin filaments could be likewise wound and unwound by the actin filaments (Jarosch and Foissner 1983; see also Matsumura and Hatano 1978). ATP-splitting, phosphate transfer and ion-displacements will accompany these winding motions too. However, many associated proteins are probably passively wound along the cytoskeletal elements. Their enzymatic work thus may appear energetically coupled with the ATP-hydrolysis. For further details see Jarosch and Foissner (1982, 1983) and the contributions on pp. 000 of this volume.

REFERENCES

Abdel-Monem, M., H.-F. Lauppe, J. Kartenbeck, H. Dürwald, and H. Hoffmann-Berling (1977): Enzymatic Unwinding of DNA III. Mode of Action of Escherichia coli DNA unwinding Enzyme. J. Mol. Biol. 110, 667-85.

Amos, W. B. (1975): Contraction and Calcium Binding in the Vorticellid Ciliates. In: Molecules and Cell Movement (eds. Sh. Inoué, R. E. Stephens) 411-36, Raven Press, New York.

Bauer, W. R., F. H. C. Crick, and J. H. White (1980): Überspiralige Formen der Erbsubstanz (der DNS). Spectrum der Wissenschaft, Sept. 1980, 25-36 (Sci. Amer. July 2980).

Franke, W. W., D. L. Schiller, and Ch. Grund (1982): Protofilaments and Annular Structures as Intermediates during Reconstitution of Cytokeratin Filaments in vitro. Biol. Cell 46, 257-68.

Fraser, R. D. B. and T. P. MacRae (1973): Conformation in Fibrous Proteins. Academic Press, New York, London.

Geisler, N., E. Kaufmann, and K. Weber (1982): Proteinchemical Characterization of Three Structurally Distinct Domains along the Protofilament Unit of Desmin 10 nm filaments. Cell 30, 277-86.

Hoffmann-Berling, H. (1958): Der Mechanismus eines neuen, von der Muskelkontraktion verschiedenen Kontraktionszyklus. Biochim. Biophys. Acta 27, 247-58.

Inesi, G. (1981): The Sarcoplasmic Reticulum of Skeletal and Cardiac Muscle. In: Cell and Muscle Motility, Vol. 1, 63-97.

Jarosch, R. (1964): Screw-mechanical basis of protoplasmic movement. In: Primitive motile systems in Cell Biology (eds. R. D. Allen, N. Kamiya), 599-622, Academic Press, New York, London.

Jarosch, R. (1972): The participation of rotating fibrils in biological movements. Acta Protozoologica 11, 23-36.

Jarosch, R. (1979): The Torsional Movement of Tropomyosin and the Molecular Mechanism of the Thin Filament Motion. In: Cell Motility: Molecules and Organization (eds. S. Hatano, H. Ishikawa, H. Sato), 291-319, University of Tokyo Press.

Jarosch, R. and I. Foissner (1982): A rotation model for microtubule and filament sliding. Eur. J. Cell Biol. 26, 295-302.

Jarosch, R. and I. Foissner (1983): The rotation model for filament sliding as applied to the cytoplasmic streaming. In: The application of laser light scattering to the study of biological motion (eds. J. C. Earnshaw and M. W. Steer), 545-58, Plenum Publ. Corp.

Krishnan, N., I. R. Kaiserman-Abramof, and R. J. Lasek (1979): Helical substructure of neurofilaments isolated from Myxicola and squid giant axons. J. Cell Biol. 82, 323-35.

Kuntz, I. D., G. M. Crippen, P. A. Kollman, and D. Kimelman (1976): Calculation of Protein Tertiary Structure. J. Mol. Biol. 106, 983-94.

Lazarides, E. (1981): Intermediate Filaments—Chemical Heterogeneity in Differentiation. Cell 23, 649-50.

Matsumura, F. and S. Hatano (1978): Reversible superprecipitation and bundle formation of plasmodium actomyosin. Biochim. Biophys. Acta 533, 511-23.

McLachlan, A. D. and M. Stewart (1975): Tropomyosin Coiled-coil interactions: evidence for an unstaggered structure. J. Mol. Biol. 98, 293-304.

McLachlan, A. D. and M. Stewart (1982): Periodic Charge Distribution in the Intermediate Filament Proteins Desmin and Vimentin. J. Mol. Biol. 162, 693-98.

Macnab, R. M. (1983): An entropy-driven engine—the bacterial flagellar motor. In: Biological structures and coupled flows (eds. A. Oplatka and M. Balaban), 147-60. Academic Press, Balaban Publishers.

Nagai, R. and T. Hayama (1979): Ultrastructure of the endoplasmic factor responsible for cytoplasmic streaming in Chara internodial cells. J. Cell Sci. 36, 121-36.

Ohashi, M. and Y. Nonomura (1979): Disappearance of smooth muscle thick filaments during K^+ contraction. Cell Struct. Function 4, 325-29.

Pauling, L. and R. B. Corey (1953): Compound helical configurations of polypeptide chains: Structure of proteins of the α-keratin type. Nature (London) 171, 59.

Phillips, Jr., G. N., J. P. Fillers, and C. Cohen (1980): Motions of Tropomyosin. Biophys. J. 1980, 485-502.

Porter, K. R. and J. B. Tucker (1981): The Ground Substance of the Living Cell. Scient. Amer. 244, 57-67.

Salisbury, J. L. (1983): Contractile flagellar roots: The role of Calcium. J. Submicrosc. Cytol. 15, 105-110.

Stone, D., J. Sodek, P. Johnson, and L. B. Smillie (1975): Proc. IX FEBS Meeting (Budapest) 31, 125-36.

Weis-Fogh, T. and W. B. Amos (1972): Evidence for a new mechanism of cell motility. Nature (London) 236, 301-04.

29 Microtubules and Growth of Myocytes in Rat Heart

L. Rappaport, J. L. Samuel,
B. Bertier, F. Marotte, K. Schwartz

INTRODUCTION

In heart muscle, microtubules have been clearly visualized (Hatt et al. 1970; Ferrans and Roberts 1973; Goldstein and Entman 1979; Samuel et al. 1983), but to date their role is poorly defined. In skeletal muscle, they are believed to be involved in cell growth (Cartwright and Goldstein 1982) and in the assembly of sarcomeres during myogenesis (Toyama et al. 1982).

In heart muscle, the contractile efficiency varies depending on the physiological state of the animal. In rat, hypothyroidism and chronic pressure overload induce, respectively, an atrophy and an hypertrophy of myocytes. Both induce a preferential synthesis of the V3 isoform of myosin (see for review Schwartz et al. 1982). The functional implications of these events are important since the size of myocytes reflects, more or less, the number of contractile units and since V3 isomyosin has an ATPase activity that is several times less than the V1 isoform (Hoh et al. 1978; Lompre et al. 1981; Litten et al. 1982). Moreover, the maximal speed of shortening of cardiac muscle was recently shown to be correlated to the myosin isozymic composition (Schwartz et al. 1981). Thus, isomyosins appeared to us to be good markers of the physiological and contractile states of myocardium. Using immunofluorescence labeling of tubulin and isomyosins in isolated cells, we looked for the changes in the microtubular pattern within cardiocytes from rats in different physiological states.

METHODS

Details of many of the models and methods utilized have been provided elsewhere (Samuel et al. 1983a and b). Most of the experiments were carried out on hearts from 3-week old rats. Growth of heart muscle was stimulated by pressure overload induced by reducing the size of the aorta (Bugaisky and Zak 1979). Atrophy, secondary to thyroxine deficiency, was induced by the administration of an iodine free diet containing P.T.U. (propyl, thiouracil), during pregnancy and nursing.

Myocytes were isolated by a modified Langendorff's method, purified on Ficoll, and labeled by an immunofluorescence technique with specific antibodies raised against each of the two isomyosins, V1 and V3 (Schwartz et al. 1982), or against brain tubulin (Samuel et al. 1983a). Antibodies were affinity purified and their specificity tested. Before labeling, myocytes were permeabilized with 1% Triton-X-100 in a 10mM EGTA buffer and were fixed with 3.7% formaldehyde.

RESULTS

Heart myocytes were isolated and purified from rats which were either normal, thyroxine deficient or with aortic stenosis. Myocytes were examined for their content of isomyosins and in their microtubule pattern by immunofluorescence microscopy.

Labeling of Myocytes with Antimyosins

Antibodies that specifically distinguish (V1 and V3) isomyosins in situ in myocytes permitted the discrimination of myosin isozymes in single myocyte (Samuel et al. 1983). In a population of adult cardiac myocytes, some cells were shown to be labeled with either one of the two antibodies, a-V1 or a-V3 myosins, or with both. Within the cell, the fluorescence was uniformly distributed and showed the typical striation pattern of myofibrils and sarcomeric structures. Fibers reacting against both antibodies exhibited the same overall morphology and periodicity (Fig. 29.1).

The labeling of a population of heart myocytes strongly depends on the physiological state of the rat (Table 29.1). When rats were 1 to 3-weeks old and thyroxine-deficient, all the cardiac myocytes reacted only with the a-V3 myosin immunoglobulins. These results enhance those from Hoh et al. (1978), who showed by an electrophoretic approach that thyroxine deficiency induces an accumulation of the V3 isomyosin in heart. In agreement with previous electrophoretic analysis (Lompre et al. 1981), we found that the whole myocyte population of

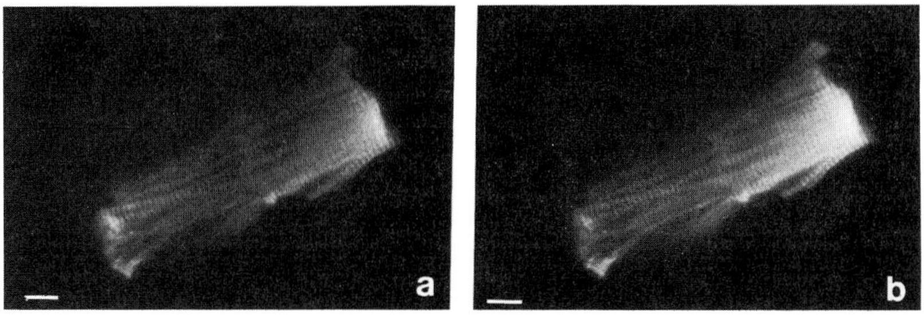

Fig. 29.1. Double indirect immunofluorescence micrograph of a myocyte isolated from a 8-week old rat heart. Incubation was performed with anti-V1 myosin labeled with rhodamin conjugated antibodies (a) and with anti-V3 myosin labeled with fluorescein conjugated antibodies (b). The staining was positive with both antibodies (Bar = 10 μm).

Table 29.1. Double labeling with antimyosin immunoglobulins of cardiac myocytes isolated from rats in different physiological states

Experimental model in rat		Myocyte	Labeling of myocytes with a-myosins IgG_S
Control (3-week old)		normal	100% a-V1
T_4-deficiency (3-week old)		atrophied	100% a-V3
Pressure overload	7 days	normal	75% a-V1
			25% a-V3 + a-V1
	3 weeks	hypertrophied	100% a-V3 + a-V1

3-week old normal rats reacted only with the a-V1 myosin immunoglobulins. However, when aortic stenosis was performed in this type of rat, myocytes reacting with both a-V3 and a-V1 immunoglobulins appeared as soon as 3 days after surgery and their percentage increased rapidly with time. Meanwhile the weight of the heart muscle was strongly increased (Bugaisky and Zak 1979).

Microtubule Pattern

Microtubule pattern was observed in these 3 experimental models (Fig. 29.2).

In Normal Rats

In 3-week old rats (as in 2 to 15-233k old rats) microtubules were organized mainly around the nuclei, with concentrations at the poles and the cone and extensions in the cytoplasm as loosely organized loops (Fig. 29.2a, b). This network was highly sensitive to depolymerizing and polymerizing agents, sch as colchicine, cold, or taxol.

In Thyroxine-Deficient Rats

Growth of myocytes in thyroxine-deficient rats was clearly delayed (Fig. 29.2c). Myocytes were very small and most of them showed only one nucleus. Microtubules were mainly concentrated around the nucleus and were underdeveloped throughout the cytoplasm, where they appeared as straight rod-shaped filaments.

In Aortic-Stenotic Rats

Soon after aortic stenosis a significant population of myocytes exhibited a modified microtubule pattern (Fig. 29.2d). Microtubules appeared densified and organized in thick arrays parallel to the long axis of the cell. The size of the myocytes was not significantly modified at that time, but both an increase in the weight of the heart (not shown, and Bugaisky et al. 1979) and a shift in the isomyosin pattern, that occured concomitantly within the myocytes, afforded proofs of the adaptational processes that happen soon after induced growth. The reorganized pattern of microtubules was no longer present several weeks after surgery, when hypertrophy of myocytes was significant, and that all the cells contained V3-isomyosin.

DISCUSSION

By means of immunofluorescence, microtubules were visualized together with isomyosins in cardiocytes isolated from rats in different physiological states.

All the criteria i.e., microtubule pattern, size of cells, and isomyosin content, differ significantly depending on the

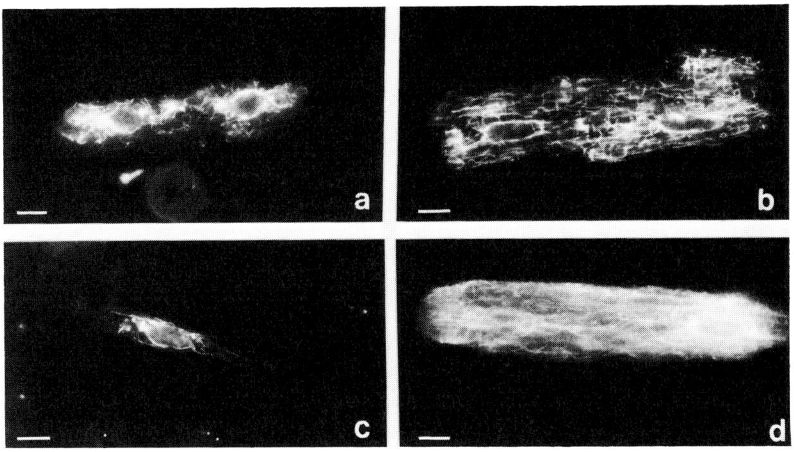

Fig. 29.2. Fluorescence micrographs of microtubules in cardiac myocytes isolated from rats in different physiological states. (a) and (b) are myocytes from normal 2 and 4-week old rats. (c) is a myocyte for a 2-week old, thyroxine-deficient rat. (d) is a myocyte from a 4-week old rat, heart being excised 7 days after aortic stenosis.

origin of myocytes. However, no relationship could be visualized between them: The same pattern of microtubules was observed in the two types of myocytes differing in both their size and isomyosins content (3-week old rats and rats several weeks after aortic stenosis). Atrophied myocytes (purified from thyroxine-deficient rat hearts) as well as hypertrophied myocytes, contained V3-myosin but in contrast, the former showed poorly developed microtubules. The atrophied cells had, in fact, the appearance of myocytes from very young rats (7 to 10 days old) which contain V1 myosin. Injection of thyroxine quickly stimulated the growth of myocytes and induced both a shift in the type of neosynthesized isomyosin and in the organization of microtubules to a "normal" pattern. We conclude that the microtubule pattern does not depend on the size of the myocytes or of their isomyosin content.

Thick arrays of microtubules organized parallel to the axis of myocytes were, however, seen soon after growth was experimentally induced. Such a reorganization may be involved in the adaptational processes occuring in myocytes during the onset of induced growth. This result is in full agreement with studies from Cartwright and Goldstein (1982) indicating, in skeletal muscle, an increase in the number of microtubules during the period of intensive growth. On the other hand, a reorganization of the microtubules in thick arrays was described in myocytes cultured in the presence of c-AMP and was associated with a decrease in contractility of cells (Nath et al. 1978). In our model, of hypertrophied heart, the decrease in contractility was clearly shown to be mainly secondary to a neosynthesis of V3-myosin. It may be hypothesized, however, that microtubules could in some conditions constitute an additional factor in the regulation of contractility in heart.

We conclude that the differences in the organization of microtubule between normal and thyroxine-deficient myocytes essentially reflects some delay in the development of the cells. No relationship between the microtubules and isomyosins patterns was identified in different types of myocytes. Microtubules could be implicated in the early cellular events occuring during acute growth of heart.

ACKNOWLEDGMENTS

Authors are grateful to Dr. Sartore and Dr. Schiaffino (Padova) for the gift of anti-V3 myosin antibodies, and to Dr. Bugaisky for providing young rats with aortic stenosis.

REFERENCES

Bugaisky L.B. and Zak R. Cellular growth of cardiac muscle after birth. Tex. Rep. Biol. Med. 39 123-38 (1979).

Cartwright J. Jr. and Goldstein M. A. Microtubules in soleus muscles of the post natal and adult rat. J. Ultrastr. Res. 79:74-84 (1982).

Ferrans V.J. and Roberts S.C. Intermyofibrillar nuclear myofibrillar connections in human and canine myocardium. An ultrastructural study. J. Mol. Cell. Cardiol. 5:247-58 (1973).

Goldstein M.A. and Entman M.L. Microtubules in mammalian heart muscle. J. Cell. Biol. 80:183-95 (1979).

Hatt P.Y., Berjal G., Moravec J. and Swynghedauw B. Heart failure. An electron microscopy study of the left ventricular papillary muscle in aortic insufficiency in the rabbit. J. Mol. Cell. Cardiol. 1:235-47 (1970).

Hoh J.H., Mc Grath P.A., Hale P.T. Electrophoretic analysis of multiple forms of rat cardiac myosin. Effect of hypophysectomy and thyroxine replacement. J. Mol. Cell. Cardiol. 10:1053-76 (1978).

Litten R.Z., Martin B.J., Low R.B., Alpert N.R. Altered myosin isozyme patterns from pressure-overloaded and thyroxic hypertrophied rabbit hearts. Circ. Res. 50: 856-64 (1982).

Lompré A.M., Mercadier J.J., Wisnewsky C., Bouveret P., Pantaloni C., d'Albis A., Schwartz K. Species and age dependent changes in the relative amount of cardiac myosin isoenzymes in mammals. Develop. Biol. 84:286-90 (1981).

Nath K., Shay J.W. and Bollen A.P. Relationship between dibutyryl cyclic AMP and microtubule organization in contracting heart muscle cells. Proc. Natl. Acad. Sci. USA 75:319-23 (1978).

Samuel J.L., Schwartz K., Lompré A.M., Delcayre C., Marotte F., Swynghedauw B. and Rappaport L. Immunological quantitation and localization of tubulin in adult rat heart isolated myocytes. Eur. J. Cell. Biol. 31:99-106 (1983a).

Samuel J.L., Rappaport L., Mercadier J.J., Lompré A.M., Sartore S., Triban C., Schiaffino S. and Schwartz K. Distribution of myosin isozymes within single cardiac cells. Circ. Res. 52:200-209 (1983b).

Schwartz K., Lecarpentier Y., Martin J.L., Lompré A.M., Mercadier J.J., Swynghedauw B. Myosin isoenzymic distribution correlates with speed of myocardial contraction. J. Mol. Cell. Cardiol. 13:1071-75 (1981).

Schwartz K., Lompré A.M., Bouveret P., Wisnewsky C., Whalen R.G. Comparisons of rat cardiac myosins at fetal stages in young animals and in hypothyroid adults. J. Biol. Chem. 257:14412-18 (1982).

Toyama Y., Forry-Schaudies S., Hoffman B. and Holtzer H. Effects of taxol and colcemid on myofibrillogenesis. 79: 6556-60 (1982).

30 Cell Physiological and Morphological Analyses on Axopodial Contraction in Heliozoa

Yoshinobu Shigenaka and Tatsuomi Matsuoka

INTRODUCTION

As implicated by its genus name, a large heliozoan Echinosphaerium is characterized by hundreds of needlelike cell extensions called axopodia. Every axopodium is well known to be maintained by an axoneme which is composed of a double-helical or dodecagonal array of microtubules (Roth et al., 1970). Once certain external stimuli are applied to the cell, on the other hand, the axonemal microtubules are apt to be disintegrated easily although the reaction is reversible (Roth and Shigenaka, 1970; Shigenaka et al., 1971). At this point, it should be noticed that these features are compatible with each other, because the axopodia should play various fundamental cell functions such as cell locomotion (Suzaki et al., 1978), cell fusion (Shigenaka et al., 1979), food intake (Suzaki et al., 1980) and so on.

During the process of our serial investigations, moreover, the axopodia were found to contract quite rapidly (Suzaki et al., 1980; Shigenaka et al., 1982). However, the mechanism of this phenomenon still remained unknown. The present study aimed to study the axopodial contraction especially from cell physiological and ultrastructural viewpoints so that the mechanism might be clarified.

AXOPODIAL CONTRACTION AND ITS INDUCERS

The axopodial contraction was usually observed to occur instantaneously when the prey attached to the tip of an axopodium (Suzaki et al., 1980). In this case, the axopodia always contracted to less than one-third of their initial length, which drew the prey toward the body surface. Complete contraction of axopodia was observed only when the axopodia had previously elongated to their full length. When recorded with a 16 mm cine camera, the velocity of axopodial contraction was measured to be more than 5 mm/sec (Shigenaka et al., 1982).

In addition to the prey organisms, the axopodial contraction was induced with various factors such as carbon particles, carmine particles coated with albumen, anion exchange resin particles, poly-L-lysine coated glass particles, and electrical stimulation (1.0 V/5 mm, 1 sec, rectangular pulse) as reported by Shigenaka et al. (1982). Furthermore, cation exchange resin particles and glass or metal needles never resulted in such contraction. These results suggest that the axopodial contraction might be triggered in collaboration with activation of chemoreceptors and/or surface negative charge on the axopodial surface membranes.

CELL PHYSIOLOGICAL ANALYSIS ON AXOPODIAL CONTRACTION

Using the electrical stimulation, the axopodial contraction has been suggested to occur with Ca^{2+} influx through the axopodial membrane, depending on the extracellular Ca^{2+} concentrations (Shigenaka et al., 1982). In the present study, some further experiments were carried out using active carbon particles as inducers of contraction, especially to ascertain whether only the Ca^{2+} influx induced the axopodial contraction. As the results, it was found that the extracellular Ca^{2+} ions might be necessarily required to induce the axopodial contraction, because the contraction occurred more conspicuously at the higher concentrations (10^{-3} to 10^{-4} M Ca^{2+}) as compared with the lower ones (10^{-6} to 10^{-7} M Ca^{2+} in the presence of 1 mM EGTA). Moreover, the 10^{-4} to 10^{-5} M La^{3+} ions were found to inhibit the axopodial contraction not completely but to some extent. These results suggest that the intracellular Ca^{2+} ions might be also required to induce the axopodial contraction in addition to the extracellular ones.

Shigenaka et al. (1982) have suggested that the tubular X-body might be transformed into the granular one by the Ca^{2+} influx through the axopodial membranes, and thereby, that the X-body is indispensable for the rapid contraction and the folding of axopodia. As to the first case, the present study showed that the intracellular Ca^{2+} ions are also effective on the axopodial contraction. Consequently, the second case was re-examined using a specific dynein ATPase inhibitor, etythro-9-(2-hydroxy-3-nonyl)adenine called EHNA. At the concentrations varying 2×10^{-4} to 4×10^{-4} M, the EHNA was found to inhibit the axopodial contraction to some extent, showing that the faint contraction occurred continuously and the carbon particles attached to the axopodial surface were conveyed very slowly to the cell body surface. These results suggest that the dynein ATPase activity might also be concerned with the axopodial contraction.

Furthermore, the cytochalasin B experiments revealed that the reagent inhibited the axopodial contraction considerably at the concentrations of 2×10^{-4} to 8×10^{-5} M, although the solvent dimethyl sulfoxide (DMSO) did not inhibit the contraction even at 4%. This result suggests that the contraction also occurs in collaboration with actin or actin-like proteins.

LIGHT AND ELECTRON MICROSCOPICAL OBSERVATIONS

When the active carbon particles were added, the heliozoan cell captures them so rapidly by the axopodial contraction (Fig. 30.1a) and convey them toward the cell body surface (Fig. 30.1b).

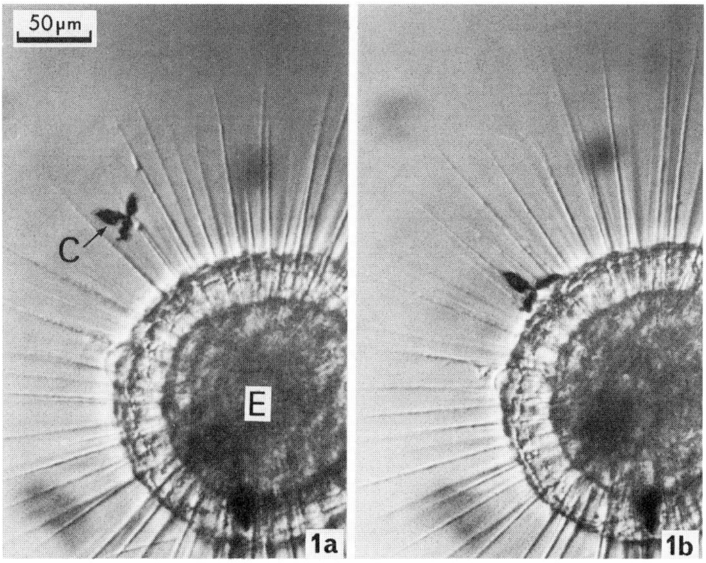

Fig. 30.1 a and b.

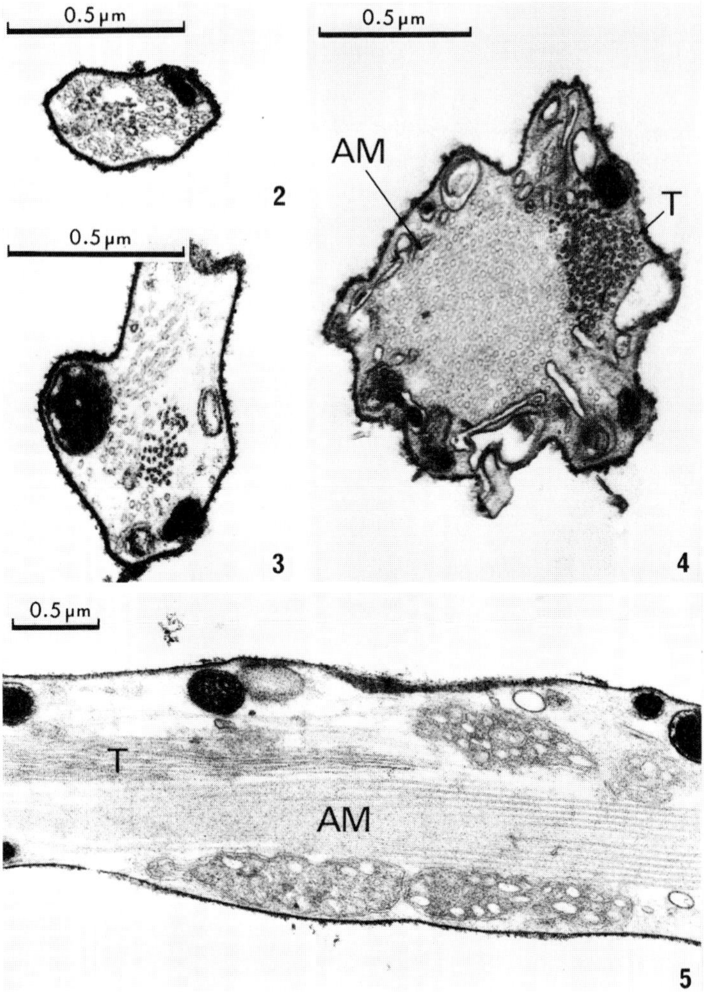

Figs. 30.2, 30.3, 30.4 and 30.5.

To know the entities inducing such contraction, the ultrathin sections were examined carefully especially in axopodia. As illustrated in Figures 30.2 to 30.5, the control organisms showed that the tubular X-body was unevenly distributed in parallel with axonemal microtubules but running through the tip to the base of axopodium.

Once the active carbon particle was captured by an axopodium, however, the axopodium contracted quite rapidly by the granulation of X-bodies and the simultaneously occurred destruction of axonemal microtubules (Figs. 30.6 and 30.7). Therefore, the cytochalasin B-treated cells (0.2 mM, 2 hrs) were examined electron microscopically, revealing that the tubular X-body was transformed to the granular one to make some coagulations toward the axopodial base (Figs. 30.8 and 30.9) and also to produce the massive group just under the cell surface, that is, in the ectoplasm. In this case, the axopodia still stood as they were. These results suggest that the X-bodies might contain actin or actin-like proteins as the skeletal elements of them.

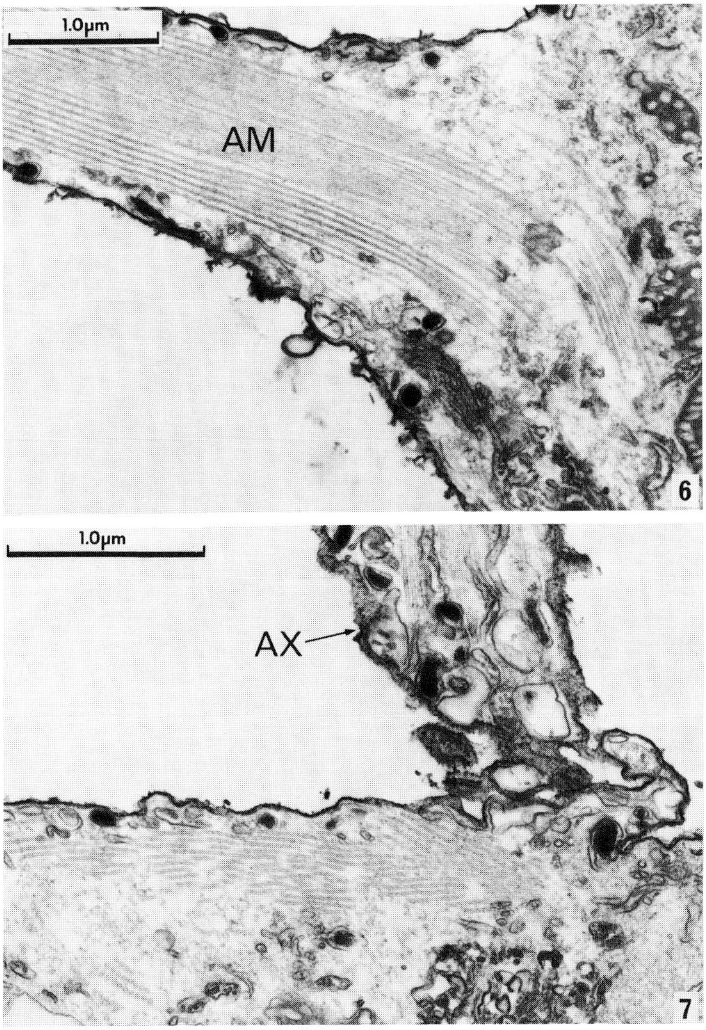

Figs. 30.6 and 30.7

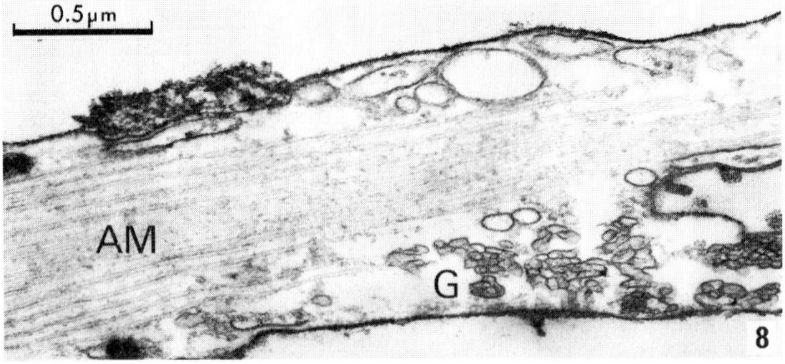

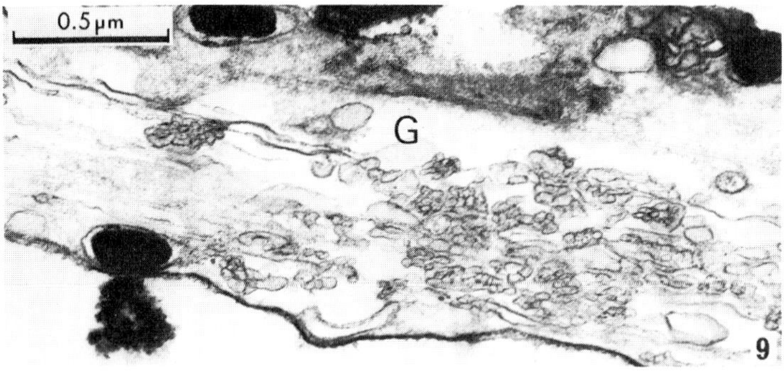

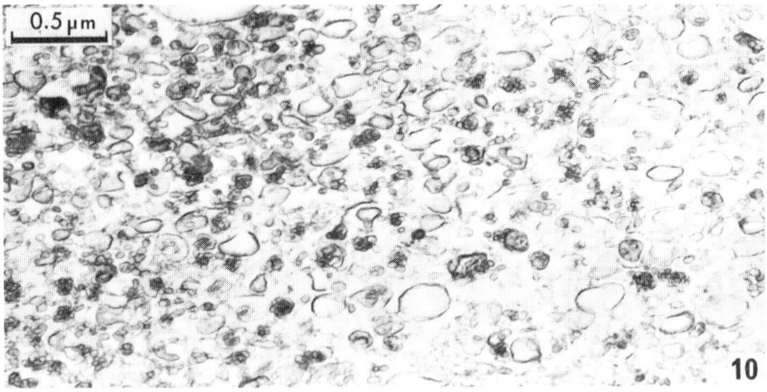

Figs. 30.8, 30.9 and 30.10.

REFERENCES

Roth, L.E. and Shigenaka, Y. (1970). J. Ultrastruct. Res., 31:356-74.

Roth, L.E., Pihlaja, D.J. and Shigenaka, Y. (1970). J. Ultrastruct. Res., 30:7-37.

Shigenaka, Y., Roth, L.E. and Pihlaja, D.J. (1971). J. Cell Sci., 8:127-52.

Shigenaka, Y., Watanabe, K. and Suzaki, T. (1980). Annot. Zool. Japon., 53:103-19.

Shigenaka, Y., Toyohara, A., Suzaki, T. and Watanabe, S. (1979). Cell Struct. Funct., 4:199-207.

Shigenaka, Y., Yano, K., Yogosawa, R. and Suzaki, T. (1982). In: Biological Functions of Microtubules and Related Structures (H. Sakai, H. Mohri and G. G. Borisy, eds.), pp. 105-14. Academic Press, Tokyo.

Suzaki, T., Shigenaka, Y. and Takeda, Y. (1978). Cell Struct. Funct., 3:209-18.

Suzaki, T., Shigenaka, Y., Watanabe, S. and Toyohara, A. (1980). J. Cell Sci., 42:61-79.

31 Approaches to Determine the Function of Nonerythrocyte Spectrin

Paul Mangeat and Keith Burridge

During the past two years a number of reports have documented the presence of proteins highly related to red blood cell spectrin in nonerythroid cells (Goodman et al., 1981; Glenney et al., 1982a,b; Repasky et al., 1982; Bennett et al., 1982; Burridge et al., 1982; Kakiuchi et al., 1982). Since these proteins have been discovered at the same time in different laboratories, they have been named differently; fodrin (Levine and Willard, 1981), DBPI for calmodulin binding protein (Davies and Klee, 1981), calspectin (Kakiuchi et al., 1982) and brain spectrin (Bennett et al., 1982; Burridge et al., 1982) all refer to the same high molecular weight protein doublet present in brain tissue; a similar molecule is present in most nonerythroid cells. TW 260/240 refers to a specific spectrin-like protein uniquely distributed in the terminal web of the intestinal epithelium (Glenney et al., 1982a). As with red blood cell spectrin, these proteins are made up of two nonidentical high molecular weight subunits behaving in native form as a tetramer $\alpha_2 \beta_2$. They are calmodulin binding proteins and bind F-actin (see all above references).

Although the structural and immunological similarities are now well established between the nonerythrocyte spectrins and the red blood cell protein, the function of nonerythrocyte spectrin has not yet been determined. Initial efforts in this direction showed that in addition to the above quoted properties, nonerythrocyte spectrin could bind to inverted red blood cell membranes depleted of spectrin and actin and could complete in this binding with erythrocyte spectrin (Bennett et al., 1982; Burridge et al., 1982). That this binding could be specifically inhibited by ankyrin antibodies (Burridge et al., 1982) suggested that the nonerythrocyte molecule possesses a preserved ankyrin binding site. In turn, this suggests that ankyrin related molecules might exist in nonerythroid tissues and that these might anchor the spectrin molecule to the cell membrane.

To test this latter hypothesis, one experimental approach was designed in our laboratory which consisted of developing an homogeneous nonerythroid model in which one could perform binding experiments similar to those that have led to our present knowledge of how erythrocyte spectrin interacts with the erythrocyte plasma membrane. Using a cultured cell line (Hela cells) we have been able to demonstrate specific binding between Hela spectrin and a defined Hela membrane fraction (Mangeat and Burridge, 1983). Briefly, we have purified Hela spectrin (Fig. 31.1) starting from 30-40 g of Hela S3 cells grown in suspension using an identical protocol developed for brain spectrin (Burridge et al., 1982). In parallel, the crude spectrin and actin depleted membranes were purified by discontinuous sucrose gradient sedimentation and three different membrane fractions were characterized. Of these, only the denser one demonstrated reassociation properties when incubated with Hela spectrin (Fig. 31.1). Specific binding was demonstrated by the failure of three other purified Hela proteins to reassociate with this membrane fraction. Two of these proteins (different forms of Hela filamin) are potent actin binding proteins, ruling out that the observed spectrin-membrane reassociation was only due to traces of actin still present in the membrane fraction. The nature of the membrane binding site remains unknown, but so far does not seem to be an ankyrin-related protein since no immunologically cross-reactive protein has been identified. Furthermore, human erythrocyte spectrin failed to bind to this Hela membrane fraction.

A second experimental approach aimed at understanding the function of nonerythrocyte spectrin has consisted of looking at what happens to a cell defective in spectrin. One way to obtain such a cell is to develop clones that have a mutation in spectrin. Another way is to create cells temporarily lacking spectrin by microinjection of specific antibodies against spectrin into live cells such that the spectrin is precipitated by the antibodies and cleared from its normal cellular location. This second approach has been used (Mangeat and Burridge, submitted for publication) and should provide a valuable tool for probing the function of spectrin in live cells. In short, nonerythrocyte spectrin has been quantitatively and selectively precipitated within living fibroblasts or epithelial cells such as Hela or MDBK cells (Fig. 31.2). This intracellular aggregation of spectrin was achieved by microinjection of either a monoclonal antibody (IgM) directed against the α-chain of

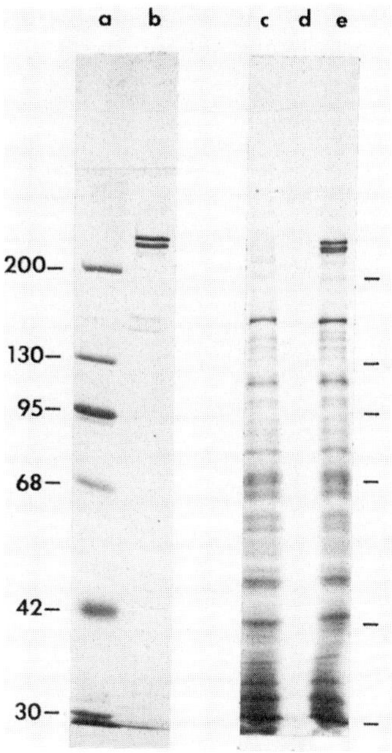

Fig. 31.1. Binding of Hela spectrin to a specific Hela membrane fraction. Analysis by electrophoresis on a 10% polyacrylamide gel in presence of sodium dodecyl sulfate. lane a: molecular weight markers (expressed on kilodaltons on the left); lane b: purified Hela spectrin; lane c: specific Hela membrane fraction pelleted through a sucrose cushion after incubation without any Hela spectrin added; lane d: Hela spectrin incubated in absence of membrane does not sediment through a sucrose cushion; lane e: Hela spectrin cosediments through a sucrose cushion after incubation with the Hela specific membrane fraction shown in lane c.

nonerythrocyte spectrin or an affinity-purified polyclonal antiserum raised against the bovine brain protein. Comparing the effect of the two types of antibody, the affinity-purified polyclonal antibody resulted in more compact aggregates of spectrin and these aligned more prominently with the microfilament bundles. These structures appeared to be unaffected by the aggregation of spectrin. We conclude that the integrity of the actin microfilament bundles does not require nonerythrocyte spectrin and that most probably these structures are linked at their termini to the membrane through proteins other than nonerythrocyte spectrin, e.g., vinculin (Geiger, 1979) or talin (Burridge and Connell, 1983). No effect of the intracellular spectrin precipitation was observed on the distribution of coated vesicles or microtubules. The aggregation of the nonerythrocyte spectrin, however, did affect the distribution of the vimentin type of intermediate filaments in most of the injected cell types. In fibroblasts, these became more distorted and condensed but did not collapse around the nucleus as occurs following microtubule disruption induced by colchicine treatment. The effect was much more dramatic in epithelial cells like Hela cells (Fig. 31.2). The clumped intermediate filaments were frequently seen to coincide with regions of aggregated spectrin. The distortion of intermediate filaments was specific for microinjection of antibodies directed against

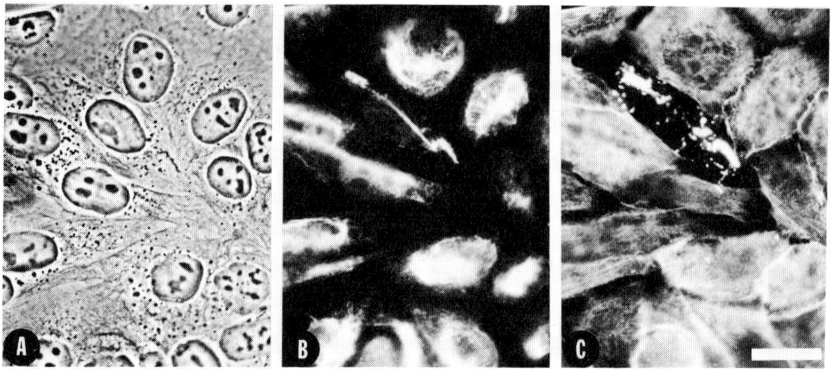

Fig. 31.2. Immunoprecipitation of Hela spectrin in Hela cells following microinjection of affinity-purified rabbit antipig brain spectrin. Cells were fixed and permeabilized 3 h after microinjection. They were then stained with antivimentin type intermediate filaments antibody (monoclonal JLB 7) and with rabbit antibrain spectrin antibody followed by rhodamine-labeled goat antimouse IgG and fluorescein-labelled goat antirabbit IgG. A: phase contrast micrograph; B: vimentin-type intermediate filament distribution; C: Hela spectrin distribution. Note that the microinjected cell is easily recognizable by the presence of the bright precipitates in contrast to the more diffuse distribution of spectrin in the surrounding uninjected cells. The post fixation staining with antibrain spectrin antibody allows not only the visualization of the normal spectrin distribution in uninjected cells, but also demonstrates that in the injected one all the detectable spectrin has been precipitated and concentrated into a few compact aggregates. Note in the injected cell the drastic change in intermediate filament distribution (B) which now lines up with the precipitated Hela spectrin-antibody complex, which even can be detected by phase contrast (A).

spectrin and no distortion was seen following microinjection of an irrelevant antibody (rhodamine-labeled goat antimouse IgG), or of rhodamine-labeled vinculin and, more convincing, after microinjection of the monoclonal anti nonerythrocyte spectrin antibody in a gerbil fibroma cell line where the antibody was unreactive. We concluded that there is some association (either direct or indirect) between intermediate filaments and nonerythrocyte spectrin and that possibly the nonerythrocyte spectrin is involved in linking intermediate filaments to the plasma membrane. Alternatively, intermediate filaments might function in the clearing of immune antigen-antibody complexes. If this is so, a similar phenomenon should be observed after microinjection of other antibodies against cellular proteins provided they also induce a quantitative precipitation of the antigen.

So far we have not detected any effect of spectrin precipitation on cell shape, cell motility, or cell division. In collaboration with Dr. K. Jacobson, we have not found spectrin precipitation to affect the lateral mobility of several surface antigens in different cell types, nor have we found an effect on the antibody-induced aggregation of β-2 microglobulin on the surface of live cells or on the capacity of macrophages to phagocytose latex beads.

ACKNOWLEDGMENTS

The authors thank Dr. J. Lin (Cold Spring Harbor Laboratory) for his generous gift of monoclonal antibody JLB7. This work was supported by a grant from NIH (GM 29860) and from the Muscular Dystrophy Association.

REFERENCES

Bennett, V., Davis, J., and Fowler, W. E. Brain spectrin, a membrane associated protein related in structure and function to erythrocyte spectrin. Nature (Lond.) 299, 126-31 (1982).

Burridge, K., and Connell, L. A new protein of adhesion plaques and ruffling membranes. J. Cell. Biol. 97, 359-67 (1983).

Burridge, K., Kelly, T., and Mangeat, P. Non-erythrocyte spectrins: actin-membrane attachment proteins occuring in many cell types. J. Cell Biol. 95, 478-86 (1982).

Davies, P.J.A., and Klee, C.B. Calmodulin-binding proteins: a high molecular weight calmodulin-binding proteins from bovine brain. Biochemistry International 3, 203-12 (1981).

Geiger, B. A 130K protein from chicken gizzards; its localization at the termini of microfilament bundles in cultured chicken cells. Cell 18, 193-205 (1979).

Glenney, J.R., Glenney, P., Osborn, M., and Weber, K. An F-actin and calmodulin-binding protein from isolated intestinal brush borders has a morphology related to spectrin. Cell 28, 843-54 (1982a).

Glenney, J.R., Glenney, P., and Weber, K. Erythroid spectrin, brain fodrin, and intestinal brush border proteins (TW-260/240) are related molecules containing a common calmodulin-binding subunit bound to a variant cell type-specific subunit. Proc. Natl. Acad. Sci. USA 79, 4002-05 (1982b).

Goodman, S.R., Zagon, I.S., and Kulikowski, R.R. Identification of a spectrin-like protein in nonerythroid cells. Proc. Natl. Acad. Sci. USA 78, 7570-74 (1981).

Kakiuchi, S., Sobue, K., Kanda, K., Morimoto, K., Tsukita, S., Tasukita, S., Ishikawa, H., and Kurokawa, M. Correlative biochemical and morphological studies of brain calspectrin: a spectrin-like calmodulin-binding protein. Biomedical Research 3, 400-10 (1982).

Levine, J., and Willard, M. Fodrin: axonally transported polypeptides associated with the internal periphery of many cells. J. Cell. Biol. 90, 631-43 (1981).

Mangeat, P., and Burridge, K. Binding of Hela spectrin to a specific Hela membrane fraction. Cell Motility. In press (1983).

Repasky, E.G., Granger, B.L., and Lazarides, E. Widespread occurence of avian spectrin in non-erythroid cells. Cell 29, 821-33 (1982).

32 Role of Polyamines in Cell Division of Amphibian Eggs

C. Oriol-Audit, N. J. Grant, C. Aimar

INTRODUCTION

The polyamines spermine and spermidine have been implicated in cytokinesis of many diverse types of cells. The polyamines may act by causing an alignment of the cortical actin filaments which form the contractile ring (Oriol-Audit, 1980). This hypothesis is supported by experiments demonstrating that the polyamines induce polymerization of actin in vitro (Oriol-Audit, 1978). In Amoeba proteus, microinjection of spermine promotes formation of cleavage furrows containing aligned actin and myosin filaments (Gawlitta et al., 1981). As well, inhibition of polyamine biosynthesis in cultured mammalian cells causes a decrease in the intracellular polyamine levels and inhibits cytokinesis (Sunkara et al., 1979).

The work presented here demonstrates that spermine and spermidine can induce early furrowing in Xenopus laevis eggs; moreover, spermine induces polymerization of egg actin into bundles.

MATERIALS AND METHODS

Microinjection of Polyamines into Xenopus Eggs

Unfertilized eggs of X. laevis were collected from females injected with 300 I.U. chorionic gonadotrophin. Jelly coats were removed with 1% L-cysteine in Steinberg's solution (58 mM NaCl, 0.67 mM KCl, 0.34 mM $CaCl_2$, 0.83 mM $MgSO_4$, 5 mM Tris) adjusted to pH 8.0. Dejellied eggs were washed and observed in Steinberg's solution at pH 7.2-7.4. The eggs were activated by pricking with a glass needle or by microinjections. The polyamines (putrescine 2HCl, spermidine 3HCl, spermine 4HCl; Sigma Chemical Co.) were diluted to the desired concentrations from 100 mM stock solutions in 10 mM Tris pH 7.5. 100 nl of polyamine solution were pressure injected using a calibrated glass micropipette into the animal hemisphere of eggs. The % acceleration of the first division was determined by using the times for 50% the eggs in each group (about 25 eggs) to divide, and then comparing the experimental groups with prick-activated control groups.

A total of 60 eggs from four females were injected with ^{14}C-spermine solution for the localization experiments (CEA, France). Each egg received 100 nl (less than 10% egg volume) of a 20 mM ^{14}C-spermine solution (specific activity: 25 mC/mmole). Cortices were isolated manually from 20 injected eggs at the time of the first furrow formation (40 min postactivation for this experiment). The isolation medium contained 40 mM TES, 60 mM PIPES, 10 mM EGTA, 20 mM NaCl, 5 mM spermine, pH 7.5. The radioactivity of the cortex was determined. The correction for quenching was made using external ^{14}C-standard (LKB-Wallace). The volume of the cortex was assumed to represent 1% of the total egg volume.

Deproteinized extracts were prepared by boiling eggs in 15% TCA for 15 min. After neutralization and centrifugation of the extracts, polyamines were separated by cellulose-TLC using butanol-1, water, pyridine, glacial acetic acid (20:10:5:10) as the solvent. The polyamine spots were removed from the chromatograms and their radioactivity was determined.

Xenopus Actin Isolation

Dejellied eggs were lysed in a Kontex homogenizer with 2 μl/egg of a KI solution (600 mM KI, 6 mM Na thiosulfate, 10 mM Tris, 1 mM ATP, 5 mM 2-mercaptoethanol, pH 7.4). The homogenate was stirred for 10 min at 4°C, and centrifuged at 27,000 g for 10 min. The supernatant was dialyzed against a polymerization solution (100 mM KCl, 1 mM $MgCl_2$, 2 mM HEPES, 0.1 mM $CaCl_2$, 0.1 mM NaN_3, 1 mM ATP, 2 mM 2-mercaptoethanol, pH 7.0). After centrifugation at 200,000 g for 1 h, the supernatant was discarded and the pellet was homogenized and dialyzed against a depolymerization solution (2 mM Hepes, 0.1 mM $CaCl_2$, 0.1 mM NaN_3, 1 mM ATP, 0.5 mM 2-mercaptoethanol, pH 7.0). After centrifugation at 200,000 g for 1 h, the supernatant (S_2) was dialyzed against the depolymerization solution containing 100 mM KCl and centrifuged at 200,000 g, 1 h. The supernatant (S_3) containing actin free of α-actinin was filtered through an Ultrogel AcA 44 column. This fraction was then polymerized with spermine.

RESULTS AND DISCUSSION

Effects of the Polyamines on Division of Amphibian Eggs

X laevis eggs injected with spermine or spermidine developed cleavage furrows earlier than control eggs mechanically activated either by pricking or by injection of buffer solutions. Estimated intracellular concentrations of 0.5 mM spermine or 1.5 mM spermidine accelerated the time of the first division by 60% (Fig. 32.1). However, putrescien which is the precursor of spermine and spermidine biosynthesis had no effect even at concentrations as high as 5 mM. Although these polyamines have been identified in X. laevis eggs (Russell, 1971), changes in their concentrations during the division cycle have not been examined. In sea urchin eggs and in dipteran eggs, increased levels of spermidine or spermine have been correlated with cleavage (Kusunoki & Yasumasu, 1976; Lundquist et al., 1983).

When ^{14}C-spermine was injected into X. laevis eggs, a fraction of the injected spermine was found in the egg cortex

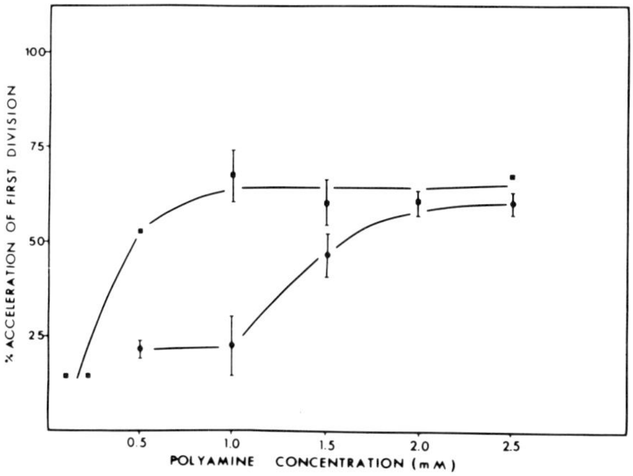

Fig. 32.1. Effects of microinjected spermine (□) or spermidine (○) on the first division cycle of unfertilized Xenopus eggs. The abscissa is the estimated intracellular concentration of the injected polyamine. Error bars represent S.E.M. (n = 4).

and the concentration was estimated to be about 3 mM (range 1 to 6 mM). The variation in cortical spermine content may arise from a rapid exchange with the unlabelled spermine in the isolation medium during the preparation of the cortex. If the polyamines bind to the actin localized in the egg cortex (Franke et al., 1976), 3 mM spermine would be enough to induce supramolecular forms of actin (Oriol-Audit, 1982; Grant et al., 1983).

In contrast to A. proteus where injection of spermine caused cytokinesis within 1 minute (Gawlitta et al., 1981), the period preceding furrow induction in X. laevis eggs is relatively long, occurring in a minimum of about 30 min. Since the timing of cleavage of the amphibian egg is controlled by a cytoplasmic clock (Hara et al., 1980), the polyamine effects on cytokinesis

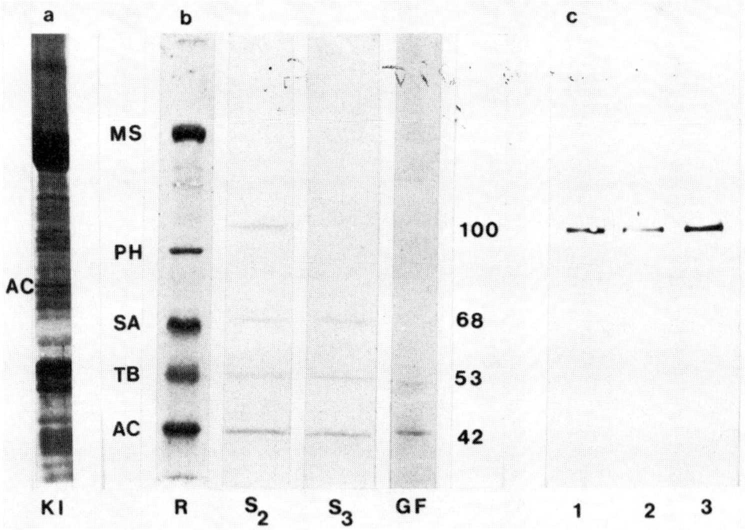

Fig. 32.2. a, b. Analysis of actin extracts during purification steps by 11% (a) and 7% (b) PAGE (Laemmli, 1970. a. KI: crude extract. B. R: standard reference proteins, MS: myosin, PH: phosphorylase, SA: serum albumin, TB: tubulin, AC: actin; S_2: supernatant of the first polymerization-depolymerization cycle; S_3: supernatant of the following polymerization step; GF: gel-filtered S_3.
c. Electrophoretic blotting of egg extracts (Towbin et al., 1979). Egg α-actinin was detected using platelet α-actinin antibody. 1 and 2: two different concentrations of S_2; 3: control, platelet α-actinin.

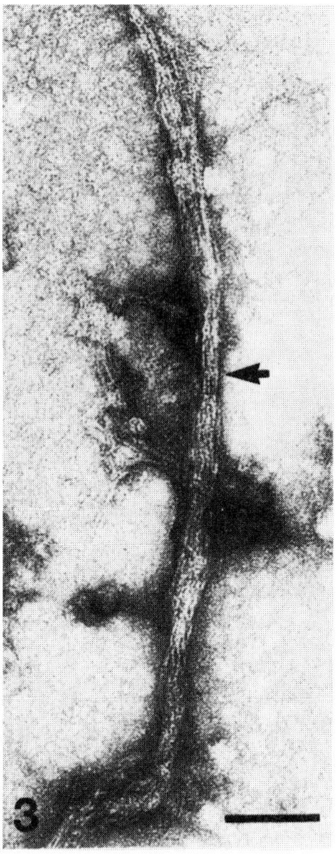

Fig. 32.3. Negative-stained actin bundle (prepared according to Grant et al., 1983). Polymerization of the gel filtered actin fraction (protein concentration: 0.125 mg/ml) was induced by 3 mM spermine. Bar: 0.1 μm; 115,000×.

may be modulated by cytoplasmic factors. The latency probably doesn't involve polyamine modifications as analysis of chromatograms of polyamine extracts of ^{14}C-spermine injected eggs indicated that metabolic transformation of spermine into its precursors, spermidine or putrescine does not occur during this period.

Effect of Spermine on Isolated Egg Actin

After an actin polymerization-depolymerization cycle of the egg extract (KI), the supernatant (S_2) contained four major proteins with M_r of 100K, 68K, 53K and 42K (Fig. 32.2a, 32.2b). The 42K protein migrates with muscle actin and the 100K protein is α-actinin (see immunoblot on Fig. 32.2c). The latter could be separated from the actin extract by another polymerization step (S_3, Fig. 32.2b). After gel filtration of the S_3 fraction, an extract containing actin and a trace of 53K protein was obtained (GF, Fig. 32.2b). Addition of 3 mM spermine to this fraction induced formation of actin bundles (Fig. 32.3). These bundles had some paracrystalline regions showing a parallel arrangement of filaments, but were more sinuous than bundles induced from muscle actin under the same conditions (Grant et al., 1983), perhaps due to the presence of the 53K protein. Clarke & Merriam (1978) also found 68K and 53K proteins in their actin extracts of X. laevis. In eggs of another amphibian, Rana pipiens, the 53K protein has been characterized as a glycoprotein associated with the egg cortex, and the 68K protein apparently coextracts with actin in the presence of 1 mM $CaCl_2$ (Smith & Burgess, 1981).

CONCLUSIONS

Spermine and spermidine clearly induced early furrow formation in X. laevis eggs, in agreement with the rapid induction of cytokinesis in A. proteus by spermine (Gawlitta et al., 1981). This effect is compatible with the fluctuations in polyamine concentrations observed during the cleavage cycle of other eggs (Kusunoki & Yasumasu, 1976; Lundquist et al., 1983) and the polyamine dependence of cytokinesis in mammalian cells (Sunkara et al., 1979).

Actin isolated from these amphibian eggs formed bundles in the presence of spermine. If this polymerization also occurs in vivo when spermine becomes associated with the egg cortex, the polyamine-actin interaction may well be essential for cytokinesis as previously postulated (Oriol-Audit, 1980). This interaction could play an important regulatory role during early embryonic development.

ACKNOWLEDGMENTS

We thank H. Gerbail for her excellent technical assistance and Y. Gache for α-actinin antibody. This project was supported by INSERM CRL 822013 (C.O.-A.), MDAC (N.J.G.) and CNRS ATP 03195 (C.A.).

REFERENCES

Clark, T. G., and Merriam, R. W.: Actin in Xenopus oocytes. I. Polymerization and gelation in vitro. J. Cell Biol. 77, 427-38 (1978).

Franke, W.W., Rathke, P.C., Seib, E., Tredelenburg, M.F., Osborn, M., & Weber, K.: Distribution and mode of arrangement of microfilamentous structures and actin in the cortex of the amphibian oocyte. Cytobiology 4, 111-30 (1976).

Gawlitta, W., Stockem, W., & Weber, K.: Visualization of actin polymerization and depolymerization cycles during polyamine-induced cytokinesis in living Amoeba proteus. Cell Tissue Res. 215, 249-61 (1981).

Grant, N.J., Oriol-Audit, C., & Dickens, M.J.: Supramolecular forms of actin induced by polyamines: an electron microscopic study. Europ. J. Cell Biol. 30, 67-73 (1983).

Hara, K., Tydeman, P., & Dirschner, M.: A cytoplasmic clock with the same period as the division cycle in Xenopus eggs. Proc. Natl. Acad. Sci USA 77, 462-66 (1980).

Kusunoki, S., & Yasumasu, I.: Cyclic change in polyamine concentrations in sea urchin eggs related with cleavage cycle. Biochem. Biophys. Res. Commun. 68, 881-85 (1976).

Laemmli, U.K.: Cleavage of structural proteins during the assembly of the head of bacteriophage T_4. Nature 227, 680-85 (1970).

Lundquist, A., Löwkvist, B., Linden, M., & Heby, O.: Polyamines in early embryonic development: their relationship to nuclear multiplication rate, cell cycle traverse, and nucleolar formation in a dipteran egg. Develop. Biol. 95, 253-59 (1983).

Oriol-Audit, C.: Polyamine-induced actin polymerization. Europ. J. Biochem. 87, 371-76 (1978).

Oriol-Audit, C.: Induction of cytokinesis by interaction between actin and polyamines. Biochimie 62, 713-14 (1980).

Oriol-Audit, C.: Actin-polyamines interaction relationship between physicochemical properties and cytokinesis induction. Biochem. Biophys. Res. Commun. 105, 1096-1101 (1982).

Russell, D.H.: Putrescine and spermidine biosynthesis in the development of normal and anucleolate mutants of Xenopus laevis. Proc. Natl. Acad. Sci. USA 68, 523-27 (1971).

Smith, D.S., & Burgess, D.R.: Actin in the noncontractile cortex of immature frog oocytes is present in polymerized form. J. Cell Biol. 91, 186a (1981).

Sunkara, P.S., Rao, P.N., Nishioka, K. & Brinkley, B.R.: Role of polyamines in cytokinesis of mammalian cells. Exp. Cell Res. 119, 63-68 (1979).

Towbin, M., Staehelin, T., & Gordon, J.: Electrophoretic transfer of proteins from polyacrylamide gels to nitrocellulose sheets: procedure and some applications. Proc. Natl. Acad. Sci. USA 76, 4350-54 (1979).

PART III

INTERACTIONS OF CYTOSKELETAL ELEMENTS WITH OTHER CELL STRUCTURES

33 Actin Capping and Severing by Gelsolinlike Proteins

Harriet E. Harris

In nonmuscle cells, the major microfilament protein, actin, can undergo substantial structural and functional changes, depending on the physiological state of the cell. A large number of actin-binding proteins has now been described which, on account of their properties in vitro and, in some instances, their cytoplasmic locations, are thought to determine the behavior of actin and thus have a crucial role in the regulation of cell motility (1,2,3). One such protein is gelsolin, first isolated from macrophages, which solvates actin gels in the presence of calcium, and may thus regulate the consistency of the cytoplasm (4). This paper describes some properties of a protein closely analogous to gelsolin, which is isolated from blood plasma.

A GELSOLINLIKE PROTEIN FROM PIG AND HUMAN PLASMA

Actin filaments are rapidly disrupted in blood plasma or serum (5,6,7). This phenomenon is due to the action of a protein of molecular weight 92,000 (8,9) which is present in plasma at a relatively high concentration, about 0.1 mg/ml^{-1} (7,10). The protein interacts with F-actin to produce an inhibitor of pancreatic DNase I and, under appropriate solution conditions (15% glycerol in the absence of ATP), DNase I inhibition is linear with amount of plasma protein, a property which is used as a quantitative assay for the plasma factor (11).

When this assay was applied to a cytoplasmic extract (lysates of washed platelets), a protein similar both in activity and size to the plasma factor was detected (B. Pope and H. E. Harris, unpublished). Comparative peptide maps of the platelet and plasma proteins, obtained by cyanogen bromide, chymotrypsin or V8 protease cleavage, showed that the two proteins were indistinguishable by this criterion (B. Pope, unpublished and (12)). Furthermore, Yin et al. have demonstrated immunological cross-reactivity between a 91,000 M_r polypeptide in human serum and macrophage gelsolin. Antiserum to cytoplasmic gelsolin also cross-reacts with proteins of similar M_r in a wide range of cell types (13). In fact, gelsolinlike proteins appear to be ubiquitous in vertebrate tissues (14).

The similarities now apparent in molecular properties (8,15), amino acid composition, tissue distribution and many functions between the plasma factor and gelsolin justify the application of the name gelsolin to both proteins. The terms plasma actin depolymerizing factor (ADF) (7,11) and brevin (9) have previously been used for plasma gelsolin.

An advantage of working with plasma gelsolin, rather than the cytoplasmic analogues, is that it is more readily purified in reasonable quantities, and may be obtained free of actin. This paper describes studies on the effects of purified pig or human plasma gelsolins on rabbit skeletal muscle actin in vitro. The aim of this work is to define in detail the molecular interactions between these two proteins. The system provides a model for the regulation of the viscosity of the cytoplasm by gelsolin.

THE ACTION OF PLASMA GELSOLIN ON PURIFIED ACTIN

Interaction with Monomeric Actin

There are several lines of evidence for a very high affinity interaction between plasma gelsolin and actin. Plasma gelsolin binds very tightly to actin affinity columns and can be eluted only under severe conditions, e.g., with denaturing agents or at very low pH. Discrete complexes of plasma gelsolin and actin, prepared by mixing the two components in approximately stoichiometric ratios, may be isolated by gel filtration. Complexes may be isolated at plasma gelsolin concentrations of about 10^{-8}M, implying that the affinity for actin is very high.

The affinity of plasma gelsolin for actin may be quantitated using actin labelled with the fluorescent probe 7-chloro-4-nitro-2,1,3-benzoxadiazole (NBD-actin) (16). The fluorescence of this probe is enhanced about twofold when monomeric actin interacts with gelsolin. The technique is sufficiently sensitive for titrations to be performed at nanomolar concentrations. The dissociation constant has thus been determined at 20 nM NBD-actin, and found to be very low: about 4nM in 10^{-4}M Ca^{2+} in an F-actin buffer (but below the critical concentration of actin). Under the same conditions, but at $<10^{-8}$M Ca^{2+}, binding was reproducibly weaker, but only by about tenfold (K_D 40 nM). Thus under approximately physiological conditions, plasma gelsolin complexes very tightly with monomeric actin.

Interaction with F-Actin

Plasma gelsolin rapidly shortens actin filaments. At low mole ratios with respect to actin, it causes a marked drop in the high shear viscosity of actin (17). The lowered viscosity correlates with shortening of filaments as observed in negatively stained electron micrographs; filaments become progressively shorter as gelsolin is increased (Fig. 33.1). In contrast to the effect of cytoplasmic (macrophage) gelsolin (4,15), Ca^{2+} concentration has little effect on either the viscosity decrease or filament length distribution (Fig. 33.2).

Sedimentation experiments using radiolabelled plasma gelsolin show that it binds with high affinity to actin filaments. To determine the mode of binding, fragments of the highly stable actin bundles from the acrosomal processes of *Limulus* sperm were used to nucleate the growth of bundles of actin filaments of uniform polarity. When incubated with G-actin under polymerizing conditions, such nuclei exhibit growth of a bundle of long filaments from one end and much shorter filaments from the other (18). Decoration with myosin subfragment-1 (SF-1) confirms that these correspond respectively to the fast-growing (barbed) and slow-growing (pointed) ends of the actin filaments (17). When the nuclei were preincubated with plasma gelsolin, before the addition of G-actin, filament growth at the barbed end was completely blocked. Plasma gelsolin thus binds with high affinity to the fast-growing ends of actin filaments and caps them, inhibiting further monomer addition at this end.

Proteins which, like plasma gelsolin, block the barbed ends of filaments, fall into two functional classes. The microvillus protein villin (18,19) and *Dictyostelium* severin (20) sever filaments along their length as well as capping the ends. However, capping and severing functions are not obligatorily linked, since proteins from *Acanthamoeba* (21) and brain (22) cap the barbed ends of filaments but do not sever. Since the

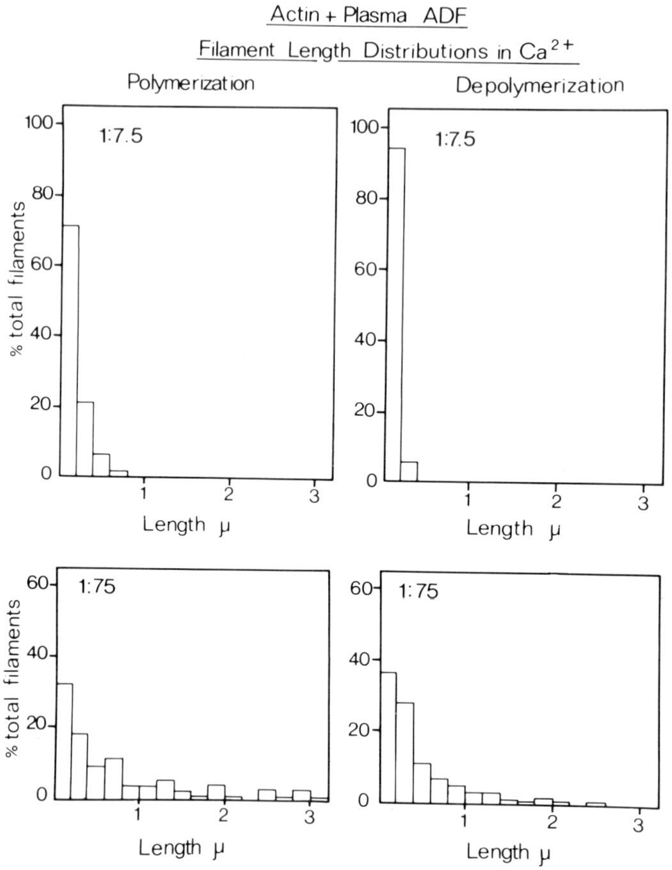

Fig. 33.1. Length distributions of actin filaments in the presence of plasma gelsolin (ADF). Left: filaments assembled in the presence of plasma gelsolin at 1-7.5 or 1-75 molar ratios to actin. Right: preformed filaments treated with plasma gelsolin under the same conditions.

action of plasma gelsolin on actin filaments is very rapid, it seemed probable that its shortening effect is due to a direct severing action (19). This was tested with actin bundles nucleated by Limulus sperm acrosomal processes (18). Bundles with a growth of filaments at both ends were placed on electron microscope grids and washed gently with plasma gelsolin solution. Filaments were cleaved off both ends of the bundles; high enough concentrations of plasma gelsolin removed all free actin. (The bundles themselves are resistant to plasma gelsolin). Since high-affinity binding to filaments was demonstrated at only one filament end, cleavage of filaments from both ends implies a direct attack along the sides of filaments. Whether gelsolin exhibits an initial, low-affinity binding before cleavage, or takes advantage of a spontaneous "breathing" of filaments to penetrate between actin monomers and cap one end is not known.

Plasma gelsolin thus severs actin filaments along their length, and binds tightly to the fast-growing end.

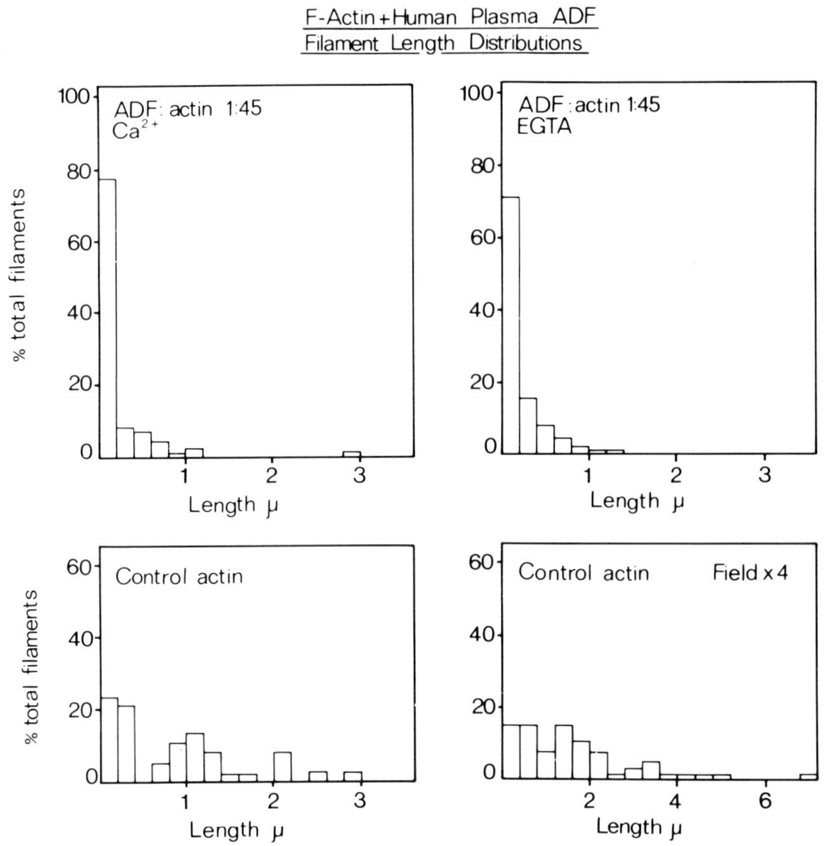

Fig. 33.2. Length distributions of actin filaments in the presence of plasma gelsolin (ADF) at 10^{-4}M (upper left) and 10^{-8}M Ca^{2+} (upper right). Plasma gelsolin-actin was 1-45. Lower histograms: lengths of control filaments without gelsolin measured in the same size of field as samples (left) or in a field with 4 times the area (right).

Effects on the Distribution of Actin Between Monomers and Filaments

All solutions of actin filaments contain an equilibrium (or steady state) concentration of monomers. To quantitate the effects of plasma gelsolin on the distribution of actin between monomers and filaments, fluorescently labelled actin was used. Actin covalently linked to N-pyrenyl-iodacetamide shows a 20-fold fluorescence increase on filament formation (24). This form of actin was used to determine the effect of plasma gelsolin on the critical concentration (or steady-state monomer concentration) (17). As little as 2nM plasma gelsolin sharply increases the critical concentration, both in the presence and absence of Ca^{+}. This is the behavior expected for treatment of actin with a protein which caps actin filaments at the end with the lower critical concentration (25,26,27), that is, the barbed end of myosin SF-1 decorated filaments (23).

The results with pyrenyl-actin are, however, more complex than predicted by this simple interpretation. The apparent critical concentration, as determined by this technique, increases progressively with plasma gelsolin concentration (17). Three explanations seem plausible: (1) The free monomer concentration may increase in proportion to the number of free filament ends; this behavior has been observed with severin (20). (2) The solution monomer concentration may be enhanced by a contribution from monomer-plasma gelsolin complexes (which are known to exhibit G-actinlike fluorescence spectra). (3) Monomers at the ends of filaments, adjacent to the capping protein, may be "silent", i.e., have G-actin fluorescence. These alternatives cannot be distinguished at present. Much more information is required on the distribution of species (free filaments, capped filaments, free monomer and small monomer-gelsolin complexes) in equilibrium mixtures of actin and plasma gelsolin.

The kinetics of filament disassembly have been analyzed by monitoring the fluorescence decrease which follows the addition of plasma gelsolin to pyrenyl-F-actin (17). This produces a rapid drop in fluorescence followed by a slower decline over several minutes. The simplest interpretation is that the rapid drop corresponds to severing (for which it must be assumed that actin subunits at the ends are "silent"), and that this is followed by a slower loss of monomers from the free ends of the newly severed and capped filaments to establish a new (higher) critical concentration. Qualitatively similar effects are obtained in the presence and absence of Ca^{2+}, although the magnitude of the fluorescence changes may be slightly smaller at $<10^{-8}$M Ca^{2+}. However, in both Ca^{2+} and EGTA, complete titration of pyrenyl-F-actin with plasma gelsolin yields a final product with a monomerlike fluorescence spectrum at plasma gelsolin to actin ratios of 1-1.

These experiments with fluorescently-labelled actin are thus consistent with the evidence for capping and severing activities implied by more direct, electron microscopic evidence but suggest, in addition, that these activities are relatively insensitive to Ca^{2+} concentration.

Effects of Plasma Gelsolin on the Assembly of Pyrenyl G-Actin

When monomeric actin at low salt concentrations is induced to assemble by addition by KCl and $MgCl_2$, there is a pronounced time lag before the rate of filament formation reaches a maximum. Plasma gelsolin, mixed with G-actin before initiation of assembly, abolishes the lag, markedly increases the assembly rate, but decreases the final extent of assembly. This effect may be monitored by either high shear viscometry or, preferably, by the fluorescence increase of pyrenyl-actin (17). The acceleration of assembly suggests that plasma gelsolin facilitates nucleation of filaments, while the lower final signals are as expected for the formation of shorter, capped filaments.

The dependence of the first-order rate constant for assembly (k_{obs}) on plasma gelsolin concentration is shown in Figure 33.3. In 10^{-4}M Ca^{2+}, k_{obs} rises steeply at lower plasma gelsolin concentrations, but shows a plateau at very high rates. Except at the highest concentrations, the data fit the relationship

$$k_{obs} \propto [\text{gelsolin}]^{\frac{1}{2}}$$

which is the theoretical relationship for actin assembly onto a fixed number of nuclei (24). In contrast to its capping and severing activities, the ability of plasma gelsolin to nucleate actin polymerization exhibits a marked Ca^{2+} sensitivity; at $<10^{-8}$M Ca^{2+}, k_{obs} is always much lower than at high Ca^{2+}, though it increases linearly with plasma gelsolin concentration (Fig. 33.3). The dependence of k_{obs} on Ca^{2+} concentration at pH 7.0 shows that there is a shift in the nucleating capacity of plasma gelsolin between 10^{-6} and 10^{-7}M Ca^{2+} (Fig. 33.4). This is in the range expected for Ca^{2+} regulation of a cytoplasmic protein, but in plasma, the protein will always be fully saturated

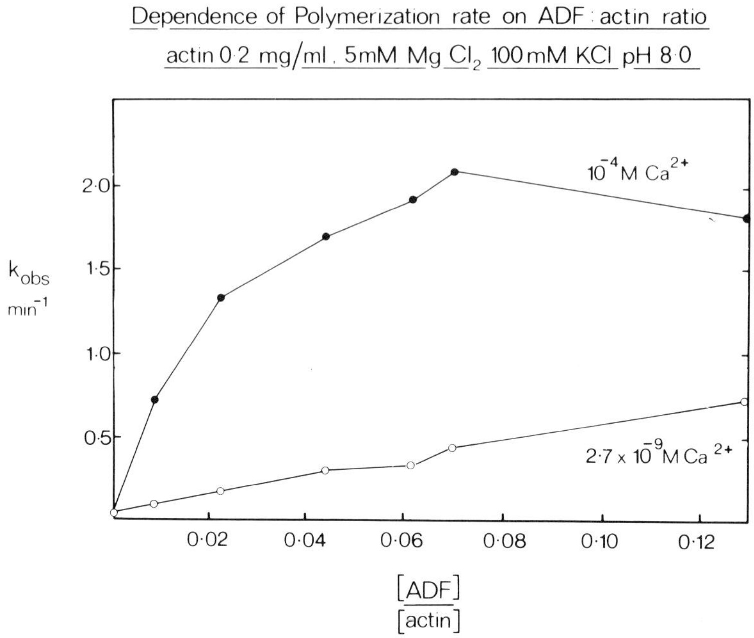

Fig. 33.3. Dependence of the first-order actin assembly rate (k_{obs}) on plasma gelsolin (ADF) concentration, in the presence and effective absence of Ca^{2+}.

with Ca^{2+}. Measurements of $^{45}Ca^{2+}$ binding to plasma gelsolin show that Ca^{2+} binds with a dissociation constant between 10^{-6} and 10^{-7}M, that is, with the right affinity for this Ca^{2+} binding site to regulate the nucleating activity of plasma gelsolin.

MODELS FOR PLASMA GELSOLIN ACTIN

The fact that the different activities of plasma gelsolin described in the previous section—capping, severing and nucleation of actin filaments—exhibit very different degrees of Ca^{2+} sensitivity, implies that there are different modes of interaction between plasma gelsolin and actin, and that these are not equally affected by Ca^{2+} concentration. One plausible model for the action of plasma gelsolin is that there is a high-affinity, Ca^{2+}-insensitive interaction with a single actin monomer. A similar interaction with filaments would account for Ca^{2+}-independent actin capping and severing. If such a complex

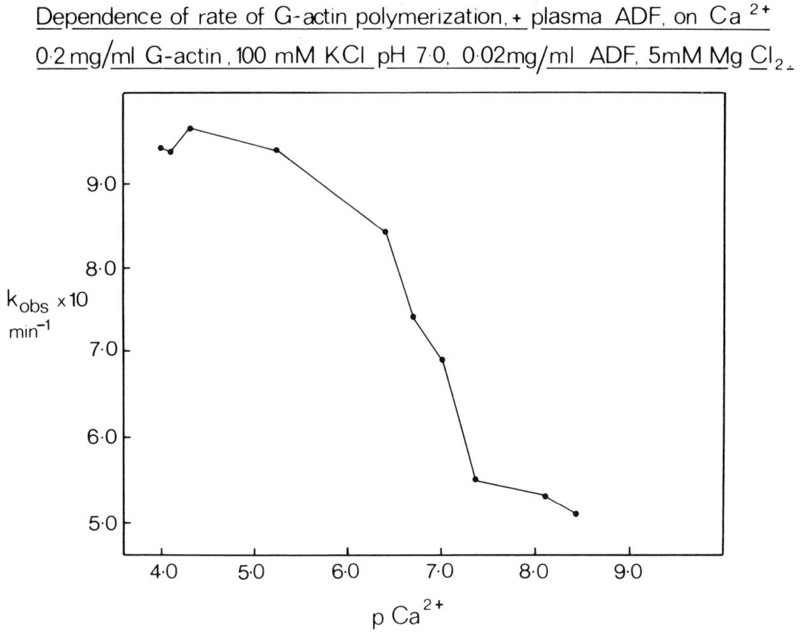

Fig. 33.4. Dependence of first-order actin assembly rate (k_{obs}) on Ca^{2+} concentration, in the presence of plasma gelsolin (ADF).

of monomer with gelsolin is not an effective nucleator of assembly, but nucleus formation requires the binding of a second actin monomer, this time in a Ca^{2+}-sensitive interaction, nucleation would exhibit Ca^{2+} dependence. There is some evidence that the behavior of platelet gelsolin conforms to this model (12), and macrophage gelsolin has been shown to bind two actin monomers per molecule (29). An alternative hypothesis is that Ca^{2+} does not affect the stoichiometry of actin binding to plasma gelsolin, but alters the conformation of an actin-gelsolin complex, so that nucleation is promoted. These two models cannot be distinguished for plasma gelsolin on present evidence, but may be tested by determining precise stoichiometries and affinities for the binding of plasma gelsolin to G- and F-actins.

The ability of plasma gelsolin to shorten actin filaments in both the presence and absence of Ca^{2+} contrasts with the behavior of macrophage gelsolin (29), and this presents a paradox. The discrepancy may be due to a real difference between the functional properties of the intra- and extracellular proteins, despite close structural similarities. For example, plasma gelsolin is unlikely to be Ca^{2+}-regulated in blood, where Ca^{2+} levels are always saturating, and it may therefore have lose some Ca^{2+} dependence. However, we cannot at present eliminate the possibility that an apparent dissimilarity results from differences in technique used in different laboratories. Further quantitation of the interactions of plasma and cytoplasmic gelsolins with actin will help to establish the extent of the similarities between these proteins. In addition, this information will help us to interpret the relationship between the in vitro properties of gelsolins and their functions in plasma and in the cell.

SUMMARY

The protein plasma gelsolin, also known as plasma actin depolymerizing factor or brevin, shows close structural and functional similarities to macrophage gelsolin. It binds with high affinity to both actin monomers and filaments, and shortens filaments. The production of shorter filaments is due to its ability both to cap filaments at the barbed end and to sever filaments. In addition, it is an efficient nucleator of assembly, and this latter activity is greatly enhanced by micromolar calcium ion concentrations.

ACKNOWLEDGMENTS

I wish to thank Dr. Alan Weeds, to whom I am indebted for his collaboration and encouragement during the course of this work. We are most grateful to John Gooch for preparing plasma gelsolin.

REFERENCES

1. Craig, S.W. and Pollard, T.D. (1982). Trends in Biochem. Sci. 7, 88-92.
2. Korn, E.D. (1982). Physiol. Rev. 62, 672-737.
3. Weeds, A.G. (1982). Nature 296, 811-16.
4. Yin, H.L. and Stossel, T.P. (1979). Nature 281, 583-86.
5. Chaponnier, C., Borgia, R., Rungger-Brandle, E., Weil, R. and Gabbiani, C. (1979). Experientia 35, 1039-41.
6. Norberg, R., Thorstensson, R., Utter, G. and Fagraeus, A. (1979). Eur. J. Biochem. 100, 575-83.

7. Harris, H.E., Bamburg, J.R. and Weeds, A.G. (1980). FEBS Lett. 121, 175-77.
8. Harris, H.E. and Gooch, J. (1981). FEBS Lett. 123, 49-53.
9. Harris, D.A. and Schwartz, J.H. (1981). Proc. Natl. Acad. Sci. U.S.A. 78, 6798-6802.
10. Thorstensson, R., Utter, G. and Norberg, R. (1982). Eur. J. Biochem. 126, 11-16.
11. Harris, H.E., Bamburg, J.R., Bernstein, B.W. and Weeds, A.G. (1982). Anal. Biochem. 119, 102-14.
12. Markey, F., Persson, T. and Lindberg, U. (1982). Biochim. Biophys. Acta 709, 122-33.
13. Yin, H.L., Albrecht, J.H. and Fattoum, A. (1981) J. Cell Biol. 91, 901-06.
14. Snabes, M.C., Boyd, A.E. and Bryan, J. (1983). Exptl. Cell Res. 146, 63-70.
15. Yin, H.L. and Stossel, T.P. (1980). J. Biol. Chem. 255, 9490-93.
16. Detmers, P., Weber, A., Elzinga, M. and Stephens, R.E. (1981). J. Biol. Chem. 256, 99-105.
17. Harris, H.E. and Weeds, A.G. (1983). Biochemistry 22, 2728-41.
18. Bonder, E.M. and Mooseker, M.S. (1983). J. Cell Biol. 96, 1097-1107.
19. Glenny, J.R., Kaulfus, P. and Weber, K. (1981). Cell 24, 471-80.
20. Yamamoto, K., Pardee, J.D., Reidler, J., Stryer, L. and Spudich, J.A. (1982). J. Cell Biol. 95, 711-19.
21. Isenberg, G., Aebi, U. and Pollard, T.D. (1980). Nature 288, 455-59.
22. Kilimann, M.W. and Isenberg, G. (1982). EMBO J. 1, 889-94.
23. Pollard, T.D. and Mooseker, M.S. (1981). J. Cell Biol. 88, 654-59.
24. Kouyama, T., and Mihashi, K. (1981). Eur. J. Biochem. 114, 33-38.
25. Wegner, A. (1976). J. Mol. Biol. 108, 139-50.
26. Kirschner, M.W. (1980). J. Cell. Biol. 86, 330-34.
27. Wegner, A. and Isenberg, G. (1983). Proc. Natl. Acad. Sci. U.S.A. 80, 4922.
28. Oosawa, F. and Kasai, M. (1962). J. Mol. Biol. 4, 10-21.
29. Yin, H.L., Hartwig, J.H., Maruyama, K. and Stossel, T.P. (1981). J. Biol. Chem. 256, 9693-97.

34 Protein Substrates of Tyrosine Phosphokinase Encoded by V-SRC Oncogene Are Associated with Cytoskeleton and Adhesion Structures in Transformed Cells

L. Naldini, M. F. Di Renzo, G. Tarone,
F. G. Giancotti, P. M. Comoglio, P. C. Marchisio

INTRODUCTION

Transformation induced by Rous Sarcoma Virus (RSV) is triggered by the action of a single transforming gene (src) via a protein termed pp60src (Hanafusa, 1977; Erikson, 1980). This protein is a phosphokinase which phosphorylates tyrosine residues in protein substrates (Collett et al. 1980; Hunter and Sefton, 1980; Levinson et al. 1980), an unusual property shared by a number of retroviral-transforming proteins (Bishop, 1983) and some growth factor membrane receptors (Hunter and Cooper, 1981; Ek et al. 1982; Rosen et al., 1983). Hence, identification of cellular substrates for tyrosine kinases has become an issue of major importance, on the assumption that the cellular targets triggered by activated receptors and retroviral-transforming proteins might be involved in driving the proliferative or neoplastic response.

A larger body of information is available for the product of the v-src oncogene, pp60src. The transforming protein of RSV in infected cells was found in the detergent-insoluble cell fraction (Burr, 1980), the active form being associated with the plasma membrane (Courtneidge and Bishop, 1982). This fact was also consistent with the often suggested role of cytoskeleton and adhesive structures as early cellular targets in the genesis of the transformed phenotype.

By a variety of approaches based on the identification of phosphotyrosine-containing proteins, a number of candidate substrates for pp60src have been described in transformed avian fibroblasts (Radke and Martin, 1979; Erikson and Erikson, 1980; Cooper and Hunter, 1981; Cooper and Hunter, 1982; Cooper and Hunter, 1983), including molecules of defined physiologic activities such as glycolytic enzymes (Cooper et al., 1983) and vinculin (Sefton et al., 1981).

We have raised antibodies against azobenzyl phosphonate, a hapten known to cross-react with phosphorylated tyrosine residues (Ross et al., 1981), and used them to identify and to localize phosphotyrosine-containing proteins within RSV transformed mammalian fibroblasts. As the pp60src protein is known to be located in the detergent-insoluble cell fraction and assuming a key involvement of the cytoskeleton in mediating cellular transformation, we started our search for phosphotyrosine-containing proteins in the detergent-insoluble cytoskeletal cell fraction.

METHODS AND RESULTS

Production and Characterization of ABP Antibodies

Antiazobenzyl phosponate (ABP) sera were raised in rabbits immunized with keyhold limpet haemocyanin to which azobenzyl phosphonate groups had been covalently cross-linked (Tabachnik and Sobotka, 1960; Landt et al., 1978). The anti-ABP activity in serum was monitored by a solid phase radioimmunoassay (Prat and Comoglio, 1976) in which polyvinyl wells of microtiter plates were coated with bovine serum albumin (BSA) coupled to ABP groups. The specificity of ABP antibodies was tested by measuring the extent of the inhibition exerted by a variety of low or high molecular weight phosphate-containing compounds. Beside the hapten, phosphotyrosine itself, as expected, inhibited ABP antibody binding. On the contrary, the other two naturally occurring phosphoamino acids, phosphoserine and phosphothreonine, did not react significantly with these antibodies.

A number of phosphate-containing inorganic and organic compounds, including various phosphoproteins and nucleic acids were also uneffective in inhibiting ABP antibody binding. Nucleoside triphosphates provided an exception by partially inhibiting this binding.

Phosphorylation In Vitro of Detergent Extracted Cells

In order to investigate potential substrates of pp60src protein kinase, 32P-γ-ATP was added to the NP40-insoluble fraction of RSV transformed fibroblasts, under conditions allowing the kinase reaction catalyzed by pp60src (Burr et al.,

1980). After phosphorylation, labelled proteins were solubilized by ionic detergents, precipitated by ABP antibodies, separated by electrophoresis and revealed by autoradiography. Controls were performed replacing ABP antibodies with preimmune rabbit immunoglobulins.

In mouse fibroblasts transformed by the SR-D strain of RSV, a major phosphorylated component of 130 Kd and two closely migrating proteins of 70-65 Kd were identified (Fig. 34.1, lane B). Two minor bands of 85 Kd and 60 Kd were also constantly observed. The latter partially overlapped with the more radioactive faster component of the 70-65 Kd doublet. In addition, after prolonged exposure of the gels, two bands of approximately 39 and 34 Kd became detectable.

The specificity of the kinase reaction was proved in the fact that no radiolabelled proteins were precipitated from mouse 3T3 control fibroblast detergent-insoluble preparations lacking viral pp60src (Fig. 34.1, lane A).

The radiolabelled phosphate transferred to these proteins was resistant to alkali treatment, ruling out the involvement of

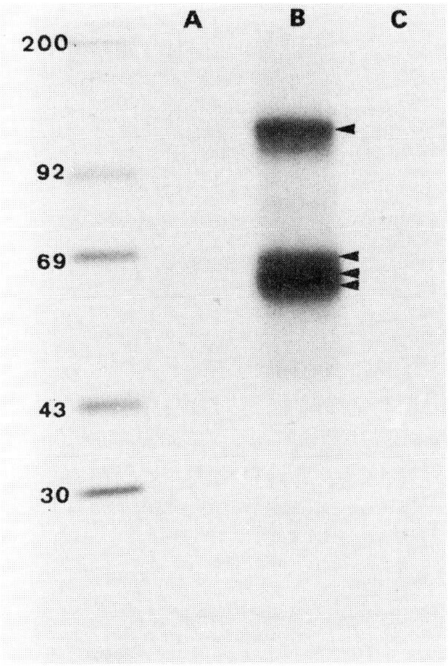

Fig. 34.1. Immunoprecipitation by ABP antibodies of phosphoproteins from the detergent-insoluble fraction of transformed fibroblasts. The detergent insoluble fractions from (A) 3T3 mouse control fibroblasts, and (B) SR-BALB mouse fibroblasts transformed by RSV, were phosphorylated in vitro with 32P-γ-ATP, solubilized and immunoprecipitated with ABP antibodies. Labelled proteins were separated by electrophoresis and exposed for autoradiography. In (C): same as (B), but ABP antibodies were replaced by preimmune rabbit immunoglobulins. A major phosphorylated component of around 130 Kd, two closely migrating proteins of 70 and 65 Kd and two minor bands of 85 and 60 were specifically precipitated from the transformed cells.

serine residue as phosphate acceptor site. Direct evidence for phosphorylation at the tyrosine residue(s) was provided by phosphoamino acid analysis. The labelled protein bands were excised from the gel, eluted and subjected to acid hydrolysis. The radiolabelled phosphate incorporated in the 130, 70-65 and 60 Kd proteins precipitated by ABP antibodies was found to comigrate with authentic phosphotyrosine in high voltage electrophoresis (Fig. 34.2).

In order to tentatively identify the phosphotyrosine-containing proteins precipitated by ABP antibodies, their electrophoretic mobility in SDS-PAGE was compared to that of vinculin (Mr 130 Kd) and pp60src (Mr 60 Kd), precipitated by specific antisera. The tyrosine-phosphorylated 130 Kd protein, immunoprecipitated by ABP antibodies comigrated with vinculin immunoprecipitated by specific antiserum from 35S-methionine metabolically-labelled detergent-extracted normal fibroblasts. However, the identify between the 130 Kd protein and vinculin was disproved since no phosphorylated vinculin was immunoprecipitated from transformed fibroblasts detergent-extracted and in vitro phosphorylated with 32P-γ-ATP, by either vinculin

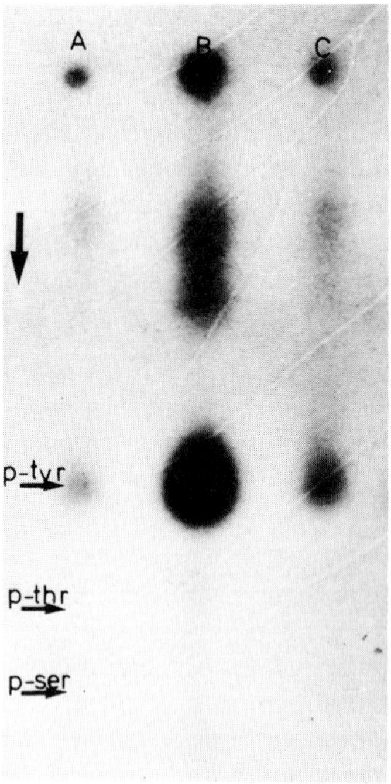

Fig. 34.2. Phosphoaminoacid analysis of in vitro phosphorylated proteins from transformed fibroblasts. Labelled proteins from a gel as in Fig. 34.1 were eluted and acid-hydrolyzed. Their phosphoaminoacids were then analyzed by high voltage electrophoresis with a mixture of unlabelled phosphoaminoacids as reference. Free phosphate was run off the plate. Lane A refers to the minor 60 Kd component, lane B to the 70-65 Kd doublet and lane C to the 130 Kd component.

antiserum or two monoclonal antibodies reacting with vinculin domains highly conserved among species.

The electrophoretic mobility of the proteins precipitated by ABP antibodies was also compared with that of in vitro phosphorylated proteins immunoprecipitated by a tumor bearing rabbit serum (TBR). Pp60src comigrated with the minor phosphorylated 60 Kd component.

Intracellular Localization of Phosphotyrosine-Containing Proteins by ABP Antibodies

The intracellular distribution of molecules cross-reacting with ABP antibodies was studied in control and RSV transformed mouse fibroblasts fixed, permeabilized and examined by indirect immunofluorescence microscopy. For immunofluorescence studies, antibodies were purified by affinity chromatography on ABP-BSA coupled to cyanogen bromide activated Sepharose 4B.

In normal fibroblasts, beside a faint cytoplasmic background, staining was observed within the nucleus. The nuclear fluorescence might be due to the above described cross-reaction of ABP antibodies with nucleoside triphosphates. In RSV transformed fibroblasts—fixed and detergent-permeabilized—ABP antibodies stained diffusely the cytoplasm and specifically decorated restricted areas of the ventral aspect of the cell, corresponding in interference reflection microscopy (IRM) to well defined dark areas representing patches of ventral membrane lying less than 10 nm from the substratum (Fig. 34.3). Moreover, the sites decorated by ABP antibodies corresponded in size, location and general morphological features to areas which were also stained by a fluorescent antivinculin serum. Consequently, the sites decorated by ABP antibodies were identified as adhesion plaques. Also, ABP antibody-decorated streaks were located at the ends of residual stress fibers, simultaneously labelled with fluorescent phalloidin. This pattern was noted particularly within areas of cells still being highly flattened. Immunostaining with ABP antibodies of areas corresponding in IRM to adhesion plaques was never observed in normal nontransformed fibroblasts, while the same cellular structures were stained by vinculin antibodies.

Specific decoration with ABP antibodies was also observed at the level of cell to cell contacts and at the tips of filopodia originating from the leading edge. This staining was never observed in nontransformed fibroblasts.

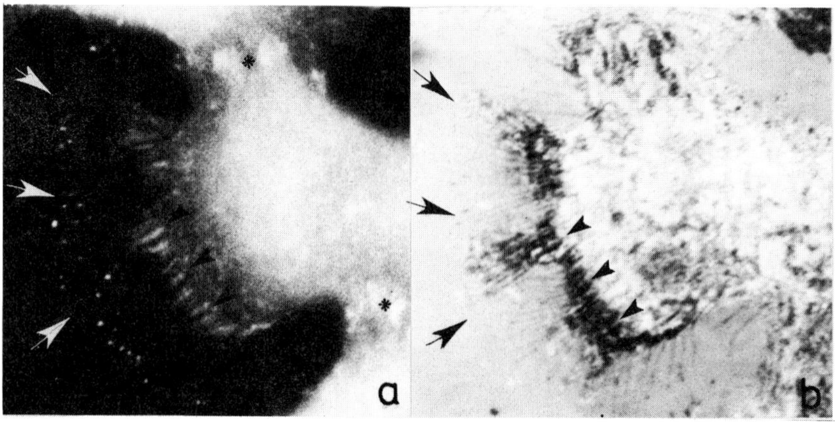

Fig. 34.3. Immunofluorescence with ABP antibodies. A RSV transformed mouse fibroblast simultaneously viewed in (A) immunofluorescence with ABP antibodies, and (B) interference reflection microscopy. The cell was permeabilized after fixation. In (A) the staining pattern is both diffusely distributed within the cytoplasm and discrete at cell edges, where restricted areas corresponding to focal contacts in interference-reflection (arrowheads) and filopodial tips (arrows) are specifically decorated. Fluorescence is more intense also at cell to cell contacts (asterisks).

Theae data suggest that in RSV transformed cells, increased phosphorylation of proteins at tyrosine residues is critically localized at cytoskeletal structures mediating cell to substratum and cell to cell contacts.

DISCUSSION

Phosphorylation of target proteins in tyrosine residues appears to be a unique feature of two families of proteins involved in the control of cellular proliferation. In fact, tyrosine phosphokinase activity has been associated both with membrane receptors mediating the growth-promoting activities of growth factors (Hunter and Cooper, 1981; Ek et al., 1982; Rosen et al., 1983) and with oncogene-encoded proteins capable of inducing a transformed phenotype in a retrovirus-infected cell (Bishop, 1983). Therefore, identifying the cellular substrates for tyrosine kinases appears to be a promising, though challenging task. It could then be possible to compare physiology and pathology and to approach crucial questions such as: (1) the existence of a final common pathway for triggering cellular proliferation controlled by either growth factors or transforming proteins, and (2) assessing the critical, quantitative or qualitative difference between their modes of action.

The cells infected by RSV are known to be transformed through the action of a single oncogene whose translation product is a tyrosine-specific protein kinase named pp60src. We have undertaken the search for its cellular substrates raising antibodies against the hapten azobenzyl phosphonate (ABP) which cross-react with phosphotyrosine and precipitate specifically tyrosine-phosphorylated proteins.

An extensive literature points to cytoskeleton organization and adhesion to substrate as likely major cellular targets of transformation. Moreover, the pp60src protein was localized in the detergent-insoluble fraction of the cell (Burr et al., 1980; Czernilofsky et al., 1980; Levinson et al., 1981) and particularly at adhesion structures (Rohrschneider and Rosok, 1983). We have then focused our investigation on the detergent-insoluble cytoskeleton of the transformed cell.

Previous work performed by immunodecoration of nitrocellulose sheets where cellular proteins from both the detergent-soluble and insoluble fractions of infected and control normal cells had been transferred after electrophoretical separation, showed the preferential association of ABP antibody-labelled material with the cytoskeletal fraction of infected cells and failed to reveal any detectable phosphotyrosyl-protein in control normal fiborblasts (Comoglio et al., 1983).

We have then applied the use of ABP antibodies to a powerful system that amplifies phosphorylation of detergent-insoluble substrates by pp60src (Gacon et al., 1982). This led to the identification of in vitro phosphorylated components, barely detectable by conventional biochemical techniques, which if proved to be true in vivo substrates, may play a significant role in the acquisition of the transformed phenotype. ABP antibodies precipitated three major detergent-insoluble phosphotyrosine-containing proteins of 130, 70 and 65 Kd, and minor components of 85, 60, 39 and 34 Kd. These proteins were substrates of the kinase reaction catalyzed in vitro by pp60src; they did not contain immunologically detectable phosphotyrosine in control untransformed fibroblasts.

The minor phosphorylated 60 Kd protein immunoprecipitated by ABP antibodies comigrated in electrophoresis with 32P-labelled pp60src immunoprecipitated from the same cell preparation with a tumor bearing rabbit serum. A major tyrosine phosphorylated molecule precipitated by ABP antibodies from detergent-insoluble fractions prepared from fibroblasts transformed by RSV has a molecular weight of 130 Kd and comigrates in electrophoresis with vinculin. A fraction of the cellular vinculin is known to be associated with detergent-insoluble structures of the plasma membrane (Geiger et al., 1980) and was found in the detergent-treated cell fractions used in this work for the kinase reaction. Since a limited amount of vinculin has been demonstrated to be phosphorylated at tyrosine residues in vivo in RSV transformed cells (Sefton et al., 1981) and this molecule and pp60src coexist with defined structures of the ventral plasma membrane—called "adhesion plaques"—(Geiger et al., 1980, Rohrschneider and Rosok, 1983) the issue whether vinculin is a significant target for pp60src has been debated (Hynes, 1982). However, the identity between vinculin and the 130 Kd phosphoprotein described in this paper was not confirmed since the latter was not immunoprecipitated by antivinculin antibodies.

Using affinity-purified ABP antibodies for immunofluorescence on fixed and permeabilized, transformed and normal fibroblasts, we observed, beside a diffuse staining of the cytoplasm of transformed cells, a discrete fluorescence pattern at the ventral surface of the transformed cell. Double immunolabelling experiments showed that the structures decorated by ABP antibodies contain vinculin and are located at the end of residual stress fibers; interference reflection microscopy of the stained cells demonstrated a correspondence between these ABP antibody-labelled structures and well defined, dark areas in IRM indicating close contacts of the ventral surface with the substrate. We have then tentatively identified these phosphotyrosyl-protein-containing structures as adhesion plaques.

Other localizations of ABP cross-reacting material in immunofluorescence microscopy were cell to cell contacts and the tips of leading edge filopodia. Again, these decorations were specific for the transformed cells.

It is then likely that cytoskeletal components of "adhesion plaques" and cell to cell contacts are involved in the transformation process, as supported by the fact that pp60src is selectively found here (Rohrschneider and Rosok, 1983). Pp60src May locally exert its kinase activity on substrates—other than vinculin—corresponding to some or all of the above described cytoskeleton-associated molecules.

ACKNOWLEDGMENTS

This work has been supported by the Italian National Research Council (C.N.R.), Progetto Finalizzato Controllo della Crescita Neoplastica.

The skillful technical assistance of M. R. Amedeo and P. Rossino is gratefully acknowledged.

REFERENCES

Bishop, J.M. (1983) Ann. Rev. Biochem., 52, 301-54.

Burr, J., Dreyfuss, G., Penman, S., and Buchanan, J. (1980) Proc. Natl. Acad. Sci. USA, 77, 3484-88.

Collett, M.S., Purchio, A.F., and Erikson, R.L. (1980) Nature, 285, 167-69.

Comoglio, P.M., Di Renzo, M.F., Tarone, G., Giancotti, F.G., Naldini, L., and Marchisio, P.C. (1983) EMBO J., in press.

Cooper, J., and Hunter, T. (1981) Mol. Cell. Biol., 1, 165-68.

Cooper, J., and Hunter, T. (1982) J. Cell Biol., 94, 287-96.

Cooper, J., and Hunter, T. (1983) J. Biol. Chem., 258, 1108-13.

Cooper, J., Reiss, N., Schwartz, R., and Hunter, T. (1983) Nature, 303, 218-23.

Courtneidge, S., and Bishop, J.M. (1982) Proc. Natl. Acad. Sci. USA, 79, 7117-21.

Czernilofsky, A., Levinson, A., Varmus, H., Bishop, J.M., Tischer, E., and Goodman, H. (1980) Nature, 287, 198-203.

Goodman, H. (1980) Nature, 287, 198-203.

Ek, B., Westermark, A., Wasteson, A., and Heldin, D.H. (1982) Nature, 295, 419.

Erikson, E., and Erikson, R.L. (1980) Cell, 21, 829-36.

Erikson, R.L., Purchio, A.F., Erikson, E., Collet, M.S., and Brugge, J.S. (1980) J. Cell Biol., 87, 319-25.

Gacon, G., Gisselbrecht, S., Piau, J.P., Fiszman, M.Y., and Fisher, S. (1982) Eur. J. Biochem., 125, 453-56.

Geiger, B., Tokuyasu, K.T., Dutton, A.H., and Singer, S.J. (1980) Proc. Natl. Acad. Sci. USA, 77, 4127-31.

Hanafusa, H., (1977) in Fraenkel-Conrat, H. and Wagner, R.R. (eds.) Comprehensive Virology, Plenum Press, NY, 10, pp. 401-83.

Hunter, T., and Cooper, J.A. (1981) Cell 24, 741.

Hunter, T., and Sefton, B. (1980) Proc. Natl. Acad. Sci. USA, 77, 1311-15.

Landt, M., Boltz, S., and Butler, L. (1978) Biochemistry, 17, 915-19.

Levinson, A., Courteneidge, S., and Bishop, J.M. (1981) Proc. Natl. Acad. Sci. USA, 78, 1624-162.

Levinson, A.D., Oppermann, H., Varmus, H.E., and Bishop, J.M. (1980) J. Biol. Chem., 255, 11793-980.

Prat, M., and Comoglio, P.M. (1976) J. Immunol. Methods, 9, 267-72.

Radke, K., and Martin, G.S. (1979) Proc. Natl. Acad. Sci. USA, 76, 5212-16.

Rohrschneider, L.R., and Rosok, M.J. (1983) Mol. Cell. Biol., 3, 731-46.

Rosen, O., Herrera, R., Olowe, Y., Petruzzelli, L.M., and Cobb, M.H. (1983) Proc. Natl. Acad. Sci. USA, 80, 3237.

Ross, A.H., Baltimore, D., and Eisen, H.N. (1981) Nature, 294, 654-56.

Sefton, B.M., Hunter, T., Ball, E.H., and Singer, S.J. (1981) Cell, 24, 165-74.

Tabachnik, M., and Sobotka, H. (1960) J. Biol. Chem. 235, 1051-60.

Tarone, G., Galetto, G., Prat, M., and Comoglio, P. (1982) J. Cell Biol., 94, 179-86.

Towbin, H., Stahelin, T., and Gordon, J. (1979) Proc. Natl. Acad. Sci. USA, 4350-54.

35 GP 85: A Plasma Membrane Glycoprotein Associated to the Triton-insoluble Cytoskeleton of BHK Cells

Guido Tarone, Riccardo Ferracini
Maria Rosa Amedeo, Paolo Comoglio

INTRODUCTION

Extraction of cells in culture with nonionic detergents solubilizes most cellular proteins, leaving insoluble structures, which retain the overall morphology of intact cells. These structures, termed detergent-resistant cytoskeletons, contain mainly nuclei and cytoskeletal filaments (1).

Since nonionic detergents do not affect protein-protein interactions (2), this method has also been applied to study membrane-cytoskeleton interactions. It has been demonstrated that cell surface glycoproteins, induced to interact in a transmembrane manner with intracellular microfilaments by aggregation with multivalent ligands, become associated to the detergent-insoluble cytoskeletons (3,4).

In this paper we report the identification of a surface glycoprotein (gp85) that remains associated to Triton-insoluble cytoskeletal structures prepared from different mammalian cells. This glycoprotein is not associated to the extracellular matrix, but rather to the cell membrane. Extraction experiments suggest a possible interaction of gp85 with actin filaments.

METHODS

Labelling of Cell Surface Proteins and Detergent Extraction

BHK21/C13 cells were grown in Dulbecco's modified Eagle's medium (DMEM) supplemented with 5% fetal calf serum, 10% tryptose phosphate broth and antibiotics. In order to selectively label surface molecules the lactoperoxidase-glucose oxidase-catalyzed radioiodination procedure was used as previously described (5). Glycoproteins were metabolically labelled by incubating cell monolayers for 15 h in complete medium containing 30 μC/ml of 3H-6D glucosamine (20 Ci/mmole Amersham).

Detergent extraction was performed by incubating either adherent or suspended cells for 20 min at 0°C in a buffer containing 0.5% Triton X-100, 20 mM Tris-HCl pH 7.4, 150 mM NaCl and 2 mM phenylmethylsulphonyl fluoride as protease inhibitor. Cells were rinsed twice in this buffer and insoluble material was dissolved in sodium dodecylsulphate (SDS) for electrophoretic analysis.

Immunofluorescence and Scanning Electron Microscopy

Cells, grown to subconfluent monolayers on 12 mm glass coverslips, were extracted with Triton X-100 under standard conditions and fixed with 3.5% formaldehyde for 20 min at room temperature. FITC-labelled phalloidin (kindly provided by Dr. Th. Wieland) was used to stain F-actin. Intermediate filaments were decorated by using affinity chromatography purified guinea pig antibodies to vimentin (a generous gift of Prof. W. Franke, Heidelberg, West Germany) and FITC-labelled antiserum to guinea pig immunoglobulins (Miles).

For scanning electron microscopy, cells were fixed with 2% glutaraldehyde and postfixed in 1% osmium tetroxide. After critical point drying and gold coating, cells were examined in the scanning electron microscope.

Immunoprecipitation Analysis

An antiserum to purified human plasma fibronectin, which cross-reacts with mouse and hamster fibronectins, has been used (6). Immunoprecipitation was performed as previously described (6). In order to identify components of the Triton-insoluble cell residues, this material was solubilized with the Triton-DOC-SDS buffer (20 mM Tris-HCl pH 7.4, 150 mM NaCl, 0.5% Triton X-100, 0.5% sodium desoxycholate, 0.1% sodium dodecylsulphate, 5 mM ethylendiamine tetraacetic acid and 2 mM phenylmethylsulphonyl fluoride as protease inhibitor). The extracts were sonicated and centrifuged at 12.000 × g for 15 min. After incubation with the appropriate antiserum, immunocomplexes were recovered by adsorption on Protein A-Sepharose beads (Pharmacia).

Electrophoresis and Fluorography

Sodium dodecylsulphate polyacrylamide electrophoresis (SDS-PAGE) was carried out in 5-15% acrylamide gradient slab gels using the procedure previously described (6).

RESULTS

Biochemical Composition of Triton-Extracted BHK Cells

As shown by scanning electron microscopy, Triton X-100 extracted adherent cells consisted in the nucleus surrounded by a fibrillar meshwork. The plasma membrane and most cytoplasmic material were removed by this treatment (Fig. 35.1A, B). Immunofluorescence with FITC-phalloidin or with antivimentin antibodies indicated that the filamentous material contained actin fibers and intermediate filaments (Fig. 35.1C, D).

Electrophoretic analysis of the detergent-insoluble material indicated that two components with molecular weight of 43K and 58K, corresponding to actin and vimentin, are the major constituents of the extracted cells (Fig. 35.2A). These results confirmed that most cytoplasmic components have been solubilized by Triton and that the insoluble residues, which will be referred to as "skeletons," consisted mainly of nuclei and cytoskeletal components.

Membrane Glycoproteins Associated with the Triton-Insoluble Cell Skeletons

When membrane proteins of adherent BHK cells were radiolabelled, a major labelled component with a molecular weight of 85K remained associated to the Triton extracted cell skeletons (Fig. 35.2C). All other cell surface membrane proteins were solubilized (Fig. 35.2B). The 85K component was also labelled with 3H-glucosamine (Fig. 35.2D), indicating that this is a surface glycoprotein (gp85) synthesized by the cell.

A surface-labelled component with a molecular weight of 220K was also found in the Triton-insoluble skeletons (Fig. 35.2C, D). This protein corresponded to fibronectin as it was selectively immunoprecipitated by specific antibodies (Fig. 35.2E). The antifibronectin serum did not recognize the gp85 (Fig. 35.2E) indicating that the latter is immunologically distinct

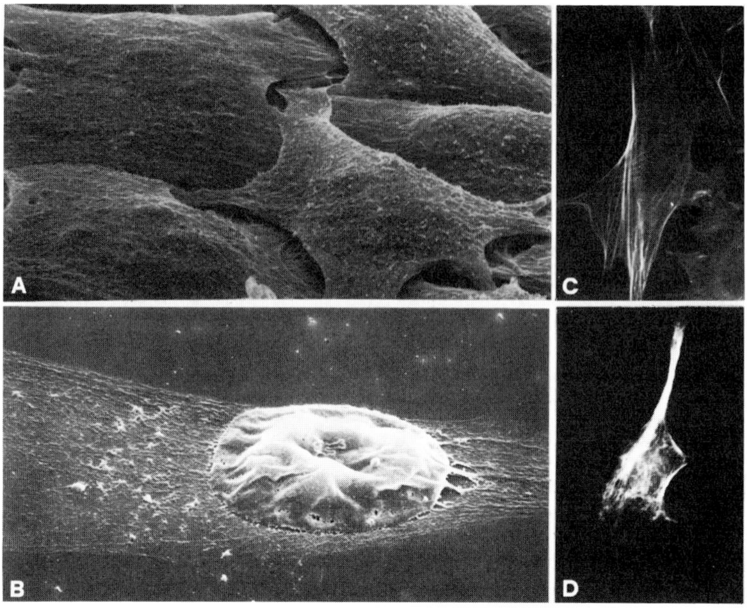

Fig. 35.1. Morphological analysis and immunofluorescence microscopy of Triton-extracted BHK cells. Scanning electron micrographs of: intact cells (A), cells extracted with 0.5% Triton in isotonic buffer for 20 min at 0°C (B). Immunofluorescence staining of Triton-extracted cells: FITC-phalloidin (C) and antivimentin antibodies followed by FITC-labelled anti-immunoglobulins (D).

from fibronectin and does not represent a proteolytic fragment of this molecule.

To test whether gp85 was associated to the matrix deposited by the cells on the culture substratum, we performed extraction experiments on suspended cells. 3H-glucosamine-labelled cells were detached from the culture dish with EDTA and subsequently extracted with Triton under standard conditions. As shown in Figure 35.2F, gp85 was still present in the Triton-insoluble material indicating that this molecule is associated with the cell membrane.

The association of gp85 with the detergent-insoluble structures was studied by using different extraction conditions. Different extraction times, from 5 min to 60 min, were tested. Most cellular glycoproteins, with the exception of gp85, were solubilized within the first five minutes of extraction. Longer extraction times did not affect significantly the amount of gp85 associated to the skeletons. Similarly three subsequent extractions with fresh buffer did not release gp85 from the skeletons, indicating that this molecule is firmly bound to the insoluble structures and its insolubility is not due to an unfavorable detergent-protein ratio.

Another set of experiments was performed by extracting the cells with Triton X-100 under different ionic conditions. Low ionic strength buffers lacking divalent cations caused almost complete solubilization of vimentin from cell skeletons (7), but did not release gp85 (Table 35.1). On the contrary, high ionic strength buffers containing divalent cations eluted a significant fraction of actin without affecting vimentin (8). Under these conditions gp 85 was partially extracted (Table 35.1), suggesting that this molecule might interact with actin filaments.

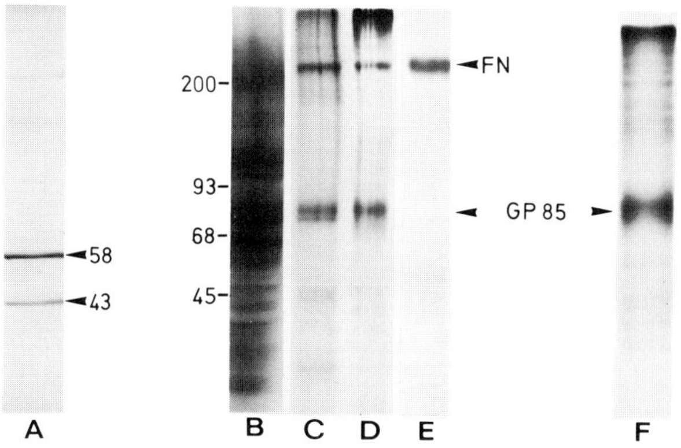

Fig. 35.2. Biochemical analysis of the Triton-insoluble skeletons of BHK cells. Cells, labelled with lactoperoxidase-catalyzed surface radioiodination or with 3H-glucosamine, were extracted with Triton under standard conditions. The insoluble skeletons were solubilized with SDS and analyzed by SDS-PAGE. The Coomassie blue stained pattern is shown in (A). Labelled material was detected by fluorography. Surface radioiodinated cells: Triton-soluble material (B), Triton-insoluble skeletons (C). 3H-glucosamine labelled cells: Triton-insoluble skeletons (D), material immunoprecipitated with antibodies to fibronectin from the Triton-insoluble fraction (E), Triton-insoluble skeletons prepared from cells in suspension (F). The mobility of proteins of known molecular weights is indicated on the right side of lane B. Arrowheads indicate the position of GP85, fibronectin (FN), vimentin (58) and actin (43).

Table 35.1

Extraction of Cell Skeletons with High of Low Ionic Strength Buffers

Ionic Strength of Extraction Buffer	% of the Triton X-100 Insoluble GP85 (*)
Isoosmotic (†)	100
Low (‡)	100
High (§)	45

(*)The values of gp85 detected in the skeletons extracted in isotonic buffer were arbitrarily considered as 100% and the amount of proteins left in the residues after extraction with other buffers is expressed as a percentage referred to this sample. Values represent the means of three independent experiments.

(†) 20 mM Tris-HCl pH 7.4, 150 mM NaCl, 0.5% Triton.

(‡)20 mM Tris-acetate pH 7.4, 1 mM EGTA, 0.5% Triton.

(§) 600 mM KCl, 4 mM Mg-acetate, 40 mM Imidazole, 1 mM cysteine 1 mM EGTA, 10 mM ATP, 0.5% Triton pH 7.

Note: 3H-glucosamine-labelled BHK cells were extracted for 20 min at 0°C with buffers indicated below, and the insoluble cell skeletons were analyzed by SDS-PAGE. The amount of gp85 was determined by cutting the electrophoretic band and counting the associated radioactivity.

Detection of GP85 in the Detergent-Insoluble Skeleton of Different Mammalian Cells

To test whether the gp85 is present in cells other than BHK fibroblasts, we analyzed the Triton-insoluble skeleton of a number of chick, mouse, rat, and human cells either normal or malignant. Subconfluent cultures were labelled with 3H-glucosamine and insoluble skeletons, prepared according to the standard protocol, were analyzed by SDS-PAGE. A glycoprotein, that comigrated with the gp85 of BHK cells, was detected as a major labelled component of the Triton-insoluble skeletons of all cells studied (Fig. 35.3) with the only exception of chick embryo cells (not shown). In NRK (rat) and HT-1080 (human) cells, in addition to the gp85 another labelled glycoprotein component, migrating with an apparent molecular weight of 140K (NRK) and 140-160K (HT-1080) (Fig. 35.3 arrow), was also detected. These components could possibly correspond to the 140K detergent-insoluble glycoprotein described by Letho and Carter (9,10).

Gp85 is, then, a membrane glycoprotein expressed in many different normal and transformed mammalian cells.

DISCUSSION

In this paper we describe a membrane glycoprotein—gp85—that is associated with detergent-insoluble skeletons of different mammalian cells in culture. This molecule is exposed at the outer cell surface as shown by its labelling with lactoperoxidase-catalyzed radioiodination. Moreover gp85 is not a serum component adsorbed at the cell surface since it can be metabolically

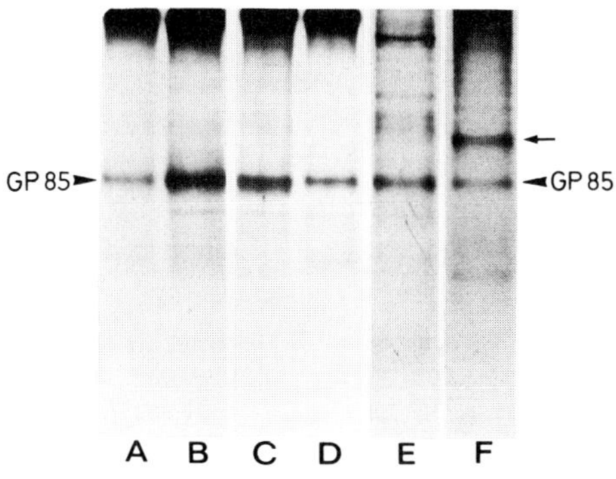

Fig. 35.3. Electrophoretic analysis of the Triton-insoluble skeletons from different cell lines. Subconfluent cell monolayers were labelled with 3H-glucosamine and Triton-insoluble skeletons were prepared according to the standard protocol. BHK hamster cells (A); B77-3T3 mouse transformed cells (B); AA6-B77-3T3 mouse cell (C); 3TP mouse cells (D); HT-1080 human sarcoma cells (E); NRK rat cells (F). Arrow indicates the position of a 140-160K component present only in HT-1080 and NRK cells.

labelled with 3H-glucosamine. In agreement with previous reports, fibronectin, a glycoprotein of the extracellular matrix, was also found to be present in the detergent-insoluble material. Using antibodies specific for fibronectin, we have shown that gp85 is not related to this molecule. Moreover, while fibronectin is mostly associated to the culture substratum (10), gp85 is part of the plasma membrane as it is linked to detergent-insoluble skeletal structures prepared from cells in suspension.

Gp85 is the only cell surface membrane glycoprotein of BHK cells that is not solubilized by Triton under the condition used. A glycoprotein with molecular weight of 140K has been described as a major cell surface component of the detergent-insoluble cytoskeleton of human fibroblasts (9,10). In this study we detected a glycoprotein with molecular weight of 140K, that can correspond to that described by Carter (9) and Lehto (10), only in the skeletons of NRK and HT-1080 cells. This component, however, was never detected in BHK cells, indicating that these cells either do not express and expose this molecule or do not incorporate it in the detergent-insoluble material.

It has been shown that antibody-induced clustering of plasma membrane glycoproteins causes their transmembrane linkage to microfilaments (11,12). After the occurrence of such interaction, the membrane proteins, that are otherwise soluble in detergents, become detergent-insoluble and remain associated to the unextracted cytoskeleton (3,4). The experiments reported in this paper indicate that ligand-induced clustering is not required for the association of gp85 with the detergent-insoluble skeletons. This surface glycoprotein is, thus, normally bound to the cytoskeleton; actin filaments might be the attachment site, as suggested by the extraction experiments.

To better understand the mechanisms of interaction of gp85 with the cell skeleton it is important to elucidate whether this molecule is integral to the membrane and spans the lipid bilayer.

ACKNOWLEDGMENTS

We would like to thank Prof. Carlo Torre of the Dept. of Forensic Medicine for the scanning electron micrographs. This work was supported by grants of the Progetto Finalizzato Controllo Crescita Neoplastica (CNR) and of the Ministry of Education.

REFERENCES

1. Osborn M. and Weber K. (1977) Exp. Cell Res. 106, 339-49.
2. Helenius A., MiCaslin D., Fried E., and Tanford C. (1979) Methods Enzymology 55:734-49.
3. Painter R. and Ginsberg M. (1982) J. Cell Biol. 92, 565-73.
4. Sheterline P. and Hopkins C. (1981) J. Cell Biol. 90, 743-54.
5. Tarone G., Galetto G., Prat M. and Comoglio P. (1982) J. Cell Biol. 94, 179-86.
6. Tarone G., Ceschi P., Prat M. and Comoglio P. (1981) Cancer Res. 41, 3648-51.
7. Traub P. and Nelson W. (1982) J. Cell Sci. 53, 49-76.
8. Bravo R., Small J., Fey S., Larsen P. and Celis J. (1982) J. Mol. Ciol. 154, 121-43.
9. Carter W. (1982) J. Biol. Chem. 257, 3249-57.
10. Lehto V.P. (1983) Exp. Cell Res. 143, 271-86.
11. Bourguignon L. and Singer S. (1977) Proc. Natl. Acad. Sci. 74, 5031-35.
12. Flanagan J. and Koch G. (1978) Nature 273, 278-81.

36 An Actin-modulating Protein from Vertebrate Smooth Muscle

H. Hinssen, J. V. Small, A. Sobieszek

INTRODUCTION

From the studies on nonmuscle cells it is well known that actin undergoes dynamic structural reorganizations during cell motility and locomotion (Lindberg et al., 1978; Korn, 1982) involving changes in its polymer state. Although these changes are, in principle, based on actins' inherent property to polymerize reversibly into filaments, the involvement of additional factors which regulate the polymerization processes is also required. In fact, several types of proteins have been detected in nonmuscle cells which affect the polymer state of actin in various ways (for reviews see Weeds, 1982; Craig and Pollard, 1982; Korn, 1982).

On the other hand, in systems with a rather invariable cytoarchitecture—like muscle cells—proteins which are supposed to influence actin dynamics may not be expected. So far, only very few muscle proteins are known which show an effect on actin polymerization, as for example β-actinin (Maruyama et al., 1977). One group of actin-binding proteins found in nonmuscle cells, specifically interacts with actin in a Ca^{2+}-dependent manner and exhibits characteristic effects on its polymer state. Because of the relative complexity of the observed effects of these proteins the term "actin-modulating protein" has been introduced for this group (Hinssen, 1981a). This class of proteins includes for example, gelsolin from macrophages (Yin and Stossel, 1980; Yin, Zaner and Stossel, 1980; Yin, Maruyama, Hartwig and Stossel, 1981), villin from intestinal brush border microvilli (Bretscher and Weber, 1980; Glenney, Bretscher and Weber, 1980; Glenney, Kaulfus and Weber, 1981), brevin from blood serum (Harris and Schwartz, 1981; Harris and Gooch, 1981), fragmin from Physarum (Hasegawa et al., 1980; Hinssen

1981a,b) and severin from Dictyostelium (Brown et al., 1982; Yamamoto et al., 1982). They are all assumed to be involved in the transformation of actin from a polymer to a nonpolymer state or to effect the gelation-solation transition of actin-containing filament networks in the cytoplasm.

In the present paper we show for the first time that actin modulators of this kind also occur in smooth muscle, a fact which raises new questions about their possible physiological functions. Here we describe the purification of an actin modulator from pig stomach smooth muscle (PSAM = pig stomach actin modulator) and the characteristics of its interaction with actin.

MATERIALS AND METHODS

Purification of Proteins

The actin modulator from pig stomach (PSAM) was purified by a method based on that developed for fragmin from Physarum (Hinssen, 1981a). Pig stomach muscle was carefully cleaned from connective tissue, minced and homogenized in 5 vol of extraction medium (20 mM KCl, 5 mM EGTA, 1 mM DTE, 1 mM $MgCl_2$, 1 mM PMSF, 1% Triton X-100, and 10 mM imidazole pH 7.0). After centrifugation for 30 min at 20,000 × g the supernatant was subjected to ammonium sulfate fractionation; the 40-55% pellet was resuspended in 0.05 M KCl, 1 mM EGTA, 1 mM DTE, 0.2 mM PMSF, 1 mM $MgCl_2$, 10 mM imidazole pH 7.5, and the ammonium sulfate rapidly removed by gel filtration on a 8 × 50 cm column of Sephadex G-25. The eluate was further purified by column chromatography using the following steps: (1) ion exchange chromatography on a 2.5 × 40 cm column of DEAE-Sepharose Cl 6B eluting with a 2-1 gradient of 0.05-0.35 M KCl, (2) gel chromatography on a 3.2 × 95 cm column of Ultrogel AcA 34, (3) chromatography on a 1.5 × 50 cm column of hydroxyapatite eluting with a gradient of 0.05-0.3 M PO_4, and (4) again chromatography on DEAE-Sepharose CL 6B, but in the presence of 6 M urea and on a 1.5 × 20 cm column. After the last purification step the modulator was renatured by dialysis and retained full activity. 1 mM EGTA was present during all chromatography steps except during hydroxyapatite chromatography. After each column, fractions were assayed for actin modulating activity by the viscometric assay described elsewhere (Hinssen, 1979, 1981a); active fractions were pooled and, if necessary, concentrated by precipitation with 65% ammonium sulfate. Skeletal muscle actin was prepared from acetone powder by the method of Spudich and Watt (1971) with a final gel filtration step on Sephadex G-150 (Rees and Young, 1967). A Ca^{2+}-sensitive actomyosin from pig stomach was prepared by the method of Sobieszek and Small (1976).

Electron Microscopy

Negative staining of protein samples was performed on carbon coated UV-irradiated 400 mesh grids using 1% uranyl acetate.

Preparation of Chick Heart Fibroblast Cytoskeletons

Chick heart fibroblasts grown on coverslips were extracted with a Triton-containing medium (Small and Celis, 1978) and the cytoskeletons were stained for actin with rhodamine-phalloidin (a gift from Prof. Th. Wieland, Heidelberg). After washing with 10 mM imidazole, 20 mM KCl, 0.2 mM $CaCl_2$ pH 7.0, the cytoskeletons were incubated with 0.2 mg/ml PSAM in the washing buffer for various times. The reaction was stopped by washing with 1 mM EGTA in 10 mM imidazole pH 7.0.

Viscometry

Viscosity was measured with Ostwald type capillary viscometers with outflow times for water of 15 and 24 sec. All measurements were made at 25°C.

Electrophoresis

SDS-electrophoresis was performed using the microslab system described by Matsudaira and Burgess (1978). For 2-dimensional electrophoresis we used the method of O'Farrell (1975) with a silver staining.

Protein Concentration

Protein was determined by either the biuret method or the dye-binding method of Bradford (1976).

RESULTS

Purification and Molecular Properties of PSAM

By the procedure described we obtained between 20 and 25 mg of purified PSAM per kg of muscle. The protein was practically homogeneous according to SDS-electrophoresis (Fig. 36.1g), and showed a single band with an apparent molar mass of $M_r \sim 85$ kd. This value was also obtained by electrophoresis of PSAM in the phosphate buffer system with a constant acrylamide concentration of 7.5% (Weber and Osborn, 1969). Gel chromatography of PSAM on a calibrated column of Ultrogel AcA 34 revealed a single component with a Stoke's radius of 3.8 nm corresponding to a molecular weight of 78 kd for a globular protein. This value was consistent with the shape of PSAM as determined from electron micrographs of rotary shadowed samples (not shown).

From the SDS-gels of the various stages of purification (Fig. 36.1) it is seen that the most effective step is the first DEAE-chromatography. The main problem thereafter was the separation of PSAM from contaminating actin to which it seemed to be complexed. This was only achieved by the second ion exchange chromatography in the presence of 6M urea.

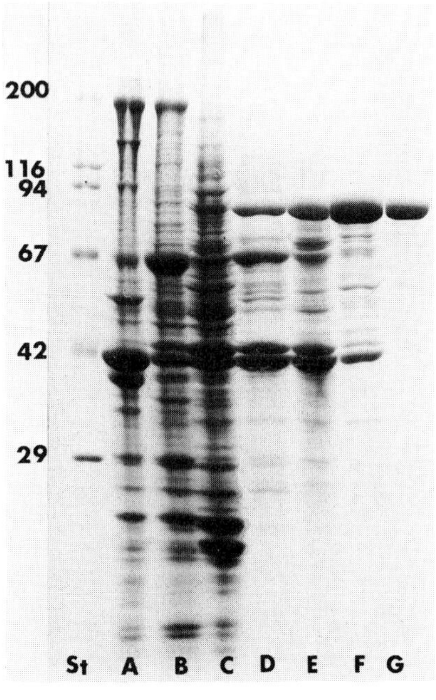

Fig. 36.1. SDS-gel electrophoresis of fractions from various purification steps of the actin modulator. St) molecular weight standards; (A) whole muscle; (B) Triton-EGTA extract; (C) 40-55% sat. ammonium sulfate pellet; (D) 1. DEAE-chromatography; (E) gel filtration; (F) hydroxyapatite chromatography; (G) 2. DEAE-chromatography.

The effect of PSAM on the polymerization of actin, as measured by a viscometric assay, increased in the various fractions by a factor of 400 during preparation with a yield of about one quarter of the total activity in the crude extract (Table 36.1). From this we estimate that there are at least 120 mg of PSAM per kg of muscle tissue. The actual content may be considerably higher because not all PSAM is extracted from the muscle as seen from the 2D electrophoresis gel in Figure 36.2d. Thus, PSAM makes up at least 0.1% of the total protein of smooth muscle.

After isoelectric focussing PSAM appears as a set of at least 5 components in the pH range from 6.0 to 6.1 (Fig. 36.2b). Indeed, it can be seen from the gels of the total muscle (Fig. 36.2a) and of the extract and residue after extraction (Fig. 36.2c,d) that this heterogeneity is not a preparation artifact because the same characteristic set of spots is present in all three gels.

Table 36.1

Purification of Pig Stomach Actin Modulator

	Protein		Activity		
Fraction	Total (mg)	Relative (%)	Specific (units/mg)	Total (units)	Yield (%)
Extract	45,000	100	0.73	32,800	100
40-55% $(NH)_4SO_4$	6,300	14	2.7	17,010	52
1. DEAE	320	0.71	39	13,480	38
Ultrogel AcA 34	162	0.36	65	10,530	32
Hydroxyapatite	68	0.15	140	9,520	29
2. DEAE	25	0.055	290	7,250	22

Note: Yield and activity of various fractions are shown for a typical preparation of PSAM starting from 1,000 g of muscle tissue. The activity was determined from the reduction in the steady state viscosity of actin when polymerized in the presence of aliquots from the fractions. One unit is defined as the amount of actin in mg for which a 50% decrease in high shear viscosity would be obtained by 1 mg protein from the respective fraction.

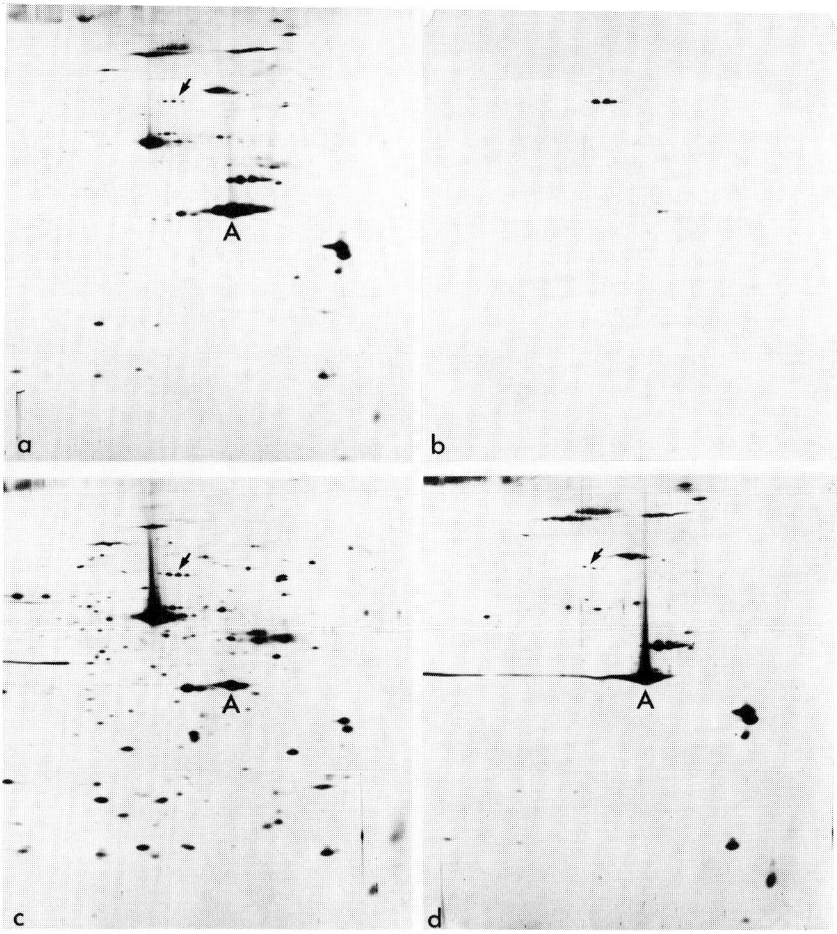

Fig. 36.2. Two dimensional electrophoresis of (a) whole muscle, (b) partially purified PSAM as recovered from hydroxyapatite chromatography, (c) crude extract, and (d) muscle residue after extraction and extensive washing. Arrows indicate the spots of PSAM; A = actin.

Effects of PSAM on the Polymer State of Actin

The pig stomach actin modulator has a very marked effect on the polymer state of actin even at very low concentration. For all experiments described here we have used skeletal muscle actin which was obtained most easily in a rigorously pure state. However, we have also tested pig stomach actin for its interaction with PSAM and found the effects to be practically identical. In Figure 36.3a some sort of overview is given of the various effects of PSAM on the polymer state of actin as revealed by viscometry. When PSAM was added at various molar ratios to G-actin before or synchronously with the salt required for polymerization the following phenomena were observed: (1) a dose-dependent enhancement of nucleation leading to a much higher rate of initial viscosity increase, (2) a reduction in steady state viscosity, also dose-dependent, and (3) a significantly slower approach to the steady state during the elongation phase. This last effect is only detectable at low molar ratios of PSAM when this delay phase is not overcome by the strong nucleation which would then reduce the time to attain steady state.

Addition of PSAM to F-actin caused a dramatic drop in viscosity with the reaction being practically complete before the first measurement was taken. Only a very slight decrease in viscosity was observed during subsequent minutes of incubation. The relative reduction in steady state viscosity can be used to quantify the effect of the modulator on actin. When this is plotted against the molar ratio of the two proteins a relationship is obtained (Fig. 36.3b) which fits to a hyperbolic curve both for addition of PSAM to G- as well as to F-actin. Measurable reductions in viscosity were already obtained at molar ratios of 1:5,000 and less. With high amounts of PSAM the viscosity approached that of G-actin. Double reciprocal plots plots linearize the relationship and we obtained for a 50% decrease in viscosity, molar ratios of 1:550 and 1:400 respectively. The effect on preformed F-actin was always somewhat less than on the polymerization of actin.

The nucleating effect of PSAM on actin polymerization was shown more quantitatively by measuring actin polymerization from the increase in absorption at 232 nm (Higashi and Oosawa, 1965; Fig. 36.4). This method does not induce any changes in the polymerization process itself and is a direct measure of the number of actin:actin bonds formed. The polymerization conditions were adjusted so that no increase in optical density was observed for at least 1 hour after addition of salt. When added to G-actin at various molar ratios PSAM caused an instant increase of absorption after the addition of salt indicating that the formation of nuclei was no longer a time-dependent reaction in the presence of PSAM. The effects of PSAM on the polymer state of actin as revealed from viscometry could result either from an increase of the amount of unpolymerized actin or from a reduction of filament length. To distinguish between these

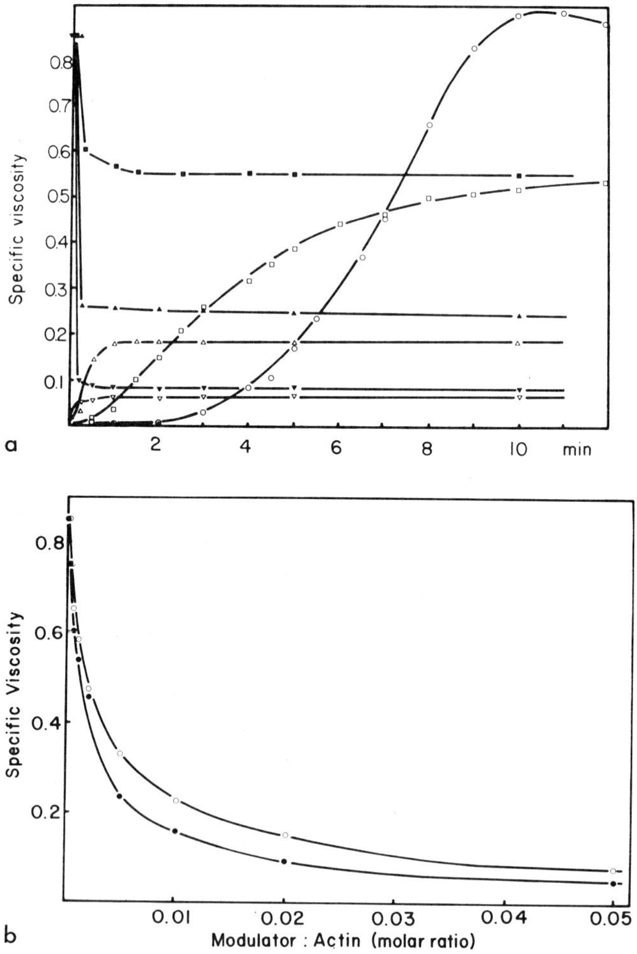

Fig. 36.3. Effects of PSAM on the polymer state of actin measured by viscometry. (a) PSAM was added at molar ratios of 1:300 (□, ■), 1:100 (△, ▲) and 1:30 (▽, ▼) to actin either before polymerization (open symbols) or to already assembled F-actin at steady state (closed symbols), and the viscosity measured at various time points after addition of salt or PSAM, respectively. (○) Control actin polymerized without PSAM. (b) Steady state viscosity of actin in the presence of PSAM as a function of the molar ratio of the two proteins. (●) PSAM added to actin before polymerization. (○) PSAM added to F-actin. Conditions: 1 mg/ml skeletal muscle actin in 1 mM ATP, 0.2 mM Ca Cl_2, 10 mM imidazole, pH 7.4. Polymerization was induced by 70 mM KCl (1), or 100 mM KCl + 2 mM $MgCl_2$ (b).

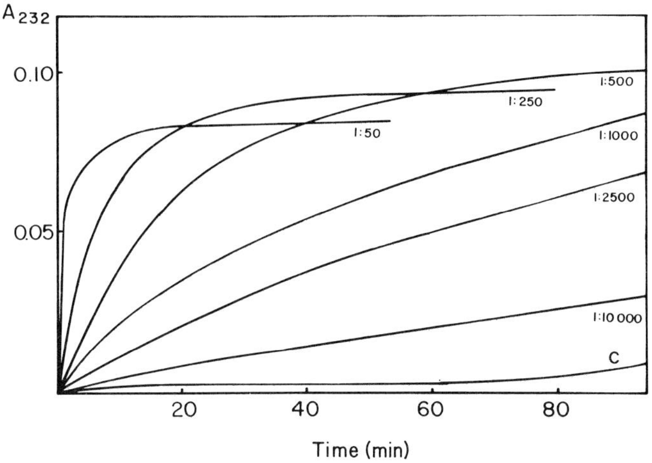

Fig. 36.4. Nucleation of actin polymerization induced by PSAM. Polymerization was followed by the increase in absorption at 232 nm. Skeletal muscle actin (0.3 mg/ml in 0.2 mM $CaCl_2$, 0.01 mM ATP and 2 mM imidazole pH 7.2) was polymerized by the addition of 0.05 M KCl. Because of low nucleation a lag phase of at least 60 min was observed under these conditions for the control (C). The other curves show the time course of polymerization when PSAM was added to actin before polymerization at the molar ratio indicated.

possibilities we have determined the critical concentration of actin in the presence of PSAM (Fig. 36.5). Under the conditions measured—that is in the presence of 2 mM $MgCl_2$—there is in fact a small increase in critical concentration from 0.015 mg/ml to 0.045 mg/ml which seems to be constant for different molar ratios down to about 1:1,000. For the 1:2,000 ratio we obtained an intermediate value and at even lower ratios it was no different from the actin control. However, it is clear that this rather small increase in the amount of unpolymerized actin cannot be responsible for the observed effects on actin viscosity. It may therefore be concluded that the primary effect of PSAM is on actin filament length. Electron microscopy of F-actin to which PSAM had been added at various molar ratios revealed a considerable shortening of filaments even at molar ratios of 1:100 (Fig. 36.6b). At higher ratios the actin was completely converted into fragments and small, nonfilamentous aggregates (Fig. 36.6c,d,e,f). The length distribution of filaments was

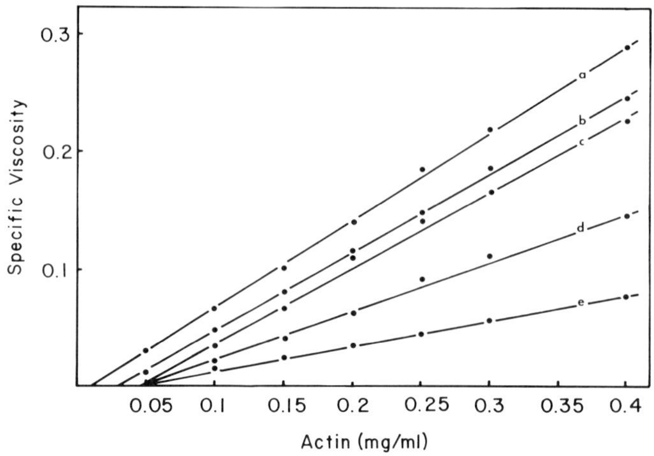

Fig. 36.5. Increase in critical concentration of actin in the presence of PSAM. Actin at various concentrations and with different amounts of PSAM added was polymerized for 15 hours at 25°C by 2 mM $MgCl_2$ and the viscosity measured. From the plots of specific viscosity against actin concentration the critical concentrations were obtained from the intercepts with the abscissa. a) actin only, b-e) actin polymerized in the presence of PSAM at molar ratios of 1:2,000 (b), 1:1,000 (c), 1:300 (d) and 1:100 (e). Conditions: Skeletal muscle actin in 0.2 mM $CaCl_2$, 1 mM ATP and 10 mM imidazole pH 7.4.

heterogeneous in all cases but the average filament length revealed a close proportionality to the ratio of actin to modulator. Moreover, the number of actin molecules per average filament was almost identical to the molar ratio of the two proteins (data not shown), indicating that one molecule of modulator was bound per filament. Similar populations of short filaments and fragments were obtained when PSAM was added to actin before polymerization.

Ca^{2+}-Dependence of PSAM–Actin Interaction

The interaction of PSAM and actin is independent of the concentrate of KCl or $MgCl_2$ used for actin polymerization and only slightly pH-dependent. However, activity is completely dependent on the presence of micromolar amounts of Ca^{2+}. At

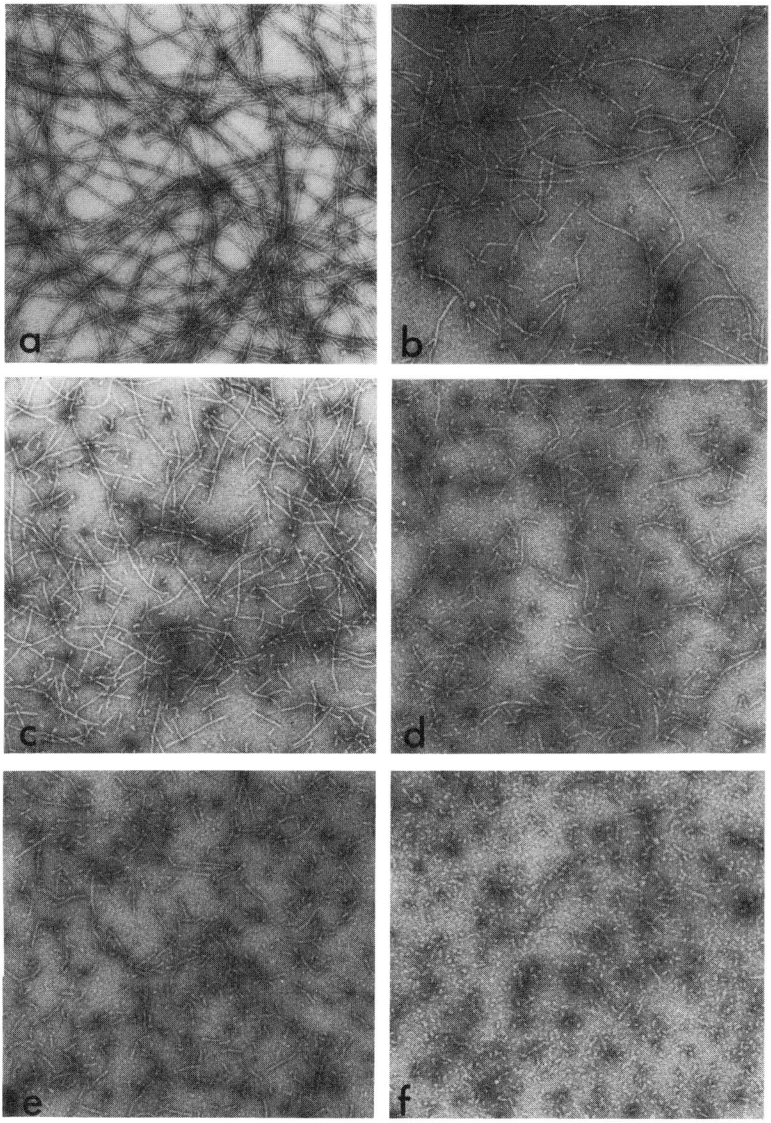

Fig. 36.6. Electron microscopy of negatively stained actin filaments after addition of various amounts of PSAM. (a) F-actin alone, b-f) F-actin with PSAM at a molar ratio of 1:100 (b), 1:50 (c), 1:20 (d), 1:10 (e) and 1:5 (f). Magnification: (a) 50,000, (b-f) 75,000.

Ca^{2+} < 10^{-6} M there is no or very little effect of PSAM as detected by either viscometry or electron microscopy.

Parallel comparison, under identical conditions, of the dependence of PSAM activity and of smooth muscle actomyosin ATPase activity on the free Ca^{2+} concentration revealed interesting differences (Fig. 36.7). Thus, whereas actomyosin ATPase was fully activated by a free Ca concentration around 10^{-6} M, maximal activity of PSAM was achieved at a 10-fold higher Ca^{2+} concentration. Accordingly, the curve of PSAM activity began to rise at 10^{-6} M Ca^{2+}, reached a maximum at 10^{-5} M and decreased again above 10^{-3} M.

From equilibrium dialysis experiments with $^{45}Ca^{2+}$ it was established that PSAM binds 2 moles of Ca in the concentration range required for activation (data not shown).

The interaction of PSAM and actin was not reversed immediately after reduction of the free Ca concentration to <10^{-7} M. When EGTA was added to a sample of actin which had been reacted with PSAM a very slow increase in viscosity was observed—about 50% within the first hour and then a steady decrease approaching equilibrium only after 2-3 days. Electron micrographs from such samples revealed a re-elongation of actin filaments.

Time Course of PSAM—Actin Interaction

As seen in Figure 36.3 the interaction between PSAM and F-actin was so rapid that its time course could not be properly followed. However, the slow reversal of the reaction after Ca^{2+} depletion enabled an approach to this problem by the following experiment: After mixing PSAM and F-actin rapidly, the reaction was stopped at different times by addition of 2 mM EGTA which instantly inactivated the modulator unbound to actin but did not effect release of modulator already bound. Thus, the effect of PSAM for short time intervals could be assessed. Figure 36.8a shows that there was already a marked reduction in viscosity after only 2 sec of modulator action, and after 10 sec the reaction was practically complete. Electron micrographs of the actin filaments at different time points revealed in fact a progressive shortening of the filaments within the first 8 sec (Fig. 36.8b,c,d).

The dynamics of interaction of the modulator on F-actin were also demonstrated on chick heart fibroblast cytoskeletons where the actin was specifically stained with rhodamine-labelled phalloidin (Fig. 36.9). With a relative excess of modulator

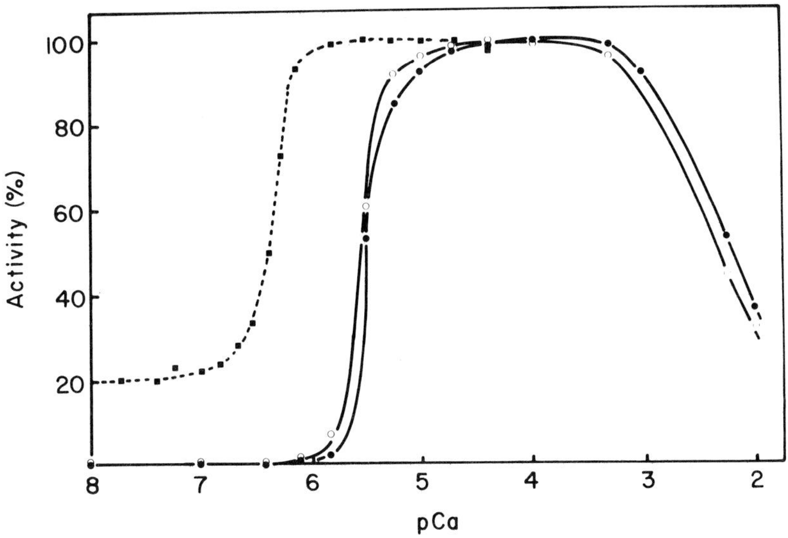

Fig. 36.7. Ca^{2+} dependence of PSAM-actin interaction. Skeletal muscle actin (1 mg/ml in 1 mM ATP and 10 mM imidazole pH 7.2) was polymerized by 100 mM KCl + 2 mM $MgCl_2$ and in the presence of various concentrations of Ca^{2+}. PSAM at a molar ratio of 1:100 to actin was added either before polymerization (○) or to F-actin (●) and from the resulting steady state viscosities relative to parallel samples containing actin only, the activity of PSAM was calculated taking the Ca^{2+} concentration with the maximum effect as 100%. The relative ATPase activity of a calcium-sensitive crude actomyosin from pig stomach was measured under identical ionic conditions and Ca^{2+} concentrations (■). Concentrations of $Ca^{2+} \leq 10^{-5}$M were adjusted with a 2 mM Ca-EGTA buffer (Sobieszek and Small 1976).

added to observe a complete degradation of the actin cytoskeleton within 3-4 min, but with an intereating differentiation for the various elements, the leading edge was most sensitive to modulator action being partly decomposed within 5 sec (Fig. 36.9b). After 30 sec the leading edge and all actin not located in the stress fibers was completely removed (Fig. 36.9c). Stress fibers seem to be relatively resistant and were gradually degraded in the following 2 min (Fig. 36.9d,e,f).

The relative resistance of the stress fibers might be due to the presence of proteins—especially tropomyosin which is not

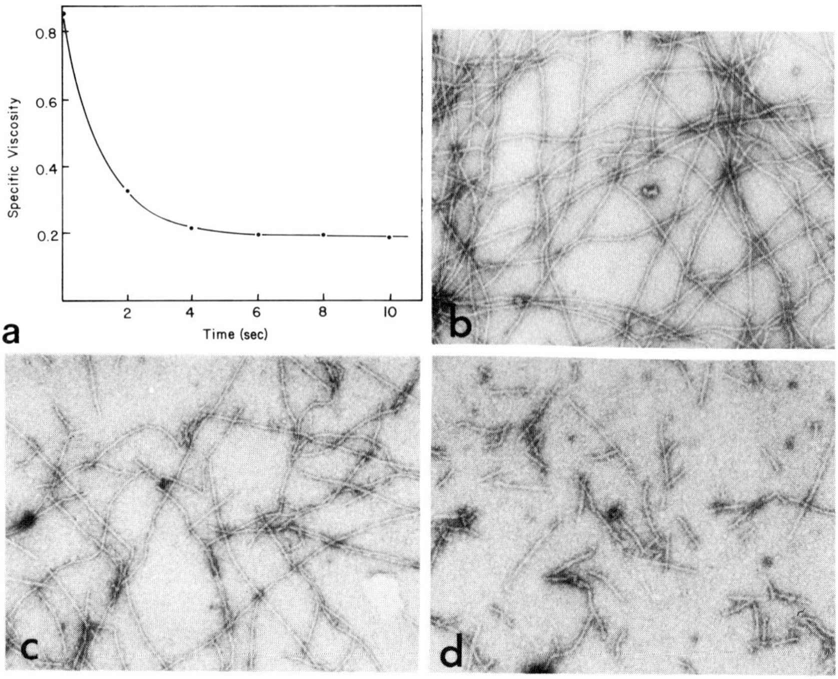

Fig. 36.8. Time course of the interaction of PSAM and F-actin. PSAM was rapidly mixed to F-actin (1 mg/ml actin in 1 mM ATP, 0.2 mM $CaCl_2$, 100 mM KCl, 2 mM $MgCl_2$, 10 mM imidazole pH 7.2) at a molar ratio of 1:50. After a given time interval the reaction was stopped by addition of 2 mM EGTA. The steady state viscosities of samples with various times of PSAM action were plotted (a), and from the time intervals of 2 and 8 sec, samples were taken for electron microscopy (c,d). From the comparison with the actin control (b), a progressive shortening of the actin filaments is revealed. Magnification: 80,000.

Fig. 36.9 (right). Decomposition of actin filaments in Triton-cytoskeletons of chick heart fibroblasts by PSAM. Coverslips with cytoskeletons specifically stained for actin with rhodamine-phalloidin, were incubated for different times with excess PSAM (0.2 mg/ml in 0.2 mM $CaCl_2$, 20 mM KCl and 10 mM imidazole pH 7.0), and the reaction was stopped by washing with 2 mM EGTA. (a) Control: cytoskeleton incubated with buffer only; (b-f) Incubation with PSAM for: 5 sec. (b), 30 sec (c), 45 sec (d), 80 sec (e), 135 sec (f). Magnification: 900.

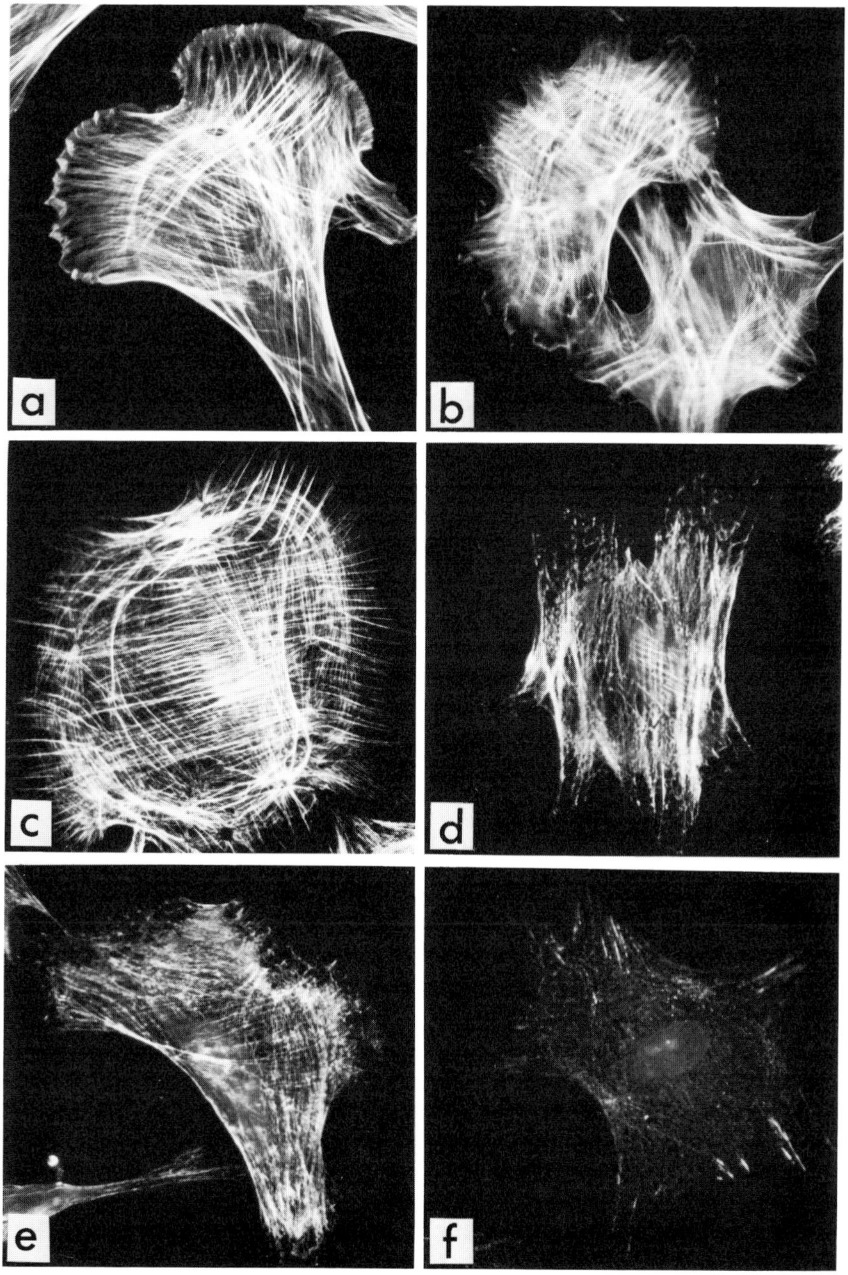
a
b
c
d
e
f

found in the leading edge and which may protect the actin from PSAM action. This is supported by the fact that after a short incubation of the cytoskeletons with skeletal muscle tropomyosin the actin in the leading edge became much more resistant to PSAM and was only degraded after several minutes (not shown).

Formation of a 2:1 Actin-PSAM Complex

Direct binding of the modulator to G-actin was demonstrated by gel chromatography. When actin was mixed with PSAM at a molar ratio of 2:1 in the presence of Ca^{2+} and this mixture applied to a gel filtration column, a single component was eluted with an apparent molecular weight of 160 kd (Fig. 36.10), indicating the formation of a stable complex of PSAM with 2 actins. Evidence for such a complex came also from electron microscopy of rotary-shadowed samples which revealed a single species of globular particles with a definitely larger diameter than the PSAM molecule alone. A 1:1 molar mixture of the two proteins yielded a heterogeneous elution profile of 3 components: one in the position of the 2:1 complex, another one in the position of the modulator alone, and a third in between, probably corresponding to a 1:1 complex. The chromatography of mixtures with higher ratios of actin (3:1 or 4:1) also lead to heterogeneous elution profiles which had always displayed a definite peak in the position of the 2:1 complex. This behavior is consistent with the formation of only one type of stable complex, namely with a 2:1 molar ratio, and indicates that the modulator has two binding sites for actin with approximately the same affinity. No complex formation was observed in the presence of EGTA.

The interaction of the 2:1 complex with actin resembled only partly that of the modulator alone, as seen from the viscosity data in Fig. 36.11. On one hand when added to G-actin the preformed complex induced nucleation and lead to the same reduction in steady state viscosity of actin. However, after addition to F-actin the complex caused no rapid drop in viscosity as seen with the modulator alone. There was only a gradual decrease in viscosity over a period of several hours and which never attained the low steady state viscosity induced by PSAM alone. An intermediate effect was observed when a 1:1 mixture of PSAM and actin was added to F-actin.

From these experiments it is concluded that the modulator when complexed with actin looses its ability to immediately shorten actin filaments. The slow viscosity decrease observed may be due to a gradual reorganization of actin into shorter

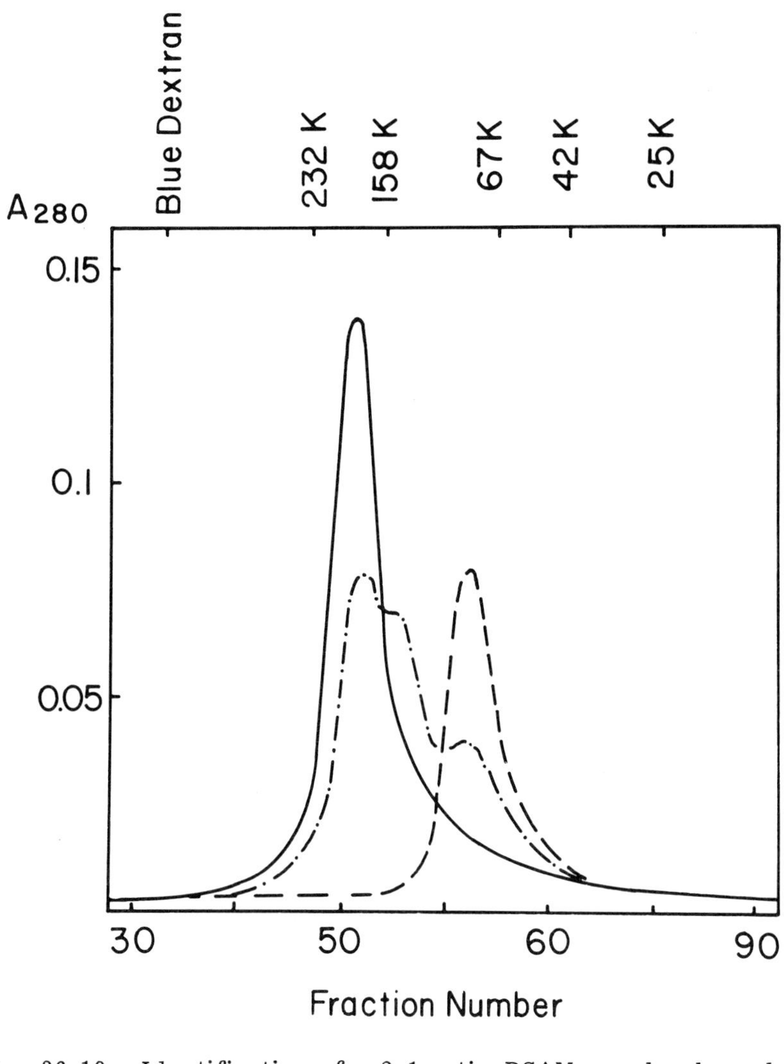

Fig. 36.10. Identification of a 2:1 actin-PSAM complex by gel chromatography. 1 ml samples were applied to a 110 × 1.5 cm column of Ultrogel AcA 34 and eluted at a flow rate of 9 ml/hour with 0.1 M KCl, 0.2 mM $CaCl_2$, 0.5 mM DTE and 10 mM imidazole pH 7.2. (---) PSAM alone (0.5 mg); a mixture of actin and PSAM (0.5 mg) at a molar ratio of 2:1 (——) and 1:1 (-·-·-), respectively. The reference scale at the top margin was obtained with a set of standard proteins (Catalase, Aldolase, Ovalbumin, Chymotrypsinogen A).

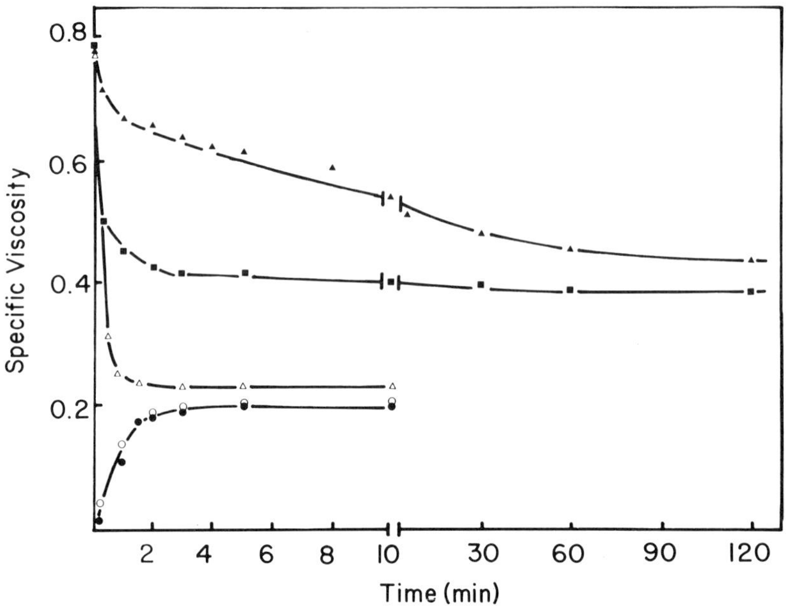

Fig. 36.11. Effect of actin-PSAM complex on the polymer state of actin. Skeletal muscle actin (1 mg/ml in 0.2 mM $CaCl_2$, 1 mM ATP, 10 mM imidazole pH. 7.4) was polymerized by 100 mM KCl + 2 mM $MgCl_2$. PSAM was added to G-actin before polymerization (○, ●) or to already assembled F-actin (△, ▲, ■) and the changes in viscosity measured. Open symbols: PSAM alone; closed symbols: mixtures of actin and PSAM of 2:1 (●, ▲) or 1:1 (■) molar ratio were added to actin

filaments caused by the combined nucleating and capping effect of the PSAM-actin complex. This assumption is supported by the observation that PSAM blocks the fast polymerizing end of actin filaments as revealed from growth experiments using myosin S-1-decorated fragments (not shown).

DISCUSSION

We have purified a protein from pig stomach smooth muscle that interacts with actin in a Ca-dependent manner. In vitro this protein modulates the polymer state of actin and its diverse effects on actin may be summarized as follows: In stoichiometric amounts PSAM forms a stable, unpolymerizable complex with two actin molecules, thereby completely inhibiting actin polymerization or—when added to F-actin—leading to a complete depolymerization. In substoichiometric amounts to actin PSAM induces the formation of short filaments no matter if it is added to actin before polymerization or to F-actin. The average filament length of actin in the presence of PSAM is dependent on the molar ratio of the two proteins. PSAM also increases the critical concentration of actin by a factor of 3-4 already at low molar ratios, it promotes nucleation of actin polymerization, but decreases the elongation rate. In addition, the activity of PSAM is completely Ca-dependent but the reversion of the reaction is very slow after Ca^{2+} is again decreased.

The following interpretation of the described effects is consistent with the experimental data: Whereas at low Ca concentrations there is no or only a very weak interaction between PSAM and actin, the modulator gains a high affinity for actin at $Ca^{2+} > 10^{-6}$ M. The stable 2:1 complex formed is a nucleus for actin polymerization but also caps the fast polymerizing end of the actin filament. PSAM has no preference for G-actin or the terminal actins of a filament, but binds with high affinity to intrafilamentous actins and—by weakening actin-actin interactions—severs the filament at this site with the modulator bound to one of the fragments.

Similar explanations have been given for the reactions of other actin modulators, for example, fragmin (Sugino and Hatano, 1982), gelsolin (Yin et al., 1981) and villin (Glenney, Kaulfus and Weber, 1981). Thus, the basic properties—severing, capping and nucleation—seem to be common to the whole group of actin modulators. The presumed severing of actin filaments distinguishes these proteins from the mere "capping proteins" (Isenberg, Aebi and Pollard, 1980; Killiman and Isenberg, 1982) which also nucleate polymerization and cap actin filaments.

The formation of short filaments from F-actin is by itself not a conclusive evidence for a direct severing or fragmenting mechanism. As has already been pointed out (Yin, Zaner and Stossel, 1980; Isenberg and Maruta, this volume) short filaments may also be formed by a reorganization of the actin after enhanced nucleation and capping. For PSAM this latter possibility was ruled out by following the time course of its interaction with F-actin very closely. As we have shown the reaction is completed within less than 10 sec and a significant shortening of the filaments is observed after only 2 sec. This cannot be explained by a reorganization process after addition of nuclei. In addition, the fast decomposition of actin structures in fibroblast cytoskeletons induced by PSAM also suggests a direct severing of actin filaments.

The barbed end of the actin filament as the capping site has already been reported for other actin modulators (Sugino and Hatano, 1982; Yin et al., 1981; Glenney, Kaulfus and Weber, 1981) and the same seems to be the case for PSAM. The decreased polymerization rate during elongation, the increase in critical concentration and the inhibition of monomer addition at the barbed end in the presence of PSAM suggest it as the preferred binding site. An additional binding of PSAM to the pointed end without interfering with monomer addition can be excluded by the observation that there is only one molecule of PSAM bound per actin filament (Hinssen, Small and Sobieszek, in press). The 3-4-fold increase in the critical concentration of actin induced by PSAM is in the same range as that reported for other barbed end cappers by other methods (Wegner and Isenberg, 1983, Tilney et al., 1983; Neuhaus et al., 1983). The fact that this increase was already obtained at PSAM to actin ratios of 1:1,000 and did not further increase with higher amounts of PSAM reflects the saturation condition where all filaments are capped. Below this ratio, part of the filaments have their barbed ends free leading to intermediate critical concentrations.

Under the conditions we have used to demonstrate the nucleation of actin polymerization by PSAM, spontaneous nucleation was extremely low and fragmentation negligible (Wegner and Savko, 1982). Thus, the initial rate of polymerization in the presence of PSAM was directly dependent on the amount of PSAM. From this and the fact that no time lag in polymerization was observed for the formation of nuclei, it is concluded that the PSAM-actin complex readily formed under G-actin conditions is the nucleus for polymerization. Normal actin polymerization requires a nucleus size of four subunits (Oosawa and Asakura, 1975; Tobacman and Korn, 1983). The strong nucleation explains the formation of short filaments when actin is polymerized in the presence of PSAM because the number of filaments is much higher than for actin alone.

In essence, the complex phenomena of the PSAM-actin interaction are based on a very simple molecular mechanism, namely the formation of a stoichiometric complex with actin. PSAM and probably the other actin modulators with severing activity are distinguished from mere capping proteins only by their ability to bind to intrafilamentous actins, whereas the affinity of the capping proteins is restricted to G-actin or to the terminal molecules of the actin filament. Both groups of proteins block only part of the binding sites of the complexed actin, making further addition of monomers possible. This property distinguishes them from other actin-complexing proteins like profilin, DNAase I and brain ADF, where a binding of actin to the complex is no longer possible.

Though all known severing proteins apparently use the same basic mechanism for interaction with actin, they reveal significant differences in other properties. First of all, they fall into two molecular weight classes, namely the 40 kd class (fragmin and severin), and the 90 kd class (gelsolin, villin, brevin or plasma ADF, and PSAM). The number of actin molecules bound is also different. Whereas fragmin binds one actin (Hasegawa et al., 1980; Hinssen, 1981a) and PSAM definitely two, an apparent number of 1.6 was found for gelsolin (Yin et al., 1981), and a complex with three actin monomers was observed for villin (Glenney, Kaulfus and Weber, 1981). On the other hand, a stable 1:1 complex was found for a gelsolinlike protein from platelets (Kurth et al., 1983). Two moles of Ca^{2+} are bound either by gelsolin (Yin and Stossel, 1980) or PSAM, while villin seems to have only one binding site (Glenney, Bretscher and Weber, 1980). In contrast to all other proteins with severing activity, villin bundles actin filaments in the absence of Ca^{2+} (Bretscher and Weber, 1980). Brevin or plasma ADF has only a limited Ca^{2+} dependence (Harris and Weeds, 1983), whereas the binding of gelsolin, PSAM, fragmin and severin is completely Ca^{2+}-dependent.

From this comparison of the different actin modulators it appears that PSAM is most closely related to gelsolin. This is in accordance with the immunological cross-reactivity of anti-gelsolin with uterine smooth muscle tissue (Yin, Albrecht and Fattoum, 1981).

Though the Ca^{2+} dependence of the actin modulators is suggestive for a possible regulatory mechanism in vivo, it was not yet clear how the activation of the modulators by Ca^{2+} might be correlated to other Ca^{2+}-dependent events in the cell, especially the regulation of contractile processes which is not fully understood as yet for nonmuscle cells. By directly comparing the activation of PSAM and the actomyosin ATPase via the myosin light chain kinase in dependence on the Ca^{2+} concentration under identical conditions, we have now been able to show that the activation of both processes is not coincident. The difference in threshold concentration of Ca^{2+} of almost one order of magnitude provides at least theoretically the possibility of a separate or sequential regulation of both reactions in the living cell.

A problem for the understanding of the possible physiological functions of PSAM is the slow reversion of the PSAM-actin interaction after decrease of Ca^{2+} below micromolar concentrations. This is again a property of all known actin modulators. The complexes of fragmin, severin and also that of the gelsolinlike protein from platelets (Bryan and Wang, 1981; Kurth et al., 1983) seem to be practically unseparable. Thus, the question arises whether these proteins might not occur mainly or exclusively in the form of actin complexes in the cell. We have shown for PSAM that its complex with actin nucleates actin polymerization and shortens filaments as well, but has lost its severing activity. A similar observation was made for fragmin (Hinssen, 1981b). The PSAM-actin complex behaves like a capping protein with no or very limited Ca^{2+} dependence. On the other hand, there is the possibility that in vivo the complex is easily separated by additional regulatory factors.

REFERENCES

Bradford, M.M. (1976) Anal. Biochem. 72, 248-54.

Bretscher, A. and Weber, K. (1980) Cell 20, 839-47.

Brown, S.S., Yamamoto, K. and Spudich, J.A. (1982) J. Cell Biol. 93, 205-10.

Craig, W. and Pollard, T.D. (1982) TIBS 7, 88-93.

Glenney, J.R., Bretscher, A. and Weber, K. (1980) Proc. Natl. Acad. Sci. USA 77, 6458-62.

Glenney, J.R., Kaulfus, P. and Weber, K. (1981) Cell 24, 471-80.

Harris, D.A. and Schwartz, J.H. (1981) Proc. Natl. Acad. Sci. USA 78, 6798-6802.

Harris, H.E. and Gooch, J. (1981) FEBS Lett. 123, 49-53.

Harris, H.E. and Weeds, A.G. (1983) Biochemistry 22, 2728-41.

Hasegawa, T., Takahashi, S., Hayashi, H. and Hatano, S. (1980) Biochemistry 19, 2677-83.

Higashi, S. and Oosawa, F. (1965) J. Mol. Biol. 12, 843-65.

Hinssen, H. (1979) In "Cell Motility, Molecules and Organization" pp. 59-85 (eds. Hatano, S., Ishikawa, S. and Sato, S.) Tokyo.

Hinssen, H. (1981a) Eur. J. Cell Biol. 23, 225-32.

Hinssen, H. (1981b) Eur. J. Cell Biol. 23, 234-40.

Isenberg, G., Aebi, U. and Pollard, T.D. (1980) Nature 288, 455-59.

Kilimann, M.W. and Isenberg, G. (1982) EMBO J. 1, 889-94.

Korn, E.D. (1982) Physiol. Rev. 62, 672-737.

Kurth, M.C., Wang, L., Dingus, J. and Bryan, J. (1983) J. Biol. Chem. 258, 10895-903.

Lindberg, U., Carlsson, L., Markey, F. and Nyström, L.E. (1978) In: The Cytoskeleton in Normal and Pathologic Processes (G. Gabbiani, ed.) Basel.

Maruyama, K., Kimura, S., Ishii, T., Kuroda, M., Ohashi, K. and Muramatsu, S. (1977) J. Biochem. (Tokyo) 81, 215-32.

Matsudeira, P.T. and Burgess, D.R. (1978) Anal. Biochem. 87, 386-96.

Neuhaus, J.M., Wanger, M., Keiser, T. and Wegner, A. (1983) J. Muscle Res. Cell Motil. 4, 507-27.

O'Farrell, D. (1975) J. Biol. Chem. 250, 4007-21.

Oosawa, F. and Asakura, S. (1975): Thermodynamics of the Polymerization of Protein, Academic Press, London.

Rees, M.K. and Young, M. (1967) J. Biol. Chem. 242, 4449-58.

Sobieszek, A. and Small, J.V. (1976) J. Mol. Biol. 101, 75-92.

Small, J.V. and Celis, J. (1978) Cytobiologie 16, 308-25.

Spudich, J.A. and Watt, S. (1971) J. Biol. Chem. 246, 4866-71.

Sugino, H. and Hatano, S. (1982) Cell Motility 2, 457-70.

Tilney, L.F., Bonder, E.M., Coluccio, L.M. and Mooseker, M.S. (1983) J. Cell. Biol. 97, 112.

Tobacman, L.S. and Korn, E.D. (1983) J. Biol. Chem. 258, 3207-14.

Wang, L. and Bryan, J. (1981) Cell 25, 637-49.

Weber, K. and Osborn, M. (1969) J. Biol. Chem. 244, 4406-12.

Weeds, A.G. (1982) Nature 296, 811-16.

Wegner, A. and Savko, P. (1982) Biochemistry 21, 1909-13.

Wegner, A. and Isenberg, G. (1983) Proc. Natl. Acad. Sci. USA 80, 4922.

Yamamoto, K., Pardee, J.D., Reidler, J., Stryer, L. and Spudich, J.A. (1982) J. Cell. Biol. 95, 711-19.

Yin, H.L. and Stossel, T.P. (1980) J. Biol. Chem. 255, 9490-93.

Yin, H.L., Zaner, K.S. and Stossel, T.P. (1980) J. Biol. Chem. 255, 9494-9500.

Yin, H.L., Maruyama, K., Hartwig, H. and Stossel, T.P. (1981) J. Biol. Chem. 256, 9693-97.

Yin, H.L., Albrecht, J.H. and Fattoum, A. (1981) J. Cell Biol. 91, 901-906.

37 Sulfogalactosylceramide: A Cytoskeleton Associated Glycolipid in EUE Cell Line

A. Giuliani, A. M. Fuhrman Conti,
L. Bolognani, B. Zalc

INTRODUCTION

For the past few years, evidences for the involvement of glycolipids in different functions of the cell membrane, such as receptors (1), ion transport (2,3), or differentiation markers (4), have shed light on the importance of this group of molecular constituents of the plasma membrane. They are generally assumed to be solely on the external leaflet of the plasma membrane (5). As a matter of fact, all the immunocytochemical studies dealing with the localization of glycolipids have reported that in normal conditions glycolipids are present only on the cell surface. More recently Raff et al. have even brought the electron microscopic evidence that the neutral glycosphingolipid, galactosylceramide, is located solely on the external surface of the oligodendrocyte in culture (6). Only in pathological conditions (such as in Gaucher's disease (unpublished data) or in the oligodendrocyte of some dysmyelinating neurological mutants such as the Jimpy mouse (7)) have we been able to localize glycosphingolipids in the cytoplasm of the cells where they accumulate.

Thus the recent report by Sakakibara et al. (8,9) on the presence of galactosylceramide on the cytoskeleton of epithelial cell lines, appears as a provocative finding. Indeed, galactosylceramide, which is the major glycolipidic constituent of myelin, has been extensively used as a surface marker for oligodendrocytes (10).

We report here on the association of an acidic glycosphingolipid—the sulfogalactosylceramide (SGC)—with cytoskeletal structures in the EUE cell line, a human heteroploide line with epithelial morphology (11).

EXPERIMENTAL PROCEDURES

Cultures

EUE were grown in Eagle's medium supplemented with 12% calf serum. The cells were synchronized as previously described (12), by selective removal of mitotic cells from an asynchronous population which was treated for 16 h with colchicine (0.02 μg/ml). For cloning, cells were seeded at 50 cells per 100 mm diameter Petri dish with 10 ml complete medium. Colonies were isolated 10-15 days after plating by the stainless steel cylinder method (13). In some cases we have also used EUE cells adapted to hypertonic medium (14).

Antibodies

Affinity purification and serological characterization of rabbit anti-SGC antibodies have been previously detailed (15). Rabbit antitubulin was a generous gift of Dr. Rappaport (Hôpital Lariboisière, Paris).

Indirect Immunofluorescence

Unfixed EUE cells grown on glass coverslips, were rinsed in 0.1 M phosphate, 0.15 M NaCl buffer pH 7.4 (PBS) prior to the incubation in the first antibody. Fixation was usually performed for 15 min. in 4% paraformaldehyde in PBS. In some cases, the cytoskeleton was stabilized by treating the cells in Pipes 45 mM, Hepes 45 mM, $MgCl_2$ 5 mM in the presence or not of 0.1% Triton X-100 prior to the fixation. Then, some cells were also submitted to either chloroform methanol (2:1) (v/v) or 98% ethanol or 0.1% periodic acid for 15 min. at room temperature, or 98% ethanol for 1 min. at 4°C prior to the incubation with the anti-SGC antibodies. After washing, the cells were incubated with affinity-purified sheep antirabbit antibodies conjugated to fluorescein (diluted 1/100 Inst. Pasteur production) as already described (16).

RESULTS AND DISCUSSION

Presence of SGC on Cytoskeletal-Like Structures on EUE Cells

When unfixed EUE cells were incubated with anti-SGC antibodies, patches of fluorescence disseminated on the cell surface

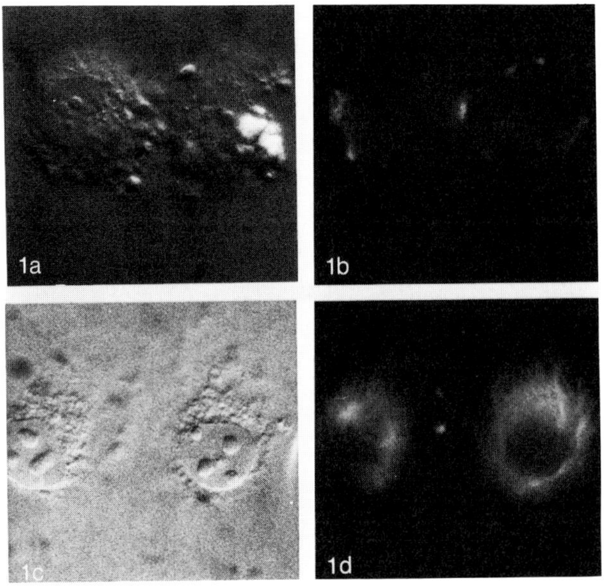

Fig. 37.1. Difference in the labelling of anti-SGC antibodies observed on unfixed (a,b) and fixed (c,d) EUE cells. (a) and (d) are differential interference contrast photographs of the cells seen in fluorescence (b,d). In (b) the fluorescence is localized to the plasma membrane, while after fixation (d) the fluorescence was mainly located in the cytoplasm on filamentous structures (× 2,500).

could be seen (Fig. 37.1b). These patches still exist when the incubation with the antibodies took place at 4°C overnight or in the presence of 0.1% azide. This suggests that this morphological distribution of SGC on the cell surface was not the consequence of an active mobilization of the lipid hapten following incubation with the antibodies. After fixation, the localization of SGC was totally different, as shown on Figure 37.1d. Indeed SGC was then seen associated to cytosolic fibrouslike structures. This staining was more intense around the nucleus and gradually decreased from the nucleus toward the plasmic membrane in a random network. Interestingly enough, in EUE cells adapted to hypertonic medium (0.25 M NaCl), the staining of SGC on these cytoskeletal-like structures was much more intense (Fig. 37.2). This correlates with our previous biochemical finding that during adaptation to hypertonic medium EUE cells increased their content in SGC (14).

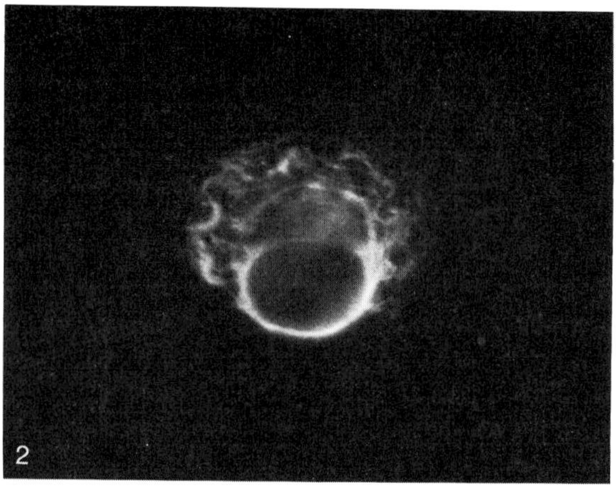

Fig. 37.2. Fixed EUE cell adapted to hypertonic medium (0.25 M NaCl) and immunostained for SGC. Note that the cell has increased in volume and that SGC is more easily visible on the cytoskeleton (× 2,500).

Evidences for the Lipidic Nature of the Labelling Observed on These Cytoskeletal Structures

If 0.1% Triton was added in the Pipes-Hepes buffer, the intracytoplasmic staining was maintained but was not localized to fibrous structures anymore. The fluorescence due to the anti-SGC antibodies was then diffuse in the cytosol. This observation was interpreted as a partial solubilization of the lipid by the detergent. The lipidic nature of the staining observed with the anti-SGC antibodies was strengthened by the fact that it disappeared after treating the cells for 15 min. at room temperature in either a chloroform-methanol (2:1) (v/v) mixture or in 98% ethanol. These solvent trreatments are indeed sufficient to extract the SGC. But, as previously shown on tissue sections (16), a 1 min. treatment in ice cold ethanol did not significantly alter the anti-SGC staining (Fig. 37.3). Similarly, exposure of the cells to 0.1% periodic acid for 15 min. at room temperature did not alter the antigen. This was expected from the fact that SGC is not susceptible to periodic acid oxidation.

The total extinction of the fluorescence after pretreatment of the cells with organic solvents at room temperature, the partial solubilization of SGC by a detergent, and the stability of the staining in conditions which either do not solubilize SGC (brief ethanol treatment at 0°C) or to which SGC is resistant (periodic acid), make many arguments in favor of the specificity of the labelling observed with the anti-SGC on the cytoskeleton of the EUE cells.

The Presence of SGC in the EUE Cells Is Genetically Linked

A striking observation was that on the total cell population, not all the cells were found SGC-positive. On a given preparation the number of positive cells very seldom exceeded 50% of the total cells (Fig. 37.4). Thus we have cloned the total population and have then been able to isolate a few clones in which all the cells were SGC-positive (Fig. 37.5). This SGC-positive phenotype was maintained aft3r several transfers. Nevertheless, it has to be noted that one of the SGC-positive clones reverted after it had been frozen.

Modification of the Cytoskeleton-Labelling with the Anti-SGC During the Cell Cycle

After fixation during the G_1 phase, although in some cells the labelling was similar to the one observed in the nonsyn-

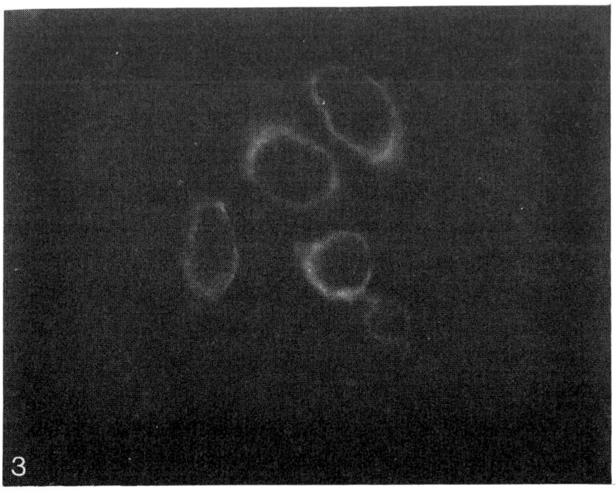

Fig. 37.3. EUE cells treated for 1 min. in ice cold ethanol prior to fixation in paraformaldehyde and incubation in anti-SGC antibodies. Note that the fluorescence, although less abundant, is still present (× 1,600).

chronized cultures, in most of the cells SGC appeared to be localized in the cytoplasm at one pole of the cell. Here again, SGC was still seen localized on reticular structures (Fig. 37.6).

During the S phase, the fluorescence could not be linked to any defined structure in the cytoplasm (Fig. 37.7). During mitosis, SGC was diffusely distributed in the cytoplasm, but the total fluorescence was quite faint (Fig. 37.8d). Staining of the cells with antitubulin antibodies (Fig. 37.8b) showing the labelling of the mitotic spindle, clearly demonstrates the dissociation of the SGC-positive cytoskeletal structures with the microtubules. This is in contradiction with what had been assumed by Nagai and Sakakibara for galactosylceramide (9). Purified tubulin or actin filaments have never been shown to contain any bound lipids. Although to our knowledge no such information exists for intermediate filaments, we suggest that in the EUE cells SGC might be bound to intermediate filaments. Indeed, morphologically the staining observed with the anti-SGC recalls mainly the one described by Franke et al. (17) for vimentin. This needs further investigation to be confirmed.

The other question raised is the physiological significance of the presence of glycolipids on these cytoskeletal structures. As stated before, glycolipids had always been assumed to be present solely on the cell surface. Indeed, on unfixed EUE cells, we have been able to localize SGC on the cell surface. Only after fixation could the anti-SGC antibodies penetrate inside the cells, and stain cytoskeletal structures which we assume are intermediate filaments. The rationale for the presence of glycolipids on the cytoskeleton is not clear. First of all, one might look (1) for the presence of other lipids on the cytoskeleton and (2) if these putative glycolipids are or not localized on the same type of filaments. Second, one might question the function of these glycolipids on the cytoskeleton. This may be too subtle to be discerned by examining culture

Fig. 37.4 (top, right). Total population of fixed EUE cell line immunostained with anti-SGC antibodies. Note that not all the cells in the population are SGC-positive. (a) Fluorescent light. (b) Same field photographed with differential interference contrast optic. Arrow heads point to SGC-positive cells (× 1,000).

Fig. 37.5 (bottom, right). Clone CS 2 immunostained with anti-SGC antibodies. Note that all the cells are SGC-positive. (a) Fluorescent light. (b) Differential interference contrast (× 2,000).

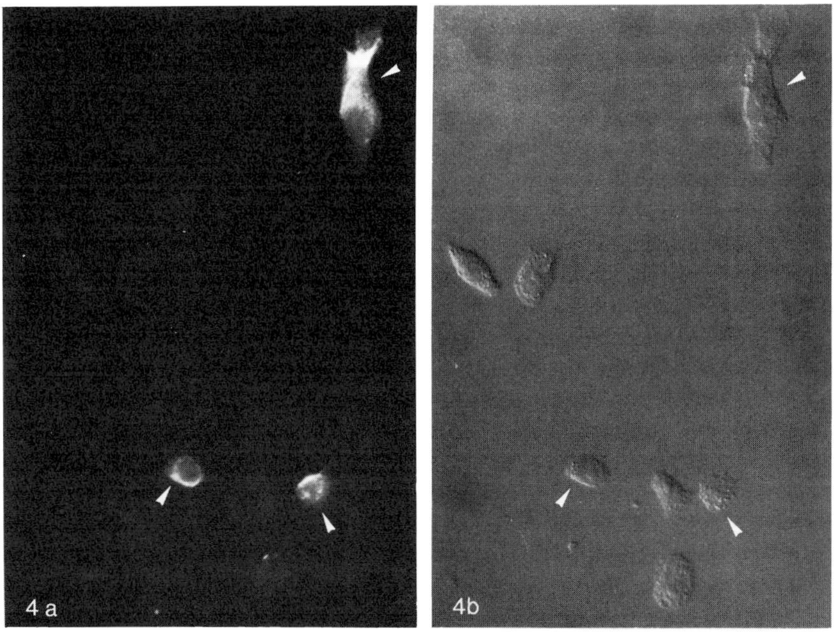
4 a
4b

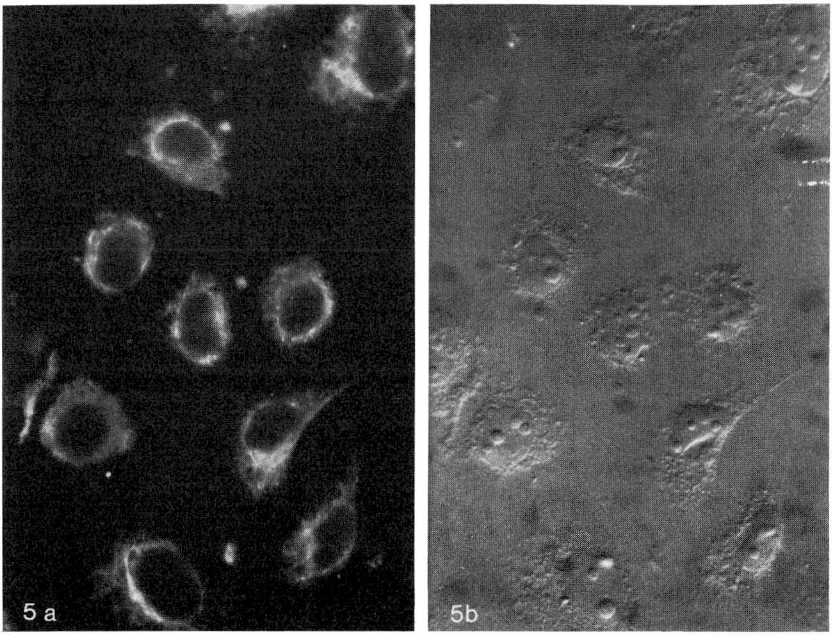
5 a
5b

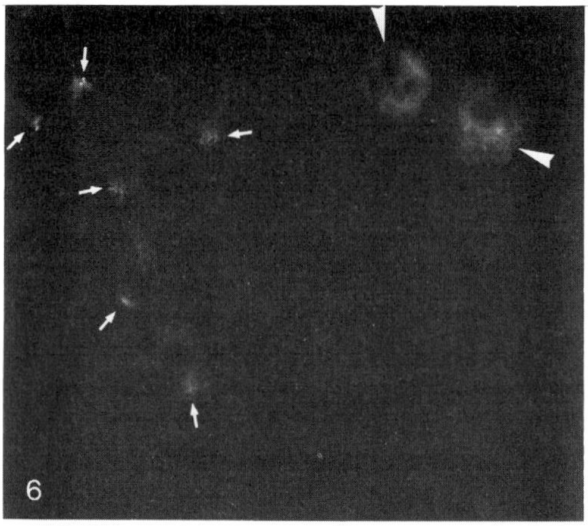

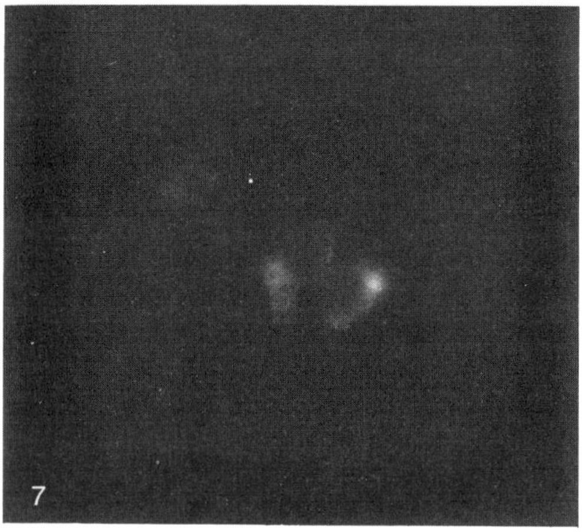

Fig. 37.6 (top). Clone CS 2 during the G_1 phase immunostained with anti-SGC antibodies. Arrow heads point to cells in which SGC is seen as a random network, while small arrows point to cells where SGC is concentrated at one pole of the cell (× 1,600).

Fig. 37.7 (bottom). Clone CS 2 during S phase immunostained with anti-SGC antibodies. The fluorescence is diffusely distributed in the cytoplasm (× 1,600).

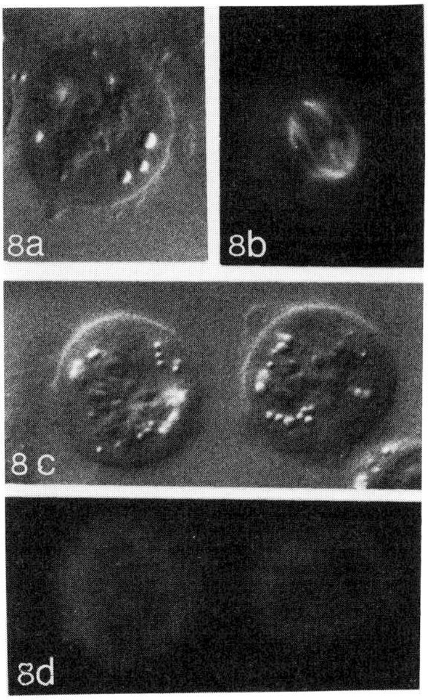

Fig. 37.8. Clone CS 2 during mitosis: (a,b) immunostained with antitubulin antiserum diluted 1/50, (c,d) immunostained with anti SGC antibodies. Note the difference of distribution of the two antigens. (a,c) Differential interference contrast. (b,d) Fluorescent light (× 2,000).

cells in a microscope. Nevertheless, one possible explanation for the presence of glycolipids (at least galactosylceramide (8,9) and SGC (this report)) on these cytoskeletal structures might be that they represent the route of transport of glycolipids from their site of synthesis (endoplasmic reticulum and Golgi apparatus) toward the plasma membrane which is their final destination.

REFERENCES

1. Kohn L.D., Consiglio E., Dewdf M.J.S., Grollman E.F., Ledley F.D., Lee G., Morris N.P.: in Proceedings Intern. Symp. on Structure and Function of Ganglioside. Mandel P. and Svennerholm L. eds. 487-504 (1979).
2. Karlsson K.A., Samuelsson B.E. and Steen G.O. Eur. J. Biochem. 46, 243-58 (1974).
3. Zalc. B., Helwig J.J., Ghandour M.S. and Sarlieve L. FEBS Letters 92, 92-96 (1978).
4. Lingwood C.A., Ng A. and Hakomori S.I. PNAS 75, 6049-53 (1978).
5. Yamakawa T. and Nagai Y. Trends in Biochemical Sciences 3, 128-32 (1978).

6. Raff M.C., Miller R.H., Noble M. Nature 303, 390-96 (1983).
7. Bologa-Sandru L., Zalc B., Herschkowitz N., Baumann N.A. Brain Res. 225, 425-30 (1981).
8. Sakakibara K., Momoi T., Uchida T. and Nagai Y. Nature 293, 76-79 (1981).
9. Nagai Y. and Sakakibara K.: in New Vistas in Glycolipid Research. Makita A., Handa S., Taketomi T. and Nagai Y. eds. Plenum Press pp. 425-43 (1982).
10. Raff M.C., Minky R., Fields K.L., Lisak R.P., Dorfman S.H., Silberberg D.H., Gregson N.A., Leibowitz S. and Kennedy M. Nature (Lond.) 274, 813-16 (1978).
11. Terni M. e Lo Monace G.B. Lo Sperimentale 108, 177-85 (1958).
12. Bettega D., Fuhrman Conti A.M., Garibaldi L., Peluchi M.T., Scaioli E. and Tallone Lombardi L. Rad. Res. 91, 457-67 (1982).
13. Puck T.T. Progress in Biophysics and Biophysical Chemistry. Burtler J.A.V. and Katz B. eds. vol. 10, p. 238, Pergamon Press, Oxford (1960).
14. Bolognani L., Fuhrman Conti A.M., Omodeo Sale M.F. Biochem. Exp. Biol. 12, 167-73 (1976).
15. Zalc B., Jacque C., Radin N.S. and Dupouey P. Immunochemistry 14, 775-79 (1977).
16. Zalc B., Monge M., Dupouey P., Hauw J.J. and Baumann n.a. Brain Res. 211, 341-54 (1981).
17. Franke W.W., Schmid E., Winter S., Osborn M. and Weber K. Exp. Cell Res. 123, 25-46 (1979).

38 Microtubule-associated Proteins in NGF-differentiated PC12 Cells

Silvia Biocca, Antonino Cattaneo, Peitro Calissano

INTRODUCTION

In nerve cells, cytoskeletal elements, besides playing functions shared with all other eukariotic cells, intervene in a crucial fashion in the process of neurite outgrowth. Little is known, however, of the precise mechanism(s) which modulates the organization and function of each cytoskeletal component in this most important differentiative neuronal event.

The clonal cell line PC12, derived from a rat pheochromocytoma tumor (Greene and Tischler, 1976), is particularly suitable to study this problem since in the presence of nanomolar concentrations of nerve growth factor (NGF) (Levi Montalcini and Booker, 1960; Levi Montalcini, 1966; Calissano et al., 1983) it ceases dividing and grows an intricate network of neurites. The possibility that NGF alters the synthesis or induces post-translational modifications of proteins constitutive of the cytoskeleton, has been recently investigated in these cells. It has been reported that NGF increases the synthesis of a high molecular weight (M_r > 300K daltons) phosphoprotein which, on the basis of immunological and physicochemical criteria, corresponds to a subspecies of a high molecular weight microtubule-associated protein (MAP 1) (Greene et al., 1983).

Studies performed in our laboratory have shown that extracts derived from NGF-differentiated PC12 cells (PC12+) stimulate the assembly of calf brain tubulin, while extracts from PC12 cells never exposed to NGF (PC12-) do not exhibit this effect (Biocca et al., 1983a). We found, moreover, that this stimulatory action is exerted by a macromolecular structure which sediments in the range between 80,000 and 100,000 × g. This structure (100K × g pellet) is also present in PC12- cells, but it contains much lower amounts of three proteins having a molecular weight of 100K, 88K and 34K daltons (Fig. 38.1). Two of these proteins (88K and the acidic 34K) are markedly reduced after incubation of PC12+ cells with colchicine for 24 hours, and increased after taxol treatment of PC12+ cells. This result suggested that their presence in the 100K × g pellet is directly connected with the state of assembly of microtubule in vivo.

In the present article we attempted to assess whether these proteins also exist in a soluble form and whether they copolymerized with taxol-induced assembled microtubules (Vallee, 1982). The protein composition of the cellular microtubules obtained by this procedure has been analyzed by 2D gel electrophoresis before and after salt extraction, as previously described to identify putative MAPs (Vallee, 1982). The results obtained show that NGF regulates the level of some proteins which, on the basis of different criteria, are tentatively identified with previously described microtubule-associated proteins (MAPs). In addition, we have further characterized the partition between soluble and particulate form of the 100K, 88K and the acidic 34K proteins which we have shown to be differentially expressed in the 100K × g pellet derived from PC12- and PC12+ cells.

MATERIALS AND METHODS

NGF was prepared as the 2.5 S form (Bocchini and Angeletti, 1969), dissolved in 2 mM acetate buffer pH 5 containing 0.5 M NaCl, filtered through a millipore and stored in sterile vials at 4°C.

PC12 cells were maintained in RPMI 1640 containing 5% foetal calf serum and 10% heat-inactivated horse serum in the presence (PC12+) or in the absence (PC12-) of 2 nM NGF. Cells were grown on collagen-coated dishes.

Cells were labeled for 20 hours with 20 μCi/ml of ^{35}S-methionine in a RPMI 1640 medium containing one tenth the normal methionine concentration and supplemented with dialyzed serum. Labeled cultures were harvested by scraping in ice cold extraction buffer (10 mM morpholino-ethane-sulfonic acid (Mes) pH 6.7, 0.1 mM EGTA, 0.1 mM $MgCl_2$, 0.1 mM phenyl-methyl-sulfonyl-fluoride (PMSF)). Cells were disrupted by sonication and centrifuged for 30 min at 80.000 × g. The 100K × g pellets were prepared from these extracts as previously described (Biocca et al., 1983a).

Taxol-facilitated assembly was performed as described (Vallee, 1982), by adjusting the 100K × g supernatants to

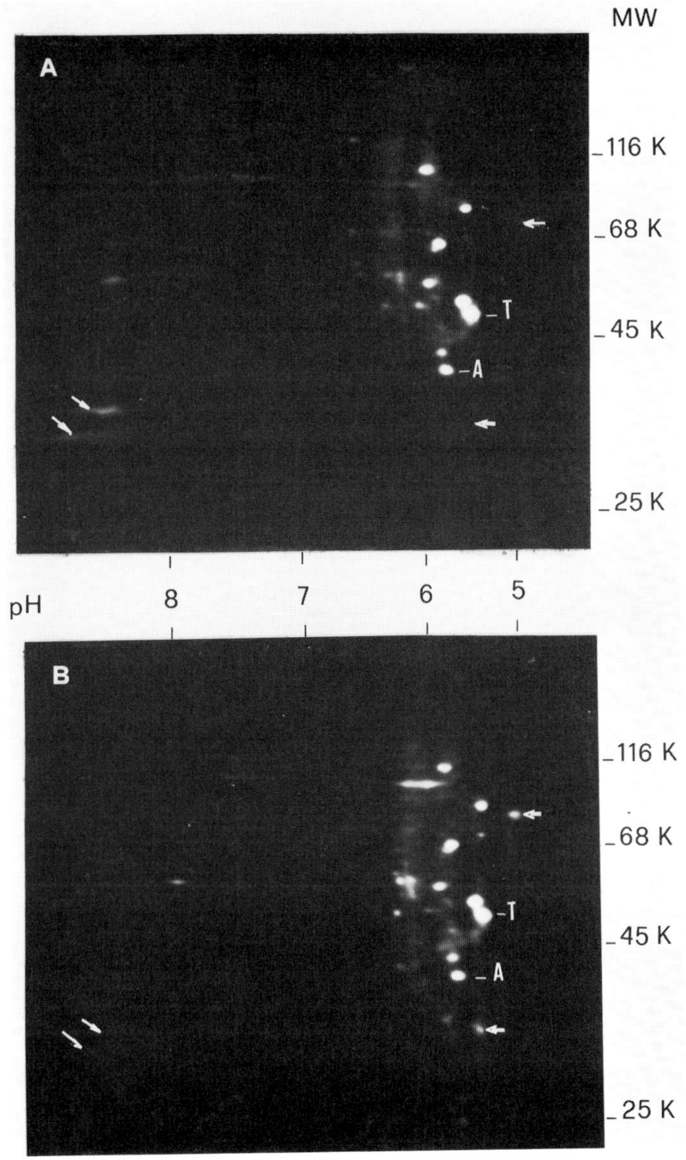

Fig. 38.1. Two-dimensional gel analysis of the 100K × g pellets. A, PC12-; B, PC12+. 90,000 c.p.m. have been applied in the first dimension in both gels, corresponding to 3.5% of the total 80,000 × g supernatants. Arrows indicate proteins selectively present or markedly enriched in one extract and absent in the other. Tubulin and actin are identified respectively by T and A. (From Biocca et al., 1983a, with the permission of EMBO J.).

contain 0.1 mM Mes, 1 mM GTP and 10 μM taxol (reassembly buffer), and by incubating them for 20 min at 37°C. The mixtures were then centrifuged and the microtubule pellet either directly dissolved in lysis buffer and analyzed by 2D gel electrophoresis (O'Farrell, 1975), or suspended in reassembly buffer (without GTP) containing 0.35 M NaCl, in order to solubilize MAPs. This material was centrifuged again and the resulting supernatants analyzed by 2D gel electrophoresis.

RESULTS AND DISCUSSION

Figure 38.2 shows the morphological appearance of PC12 cells before (2A) and after (2B) incubation with 2 nM NGF for 2 weeks. It can be seen that following incubation with NGF these cells grow long, branching neurites which, according to previous studies (Dichter et al., 1977), are electrically excitable. The studies reported in this paper have been performed with similar populations of cells since, as previously reported (Biocca et al., 1983a), such a condition is a prerequisite to detect the observed changes in microtubule-associated proteins.

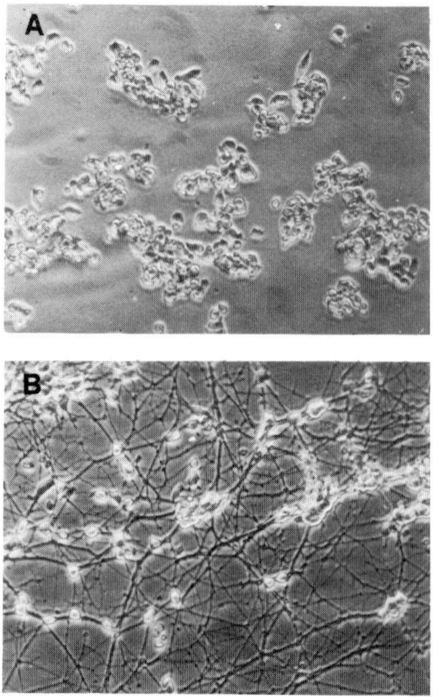

Fig. 38.2. Morphological appearance of PC12 cells grown in the absence (A) and in the presence (B) of 2 nM NGF for 15 days.

PC12- and PC12+ cells were metabolically labeled with ^{35}S-methionine and the resulting 100 K × g supernatants incubated with 10 μM taxol and 1 mM GTP at 37°C, in order to allow microtubule formation. It has been shown (Vallee 1982) that under these conditions cellular microtubules can assemble in the absence of added carrier tubulin and with the active participation of endogenous MAPs.

Figure 38.3 shows a 2D gel electrophoretic analysis of the microtubule pellets derived from PC12- (3A) and PC12+ (3B) supernatants. The two major spots are constituted by the α and β subunit of tubulin (identified with cold markers). The other more abundant protein is an acidic 34K dalton polypeptide (see arrow) similar in molecular weight and charge to a previously identified low molecular weight MAP (LMW MAP) (Berkowitz et al., 1979). This protein is equally intense in microtubule pellets derived from PC12 cells before and after differentiation. It is noteworthy to recall that the acidic 34K dalton protein is present only in the 100K × g pellet derived from PC12+ cells or PC12+ cells incubated in vivo with taxol, while being totally absent in the corresponding fraction from PC12- cells and from PC12+ cells treated in vivo with colchicine (Biocca et al., 1983a). The 88K and 100K dalton proteins present in the 100K × g pellet from PC12+ cells (Fig. 38.1) are completely absent from the microtubule pellets derived from both PC12- and PC12+ cells. Thus, it can be concluded that the acidic 34K protein is entirely soluble in PC12- cells, while in PC12+ cells becomes constitutive of a macromolecular structure favoring microtubule assembly. The 100K and 88K proteins are solely constitutive of the 100K × g pellet in PC12+ cells.

Autoradiograms derived from PC12- cells (Fig. 38.3A) show the presence of a protein with M_r of 34K and pI around 8.0 which is undetectable in differentiated PC12 cells. This protein is not only present in the taxol-induced microtubule pellet, but it is also constitutive of the 100K × g pellet from PC12- cells (Fig. 38.1). We have found (Biocca et al., 1983b) that this protein exhibits a specific affinity for single-stranded DNA. Due to this unique property it has been named 34K-single strand binding protein (34K-ssbp) and, exploiting its affinity for denatured DNA, it has been purified on a preparative scale to near homogeneity. Proteins of similar charge and molecular weight have been identified in prokariotes and in eukariotes and were shown to play a crucial role in DNA replication by facilitating its unwinding or reannealing (Kornberg, 1980). It is not yet clear whether the 34K-ssbp identified

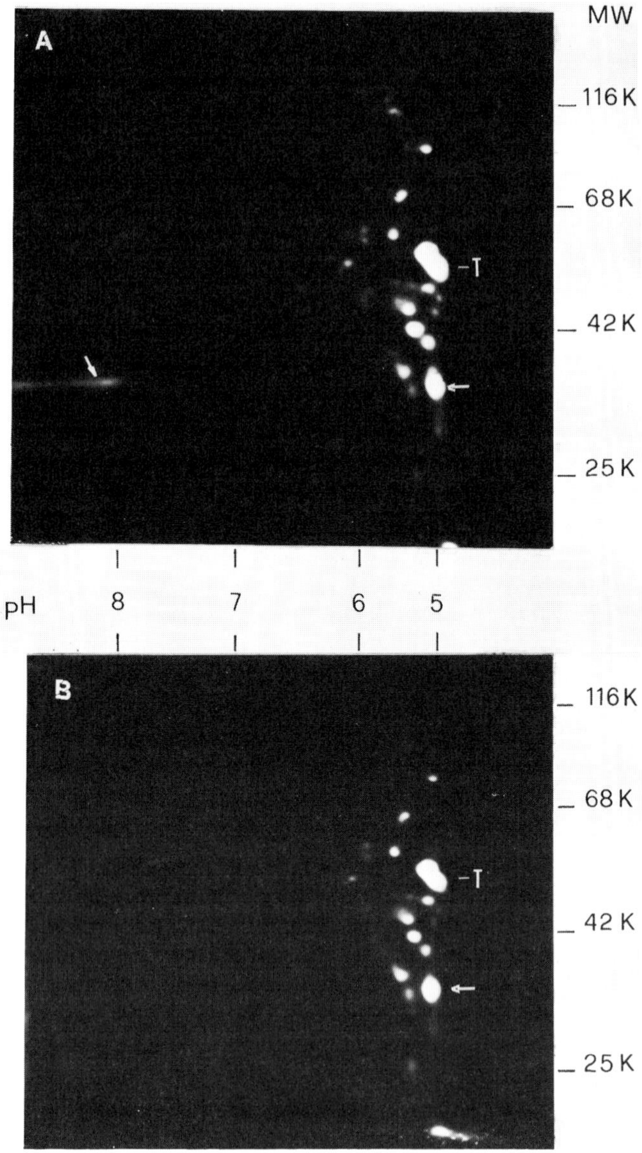

Fig. 38.3. Microtubule pellet obtained by taxol-forced assembly of 100K × g supernatants from PC12- (A) and PC12+ (B) cells. Arrows indicate the 34K ssbp (see text) of basic charge, present only in PC12- extracts and the acidic 34K protein, present in both microtubule pellets (compare the same protein in Fig. 38.1).

in our cells plays a somewhat similar function(s) or whether it has some connection with assembly, stability or structural state of microtubules. The finding that the synthesis of this protein progressively decreases with arrest of mitosis and onset of neuronal differentiation, may also point to a role of the 34K-ssbp in the synchronous unfolding of these two most important cellular events.

In conclusion, the protein pattern of the microtubule pellet derived from PC12+ cells is apparently identical to that observed in PC12- cells, with the exception of the above described 34K-ssbp (see arrow). Differences in minor components, however, can be visualized by overexposing the autoradiograms (not shown). These differences can be made readily detectable by treating the microtubule pellets with 0.35 M NaCl, a procedure which was shown to release among other proteins, the three previously identified classes of MAPs, one of high molecular weight (HMW) and two of lower molecular weight, designated tau and low molecular weight (LMW) MAPs. Such a treatment releases respectively 9% and 13% of the total microtubule proteins from PC12- and PC12+ extracts. The 30% increase of radioactivity released from PC12+ microtubule pellets might be accounted for by the presence in the salt extract of the latter of a group of proteins absent in PC12- microtubule pellets. This is illustrated in Figure 38.4 which shows the 2D gel electropherograms of the proteins released by salt extraction of microtubule pellets from PC12- (4A) and PC12+ (4B) cells. The overall patterns of the two autoradiograms are very similar, with the exception of (1) a protein of M_r > 250K and pI around 6, (2) a cluster of proteins of M_r between 60K and 70K daltons and pI 6.2-6.5, and (3) a protein of M_r 120K and pI 6.2 which are present in PC12+ extracts and absent in PC12- extracts (see arrows). The high molecular weight protein can likely be identified with the HMW MAP 1 whose level has been recently shown to be regulated by NGF treatment of PC12 cells (Greene et al., 1983). The cluster of proteins with molecular weight ranging between 60K and 70K daltons and pI 6.2-6.5 (Fig. 38.4B) has a bidimensional electrophoretic mobility very similar to that of the tau group of MAPs which, as recently shown (Pallas and Solomon, 1982; Black and Kurdyla, 1983), appears to constitute an heterogeneous group of closely related proteins, probably differing in posttranslational modifications. In this connection notice also that a streaky protein of M_r 60K daltons and more basic charge is vice versa, present only in PC12- and not in PC12+ extracts (see arrow in Fig. 38.4A). A protein of M_r 125K daltons copurifying with microtubules from HeLa cells and stimulating

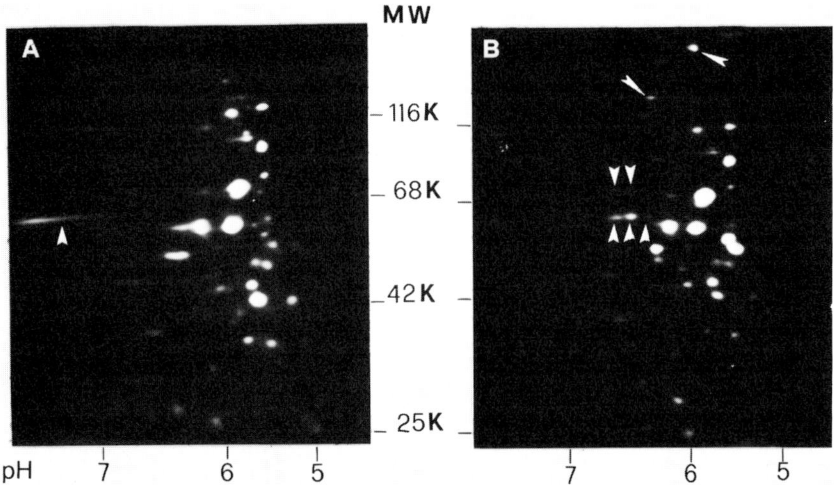

Fig. 38.4. Two-dimensional gel electrophoretic analysis of the 0.35 M NaCl released proteins from the microtubule pellet derived from PC12- (A) and PC12+ (B) extracts. 70,000 c.p.m. were applied in both gels. Arrows indicate proteins selectively present in one extract and absent in the other.

the rate and extent of tubulin polymerization, has been previously described (Bulinski and Borisy, 1980; Weatherbee et al., 1980). The 120K dalton protein stimulated by NGF might be related to this MAP.

Identification of microtubule-associated proteins (MAPs) was obtained with in vitro reconstitution experiments (Murphy and Borisy, 1975; Weingarten et al., 1975) whereby these proteins were shown to coassemble with tubulin, and in some instances, to be absolutely required for in vitro tubulin assembly. Other operative definitions of MAPs have been recently suggested by Solomon et al. (1979) and exploited by Black and Kurdyla (1983) to characterize MAPs of neurons. Accordingly, a protein will be identified as a MAP if it (a) remains associated with cytoskeletons prepared with Triton-containing microtubule-stabilizing buffer, (b) is released from cytoskeletons by incubation in microtubule-depolymerizing buffer, (c) is dependent on microtubule integrity in vivo for its presence in the cytoskeletal fraction, and (d) is able to cocycle with exogenously added tubulin in vitro. An alternative procedure for the isolation and characterization of MAPs involves the taxol-induced assembly of cellular microtubules in vitro, in the absence of exogenous tubulin (Vallee, 1982). The MAPs can be released from preformed microtubules by exposure to elevated ionic strength. These operative definitions, while being very stringent for demonstrating the association of MAPs with microtubules in situ, do not allow us to distinguish proteins regulating microtubule assembly and/or function from proteins simply associated with microtubules, but not involved in modulating their function.

In a previous article (Biocca et al., 1983a) we have reported on the effects of cellular extracts on microtubule assembly in vitro and demonstrated the existence, in NGF-differentiated PC12 cells, of a macromolecular moiety favoring polymerization of tubulin. This approach led to the identification of a set of proteins which are differentially expressed and/or partitioned in PC12- and PC12+ cells, and which would have probably escaped an analysis performed according to the above mentioned criteria for the characterization of MAPs. The aim of the present report was: (1) to further characterize the interaction of these proteins with microtubules and (2) to identify other putative MAPs in relation with the morphological differentiation induced by NGF. The data obtained show that NGF action in PC12 cells results in (a) compartmentalization of a 34K, acidic protein which is selectively enriched in the 100K × g pellet derived from PC12+ cells, while equally abundant in taxol-induced microtubule pellet from both PC12- and PC12+ extracts; (b) appearance in the same 100K × g pellet of two proteins of 100K and 88K daltons; (c) increased synthesis of a group of proteins which, on the basis of their interaction with microtubules and their electrophoretic mobility, may be identified with previously described MAPs; (d) inhibition of the synthesis of a 34K protein with positive charge, which is found in the 100K × g pellet, but exhibits a specific affinity for single-stranded DNA whose function remains to be elucidated.

Future studies performed with immunological and biochemical methods will allow us to assess the intracellular distribution and the role of these proteins in relation to the highly organized spatial pattern of microtubules in axons and dendrites of differentiated neurons.

REFERENCES

Berkowitz S.A., Katagiri J., Binder H.D. and Williams R.C. (1977) Biochemistry 16, 5610-17.

Biocca S., Cattaneo A. and Calissano P. (1983) The EMBO J. 2, 643-48.

Biocca S., Cattaneo A. and Calissano P. (1983) Proc. Natl. Acad. Sci. USA, in press.

Black M.M. and Kurdyla J.T. (1983) J. Cell Biol. 97, 1020-28.

Bocchini V. and Angeletti P.U. (1969) Proc. Natl. Acad. Sci. USA 64, 784-94.

Bulinski J.C. and Borisy G.G. (1980) J. Biol. Chem. 255, 11570-576.

Calissano P., Cattaneo A., Aloe L. and Levi-Montalcini R. (1983) In: "Hormonal proteins and peptides" (C.H. Li ed.) 12, in press.

Dichter M.A., Tischler A.S. and Greene L.A. (1977) Nature 268, 501-04.

Greene L.A. and Tischler A.S. (1976) Proc. Natl. Acad. Sci. USA 73, 2424-28.

Greene L.A., Liem R.K.H. and Shelanski M.L. (1983) J. Cell Biol. 96, 76-83.

Kornberg A. (1980) In: "DNA replication," 278-317.

Levi-Montalcini R. and Booker B. (1960) Proc. Natl. Acad. Sci. USA 46, 373-84.

Levi-Montalcini R. (1966) Harvey Lect. 60, 217-59.

O'Farrell (1975) J. Biol. Chem. 250, 4007-21.

Pallas D. and Solomon F. (1982) Cell 30, 407-14.

Solomon F., Magendantz M. and Salzman A. (1979) Cell 18, 431-38.

Vallee R.B. (1982) J. Cell. Biol. 92, 435-42.

Weatherbee J.A., Luftig R.B. and Weihing R.R. (1980) Biochemistry 19, 4116-23.

Weingarten M.D., Lockwood A.H., Hwo S.Y. and Kirschner M.W. (1975) Proc. Natl. Acad. Sci. USA 72, 1858-62.

39 Cytoarchitectural Changes Associated with Fibroblast Locomotion: Involvement of Ruffling and Microtubules in the Establishment of New Substrate Contacts at the Leading Edge

J. V. Small, G. Rinnerthaler, Z. Afnur, B. Geiger

INTRODUCTION

The locomotion of fibroblastic cells over a planar substratum involves two distinct, repetitive processes: (1) a net advance of a leading zone or "leading lamella" and (2) a subsequent forward retraction of the trailing cell tail into the cell body (for review, see Abercrombie, 1980). The advance of the leading lamella is not a smooth, continuous process but very irregular in nature (Abercrombie et al., 1970a,b), attributable to the irregular, active movements of a thin cytoplasmic sheet or lamellipodium (Abercrombie et al., 1970b) of some several microns in width at the cell front. These irregular motions of the thin lamellipodium take the forms of phases of protrusion, standstill and withdrawal that occur both close and parallel to, as well as away from the substratum. The latter, upward motion of the lamellipodium, involving normally its backward retreat into the cell body, has been termed "ruffling" (Ingram, 1969; Abercrombie et al., 1970b). According to studies of the dynamic changes in cell-to-substrate contacts during cell locomotion, undertaken using interference reflection microscopy, the lamellipodium serves the primary requirement of generating new contact sites in front of the cell such that directional locomotion may occur (Izzard and Lochner, 1980).

While the importance of the lamellipodium or "leading edge" as a motile organelle is generally acknowledged, the mechanisms that underly its motility as well as those that regulate the formation of new substrate contacts are poorly understood. In an effort to shed some light on these mechanisms, we have been undertaking studies to correlate the locomotion of selected cells, recorded by cinematography, with the organization of their cytoskeleton as determined, for a known phase of locomotion, by immunochemistry. The results of the first stage in these studies described here and elsewhere (Rinnerthaler et al., in preparation) indicate firstly, that ruffling activity in one form or another, is involved in contact formation in the lamellipodium and secondly, that microtubules play a primary role in determining the sites at which stable contacts will form.

MATERIALS AND METHODS

Cells and Cinematography

Primary cultures of fibroblasts were maintained at 37°C either in Dulbeccos modified essential medium (DMEM–Gibco) containing 10% fetal calf serum (FCS–Gibco) in the presence of 5% CO_2 or in Eagles basal medium (Hanks' salts) or Leibovitz medium (Flow Laboratories) containing 10% FCS in an air atmosphere. For cinematography, cells were plated (3rd to 5th subculture) onto 12mm diameter glass coverslips and used one to three hours after spreading. The coverslips, marked with a grid pattern to facilitate cell location (Small, in preparation), were mounted into a flow-through chamber of the Dvorak-Stotler type (Dvorak-Stotler, 1971) and transferred directly to the stage of a Zeiss photomicroscope maintained at 37°C, for cinematography. Details of the perfusion chamber and heating stage will be given elsewhere (Rinnerthaler and Small, in preparation). By means of a drip feed and siphon outflow the medium in the chamber could be continuously replenished without focus change, and exchanged rapidly for any desired fixation mixture.

Typically, cells were filmed for a period of 3 to 4 minutes and then the medium exchanged, first with buffer (cytoskeleton buffer—as Small (1981), but buffered with 10 mM MES instead of PIPES) and then with a fixative mixture (0.5% to 0.3% Triton X-100, 0.5% to 0.25% glutaraldehyde in cytoskeleton buffer; for details, see Results), the whole process being filmed until and after arrest of locomotion. After 2-3 min exposure to the Triton-glutaraldehyde mixture the coverslip was removed and placed for a further 10 min in 1% glutaraldehyde and then stored in the same buffer prior to immunocytochemistry. Cell locomotion was recorded in phase contrast (40× objective) on 16mm Kodak Technical Pan Film using either a Bolex or Beaulieu camera operating at 2 to 4 frames per second.

Immunocytochemistry

Immunocytochemistry was performed on cells cultured normally in dishes as well as on those subjected to cinematography, in each case using grid-marked coverslips for cell relocation. Fixation was performed using an initial fixation in one or two Triton-glutaraldehyde mixtures (see above and also Results) for 2 min followed by 10 min in 1% glutaraldehyde (all at R.T.). Prior to application of the first antibodies, coverslips were treated with 5 mg/ml $NaBH_4$ (Weber et al., 1978) for 2 × 5 min and 1 × 10 min on ice in cytoskeleton buffer. The antibodies to vinculin and tubulin were either affinity-purified polyclonal rabbit antibodies or monoclonal antibodies from rat or mouse. The antibodies to tubulin were generous gifts from Dr. J. De Mey and Dr. J. V. Kilmartin. Labelling for actin was performed either with fluorescent phalloidin (a gift of Prof. Th. Wieland, Heidelberg) or with affinity-purified actin antibodies (raised against chicken gizzard actin in collaboration with Dr. J. De Mey, Janssen Pharamceutica, Belgium). Antibody labelling as well as washing procedures were carried out in Tris-buffered saline (155mM NaCl, 10mM Tris, 2mM $MgCl_2$, 2mM EGTA pH 7.6) at room temperature. For microscopy, coverslips were mounted either in Gelvatol 20-30 (Monsanto Polymers, St. Louis, Missouri, U.S.A.) or 50% glycerol (for multiple labelling) with added 6mg/ml n-propyl gallate to inhibit bleaching (Dr. J. Sedat, priv. communication). Second antibodies conjugated with fluorescein were purchased from Nordic; those coupled to rhodamine were prepared as described elsewhere (Geiger et al., 1981).

Multiple Imaging and Labelling

After fixation of filmed cells as well as cells not subjected to cinematography, phase contrast and interference reflection contrast images (Zeiss optics) were recorded with coverslips mounted in cytoskeleton buffer in the perfusion chamber. Thereafter the cells, whose positions were recorded, were labelled with the first two antibodies (tubulin and vinculin) and the coverslips remounted in the same chamber in 50% buffered glycerol (with added n-propyl-gallate) for immunofluorescence microscopy. The coverslips were again removed, washed free of glycerol and stained for actin using either actin antibodies or rhodamine-labelled phalloidin. Finally, the coverslips were mounted in Gelvatol to record the actin pattern.

With this method the actin pattern was superimposed on either the microtubule or vinculin patterns (depending on the fluorophore used). Since the latter patterns had already been recorded and were clearly distinct from the actin pattern, the relationships of the three components could be readily evaluated. An alternative, one step method involved the simultaneous staining of vinculin (rhodamine-labelled second antibody) and actin (rhodamine-labelled phalloidin) with the phalloidin diluted 40-fold more than normal so as to produce a clear distinction between the vinculin and actin patterns.

Immunofluorescence microscopy was carried out using a Zeiss photomicroscope fitted with epi-illumination.

Photographic Evaluation Procedures

For evaluation of the distribution of multiple components in the same cell, as determined by immunocytochemistry, and their localization relative to details recognizable by other methods (phase contrast and interference reflection microscopy), separate photographic prints were found to be unsuitable. Cell images were therefore reproduced at the same magnification on 13 × 18 cm sheet film in positive or negative contrast, as desired, such that any combination of images could be superimposed and judged on the surface of a light box. For the images obtained by immunofluorescent microscopy a further refinement involved conversion of the original black and white images into color (blue or red) using a chromatic developing procedure similar to that used earlier in color photography (Rinnerthaler and Sulzer, unpublished). An alternative method used for evaluation of immunofluorescent images entailed making double exposures (in the fluorescein and rhodamine channels) on color slide film (Kodak Ektachrome 400) utilizing a camera lock on the microscope.

RESULTS

Preservation of the Leading Edge (Lamellipodium): A Note on Fixation Procedures

The methods commonly used in immunofluorescence microscopy to preserve cell structure, and that are generally compatible with the demonstration of antibody-antigen reactions, involve fixation in methanol and acetone or in formaldehyde (see e.g. reviews by Fujiwara and Pollard, 1980, and Lazarides, 1982). These procedures, however, fail to properly preserve the delicate structure of the leading edge through the steps required for indirect antibody labelling. In our hands the most suitable method that satisfied both requirements of structural preservation and antibody accessibility into the cell was an initial, brief fixation in a glutaraldehyde-Triton X-100 mixture (see Small, 1981) followed by limited glutaraldehyde fixation. After subsequent reduction of free aldehyde groups with sodium borohydride (Weber et al., 1978) the intensity of labelling by the antibodies we have used was equal or superior (for microtubules) to that obtained with formaldehyde. The superiority of glutaraldehyde in preserving microtubular organization for fluorescence microscopy has also been noted earlier (De Mey et al., 1976; Osborn and Weber, 1982).

A further consideration, worthy of attention was the preservation of the three dimensional structure of the leading edge, or more precisely of the ruffling regions. With the glutaraldehyde-Triton mixtures containing relatively high Triton concentrations (0.5%) ruffling lamellipodia normally collapsed onto the cell or leading edge on fixation, such that the three dimensional form was partly lost. This collapse was prevented by using a sequence of Triton-glutaraldehyde mixtures prior to the final 10 min fixation in glutaraldehyde alone, namely: 0.1% Triton/0.5% glutaraldehyde, 2 min, followed by 0.5% Triton/ 0.25% glutaraldehyde, 2 min. (Examples of cells fixed in this way and then processed for multiple labelling of the cytoskeleton are shown in Figs. 39.3 and 39.6.)

Colocalization of Vinculin and Microtubules in the Leading Edge

In fibroblasts observed by immunofluorescence microscopy, two notable components were found to be present, intermittently but persistently, at the leading edge: (1) microtubules (Fig. 39.1a,b) and (2) fine vinculin-containing sites, distinct from those at the termini of the stress fiber bundles behind the leading edge (Fig. 39.1c,d). In this respect, two distinct locations in the leading edge should be distinguished—the base of the leading edge, that is its clearly delineated rear margin evident after labelling for actin, and the leading edge proper.

At the base of the leading edge, vinculin occurred in distinct spots or oblate streaks that were either associated or not associated with the termini of recognizable stress fiber

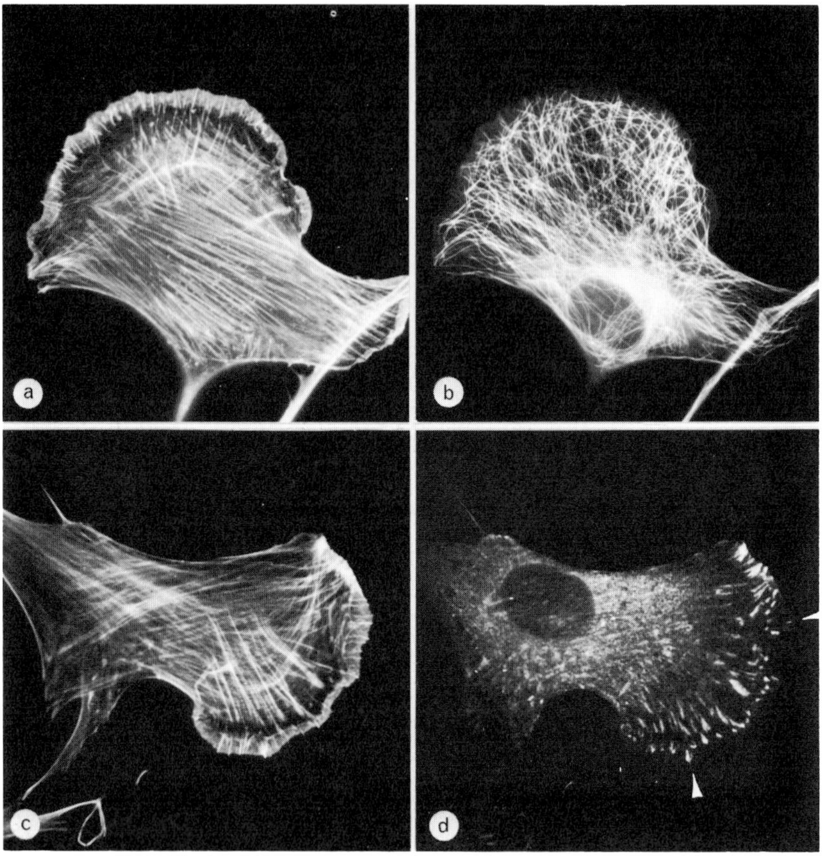

Fig. 39.1. The presence of microtubules and vinculin in the leading edge of chick heart fibroblasts. a,b: cell double-labelled for actin (a) and tubulin (b). Note penetration of microtubules into the broad convex band of actin that makes up the leading edge. c,d: cell double-labelled for actin (c) and vinculin (b). Prominent vinculin sites present in the leading edge are indicated by arrows. Note also vinculin sites at base of leading edge not associated with stress fibers. (Initial fixation: 0.5% Triton/0.25% glutaraldehyde.) Magnification: a,b × 1,035; c,d × 1,275.

bundles (Fig. 39.1c,d). The proportion of sites not associated with stress fiber bundles was highly variable and difficult to quantitate. However, it was notable that in convex, fan-shaped cells with narrow leading edges and poorly developed stress fibers, the base of the leading edge was marked by a more or less continuous punctate line of vinculin sites; that is, the presence of vinculin at the base of the leading edge was clearly not stress-fiber dependent. In cells with broader leading edges, in which actin bundles (microspikes) were more prominent, vinculin showed no particular preference for the base of the filament bundles. However, in the case that a vinculin site extended radially into the leading edge, from the base towards the tip, it was often (but not always) associated with such a bundle (Fig. 39.1c,d). Elsewhere in the leading edge proper, vinculin occurred in fine discrete spots or streaks that extended in visibility down to the limits of detectability by the immunofluorescence method.

The analysis of microtubule patterns in cells labelled for tubulin and actin revealed that almost all microtubules terminated in either of three sites: (1) the termini of stress fiber bundles, (2) the base of the leading edge, or (3) within the leading edge. Figure 39.1 (a,b) shows an example in which the microtubules show a dominant interaction with the leading edge, penetrating in some cases almost to its tip. The number of microtubules in the leading edge was highly variable and depended, as we shall see, on the motile activity of the lamellipodium. Correspondingly, the proportion of microtubule ends found in the leading edge compared to the total complement (at the base, and in the leading edge) ranged from 0% to 80% (average, 35%). For a selected number (15) of polarized cells with convex leading edges, this corresponded to a range from 0 to 40 microtubule ends terminating in the leading edge region. Electron microscopy of negatively-stained cytoskeletons confirmed the entry of microtubules into the actin meshworks in the leading edge (Fig. 39.2), as well as their association in some instances with microspikes.

By far the most interesting and significant aspect of the interactions of vinculin and tubulin with the leading edge was the colocalization of these two components. Thus, in the majority of cases, microtubules either terminated, crossed, or came very close to the vinculin sites in and at the base of the leading edge. A typical example is shown in Figure 39.3d. (This figure is also discussed in more detail below). At the base of the leading edge very few tubules were found not terminating in vinculin sites. And within the leading edge coincidence occurred in an average of 75% of cases (either tubulin in vinculin sites or vinculin sites with tubulin). The relevance of this co-localization became apparent only through correlated cinematography and immunofluorescence microscopy of the same cells. The data is presented in the following section.

Correlation of Vinculin and Tubulin Localization with Ruffling Activity in the Leading Edge

The motiliby of the lamellipodium characteristically involves folding or "ruffling" of the leading edge, generally recognized in living cells from increases in density in phase contrast. Within this term "ruffling" (Abercrombie et al., 1970b), we shall include all those forms of activity that lead to transitory increases of phase density at the leading edge, ranging from the folding of the whole extent of the lamellipodium to small, discreate increases of phase density within it.

Figure 39.4 shows an example of a cell followed by cinematography, that was fixed and then processed for immunocytochemistry for vinculin and actin. The time lapse sequence

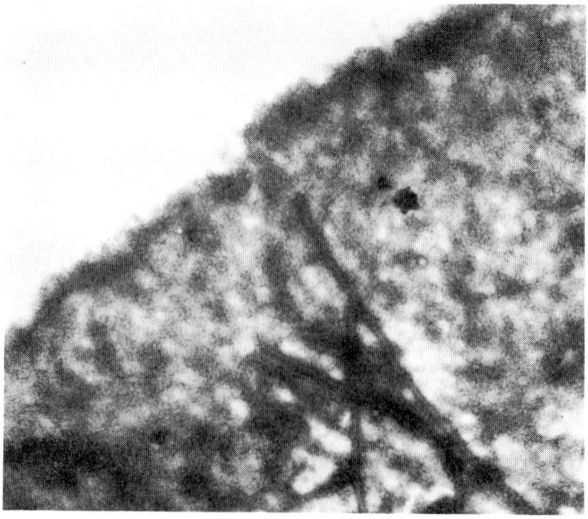

Fig. 39.2. Leading edge of fibroblasts as seen after negative staining in the electron microscope and showing penetration of microtubules into the actin meshworks. The density of the microtubules is enhanced by the addition of α-tubulin antibodies. Stained with sodium silicotungstate. Magnification: × 32,000.

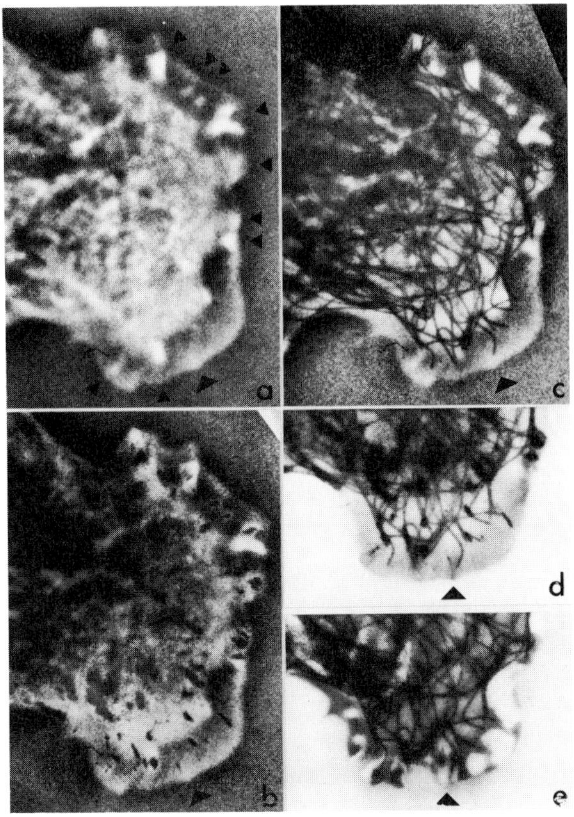

Fig. 39.3. Cell processed for msultiple imaging (see Material and Methods) and showing the distributions of the cytoskeletal components (vinculin, tubulin and actin) with respect to each other (d) and in relation to the ruffling activity of the cell (a,b,d) as well as the interference reflection contrast image (e). Ruffling activity (see a, arrowheads) appears in white in these reversed phase contrast images. Not the coincidence of the vinculin sites with ruffling activity in b (vinculin plus negative phase contrast) and the close association of microtubules with the same ruffles in c (tubulin plus negative phawe contrast). In d the vinculin and microtubule patterns are superimposed demonstrating their colocalization in the leading edge. In the same picture the actin pattern (diffuse grey), at a focus above substrate is also included to show the presence of a small "radial ruffle" (arrow), colocalized with a vinculin streak and the end of a microtubule. (This site is indicated in all pictures by a broad arrowhead). The same site in the reflection contrast image (e) shows the existence of a fine, radial close contact. See also text. Magnifications: a-c, × 3,200; d,e × 3,500.

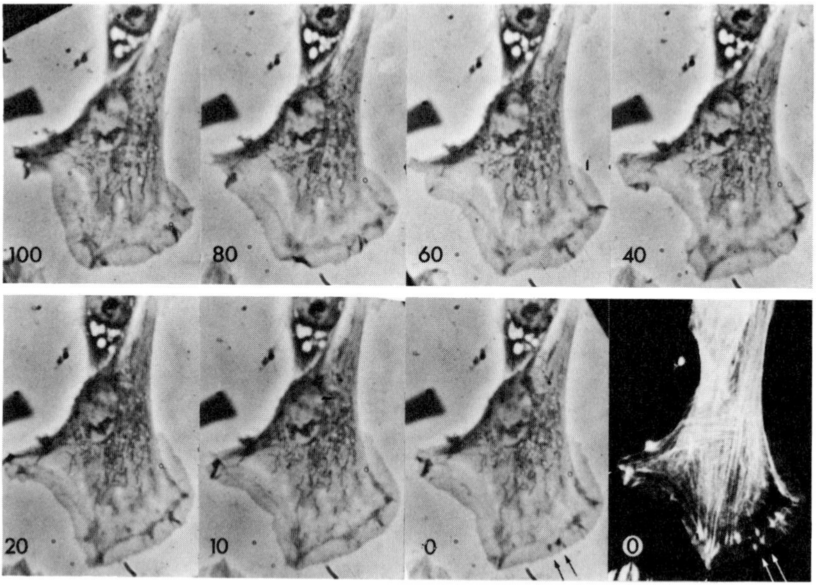

Fig. 39.4. Time lapse sequence of a cell taken from the last 100 seconds before fixation. Numbers indicate seconds before fixation and 0 the frame at the instant of fixation. (Fixation mixture: 0.5% Triton X-100/0.25% glutaraldehyde.) The fluorescence image, 0, corresponds to the vinculin pattern (intense white spots) superimposed on the actin pattern (grey) of the fixed cell. Note course of ruffling activity and in particular, the coincidence of vinculin with two "mini-ruffles" in the leading edge (arrows). These ruffles become prominent in the last 10 sec but also show precursors at an earlier stage (e.g. 80 and 60). See also text. Magnification: × 1,275.

is taken from the final period of 100 sec prior to fixation (0 sec, Fig. 39.4). The vinculin pattern (high contrast spots) is superimposed on the actin pattern of the fixed cell in Figure 39.4. What is immediately apparent is that the vinculin sites in the leading edge correspond precisely to the regions showing ruffling activity in the final seconds before the cell was fixed. In particular, a pair of small vinculin sites within the leading edge, coincides with the pair of punctate thickenings that became prominent in the final 10 seconds. Close inspection of the film revealed, in fact, that similar ruffling activity also occured in one of these sites some 100 sec earlier. Analysis of many series of this type revealed a consistent correlation between ruffling activity and the localisation of vinculin, either beneath, at the base of, or to the side of a ruffle. Interesting exceptions were, however, found. In those cases where a ruffle was formed for the first time in the last 10-20 sec of movement, vinculin was generally absent. In other words, a time of greater than about 20 secs appeared to be required for the laying down of vinculin at these sites. The clear association of vinculin sites with ruffling activity is shown also in Figure 39.3c for a cell not followed by cinematography but processed for multiple imaging and labelling (see Methods).

Equally interesting correlations were found with respect to the microtubule patterns (Figs. 39.3 and 39.5). Referring first to Figure 39.3 (a,b,e) the penetration of microtubules into the leading edge was associated in more or less all instances with ruffling activity, in one form or another. This was clearly demonstrated by superimposing the microtubule pattern on the phase contrast image (Fig. 39.3b). After staining for actin and focusing a little above the substrate, the ruffling regions may also be identified; for the cell in Figure 39.3 this revealed the correlation of one microtubule with a raised "radial ruffle" (Fig. 39.3d) that was not detectable in the phase contrast picture. From correlations by cinematography the same relationship was found, but with some additional aspects (Fig. 39.5). Thus, when the microtubule pattern was superimposed on the phase contrast images of the different time lapse pictures, it did not necessarily match best with final frame before fixation. Instead, where distinct changes in the ruffling patterns occurred, the microtubules often matched the centers or limits of ruffling activity exhibited 40 to 60 seconds prior to fixation (Fig. 39.5). This was taken to imply that the microtubule distribution with respect to a particular region in the leading edge was stable for at least this time period. In other examples (Rinnerthaler et al., in preparation) we have found that for ruffles formed

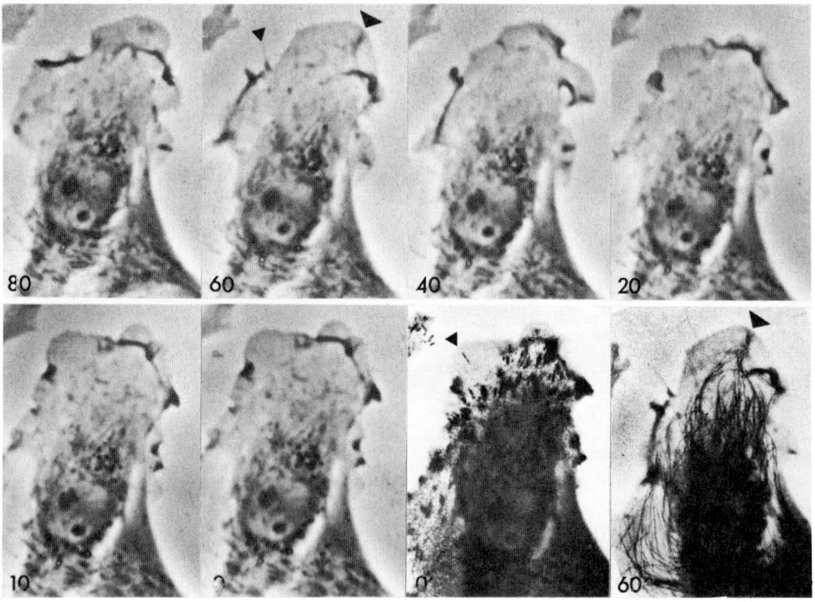

Fig. 39.5. Time lapse sequence, as Fig. 39.4, showing cell with leading edge that exhibits ruffling, retraction and protusion in the final 80 sec before fixation. The superimposition of the vinculin pattern on the phase contrast picture at fixation (0') shows again the correspondence between vinculin localization at the cell front and ruffling activity. Note also vinculin streak in rapidly expanded lamella (arrowhead) that corresponds to microspike extended 60 sec (60) before fixation. Superimposition of the microtubule pattern of the fixed cell on the time lapse pictures showed that it fitted best to the pattern of ruffling activity seen 60 seconds before fixation (60'). The main focus of microtubule ends is on the radial ruffle indicated by the large arrowhead. (Fixation mixture: 0.5% Triton X-100/ 0.25% glutaraldehyde.) Magnification: × 1,700.

in the terminal 10 to 20 seconds and in which vinculin was absent, microtubules occurred in, or at the base of the ruffle. One conclusion from these time lapse studies is thus that microtubules occur prior to vinculin in the ruffling zones.

Other features of the cytoarchitectural associations are also worth noting. As shown in Figure 39.3 there is a close correlation between the vinculin sites in the leading edge and the grey regions in the interference reflection pattern, taken to correspond to regions of close contact. Also, the coassociation between microtubules and vinculin sometimes reveals itself, rarely but dramatically, in cases where small microtubule fragments occur in the leading edge. In some of these instances the microtubule fragment precisely matches the vinculin site in size and position (Fig. 39.6). However, microtubules and vinculin may also be associated with the same ruffling activity without coming into direct contact (Fig. 39.3, arrow). Finally, vinculin association does not appear to be exclusively associated with ruffling activity in the lamellipodium. A case in point is illustrated in Figure 39.5 in the region of the expanding lamellum, free of ruffling activity. In this instance a vinculin streak appears in the position of a filopodial extension that first projected rapidly from a phase dense patch at the cell margin, and then attached to the substrate about 1 min before the cell was stopped. Subsequently, the lamellipodium expanded to fully incorporate this filopodium (see 0 sec picture).

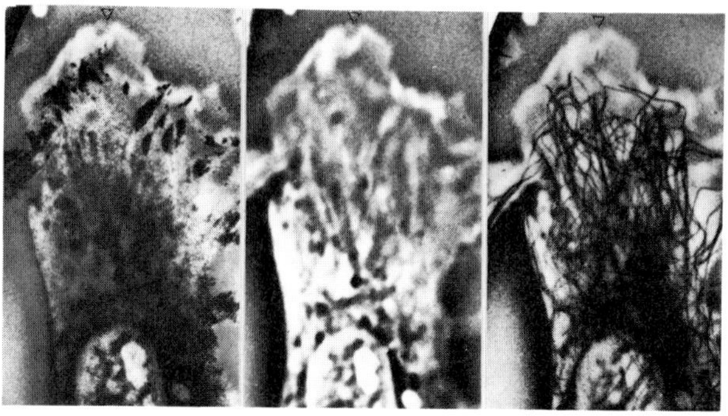

Fig. 39.6. Colocalization of microtubule fragment with vinculin "precursor site" in a ruffle (open triangle). Center, negative phase contrast picture (white ruffles); right and left, vinculin and tubulin patterns, respectively, superimposed on phase image. Magnification: × 2,320.

Effect of Microtubule Inhibitions on Vinculin Distribution

As shown by Vasiliev and Gelfand (1976) and De Brabander and coworkers (1977), microtubule distribution leads, in cultured cells, to the loss of directed cell locomotion, but not the ability to spread on a substrate. Treatment of chick heart fibroblasts with 10 μg/ml Nocodazole (De Brabander et al., 1976) results in the complete loss of formed microtubules from the spreading regions after 10 min, and of microtubules altogether by about 40 min. The course of subsequent morphological changes (Vasiliev and Gelfand, 1976; De Brabander et al., 1977) includes the spasmodic extension and retraction of different parts of the cell periphery. After doubling labelling for vinculin and actin, it is found that vinculin is still present in the lamella regions. However, although these regions, now of limited extent, appeared normal with respect to actin staining, it is noteworthy that mature focal contacts with associated stress fibers were not formed behind them (data not shown).

DISCUSSION

Because of the heterogeneity in lamellipodial activity, not only from one cell to another, but along one and the same leading edge (Abercrombie et al., 1970a), any meaningful structural correlations aimed at yielding details of the motile process must be made on cells whose history of movement is known. By their nature, such studies are extremely time consuming and have been undertaken all too infrequently (Price, 1968; Heaysman, 1973; Buckley, 1974; Heath and Dunn, 1978; Izzard and Lochner, 1980; Chen, 1981; Herman et al., 1981; Tosney and Wessels, 1983). In the present study we have focussed attention, for the first time, on the rearrangements of two components with locomotion whose interaction with the leading edge was hitherto not fully appreciated—namely, microtubules and vinculin.

Perhaps the most striking and unexpected result was the intimate involvement of ruffling activity in the formation of new, vinculin-containing substrate contacts. New contacts could also be formed by the extension of filopodia, just as Izzard and Lochner (1980, Fig. 39.6) describe (see also Fig. 39.5), but in the cells we observed, contact formation beneath ruffles was far more common. Since ruffles are not detectable by interference reflection microscopy, their association with contact formation was earlier overlooked. As emphasized in the Results and intimated in the studies of Abercrombie et al. (1970b) ruffling is expressed in a variety of forms. Apart from "typical" ruffles formed from the total backward folding of the lamellipodium or upward protrusion from its base, we draw attention to two additional types. The first forms from an apparent sideways folding of the leading edge causing a radial increase in density. Such "radial ruffles" (see e.g. Fig. 39.3) include, or are bordered by microspike bundles (unpublished results). The second type, that we would prefer to call "mini-ruffles" (Figs. 39.3 and 39.4), corresponds in morphology to the "microcolliculi" described as the only ruffle type in Xenopus epidermal cells (Bereiter-Hahn et al., 1981) and appears to arise from localized bulging of the lamellipodium.

At the limit of sensitivity of the methods used here, there was a good coincidence between the vinculin sites in the leading edge and regions of close contact, as recognized by interference reflection microscopy. It is therefore reasonable to presume that these vinculin sites constitute potential precursors to the mature focal contacts associated with microfilament bundles (see review by Geiger, 1983). The presence of vinculin in close-contact regions was also earlier recognized (Geiger, 1980). From the current data we cannot say how the transition from nascent vinculin sites in the leading edge to mature focal contacts occurs. However, we suggest that the vinculin sites at the base of the leading edge, not associated with stress fibers, represent a transition stage between a leading edge site and a focal contact proper. In other words, the sites within the leading edge mark the new limits of the close contact region (Izzard and Lochner, 1980) and the subsequent base of the leading edge. Once the base has been established at this level the sites presumably undergo further development which then leads to the formation of stress fibers, or equivalent actin filament arrays.

From the experiments with nocodazole, the laying down of vinculin in the leading edge is apparently a spontaneous process, not dependent on microtubules. However, the coordinated differentiation of contact sites in the leading edge into focal contacts behind it, necessary for directional locomotion, is apparently uncoupled in the absence of microtubules. This leads to the random development of stable and active cell edges in cells treated with microtubule inhibitors (Vasiliev and Gelfand, 1976; Vasiliev, 1982). De Brabander et al. (1977) concluded that the requirement of microtubules for polarized locomotion and other functions may be explained by their performing a signal transducing role, whereby they influence the activity of the microfilament system in response to signals from the cell center. Consistent with this idea, the present findings now indicate a route by which microtubules exert their influence on the actin-based motile apparatus. As we show, microtubules colocalize with vinculin in the leading edge and further, appear to preempt vinculin in ruffling zones, in which the latter is subsequently laid down. It is tempting to speculate that microtubules determine the sites of ruffling activity and thereby the sites at which stable contacts will be made, but we have no direct proof that this is the case. Neither is it possible to say whether tubulin directly or indirectly, via associated proteins, is capable of binding to vinculin. These are questions that still remain open. However, the present data points clearly to the existence of a direct interaction between the microtubule and microfilament systems and a possible route whereby microtubules may influence processes occuring at the cell membrane.

ACKNOWLEDGMENTS

The authors are indebted to Ms. M. Hattenberger and Mrs. H. Kunka for excellent technical assistance and Mr. G. Sulzer for the invaluable developments in photography that made proper data analysis and presentation possible. We also thank Dr. J. De Mey for discussion and for providing some of the antibodies and Mrs. G. McCoy for typing. This work was supported by a grant from the Austrian Research Council.

REFERENCES

Abercrombie, M. (1980). The crawling movement of metazoan cells. Proc. R. Soc. Lond. B 207, 129-47.

Abercrombie, M., Heaysman, J.E.M. and Pegrum, S.M. (1970a). The locomotion of fibroblasts in culture. I. Movements of the leading edge. Expl. Cell Res. 59, 393-98.

Abercrombie, M., Heaysman, J.E.M. and Pegrum, S.M. (1970b). The locomotion of fibroblasts in culture. II. "Ruffling." Expl. Cell Res. 60, 437-44.

Bereiter-Hahn, J., Strohmeier, R., Kunzenbacher, I., Beck, K. and Vöth, M. (1981). Locomotion of Xenopus epidermis cells in primary culture. J. Cell Sci. 52, 289-311.

Buckley, I.K. (1974). Subcellular motility: A correlated light and electron microscopic study using cultured cells. Tissue Cell 6, 1-20.

Chen, W.-T. (1981). Mechanism of retraction of the trailing edge during fibroblast movement. J. Cell Biol. 90, 187-200.

De Brabander, M., De Mey, J., Van de Veire, R., Aerts, F. and Geuens, G. (1977). Microtubules in mammalian cell shape and surface modulation: an alternative hypothesis. Cell Biol. Int. Rep. 1, 453-61.

De Mey, J., Hoebeke, I., De Brabander, M., Geuens, G. and Joniau, M. (1976). Immunoperoxidase visualization of microtubules and microtubular proteins. Nature (Lond.) 264, 273-75.

Fujiwara, K. and Pollard, T.D. (1980). Techniques for localizing contractile proteins with fluorescent antibodies. In "Current Topics in Developmental Biology" Vol. 14, pp. 271-96. Academic Press.

Geiger, B. (1981). Transmembrane linkage and cell attachment: the role of vinculin. In "International Cell Biology 1980-1981" (ed. by H.G. Schweiger), pp. 761-73. Springer Verlag.

Geiger, B. (1983). Membrane-cytoskeleton interaction. Biochim. Biophys. Acta 737, 305-41.

Geiger, B., Dutton, A.H., Tokuyasu, K.T. and Singer, S.J. (1981). Immunoelectron microscopic studies of membrane-microfilament interactions. The distributions of α-actinin, tropomyosin and vinculin in intestinal epithelial brush border and in chicken gizzard smooth muscle. J. Cell Biol. 91, 614-28.

Heath, J.P. and Dunn, G.A. (1978). Cell to substratum contacts of chick fibroblasts and their relation to the microfilament system. A correlated interference-reflexion and high-voltage electron-microscope study. J. Cell Sci. 29, 197-212.

Heaysman, J.E.M. and Pegrum, S.M. (1973). Early contacts between fibroblasts: an ultrastructural study. Expl. Cell Res. 78, 71-78.

Herman, I.M., Crisona, N.J. and Pollard, T.D. (1981). Relation between cell activity and the distribution of cytoplasmic actin and myosin. J. Cell Biol. 90, 84-91.

Ingram, V.M. (1969). A side view of moving fibroblasts. Nature (Lond.) 222, 641-44.

Izzard, C.S. and Lochner, L.R. (1980). Formation of cell-to-substrate contacts during fibroblast motility: an interference-reflexion study. J. Cell Sci. 42, 81-116.

Lazarides, E. (1982). Antibody production and immunofluorescent characterization of actin and contractile proteins. In "Methods in Cell Biology" Vol. 24, pp. 313-31. Academic Press.

Osborn, M. and Weber, K. (1982). Immunofluorescence and immunocytochemical procedures with affinity purified antibodies: tubulin-containing structures. In "Methods in Cell Biology" Vol. 24, pp. 97-132. Academic Press.

Price, Z.W. (1972). A three-dimensional model of membrane ruffling from transmission and scanning electron microscopy of cultured monkey kidney cells ($LLCMK_2$). J. Microsc. 95, 493-505.

Small, J.V. (1981). Organization of actin in the leading edge of cultured cells: influence of osmium tetroxide and dehydration on the ultrastructure of actin meshworks. J. Cell Biol. 91, 695-705.

Tosney, K.W. and Wessells, N.K. (1983). Neuronal motility: the ultrastructure of veils and microspikes correlates with their motile activities. J. Cell Sci. 61, 389-411.

Vasiliev, J.M. (1982). Spreading and locomotion of tissue cells: factors controlling the distribution of pseudopodia. Phil. Trans. R. Soc. Lond. B. 299, 159-67.

Vasiliev, J.M. and Gelfand, I.M. (1976). Effects of colcemid on morphogenetic processes and locomotion of fibroblasts. In "Cell Motility" (ed. by R. Goldman, T. Pollard and J. Rosenbaum), pp. 279-303. Cold Spring Harbor Laboratory, 1976.

Weber, K., Rathke, P.C. and Osborn, M. (1978). Cytoplasmic microtubular images in glutaraldehyde-fixed tissue culture cells by electron microscopy and by immunofluorescence microscopy. Proc. Natl. Acad. Sci. 75, 1820-24.

40 Possible Mechanism of Regulation of Membrane-cytoskeletal Interactions in Erythrocytes

Philip S. Low and Kenneth C. Appell

INTRODUCTION

The cytoskeleton of the human erythrocyte membrane is largely responsible for control of the norphology and deformability of the cell (Agre et al., 1982; Chang et al., 1979; Goodman et al., 1982; Liu et al., 1982; Lux, 1979; Mohandas et al., 1982; Shohet, 1979; Tchernia et al., 1981; Yu et al., 1973). The cytoskeleton is primarily composed of the proteins spectrin (bands 1 and 2), actin (band 5), band 4.1 and band 4.9. The cytoskeleton is attached to the membrane via ankyrin (bands 2.1-2.6), which has a high affinity site for both the β subunit of spectrin and the cytoplasmic domain of the integral membrane protein, band 3 (Bennett, 1982; Branton et al., 1981; Yu and Goodman, 1979). Thus, the ankyrin-band 3 complex provides the required communication of the cytoskeletal network to the surrounding lipid envelope.

Regulation of ankyrin-band 3 interactions has been a matter of some confusion. Although there are ∿ 1.2×10^6 copies of band 3 per cell (Knauf, 1979), there are only ∿ 10^5 ankyrin binding sites on the red cell membrane (Goodman et al., 1982; Hargreaves et al., 1980; Bennett and Stenbuck, 1979). Thus, a majority of the band 3 proteins of the erythrocyte membrane apparently do not participate in ankyrin binding. Curiously, those copies of band 3 which apparently do not bind ankyrin in situ can be extracted and shown to exhibit normal ankyrin affinity when reconstituted into liposomes (Hargreaves et al., 1980) or when mixed with ankyrin in aqueous buffer (Bennett and Stenbuck, 1980). However, even this reconstituted fraction participates only partially in ankyrin binding. Thus, it would appear that a small but relatively constant fraction of any band 3 population can assume the ability to bind ankyrin. This behavior is obviously the type expected of a protein which exists in equilibrium between high affinity and low affinity forms. However, whether the putative equilibrium involves a subunit association or a conformational rearrangement cannot be predicted from the data. In fact, either explanation is fully consistent with the observed negative cooperativity of ankyrin binding to band 3 (Agre et al., 1981; Bennett and Stenbuck, 1980).

Since we have been characterizing a conformational equilibrium in the cytoplasmic domain of band 3 (Appell and Low, 1981; Appell and Low, 1982), we decided to investigate whether this structural equilibrium might be involved in the regulation of ankyrin binding. The following account summarizes some of the evidence for a conformational equilibrium in the cytoplasmic domain of band 3 and then presents data suggesting that the structural equilibrium might regulate ankyrin binding affinity.

THE ISOLATED CYTOPLASMIC FRAGMENT OF BAND 3 UNDERGOES TWO pH DEPENDENT STRUCTURAL TRANSITIONS

To study the structural properties of the cytoplasmic domain of band 3, we have cleaved the domain from the membrane and have isolated it in relatively pure form (Appell and Low, 1981). Figure 40.1 shows the data from three of several studies which demonstrate the existence of a conformational equilibrium among native states of this region of band 3. In Figure 40.1A the intrinsic fluorescence at 335 nm ($\lambda_{\underline{ex}}$, 290 nm) of the isolated cytoplasmic fragment dissolved in 50 m$\underline{M}$ sodium phosphate, 50 mM sodium borate, 70 mM NaCl, at 24°C is plotted as a function of pH. As the solution pH is elevated, the intensity of the intrinsic fluorescence increases, but in a complex manner. Curiously, the data could not be easily fitted to a single titration curve, regardless of the Hill coefficient. Instead, it was found that the data could be well described by two independent but partially overlapping titration curves. The derived apparent pKas are 7.2 ± 0.1 and 9.2 ± 0.07. The low pH transition contributes slightly more than 25% of the total fluorescence increase, suggesting the conformational change triggered by this ionization has a lesser effect on the domain's tryptophan residues than the high pH transition. The horizontal line in the figure describes the behavior of the cytoplasmic fragment in either 8 M urea or after heat denaturation. Clearly, the native conformation of the fragment is essential for both pH-

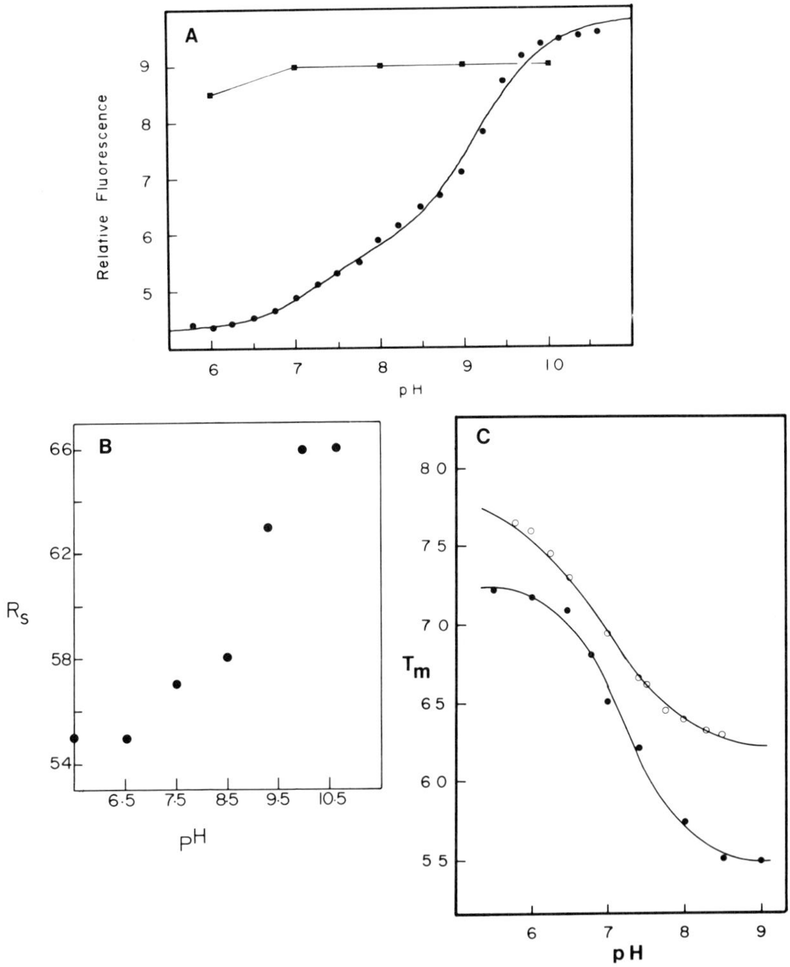

Fig. 40.1. The effect of pH on the (A) intrinsic fluorescence (λ_{ex} = 290 nm, λ_{em} = 335 nm), (B) Stokes radius, and (C) denaturation temperature of the 40,000-dalton cytoplasmic fragment of band 3 dissolved in 50 mM sodium phosphate, 50 mM boric acid, and 70 mM NaCl.

dependent structural changes. Titration of the fragment at 20°, 23°, 25°, 27°, 30°, 33°, 37° and 40°C also invariably yielded two distinct transitions. The ΔpKa/°C values were estimated to be -0.019 and -0.031 for the lower and higher transitions, respectively.

Figure 40.1B shows the dependence of the fragment's Stokes radius on pH, as measured on a Sephacryl S-300 column calibrated with ovalbumin, bovine serum albumin, aldolase, catalase and apoferritin. Again, the data appear to describe a double titration more accurately than a single titration. However, the number of measurements conducted are insufficient to draw any quantitative conclusions. The change in Stokes radius is very large, indicating that the pH-triggered structural rearrangements are substantial.

Figure 40.1C (open circles) demonstrates how the thermal stability of the cytoplasmic domain of band 3 is influenced by pH. Unfortunately, the calorimeter either is incapable of resolving the two overlapping conformational transitions or one of the structural changes has no effect on the domain's thermal stability. Regardless, the effect of pH on the net calorimetric properties of the fragment is profound, leading to a change in thermal stability of a magnitude rarely observed over the physiological pH range. The computer-derived apparent pKa is 6.9 and the Hill coefficient is 0.8.

Other studies from our laboratory (to be published) suggest that the above conformational transitions involve the rotation of subdomains of band 3 at a hinge region. Thus, no change in the CD spectrum between 195 and 250 nm is observed between pH 6 and 10. Furthermore, the cytoplasmic domain remains a dimer over this entire pH range, as long as the temperature is kept below Tm, i.e., the denaturation temperature. And finally, Förster energy transfer measurements indicate that the cluster of tryptophans in the cytoplasmic domain (residues 76, 82, 95 and 106) separates from cysteine 202 as pH is elevated.

THE CYTOPLASMIC AND MEMBRANE-SPANNING DOMAINS OF BAND # CAN BE INDEPENDENTLY REGULATED

The closed circles in Figure 40.1C show the effect of pH on thermal stability of the cytoplasmic domain of band 3 in situ, i.e., uncleaved and on the membrane. The apparent pKa and Hill coefficient were 7.2 and 1.0, respectively. Importantly, this behavior is not grossly different from the calorimetric behavior of the isolated cytoplasmic fragment and this suggests two important conclusions. Firstly, the conformational equilibrium which occurs in the isolated cytoplasmic fragment of band 3 over the physiological pH range also occurs in the cytoplasmic domain of band 3 on the intact membrane. And secondly, the cytoplasmic domain of band 3 must be structurally independent from the membrane-spanning domain of band 3; i.e., separation of the two domains by proteolysis has little effect on the behavior of the cytoplasmic domain. Thus, this region of band 3 can be independently regulated without perturbing the properties of anion transport through the membrane-spanning domain of band 3. The structural independence of the two major domains of band 3 has been documented much more extensively elsewhere (Appell and Low, 1982).

THE pH DEPENDENCE OF ANKYRIN BINDING TO BAND 3

The demonstration of multiple native conformations of the cytoplasmic domain of band 3 immediately suggests an hypothesis to account for the nonparticipation of part of the band 3 population in ankyrin binding. Thus, ankyrin may exhibit a high affinity for only one conformation of band 3 and not associate with the other two populations. Since physiological pH is near pH 7.4, the relevant structural equilibrium is likely the one characterized by the lower pKa, i.e., pKa $\sim$7.2. Thus, if ankyrin interacted with only the low pH form of band 3, then at physiological pH only $\sim$40% of the band 3 molecules would serve as ankyrin sites. Since this value is within the range of published estimates for the percentage of band 3 molecules which participate in ankyrin binding (Goodman et al., 1982; Sheetz, 1979; Bennett and Stenbuck, 1979; Hargreaves et al., 1980; Nigg and Cherry, 1980), we decided to examine the pH dependence of ankyrin binding experimentally. Human red cell ghosts were depleted of spectrin and actin by incubation at 37°C for 30 min. in 0.5 mM EDTA, 50 μg/ml phenylmethylsulfonyl fluoride (PMSF), pH 8. The resulting inside-out membrane vesicles were washed in 5 volumes of 50 mM sodium phosphate, 50 mM boric acid, 70 mM KCl, 2 mM $MgCl_2$, 7 mM ATP, 0.2 mM dithiothreitol, 50 μg/ml PMSF, 2 μg/ml pepstatin A, and 200 U/ml penicillin G, pH 7.15, and then resuspended in 30 volumes of the above buffer preadjusted to various pHs between pH 6 and 10. After incubation at 24°C for 16 hrs., the samples were pelleted at $\sim$40,000 × g for 45 min. and the pH of each supernatant was measured and recorded as the pH of incubation. The pellet was washed 1× in 0.5 mM EDTA, 50 μg/ml PMSF, pH 8 and then dissolved in a sodium dodecylsulfate solubilization buffer. The dissolved membrane samples were loaded onto a 5% polyacrylamide slab gel and separated electrophoretically according to the procedures of Fairbanks et al. (1981). The developed gel is shown in Figure 40.2A and the ratio of the ankyrin:band 3 staining intensity is plotted as a function of pH in Figure 40.2B. The data clearly show that less ankyrin is bound to the membrane at high pH than low pH. The insensitivity of some of the ankyrin to elution at alkaline pH probably derives from the extraordinarily slow rate of band 3-ankyrin dissociation. Assuming a binding constant (K_D) of 5×10^{-9} M and a half-time ($t_{\frac{1}{2}}$) of band 3-ankyrin association of ∿30 min at protein concentrations near 10^{-7} M (Bennett and Stenbuck 1980), the halftime of ankyrin dissociation can be estimated to be near 10 hrs. Therefore, ankyrin dissociation must have been incomplete at the conclusion of our 16 hr. incubation. Furthermore, some of the ankyrin appearing on the gels may have been trapped in resealed membrane vesicles. Regardless, the computer-derived apparent pKa of binding is ∿7.7 ± 0.7 which is not grossly different from the conformational pKa of ∿7.2, especially when one considers the inherent difficulties of accurately quantitating gel staining intensities. The loss of ankyrin binding at higher pHs is not due to increased proteolysis, since neither the major proteolytic products of ankyrin (a 72,000-dalton water-soluble, spectrin-binding fragment, and a ∿140,000-dalton band 3-binding fragment) nor of band 3 (a 40,000-dalton cytoplasmic fragment and a 55,000-dalton membrane-spanning fragment) are observed in any of the incubated samples (Fig. 40.2A). Therefore, the pH-dependent loss of membrane-bound ankyrin at elevated pHs must reflect a reduced affinity of ankyrin for the less protonated conformations of band 3. This observation is consistent with the ankyrin purification step employed by Hargreaves et al. (1980), where ankyrin is rapidly extracted from erythrocyte membranes in 1 M KCl by raising the pH from 6.75 to 7.6.

We, therefore, propose that all copies of band 3 are capable of binding ankyrin and that the solution pH (and perhaps other solutes) regulates the fraction of band 3 molecules which actually participate in ankyrin binding. Since at physiological pH the number of participating band 3 molecules closely matches the total number of ankyrin molecules in each cell, any major change in the band 3 conformational equilibrium should significantly influence the binding of cytoskeletal proteins

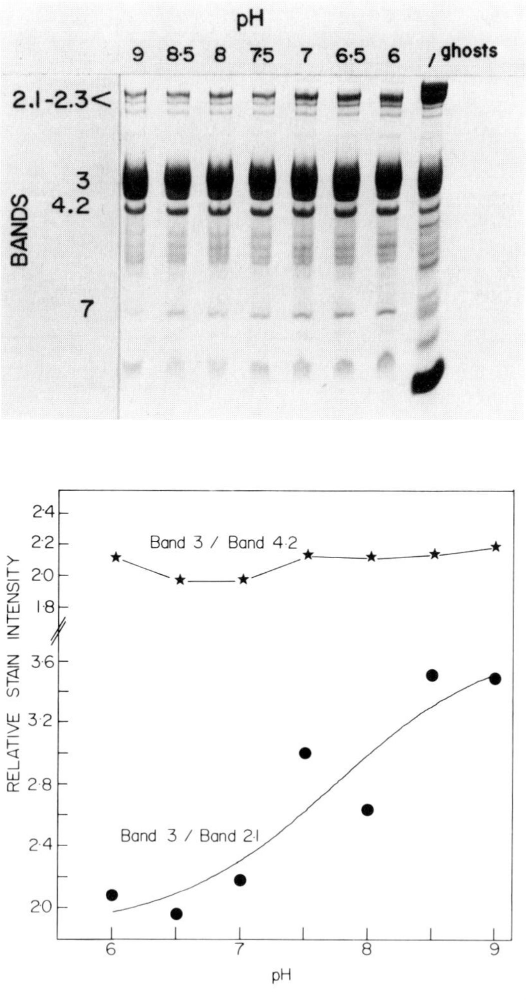

Fig. 40.2. (A) Polyacrylamide slab gel electrophoresis in 0.2% sodium dodecylsulfate of spectrin/actin-depleted human erythrocyte membrane vesicles incubated as described in the text at the pHs shown above. A few relevant protein bands are labelled. Intact erythrocyte membranes, containing the normal complement of ankyrin, spectrin and actin, were separated electrophoretically in the lane on the far right. (B) The ratio of the integrated staining intensities of band 3:ankyrin as a function of pH. The Coomassie blue-stained gels were scanned on an E-C Apparatus Corp. densitometer. The ratio of the band 3:band 4.2 staining intensities is also plotted as a function of pH as a control.

to the membrane. It is, therefore, not inconceivable that the observed pronounced dependence of red cell filterability on pH between pH 6.6 and 8 (Gummelt et al., 1981; LaCelle, 1969) might in part derive from the influence of pH on ankyrin-band 3 interactions.

ACKNOWLEDGMENTS

We thank Janet Keyl for typing the manuscript and acknowledge the support of a grant from the NIH (GM24417) in this work.

REFERENCES

Agre, P., Orringer, E.P., Chui, D.H.K. and Bennett, V. A molecular defect in two families with hemolytic poikilocytic anemia. J. Clin. Invest. 68, 1566-76, 1981.

Agre, P., Orringer, E.P. and Bennett, V. Deficient red-cell spectrin in severe, recessively inherited spherocytosis. N. Engl. J. Med. 306, 1155-61, 1982.

Appell, K.A. and Low, P.S. Partial structural characterization of the cytoplasmic domain of the erythrocyte membrane protein, band 3. J. Biol. Chem. 256, 11104-111, 1981.

Appell, K.A. and Low, P.S. Evaluation of structural interdependence of membrane-spanning and cytoplasmic domains of band 3. Biochemistry 21, 2151-57, 1982.

Bennett, V. and Stenbuck, P.J. The membrane attachment protein for spectrin is associated with band 3 in human erythrocyte membranes. Nature London 280, 468-73, 1979.

Bennett, V. and Stenbuck, P.S. Association between ankyrin and cytoplasmic domain of band 3 isolated from the human erythrocyte membranes. J. Biol. Chem. 255, 6424-32, 1980.

Bennett, V. The molecular basis for membrane-cytoskeleton association in human erythrocytes. J. Cell Biochem. 18, 49-65, 1982.

Branton, D., Cohen, C.M. and Tyler, J. Interaction of cytoskeletal proteins on the human erythrocyte membrane. Cell 24, 24-31, 1981.

Chang, K., Williamson, J.R. and Zarkowsky, H.S. Effect of heat on the circular dichroism of spectrin in hereditary pyropoikilocytosis. J. Clin. Invest. 64, 326-28, 1979.

Fairbanks, G., Steck, T.L. and Wallach, D.F.H. Electrophoretic analysis of the major polypeptides of the human erythrocyte membrane. Biochemistry 10, 2606-17, 1971.

Goodman, S.R., Schiffer, K.A., Casoria, L. and Eyster, M.E. Identification of the molecular defect in the erythrocyte membrane skeleton of some kindreds with hereditary spherocytosis. Blood 60, 722-84, 1982.

Gummelt, M., Unger, J., Scheven, C.H. and Geyer, G. Deformability of erythrocytes incubated at various pH-values. Acta Biol. Med. Germ. 40, 379-83, 1981.

Hargreaves, W.R., Giedd, K.N., Verklei, A. and Branton, D. Reassociation of ankyrin with band 3 in erythrocyte membranes and in lipid vesicles. J. Biol. Chem. 255, 11965-972, 1980.

Knauf, P. Erythrocyte anion exchange and the band 3 protein: transport kinetics and molecular structure. Curr. Topics Membr. Transp. 912, 249-363.

LaCelle, P.L. Alteration of deformability of the erythrocyte membrane in stored blood. Transfusion 9, 238-45.

Liu, S.C., Palek, J. and Prchal, J.T. Defective spectrin dimer-dimer association in hereditary elliptocytosis. Proc. Natl. Acad. Sci. USA 79, 2072-76, 1982.

Lux, S.E. Spectrin-actin membrane skeleton of normal and abnormal red blood cells. Semin. Hematol. 16, 21-51, 1979.

Mohandas, N., Clark, M.R., Heath, B.P., Rossi, M., Wolfe, L.C., Lux, S.R. and Shohet, S.B. A technique to detect reduced mechanical stability of red cell membranes: relevance to elliptocytic disorders. Blood 59, 768-74, 1982.

Nigg, E.A. and Cherry, R.J. Anchorage of a band 3 population at the erythrocyte cytoplasmic membrane surface:

protein rotational diffusion measurements. Proc. Natl. Acad. Sci. USA 77, 4702-4706, 1980.

Sheetz, M.P. Integral membrane protein interaction with triton cytoskeletons of erythrocytes. Biochim. Biophys. Acta 557, 122-34, 1979

Shohet, S.B. Reconstitution of spectrin deficient spherocyte membranes. J. Clin. Invest. 64, 483-94, 1979.

Tchernia, G., Mohandas, N. and Shohet, S.B. Deficiency of skeletal membrane protein band 4.1 in homozygous hereditary elliptocytosis. Implications for erythrocyte membrane stability. J. Clin. Invest. 68, 454-60, 1981.

Yu, J., Fischman, D.A. and Steck, T.L. Selective solubilization of proteins and phospholipids from red blood cell membranes by nonionic detergents. J. Supramol. Struct. 1, 233-48.

Yu, J. and Goodman, S.R. Syndeins: the spectrin binding protein(s) of the human erythrocyte membrane. Proc. Natl. Acad. Sci. USA 76, 2340-44, 1979.

41 Centripetal Movements of Actin and Cell Surface Receptors during Fibroblast Locomotion

Julian P. Heath

BEHAVIOR AND STRUCTURE OF ARCS IN CRAWLING FIBROBLASTS

Arcs are curvilinear phase-dense fibers that form close to the lamellar margin and then travel centripetally at 1 to 3 $\mu m\ min^{-1}$ relative to the substratum before disappearing in the perinuclear region (Fig. 41.1). Arc formation is specifically associated with protrusion of the lamellar margin; it occurs in respreading cells, during normal locomotion, and after tail detachment, but is absent in polygonal cells and during contact inhibition of locomotion (1,2).

By SEM, arcs are seen as curved ridges in the dorsal surface of the lamella (Fig. 41.2). TEM of vertical sections show these ridges to be bundles of 200 to 500 actin microfilaments (Fig. 41.3). But arcs are not discrete structures; they are an integral part of a submembranous sheath of circumferentially-oriented microfilaments that characterizes actively motile fibroblasts. This sheath, called the dorsal cortical microfilament sheath (DCMS), is normally 5 to 10 filaments deep and is separate from the adhesion-linked stress fibers. Arcs appear to develop from a local S-shaped folding of the DCMS which subsequently flattens, forming an arc visible by light microscopy. Arcs and the DCMS stain with antibodies to actin, myosin, α-actinin and vinculin, but the spatial arrangements of these proteins remain to be determined. However, the pseudosarcomeric organization of microfilaments and dense bodies found in arcs (3) suggests that DCMs movements may be generated by a sliding filament mechanism.

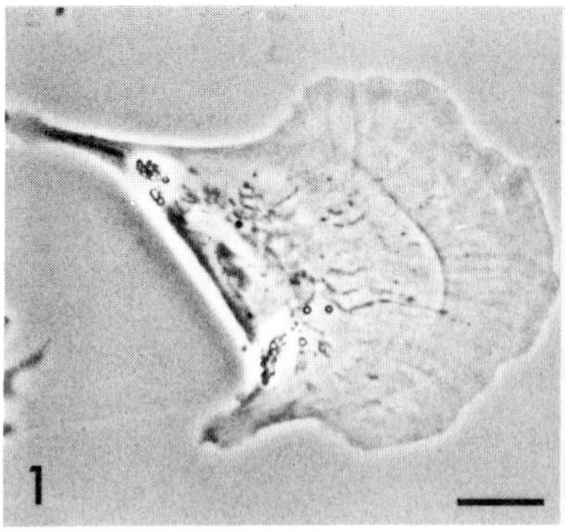

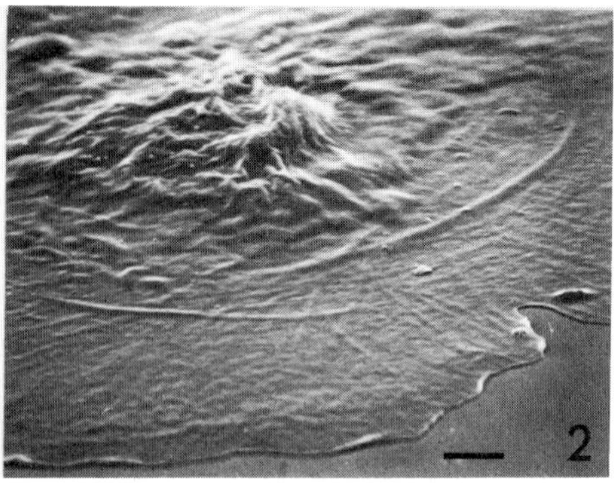

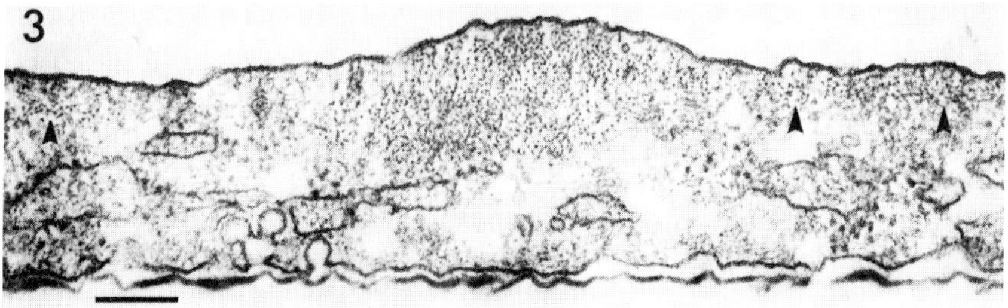

Fig. 41.1. Phase contrast micrograph of a chick hearr fibroblast with an arc in its leading lamella. Scale bar = 10 μm.

Fig. 41.2. Scanning electron micrograph showing two arcs in a respreading chick fibroblast. Scale bar = 2 μm.

Fib. 41.3. Transmission electron micrograph of a vertical section through the lamella of a human fibroblast showing an arc and the dorsal cortical microfilament sheath (arrows). Scale bar = 200 nm.

CAPPING OF CELL SURFACE RECEPTORS

Behaviorally, arc movement has much in common with particle transport and the discovery (2) that the two events occur in a coordinated manner indicated that arcs, and the DCMs, could be involved in the mobility of surface receptors and their ligands.

This possibility was tested directly by using a rabbit antiserum to chick cell surfaces—antiCSN. Chick fibroblasts are uniformly stained by antiCSN which, additionally, behaves as a monovalent ligand since there is no redistribution of anti-CSN on translocating cells even during arc formation and movement (3). This shows that there is no interaction of the DCMS with uncross-linked receptors. However, when antiCSN-labelled cells are incubated in antiIgG, the receptors are patched and, moreover, become attached to the detergent-resistant cyto-skeleton. On translocating or respreading fibroblasts, the patches are cleared centripetally from actively protruding lamellae and within 15 min have collected in a cap over the nucleus and perinuclear zone. There is no directed or rapid movement of patches on stationary cells or on the nonprotruding processes of moving cells. By taking SIT video recordings it was found that patches move back as a unit, maintaining their relative positions until reaching the rear of the lamella. On arc-forming cells patches were always more dense over arcs than elsewhere on the lamella and these patches moved rearwards coordinately with the arcs for many microns until they reached the perinuclear zone (Fig. 41.4).

Other surface receptors behave in a similar manner with respect to arcs. To date these include receptors for concanavalin A, anti-β_2 microglobulin, and cholera toxin which binds to gangliosides.

CONCLUSIONS

This study has shown that during the capping process on crawling fibroblasts, patched cell surface receptors move coor-

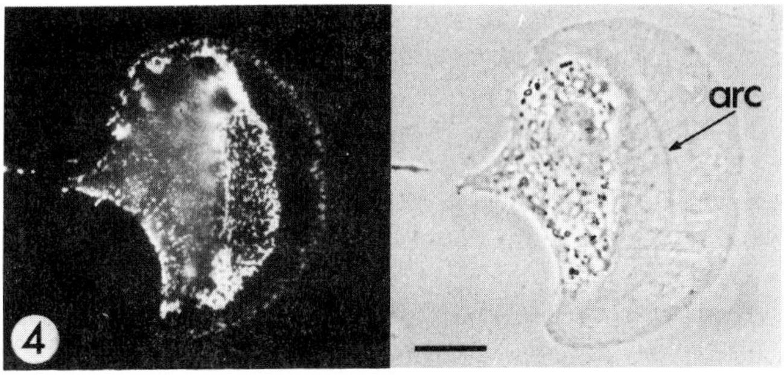

Fig. 41.4. Paired immunofluorescence and phase contrast micrographs showing the coordinate centripetal movement of an arc and patched antiCSN receptors. The anterior band of patches is on the ventral cell surface and does not move rearwards. This cell was incubated in antiCSN for 10 min at 37 C, washed and then incubated in fluoresceinated antirabbit Ig for 6 min at 37 C before fixation. Scale bar = 10 μm.

dinately with submembranous bands of actin microfilaments. The data support the conclusion that receptor movement is mediated by the DCMS rather than by other mechanisms such as lipid flow or surface waves. The nature of the interaction between the DCMS and receptor patches is still largely a matter for speculation, but it must be fairly nonspecific to allow different glycoprotein and glycolipid receptors to behave similarly.

The behavior of patches points to a bulk flow of DCMS material beneath the dorsal surface during movement rather than minor displacements in a stationary DCMS such as might be caused by longitudinal or transverse waves. Microinjection of labelled actin may resolve this question. The bulk flow of the DCMS could be driven by sliding filaments, given the circumferential orientation of the DCMS. DCMS material would be disassembled at the rear of the lamella and recycled for assembly at the lamellar margin, a process that could drive lamellar extension. In this model, any receptor that as a result of patching formed an attachment to, or was constrained to move with, the DCMS would be swept backwards on the lamella. Although there is a small amount of clearance of patches from the ventral surface, the majority of ventral patches appear to be immobile and are frequently left behind on the substratum as the cell moves on. Thus, there is not yet sufficient evidence for a centripetal flow of actin at the ventral cell surface, which if present must necessarily be complex because of the involvement of the ventral cortical actin in adhesion and stress fiber formation.

REFERENCES

Heath, J.P. Arcs: Curved microfilament bundles beneath the dorsal surface of the leading lamellae of moving chick embryo fibroblasts. Cell Biol. Int. Reps. 5, 975-80 (1981).

Heath, J.P. Behaviour and structure of the leading lamella in moving fibroblasts. I. Occurrence and centripetal movement of arc-shaped microfilament bundles beneath the dorsal cell surface. J. Cell Sci. 60, 331-54 (1983).

Heath, J.P. Direct evidence for microbilament-mediated capping of surface receptors on crawling fibroblasts. Nature 302, 532-34 (1983).

42 Troponin in Normal and Regenerating Fibers in the Anterior Latissimus Dorsi Muscle of the Chick

Yutaka Shimada, Mitsutada Miyazaki, Naoji Toyota

INTRODUCTION

Extrafusal fibers of the anterior latissimus dorsi (ALD) muscle of the adult chicken can be called "slow tonic" fibers to distinguish from "slow twitch" fibers (Hess, 1970), and have long been considered to be composed of fibers exhibiting a homogeneity in structural, functional and histochemical properties (Ginsborg, 1960; Hess, 1970). In recent years, however, observations by histochemistry of myofibrillar ATPase staining and immunohistochemistry using antibodies against myosin have suggested that this muscle is a mixture of fiber types; it has two subtypes of slow tonic fibers and a small percentage of fast twitch fibers (Pierobon Bormioli et al., 1980; Toutant et al., 1980; Hikida and Wang, 1981).

The present study was undertaken to investigate the distribution of the regulatory proteins of troponin (TN) in both extra- and intrafusal fibers of the ALD. For this purpose, we have utilized an immunohistochemical approach using antibodies specific for breast and ventricular subunits of TN (Toyota and Shimada, 1981, 1983) combined with myofibrillar ATPase staining. Further, with the use of these antibodies we examined the problem of what types of muscle fibers are produced in the ALD and posterior latissimus dorsi (PLD) during regeneration after cold injury.

MATERIALS AND METHODS

Normal (ADL) and regenerating (ALD and PLD) muscles of the adult chicken were quickly frozen in isopentane cooled to -160°C with liquid nitrogen. Transverse sections were reacted

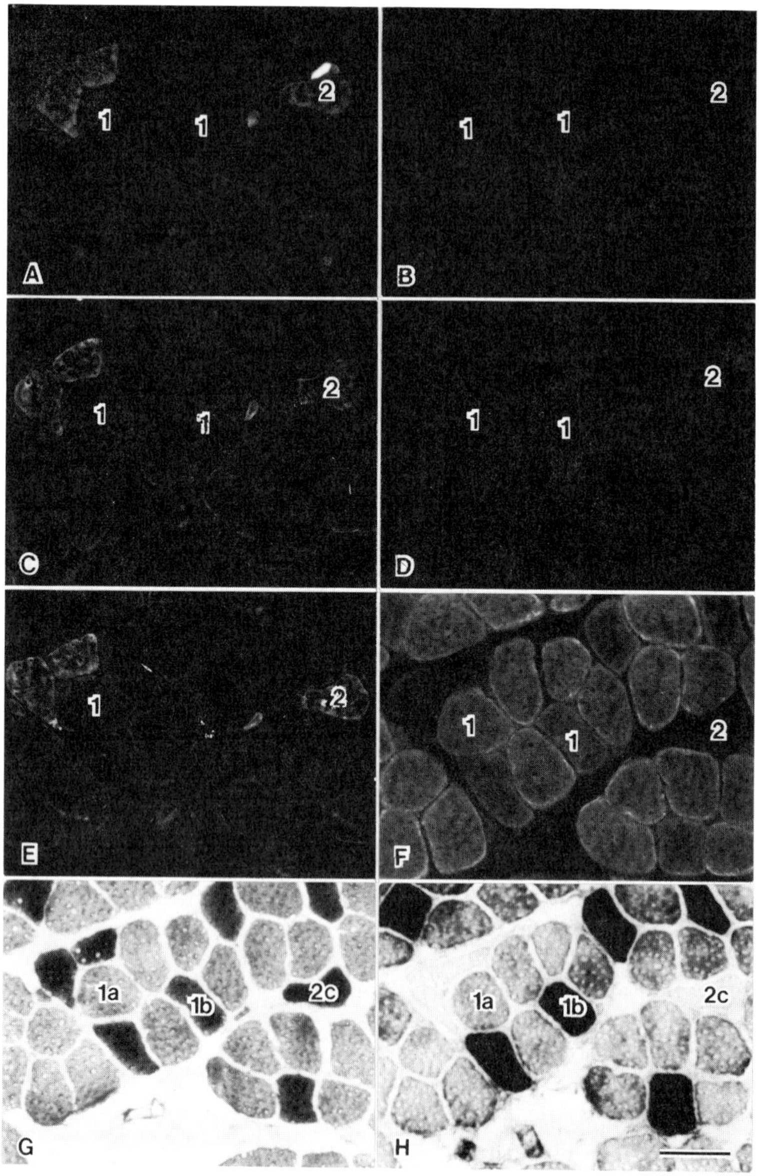
A
1
1
2
B
1
1
2
C
1
1
2
D
1
1
2
E
1
2
F
1
1
2
G
1a
1b
2c
H
1a
1b
2c

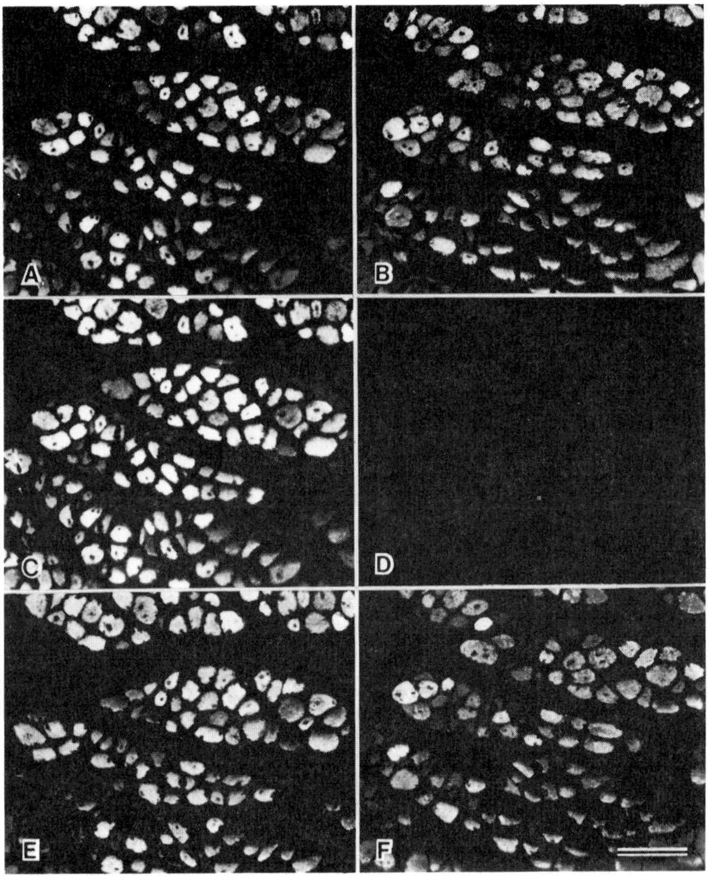

Figs. 42.1 and 42.2. Serial cross sections of the normal (Fig. 42.1) and regenerating (Fig. 42.2) fibers in the ALD muscle. Stained with antibodies against breast TN-T (A), TN-I (C) and TN-C (E), and ventricular TN-T (B), TN-I (D) and TN-C (F). Myofibrillar ATPase activity after alkaline (Fig. 42.1G) and acid (Fig. 42.1H) preincubation. See text for details. Magnification: × 120. Scale bar = 100 μm.

with antibodies against TN components (T, I, and C) of breast and ventricular muscles from the adult chicken. They were treated with FITC-labeled second antibody and examined under a Zeiss microscope equipped with epifluorescence. Sections serial to those used for immunofluorescence were processed for the histochemical demonstration of myofibrillar ATPase activity after alkaline (pH 10.4) and acid (pH 4.3) preincubation. For regeneration experiments, a focal cold injury was brought about on the ALD and PLD muscle with a liquid-nitrogen cooled brass rod.

RESULTS

Normal Muscle Fibers

Four distinct categories of extrafusal muscle fibers were distinguished on the basis of their differential reactivity with the antibodies. The predominant population of fibers (>95%) was stained only with antiventricular TN-C; they were not reactive with any other antibodies prepared (1, Fig. 42.1A-F). These fibers can be considered to be slow tonic fibers, because their myofibrillar ATPase activity was either alkali- and acid-labile (1a, Fig. 42.1G and H) or alkali- and acid-stable (1b, Fig. 42.1G and H) (Ovalle 1978).

The second group of fibers (<5%) reacted with antibodies directed against breast TN-T, -I, and -C, but did not react with any of the antibodies against the ventricular TN components (2, Fig. 42.1A-F). Since these fibers showed high myofibrillar ATPase activity after alkaline preincubation and no myofibrillar ATPase activity after acid preincubation (2c, Fig. 42.1G and H), they seem to correspond to fast twitch fibers (Burke et al., 1971).

The third type of fibers (<1%) was labelled not only with antibodies against all of the three components of breast TN, but also with antibodies raised against ventricular TN-T and -C. They were not reactive with antiventricular TN-I antibody. The remainder of the fibers (<1%) in the ALD had a positive response to antibodies raised against ventricular TN-T and -C, but a negative reaction to those against all three breast TN components and ventricular TN-I. Since myofibrillar ATPase activity of both fiber types was alkali- and acid-labile, they are regarded as slow tonic fibers.

In intrafusal muscle fibers of the ALD, the same four types of fibers as those found in extrafusal fibers were observed.

Table 42.1. Normal and Regenerating Fiber Types in the ALD Muscle of the Adult Chicken

Fiber Type		Extrafusal Fibers 1	Extrafusal Fibers 2	Extrafusal Fibers 3	Extrafusal Fibers 4	Intrafusal Fibers, Equatorial Region 1	Intrafusal Fibers, Equatorial Region 2	Intrafusal Fibers, Equatorial Region 4	Intrafusal Fibers, Polar Region 3	Intrafusal Fibers, Polar Region 4	Regenerating Extrafusal Fibers 3
Antibody against	Breast TN-T	-	+	+	-	-	+	-	+	-	+
	Breast TN-I	-	+	+	-	-	+	-	+	-	+
	Breast TN-C	-	+	+	-	-	+	-	+	-	+
	Ventricular TN-T	-	-	+	+	-	-	+	+	+	+
	Ventricular TN-I	-	-	-	-	-	-	-	-	-	-
	Ventricular TN-C	+	-	+	+	+	-	+	+	+	+
Myo-fibrillar ATPase	Alkaline (pH 10.4)	low or high	high	low	low						
	Acidic (pH 4.3)	low or high	no	low	low						

Within four types of intrafusal fibers, three were found in the equatorial zone and two were seen in the encapsulated polar region (not shown). These results indicate that there appears to be some variation in the stainability of fibers along their length.

Regenerating Muscle Fibers

After the muscle was subjected to cold injury, new cells could be observed in the intermediate region between the outer necrotic area and the underlying uninjured muscle. These new fibers were of a smaller diameter than normal adult fibers and had a central nucleus. They were presumed to be regenerating myotubes, and exhibited different stainabilities with the antibodies from those of the adjacent uninjured tissue.

Although differenr fiber types are present in the normal ALD, all new cells arose from satellite cells in this muscle stained with antibodies against all breast TN components and ventricular TN-T and -C (Fig. 42.2A-F). The same stainabilities were also observed with all regenerating fibers in the PLD. These reactivities in regenerating fibers are the same as those seen in embryonic skeletal muscles (Toyota and Shimada, 1981).

Reactivities to the antibodies that do not normally bind uninjured muscles of each type began to appear 2 days after injury and were prominent during subsequent days. By 8 weeks, the regenerating fibers in the ALD and PLD recovered their normal diameter and characteristic stainabilities.

The results are summarized in Table 42.1.

CONCLUSIONS

The slow ALD muscle of the adult chicken contains a heterogeneity of fiber types regarding the reactivity with antiTN component antibodies in both extra- and intrafusal fibers.

Regenerating fibers, iirespective of the kinds of muscles (ALD or PLD) or the type of fibers, synthesize TN subunits possessing embryonic reactivity. It appears that during regeneration from cold injury, skeletal muscles recapture an embryonic stage and synthesize embryoniclike TN isoforms.

REFERENCES

Burke, R.E., Levine, D.N., Zajac, F.E., Tsaivis, P. and Engle, W.K. 1971. Mammalian motor units. Physiological-histochemical correlation in three types in cat gastrocnemius. Science 174:709-12

Ginsborg, B.L. 1960. Some properties of avian skeletal muscle fibers with multiple neuromuscular junctions. J. Physiol. 154:581-98.

Hess, A. 1970. Vertebrate slow muscle fibers. Physiol. Rev. 50:40-62.

Hikida, R. and Wang, R.J. 1981. Tenotomy of the avian anterior latissimus dorsi muscle. I. Effect of age on fiber-type transformation and regeneration from the stump in chicks. Am. J. Anat. 160:395-408.

Ovalle, W.K. 1978. Histochemical dichotomy of extrafusal and intrafusal fibers in an avian slow muscle. Am. J. Anat. 152:587-98.

Pierobon Bormioli, S., Sartore, S., Vitadello, M. and Schiaffino, S. 1981. "Slow" myosins in vertebrate skeletal muscle. An immunofluorescence study. J. Cell Biol. 85:672-81.

Toutant, J.P., Toutant, M.N., Renaud, D. and Le Douarin, G.H. 1980. Histochemical differentiation of extrafusal muscle fibers of the anterior latissimus dorsi in the chick. Cell Differ. 9:305-14.

Toyota, N. and Shimada, Y. 1981. Differentiation of troponin in cardiac and skeletal muscles in chicken embryos as studied by immunofluorescence microscopy. J. Cell Biol. 91:497-504.

Toyota, N. and Shimada, Y. 1983. Isoform variants of troponin in skeletal and cardiac muscle cells cultured with and without nerves. Cell 33:297-304.

43 Mechanism of Interaction of Brain Actin Fragmenting Protein with Actin Filaments

T. C. Petrucci

A calcium-dependent actin fragmenting protein with molecular weight 90,000 daltons (90K protein) has been purified to homogeneity from bovine brain (I). This protein is globular with a diameter of 7-9 nm in electron microscope and consists of a single polypeptide chain. A similar protein was detected by immunoblotting and immunofluorescence in pure cultured neurones (I). We studied the actin of brain 90K protein by viscometry, electron microscopy, sedimentation and fluorescence methods. The activity of the 90K protein from brain was very similar to that of gelsolin, an actin-binding protein originally isolated from rabbit lung macrophages (2). Actin filaments are rapidly fragmented on addition of 90K protein at Ca^{2+} greater than 10^{-7}. Substoichiometric concentration of 90K protein also effects the salt-induced polymerization of muscle G-actin as measured by viscometry.

The interaction of 90K protein with muscle actin was studied by using a pyrenyl derivative of actin that increases its quantum yield by more an order of magnitude on polymerization (3). This fluorescent probe provides a quantitative measure of incorporation of monomer into polyer in solutions undisturbed by shear stress.

In solution of muscle G-actin containing trace amounts of pyrene-labelled actin (5-6% of the total actin) the 90K protein induces polymer formation in the presence of micromolar concentration of calcium (Fig. 43.1). In buffers containing either 0.1 M KC1 or 1 mM $MgCl_2$, the 90K protein increases the initial rate of polymerization as reflected by an enhanced rate of incorporation of monomer into polymer (Fig. 43.2). However, the extent of polymerization is reduced in the presence of brain 90K protein, and the extent of this decrease is dependent on the condition of polymerization.

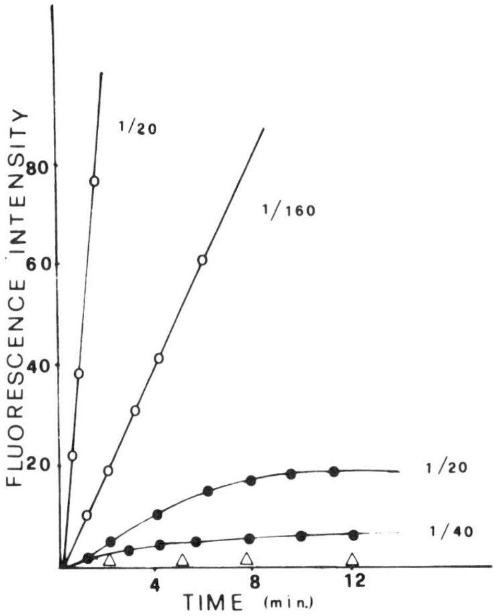

Fig. 43.1. Polymerization of pyrene-labelled G-actin nucleated by brain 90K protein added to actin at the indicated molar ratio (○), in absence (closed symbol) and in presence (open symbol) of Ca^{2+} μM. Actin alone 16 μM (△). The fluorescence enhancement was monitored (386 nm, excited 366 nm).

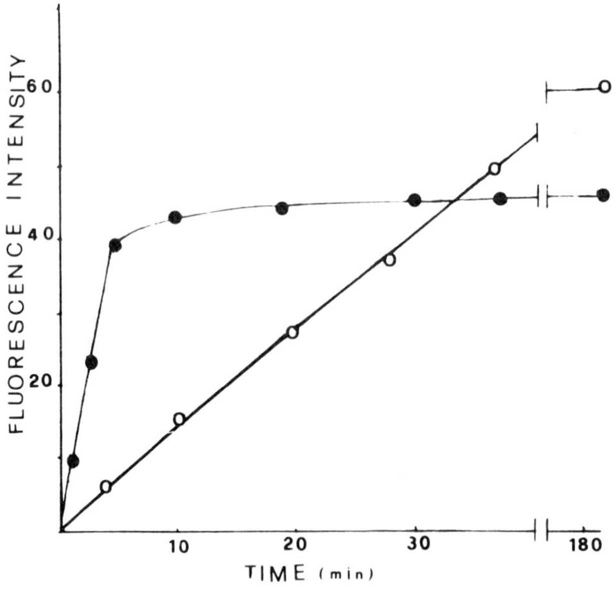

Fig. 43.2. Fluorescence intensity (in arbitrary units) at 386 nm of G-actin (16 μM-6% pyrenyl actin), monitored in the course of salt-induced polymerization into actin filaments. (○) control; (●) in presence of brain 90K protein (0.12 μM).

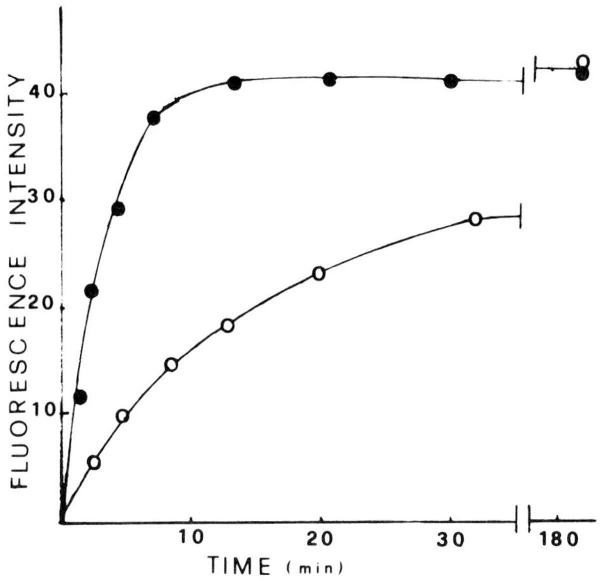

Fig. 43.3. Incorporation of pyrenyl actin into F-actin in absence (○) and in presence of 0.12 μM of brain 90K protein (●). Muscle G-actin (16 μM) was polymerized at 25°C by addition of 1 mM $MgCl_2$ and 0.1 M KCl. Pyrenyl actin (5 μl) and 90K protein (25 μl) or buffer were added to 0.5 ml of F-actin solution and the fluorescence enhancement was monitored (386 nm, excited 366 nm).

Measurements of incorporation of G-actin into F-actin were made by adding trace amounts of pyrenyl actin to the solutions of F-actin at steady state. As shown in Figure 43.3, the initial rate of fluorescence increase is very high in the presence of 90K protein and a fluorescence equilibrium is reached after ten minutes. The very rapid exchange of actin monomer into polymer (enhanced fluorescence) could be explained by an increase of free ends of actin filaments as a consequence of severing by 90K protein.

These studies confirm previous results (1) that the brain 90K protein has more than one action in vitro. The brain 90K protein a) binds to monomeric actin, b) nucleates filament assembly, and c) has the ability to sever filaments in a calcium-dependent manner.

REFERENCES

1. Petrucci T.C., Thomas C., Bray D., J. Neurochem. (1983), 40, 1507-16.
2. Lin H.Y. and Stossel T.C., J. Biol. Chem. (1981), 255, 9490-93.
3. Kouyama T. and Mihashi K., Eur. J. Biochem. (1981), 114, 33-38.

44 Immunocytochemical Localization of Cytoskeletal and Extracellular Matrix Proteins Using Semithin Sections of Quick-frozen, Freeze-dried and Plastic-embedded Tissue

D. Drenckhahn, M. Gerhardt, R. Steffens, M. Brandner, J. Wagner, C. Gerhardt, T. Schäfer, M. Prinz, K. Zinke

INTRODUCTION

Many cytoskeletal proteins are soluble in physiologic buffers such as phosphate-buffered saline (PBS) or other buffers frequently used for immunocytochemical staining procedures. It must be taken into account, therefore, that some of these proteins, in particular those associated with the microfilament system, are sensitive to artifactual removal or redistribution during the rinsing and incubation steps used for immunocytochemical staining of cryotome sections of unfixed or mildly fixed tissues. This possibility was investigated by SDS gel electrophoretic analysis of buffer supernatants (isotonic PBS, pH 7.4) which had been used for rinsing air-dried cryostate sections of chicken gizzard (unfixed or prefixed by 1% paraformaldehyde in PBS) mounted on glass slides. After incubation for 30 min. at room temperature the rinsing buffer was removed from the slides, cleared by high speed centrifugation, mixed with sample buffer and applied to a 5 to 20% polyacrylamide gradient gel (Fig. 44.1).

Comparison of Coomassie blue stained tracks with tracks of whole gizzard homogenate blotted on nitrocellulose paper and stained with antibodies to actin, myosin, α-actinin, and tropomyosin revealed the presence of these microfilament-associated proteins in the rinsing buffer supernatant. These experiments indicate that at least some of the major proteins constituting the microfilament system become partly removed from cryostate sections of unfixed and also from mildly fixed tissue during exposure to isotonic buffers. Considering the possibility that solubilization of cytoskeletal proteins might cause false negative (e.g., removal of proteins) or false positive (displacement of proteins to other cellular structures or cells) staining results we applied immunofluorescence and immunoperoxidase staining to semithin tissue sections (0.5-1 µm) of freeze-dried and plastic-embedded tissues (FDP method).

TECHNICAL NOTES

Small tissue pieces of various organs or whole cover slips with adherent tissue culture cells were immediately frozen by placing in melting isopentane or by pressing on a polished copper block cooled with liquid nitrogen. Freeze-drying was performed at 10^{-4} Torr starting with a temperature of -80°C,

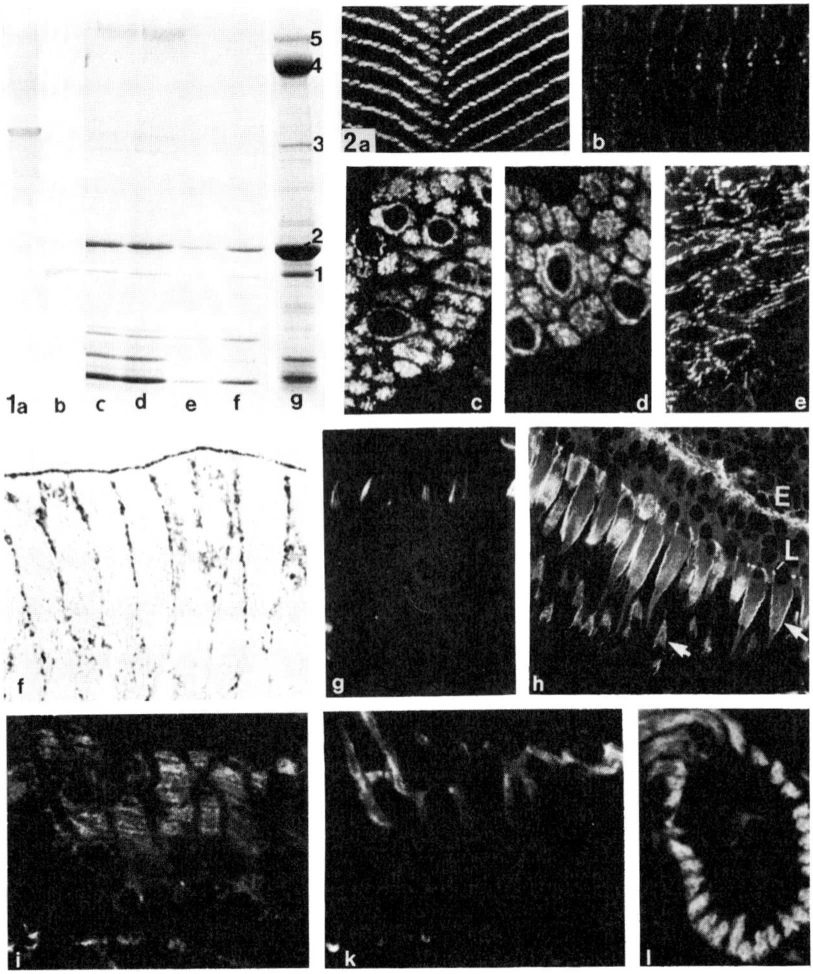

which was stepwise raised (intervals of 12 h) to 40°C, -20°C, 0°C and finally +20°C. Freeze-dried tissue samples were placed in Epon (or Araldite) containing 1.8% accelerator and were exposed to constant vacuum (10^{-3} Torr) for 48 h in order to allow complete penetration of the resin. Glass slides with mounted sections (0.5-1 μm) of plastic-embedded, freeze-dried tissue or tissue culture were placed for 2 to 12 min. in a mixture of sodium methoxide, methyl alcohol, and benzene (according to Mayor et al. 1961) in order to partially remove the resin. The sodium methoxide solution was freshly prepared from metallic sodium and water-free methyl alcohol and was used for about 4 weeks.

Fig. 44.1 (top left corner). Electrophoretic analysis of proteins extracted from cryotome sections (5 μm) of unfixed (b,c) and fixed (1% formaldehyde) (d,e) chicken gizzard tissue during 30 min of incubation with PBS. Comparison of the buffer supernatants with whole extracts of sectioned tissue (f) and immunoperoxidase-stained blots of the electrophoretograms (examples of tracks stained with antibodies to α-actinin (a) and tropomyosin (b) are shown) reveals several microfilament-associated proteins removed from the sections during the incubation period. Protein bands determined: tropomyosin (1), actin (2), α-actinin (3), myosin (4), filamin (5).

Fig. 44.2. Semithin (0.5 μm) sections of freeze-dried and plastic-embedded tissues incubated with the antibodies indicated. (a,b) Parts of skeletal muscle fibers stained with anti-α-actinin (a) and antidesmin (b). (c,d,e) Intestinal smooth muscle stained with anti-α-actinin (c), antidesmin (d) and antivinculin (e). (f) Intestinal epithelium stained with anti-α-actinin (immunoperoxidase). (g) Chicken macula statica showing fimbrinlike fluorescence confined to stereocilia. (h) Fish (Guppy) retina stained with antigizzard actin (E, external plexiform layer, L, external limiting membrane, arrows, cone and rod ellipsoids). (i,k) Tangential section of human splenic sinusoidal endothelium stained subsequently with antimyosin (i, note parallel aligned stress fibers) and antifibronectin (k). Fluorescent lines in k (annular component of the basement membrane, "ring fibers") correspond to the unstained lines in i (contact zone of the stress fibers to the extracellular matrix). (l) Splenic sinus endothelium stained with antivimentin. Magnifications: a × 1,500, b × 1,700, c × 1,200, d × 1,700, e × 1,200, f × 1,600, g × 700, h × 700, i,k × 1,000, l × 1,500.

In most instances a 12 min. exposure of 1 μm thick plastic sections to the methoxide methyl alcohol-benzene solution gave satisfactory removal of resin. After this treatment slides were transferred for 15 min. to a 1:1 (v/v) mixture of methyl alcohol and benzene (freshly prepared), then placed in acetone (20 min., two changes), and finally rinsed in PBS (30 min., two changes). Tissue sections mounted on slides and treated as indicated above were incubated with antibodies (0.1-1 mg/ml IgG in PBS) in a humidified atmosphere at 21°C for 30 min. After washing in PBS for 30 min. (3 changes) sections were processed for indirect immunofluorescence or immunoperoxidase staining (using the appropriate controls) as described (Drenckhahn et al. 1983).

RESULTS

The applicability of this method of tissue processing for the immunocytochemical localization of several cytoskeletal proteins was tested in striated and smooth muscle, the intestinal epithelium, vascular endothelial cells, the retina and auditory and vestibular hair cells. (Fig. 44.2).

In striated muscle antibodies to actin, myosin, tropomyosin, and α-actinin reacted brilliantly with the corresponding myofibrillar bands and lines where these proteins are known to be located (see Fig. 3 in Gröschel-Stewart and Drenckhahn 1982). Antivinculin displayed (as recently shown by Pardo 1983) a preferential affinity for the muscular plasma membrane at the level of the Z-lines. Antigizzard desmin gave, as expected (Lazarides 1980), an interrupted beadlike staining pattern in the level of the Z-lines.

In smooth muscle antibodies to α-actinin (gizzard) and desmin (gizzard) displayed virtually identical staining patterns which were characterized by numerous fine cytoplasmic particles (probably reflecting dense bodies) and small plasmalemmal spots or lines (probably dense areas). Antivinculin was confined to the α-actinin and desminlike plasmalemmal spots as to be expected from the studies of Geiger et al. (1982). Antibodies to actin, myosin, myosin kinase, myosin light chain, tropomyosin and filamin gave a diffuse cytoplasmic immunostaining.

Immunostaining of intestinal epithelium with antibodies to actin, myosin, tropomyosin, α-actinin, filamin, vinculin, villin, and fimbrin gave the typical distribution patterns recently described for these proteins in the brush border region (for review, see Gröschel-Stewart, Drenckhahn 1982). Antibodies to actin and α-actinin gave a beadlike interrupted staining pattern along the entire lateral and basal plasma membrane, which is similar to the pattern of microfilament-associated spot desmosomes (adhaerens plaques) described for acinar cells of various exocrine glands (Drenckhahn and Mannherz 1983).

Rod and cone inner segments of the retina of all nonmammalian species investigated contained prominent actin bundles aligned parallel to the photoreceptor long axis (probably important for retinomotoric activity, Burnside 1978). The external limiting membrane was shown to be composed of actin filaments, α-actinin and vinculin (confirmed by ultrastructural studies to be published).

The apex of vestibular and auditory hair cells contains actin, myosin (only weak reaction), α-actinin, vinculin and tropomyosin (Flock et al. 1983, Drenckhahn et al. 1983). These proteins showed a similar distribution as described for the intestinal brush border. In vestibular hair cells the actin cross-linking protein of intestinal microvilli, fimbrin, was associated with the stereocilia but not with the cuticular plate as described recently for the cuticular plate of auditory hair cells in the guinea pig (Flock et al. 1983).

Splenic sinus endothelium was chosen to study the composition of endothelial stress fibers. The stress fibers reacted with antibodies to myosin, α-actinin, actin and tropomyosin in a similar distribution pattern described for stress fibers found in cultured cells (see Drenckhahn 1983). Vinculin was located at the contact zone of stress fibers with the extracellular matrix (so-called ring fibers). On the extracellular side this contact zone was characterized by high affinity for antibodies to fibronectin, collagen type III and laminin (antilaminin bound only to frozen sections and not to the plastic sections).

CONCLUSIONS

The use of semithin plastic sections of freeze-dried and plastic-embedded tissues (FDP method) for immunocytochemical location of cytoskeletal and extracellular matrix proteins avoids diffusion artifacts and allows the immunocytochemical location of all cytoskeletal proteins so far tested (actin, tropomyosin, α-actinin, vinculin, filamin, villin, fimbrin, myosin, myosin 20k light chain, myosin light chain kinase, vimentin, desmin, prekeratins, tubulin) as well as many extracellular matrix proteins (fibronectin, collagen Types I, III, but not laminin). Further advantages of the FDP method are: (1) good structural preservation, (2) high light microscopic resolution, (3) the possibility to use serial sections for multiantibody studies in identical cells, (4) accessibility to tissues which are difficult to investigate by cryotomy (e.g., the auditory organ, tissues of the eye), and (5) the use of individual tissue blocks for unlimited periods (antigenicity was fully preserved in tissue pieces embedded nine years previously).

REFERENCES

Burnside, B.: Thin (actin) and thick (myosin-like) filaments in cone contraction in the teleost retina. J. Cell Biol. 78, 227-46 (1978).

Drenckhahn, D.: Cell motility and cytoplasmic filaments in vascular endothelium. Prog. Appl. Microcirc. (Karger, Basel) 1, 53-70 (1983).

Drenckhahn, D., U. Gröschel-Stewart, J. Kendrick-Jones, J. Scholey: Antibody to thymus myosin: its immunological characterization and use for immunocytochemical localization of myosin in vertebrate nonmuscle cells. Eur. J. Cell Biol. 30, 100-111 (1983).

Drenckhahn, D., J. Kellner, H. G. Mannherz, U. Gröschel-Stewart, J. Kendrick-Jones, J. Scholey: Absence of myosin-like immunoreactivity in stereocilia of cochlear hair cells. Nature 300, 531-32 (1982).

Drenckhahn, D., H. G. Mannherz: Distribution of actin and the actin-associated proteins myosin, tropomyosin, alpha-actinin, vinculin and villin in rat and bovine glands. Eur. J. Cell Biol. 30, 167-76 (1983).

Flock, A., A. Bretscher, K. Weber: Immunohistochemical localization of several cytoskeletal proteins in inner ear sensory and supporting cells. Hearing Res. 6, 75-89 (1982).

Geiger, B., A. H. Dutton, K. T. Tokuyasu, S. J. Singer: Immunoelectron microscope studies of membrane-microfilament interactions. Distributions of α-actinin, tropomyosin, and vinculin in intestinal epithelial brush border and chicken gizzard smooth muscle cells. J. Cell Biol. 91, 614-28 (1981).

Gröschel-Stewart, U., D. Drenckhahn: Muscular and cytoplasmic contractile proteins. Biochemistry, immunology, structural organization. Collagen Rel. Res. 2, 381-463 (1982).

Lazarides, E.: Intermediate filaments as mechanical integrators of cellular space. Nature 283, 249-56 (1980).

Mayor, H. D., J. C. Hampton, B. Rosario: A simple method for removing the resin from epoxy-embedded tissue. J. Biophys. Biochem. Cytol. 9, 909-10 (1961).

Pardo, J., J. D'Angelo Siciliano, S. W. Craig: A vinculin-containing cortical lattice in skeletal muscle: Transverse lattice elements ("costameres") mark sites of attachment between myofibrils and sarcolemma. Proc. Natl. Acad. Sci. U.S.A. 80, 1008-12 (1983).

45 Concentration of F-Actin and Myosin in Synapses and in the Plasmalemmal Zone of Axons

H. W. Kaiser and D. Drenckhahn

INTRODUCTION

Brain was one of the first nonmuscle tissues from which actomyosinlike proteins were extracted. Since then several other actin-associated proteins have been demonstrated in brain (for review, see Puszkin and Schook 1979). Previous studies have shown that actin is enriched in synaptosome fractions and in isolated postsynaptic density (Berl et al. 1973; Marotta et al. 1978; Crick 1982). Heavy meromyosin decoration demonstrated the presence of actinlike filaments in pre- and postsynaptic terminals (Le Beux and Willemot 1975; Fifkova and Delay 1982). The presence of actin in synaptic formations was further indicated by immunofluorescent staining with human autoantibodies cross-reacting with actin (Toh et al. 1976). Recent electron microscopic studies demonstrated actinlike immunoreactivity in dendritic spines and in postsynaptic densities. Since the antibodies used in those studies did not recognize presynaptic actin, it has been argued by the authors that nerve cells might contain different immunological subtypes of actin (Matus et al. 1982, Caceres et al. 1982). In the present study we applied staining with fluorescent phalloidin as a specific probe for all isoforms of actin so far known (Wulf et al. 1980) in order to determine the overall distribution of actin filaments in brain tissue. Moreover, antibody staining with anticalf thymus myosin (Drenckhahn et al. 1983) was used to colocalize F-actin with myosin, its force-gene-rating counterpart in muscle (for technical details see Drenckhahn and Kaiser 1983).

RESULTS

In the rat and guinea pig cerebellum, phalloidin and antibodies to myosin displayed a preferential affinity for the main synaptic formations, i.e., the molecular layer, the glomerula cerebellosi of the granular layer and the surface of the Purkinje cell somata. At higher magnification, the molecular layer and glomerula were seen to be crowded with numerous fluorescent particles having the size of 0.5-2 μm in diameter. The fluorescent margin of the Purkinje cell somata was frequently seen to be composed of a row of fine fluorescent particles. A similar interrupted fluorescent line was seen along the surface of motoneurons and their processes in the spinal cord.

In the guinea pig hippocampus (which was investigated by phalloidin fluorescence only), fluorescent particles showed a conspicuous alignment along the surface of the dendrites of granule cells and pyramidal cells. Extraordinarily large fluorescent particles (2-5 μm) were seen in the hippocampal mossy fiber termination zone at the dendrites of CA 4 and CA 3 pyramidal cells. This zone has been shown by electron microscopy to contain large synaptic formations consisting of particularly large spines and large complex presynaptic boutons (e.g., Frotscher et al. 1981). In the chick retina phalloidin staining resulted in a strong fluorescence of the cone pedicles which are the presynaptic part of photoreceptor synapse formations.

Apart from synaptic formations, phalloidin and thymus myosin were seen to bind to a narrow zone located close to the plasma membrane of virtually all nerve fibers. Axonal fluorescence was rather weak as compared to the strong fluorescence seen in association with synaptic structures. Axolemmal fluorescence was particularly prominent in cross-sectioned nerve fibers of the superior cerebellar peduncle and of the spinal roots.

Biochemical analysis of rat brain synaptosome fractions (prepared according to Hajos 1975) showed protein bands having the molecular weight of actin and myosin heavy chain. The identity of these protein bands with myosin and actin was demonstrated by antibody staining of gels electrophoretically blotted on nitrocellulose paper as well as by double immunofluorescent staining of synaptosome smears (Drenckhahn and Kaiser 1983).

CONCLUSIONS

The present study demonstrates the coexistence of F-actin and myosin in synaptic formations of the central nervous system and in the axolemmal zone of central and peripheral nerve fibers. As judged from the distribution pattern and intensity of fluores-

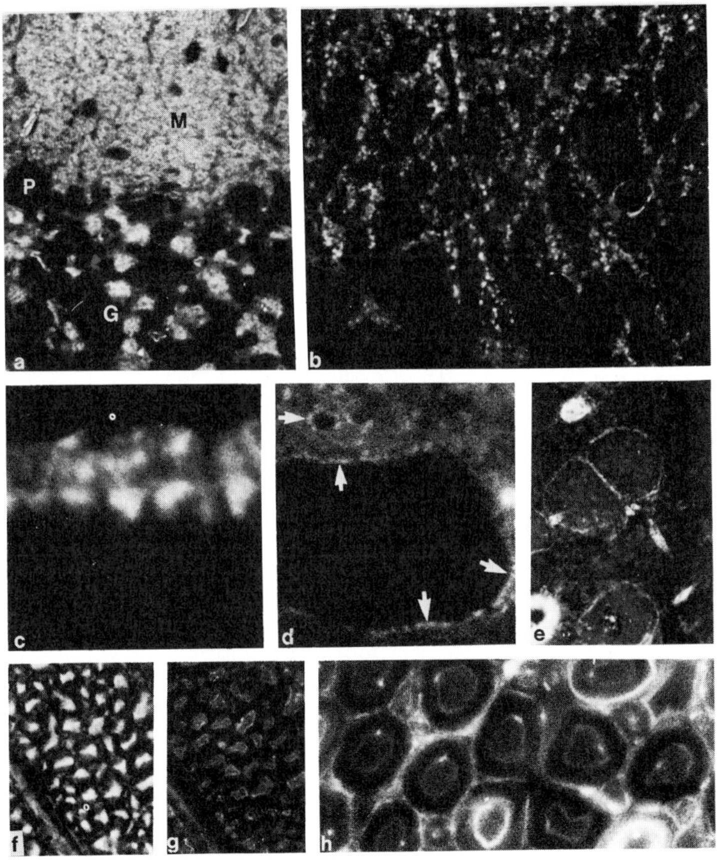

Fig. 45.1. Fixed frozen section of rat cerebellum (a), guinea pig hippocampus (b), external plexiform layer of chicken retina (c) and rat spinal cord motoneuron (d) stained with fluorescent phalloidin. Note strong actin-specific fluorescence associated with synaptic formations in the cerebellum (molecular layer, M; glomerula, G; surface of Purkinje cell somata, P), the hippocampus (mossy fiber terminating zone along CA3 pyramidal cell dendrites), the retina (cone pedicles and other structures), and the spinal cord (surface of motoneurons and their processes, arrows). (e) Rat brain stem neurons stained along their surface with antithymus myosin. (f,g) Corresponding phase contrast (f) and immunofluorescence (g)–microscopic images of cross-sectioned nerve fibers in the rat brain stem stained with antithymus myosin. Note affinity of antimyosin for the axolemmal zone of virtually all axons. (h) Rat ventral spinal root stained with fluorescent phalloidin, which binds to Schwann cells and to a narrow layer in the area of the axon membrane. Magnifications: a × 250, b × 680, c × 1,500, d × 1,200, e × 600, f,g × 700, h × 1,500.

cence, probably more than 90% of the total amount of neuronal F-actin and myosin appears to be concentrated in synapses (pre- and postsynaptic), whereas the remaining amount of actin and myosin is probably associated with the plasmalemmal zone of axons. Regarding the functional role of synaptic actin and myosin, both proteins might act to maintain the complex structure of synaptic formations (static function) as well as to play a role in dynamic events such as changes in shape and volume of dendrites in response to various experimental stimuli, as for example afferent stimulation or denervation ("synaptic plasticity," Crick 1982; Fifkova and Delay 1982).

In axons, actin and myosin have been postulated to generate the force driving axonal transport. Experimental support for this hypothesis comes from microinjection studies with DNAase I, an actin-depolymerizing pancreatic enzyme (Goldberg 1982; Isenberg et al. 1980). On the other hand, it is well documented that microtubles are involved in axonal transport. Considering the weak axolemmal fluorescence observed with fluorescent phalloidin and antimyosin which, at the ultrastructural level, apparently corresponds to the poorly developed subaxolemmal microfilament web (Hirokawa 1982), it seems more likely that actin and myosin have a structural role in supporting the axon membrane and/or controlling the lateral mobility of membrane macromolecules.

REFERENCES

Berl, S., S. Puszkin, W. J. Nicklas: Actomyosin-like protein in brain. Actomyosin-like protein may function in the release of transmitter material at synaptic endings. Science 179, 441-46 (1973).

Caceres, A., M. R. Payne, L. I. Binder, O. Stewart: Immunocytochemical localization of actin and microtubule-associated protein MAP 2 in dendritic spines. Proc. Natl. Acad. Sci. U.S.A. 80, 1738-42 (1983).

Crick, F.: Do dendritic spines twitch? Trends Neurosci. 5, 44-46 (1982).

Drenckhahan, D., U. Gröschel-Stewart, J. Kendrick-Jones, J. Scholey: Antibody to thymus myosin: its immunological characterization and use for immunocytochemical localization of myosin in vertebrate nonmuscle cells. Eur. J. Cell Biol. 29, 100-111 (1983).

Drenckhahn, D., H. W. Kaiser: Evidence for concentration of F-actin and myosin in synapses and in the plasmalemmal zone of axons. Eur. J. Cell Biol. 32 (1983).

Fifkova, E., R. J. Delay: Cytoplasmic actin in neuronal processes as a possible mediator of synaptic plasticity. J. Cell Biol. 95, 345-50 (1982).

Frotscher, M., U. Misgeld, C. Nitsch: Ultrastructure of mossy fiber endings in in vitro hippocampal slices. Exp. Brain Res. 41, 247-55 (1981).

Goldberg, D.: Microinjection into an identified axon to study the mechanism of fast axonal transport. Proc. Natl. Acad. Sci. U.S.A. 79, 4818-22 (1982).

Hajos, F.: An improved method for the preparation of synaptosomal fractions in high purity. Brain Res. 93, 485-89 (1975).

Hirokawa, N.: Cross-linker system between neurofilaments, microtubules, and membranous organelles in frog axons revealed by the quick-freeze, deep-etching method. J. Cell Biol. 94, 129-42 (1982).

Isenberg, G., P. Schubert, G. W. Kreutzberg: Experimental approach to test the role of actin in axonal transport. Brain Res. 194, 588-93 (1980).

LeBeux, Y. J., J. Willemot: An ultrastructural study of the microfilaments in rat brain by means of heavy meromyosin labelling. Cell Tiss. Res. 160, 1-68 (1975).

Marotta, C. A., P. Strocchi, J. M. Gilbert: Microheterogeneity of brain cytoplasmic and synaptoplasmic actins. J. Neurochem. 30, 1441-51 (1978).

Matus, A., M. Ackermann, G. Pehling, H. R. Byers, K. Fujiwara: High actin concentrations in brain dendritic spines and post-synaptic densities. Proc. Natl. Acad. Sci. U.S.A. 79, 7590-94 (1982).

Puszkin, S., W. Schook: The role of cytoskeleton in neuron activity. Meth. Achiev. Exp. Pathol. 9, 87-111 (1979).

Toh, B. H., H. A. Gallichio, P. L. Jeffrey, B. G. Livett, H. K. Muller, M. N. Cauchi, F. M. Clarke: Anti-actin stains synapses. Nature 264, 648-50 (1976).

Wulf, E., A. Deboben, F. A. Bautz, H. Paulstich, T. Wieland: Fluorescent phallotoxin, a tool for the visualization of cellular actin. Proc. Natl. Acad. Sci. U.S.A. 76, 4498-4502 (1980).

46 Characterization of a 40° C Red Blood Cell Structural Transition and Its Modification in Hereditary Spherocytosis

M. Minetti, M. Ceccarini,
A. M. M. Di Stasi, T.C. Petrucci, G. Scorza,
G. D. Di Nucci, V. T. Marchesi

Several spectroscopic techniques detect thermotropic transitiors in RBC membrane (see references reported in 1) at temperatures similar to those of some thermotropic membrane phenomena. However the structural changes involved in these transitions are still unclear. We observed that a spin-labelled stearic acid (16-dinyloxyl stearic acid, 16DSA) detects in RBC membrane three thermotropic breaks in its freedom of motion at 8°, 20° and 40°C. Ghosts from 6 patients with Hereditary Spherocytosis (HS) display a decrease in the temperature of 40°C transition (i.e. 36° ± 1°C versus 40° ± 0.5°C of control ghosts, Fig. 46.1).

In order to characterize the membrane component(s) involved in this transition, we treated ghosts with alkali to remove skeletal proteins. Figure 46.2 shows that skeleton-free vesicles lack the 40°C transition. Moreover a similar result can be obtained with ghosts resealed after treatment with polyclonal monospecific antispectrin antibodies (Fig. 46.2). These data suggest the involvement of skeletal proteins, and particularly spectrin, in the 40°C transition.

Spectrin presents a thermotropic conformotional change detectable by covalent spin-labelling methods on both membranes and purified spectrin preparations (2, and Fig. 46.3). The onset of this thermotropic change can be extrapolated at 40°C. After labelling of ghosts with a maleimide-analogue sulphydryl spin probe (4-MAL), the same thermotropic protein conformational change in normal and HS ghosts was observed (compare Figs. 46.3 and 46.4).

To further characterize thermotropic properties of RBC membrane we used monoclonal antibodies direct against spectrin. We tested two monoclonal antispectrin antibodies, one obtained

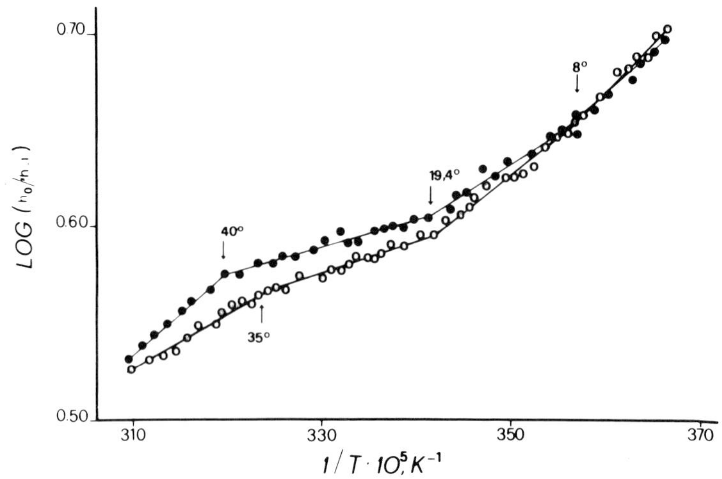

Fig. 46.1. Plot of log (h_0/h_{-1}) versus 1/T for 16DSA-labelled normal and HS ghosts. (●-●) Normal ghosts, (○-○) HS ghosts.

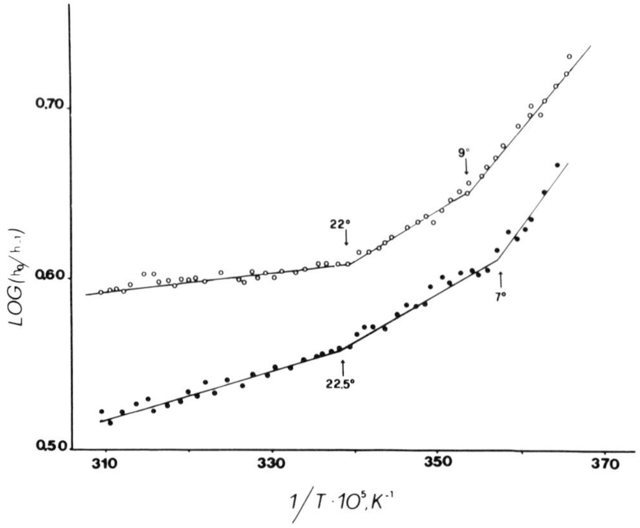

Fig. 46.2. Plot of log (h_0/h_{-1}) versus 1/T for 16DSA-labelled ghosts. (○-○) Alkali-stripped ghosts; (●-●) ghosts resealed after treatment with polyclonal antispectrin Ab.

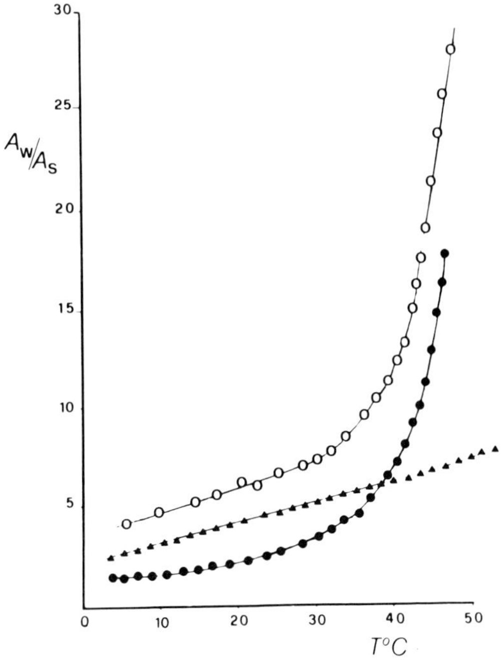

Fig. 46.3. Effect of temperature on the A_W/A_S ratio of membranes and spectrin preparations labelled with 4-MAL. (○-○) Purified spectrin; (▲-▲) alkali-stripped ghosts; (●-●) intact ghosts. A_W and A_S represent the amplitude of weakly (w) and strongly (s) immobilized components of the spectra.

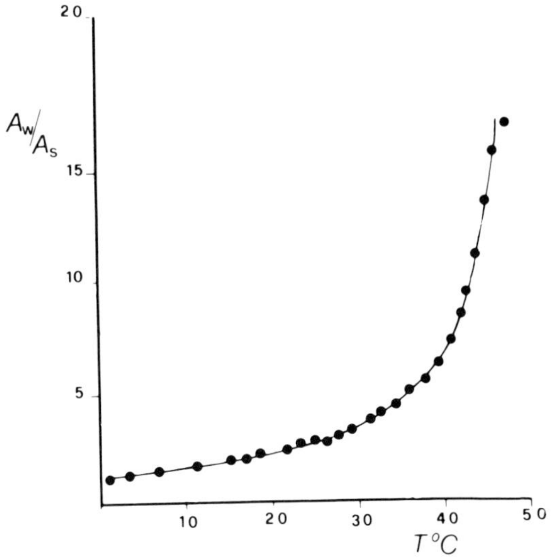

Fig. 46.4. Effect of temperature on the A_W/A_S ratio of HS ghosts labelled with 4-MAL.

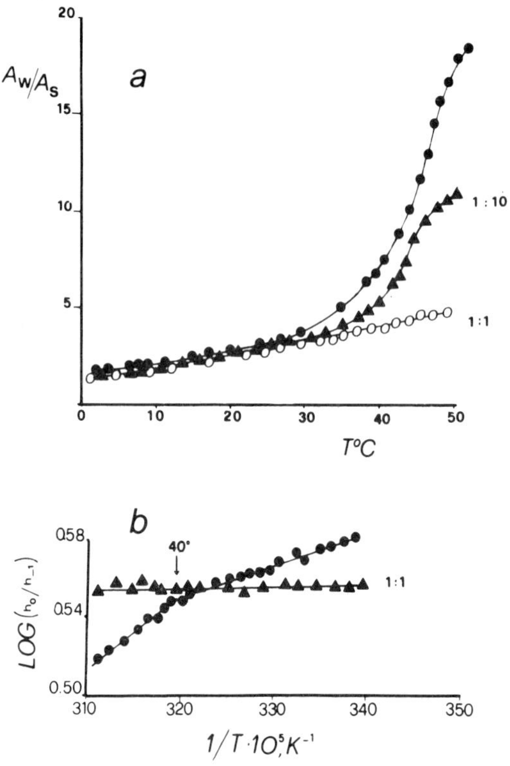

Fig. 46.5. Effect of ghosts' treatment with the autoimmune Ab on thermotropic properties detected by 4-MAL and 16DSA spin probe. (a) A_WA_S ratio as a function of temperature for (●-●) control ghosts; autoimmune Ab-treated ghosts at Ab with spectrin molar ratio 1:10 (▲ -▲) and 1:1 (○-○). (b) Plot of log (h_0/h_{-1}) versus 1/T for 16DSA-labelled ghosts (●-●) and autoimmune Ab-treated ghosts at Ab with spectrin molar ratio 1:1 (▲-▲).

from a mouse with an autoimmune disease (autoimmune Ab) and another specific for the 80K terminal fragment of the α-spectrin subunit (3). The autoimmune Ab is specific for internal domains occurring in many fragments obtained after limited proteolysis of spectrin (4). Thermotropic transitions of ghosts resealed after treatment with these antibodies were measured by the two spin probes. As shown in Figure 46.5, the autoimmune Ab inhibits the 40°C spectrin conformational change detected by

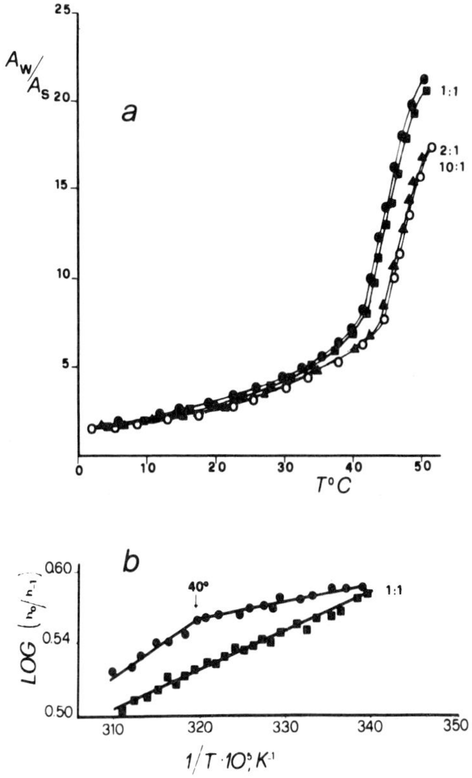

Fig. 46.6. Effect of ghosts' treatment with the anti-80K antibody on thermotropic properties detected by 4-MAL and 16DSA spin probe. (a) A_W/A_S ratio as a function of temperature for (●-●) control ghosts; anti-80K Ab-treated ghosts at Ab with spectrin molar ratio 1:1 (■ ■), 2:1 (▲-▲), and 10:1 (○-○). (b) Plot of log (h_0/h_{-1}) versus 1/T for 16DSA-labelled ghosts (●-●) and anti-80K Ab-treated ghosts at Ab with spectrin molar ratio 1:1 (■-■).

4-MAL, and effaces the 40°C thermotropic break detected by 16DSA spin label. Therefore the hydrophobic domains of spectrin recognized by this antibody seem to be involved in both the structural changes monitored by 16DSA and 4-MAL. The anti-80K antibody did not affect the 40°C spectrin conformational change monitored by 4-MAL, while it inhibited the thermotropic transition detected by 16DSA (Fig. 46.6).

Interestingly the 80K fragment is involved in α-β end-to-binding and is responsible for spectrin oligomer formation(s). Moreover, the ankyrin binding site seems to be localized near the terminal end of the β-subunit (for a review see 5). It is likely, therefore, that anti-80K treatment modifies spectrin-spectrin interaction and possibly skeleton-membrane association. We suggest that 16DSA detects a transition at the level of membrane core, due to or influenced by skeleton-membrane association. This association seems to be weakened in HS disease. Thermotropic properties of HS spectrin, as detected by 4-MAL, are apparently normal. Our results are in accordance with recent findings (6,7) showing normal spectrin but abnormal spectrin-membrane interactions in HS erythrocytes.

REFERENCES

1. Minetti, M., Ceccarini, M. (1982) J. Cell Biochem. 19, 59-75.
2. Cassoly, R., Daveloose, D., Leterrier, F. (1980) Bioch. Biophys. Acta 601, 478-89.
3. Yurchenco, P. D., Speicher, D. W., Morrow, J. S., Knowless, W. J., Marchesi, V. T. (1982) J. Biol. Chem. 257, 9102-9107.
4. Helteanu, I., Bologna, M., Speicher, D. W., Marchesi, V. T., manuscript in preparation.
5. Marchesi, V. T. (1983) Blood 61, 1-11.
6. Wolfe, L. C., John, K. M., Falcone, J. C., Byrne, A. M., Lux, S. E. (1982) N. Engl. J. Med. 307, 1367-73.
7. Goodman, S. R., Shiffer, K. A., Casoria, L. A., Eyster, M. E. (1982) Blood 60, 772-84.

47 Air-waterphase Dependent Interaction among Vinculin Molecules as a Model for Possible Vinculin Aggregation in Plasma Membranes

R. K. Meyer, P. Burn, M. M. Burger

The aim of this work was to analyze interactions between vinculin and lipid membranes to contribute to the unresolved question of how vinculin is anchored in the focal contacts of living cells (Geiger 1979; Burridge and Feramisco 1980; Geiger et al. 1980; Tokuyasu et al. 1981; Geiger 1981; Jockusch and Isenberg 1982; Birchmeier et al. 1982; Geiger 1982; Avnur et al. 1983).

Vinculin was found to insert into a lipid monolayer interface at lipid surface pressures as high as were reported for natural cell membranes. The inserted vinculin displaced the lipids and an unexpected and unusually rigid layer (similar to alpha-actinin; Meyer et al. 1982) was formed at the air-water interface, possibly by a protein-protein interaction. Such rigid layers have never been found with other proteins or glycoproteins so far. Acid treatment of vinculin eliminated this surface activity and subsequent ethanol treatment reestablished the original behavior of vinculin.

SURFACE ACTIVITY OF VINCULIN: STUDIES ON A LANGMUIR TROUGH

The surface activity of vinculin was tested measuring its insertion into the air-water interface by using a modified Langmuir trough (Fromherz 1975). A lipid monolayer was spread at the buffer surface to allow a contact zone with the Wilhelmy balance. Its surface pressure was adjusted and held constant by an electronically controlled movable surface barrier. Vinculin was injected 3 mm below the monolayer into the central part of the trough. The spontaneous insertion of vinculin molecules into the air-water interface was measured by observing the movement of the surface barrier. With this method both the final increase of the monolayer surface area (cm^2) due to protein insertion (Fig. 47.1), and the insertion rate (cm^2/min) could be recorded (Table 47.1).

Vinculin showed its surface activity only at physiological ionic strength (150 mM NaCl in 20 mM Tris-HCl pH 7.6 or phosphate-buffered saline). No insertion was observed in buffer solutions with low ionic strength (20mM Tris-HCl pH 7.6). Ca^{++} and Mg^{++} did not influence the surface activity of vinculin in these experiments. The insertion rate measured as an increase in the phospholipid monolayer area at constant pressure immediately after injection of vinculin, was highly sensitive to the surface pressure chosen. However, an insertion of vinculin into the air-water interface could be observed within a reasonable time even at surface pressures over 25 mN/m.

In subsequent experiments increasing amounts of vinculin were injected into the central part of the trough (physiological ionic strength) at a constant surface pressure of 15 mN/m.

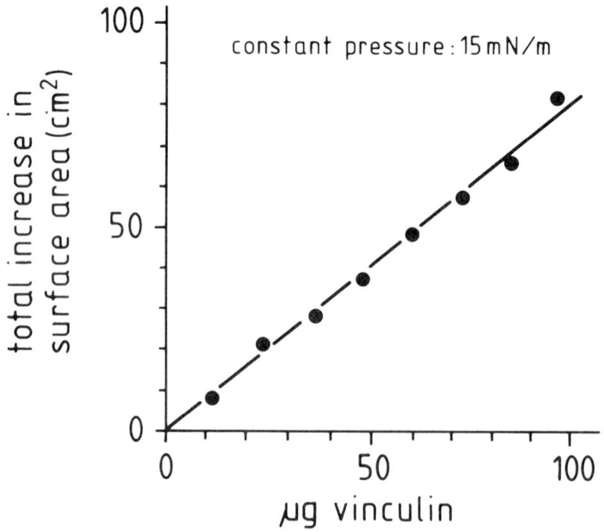

Fig. 47.1. Increase of the phospholipid monolayer area by the insertion of increasing amounts of vinculin. The monolayer was kept at a constant pressure of 15 mN/m. In subsequent experiments increasing amounts of vinculin were injected underneath the lipids. The increase in the lipid monolayer area was measured at equilibrium.

Table 47.1. Vinculin Insertion Rate Monitored by the Increase of the Phospholipid Monolayer Surface Area

Vinculin		
Native	42.8 ± 6.6[a]	cm^2/min
Acid-treated	0.5 ± 0.3[b]	cm^2/min
Acid-treated, renatured with ethanol	16.8 ± 1.7[b]	cm^2/min

Note: 100 μg vinculin were injected at a constant surface pressure of 15 mN/m. The insertion rate was measured immediately after injection of vinculin. (The standard deviation of the mean value is given for eleven (a) and for four (b) independent experiments.)

The increase in the lipid monolayer area was measured at equilibrium (Fig. 47.1). As expected, a noncooperative interaction linear relationship has been found. To verify that it was vinculin penetrating into the air-water interface, the surface layer formed was carefully moved to another buffer compartment and so removed from proteins not interacting with the surface layer. Then the material from the surface was removed by suction and concentrated by freeze-drying and analyzed on a $NaDodSO_4$ polyacrylamide gel (Fig. 47.2B). Only the 130,000 vinculin band was visible after Coomassie brilliant blue staining. The monolayer formed by vinculin was unexpectedly rigid. Both the insertion of native vinculin as well as the rigidity of the layer formed, were found to be independent of the lipid composition at the air-water interface. Both observations could also be made in the absence of any added lipids. Naturally occurring phospholipids as well as mixtures of lipid extracts from natural sources were tested. In all these experiments the lipids at the air-water interface were displaced by the inserting vinculin molecules; no lipids were found within the protein surface layer. The high viscosity of the vinculin layer is probably due to protein-protein interaction and most likely prevents mixing of lipids with proteins. On the other hand, the vinculin layer is flexible enough to take up additional vinculin molecules from the buffer phase.

Even in a high ionic strength buffer the surface activity of vinculin was lost in a reversible way after acid treatment (pH = 1.0). Acid treatment and renaturation did not degrade vinculin since no change in molecular weight could be observed after gel electrophoresis (Figs. 47.2C and 47.2D), and since the rigid behavior on the monolayer was restored after treatment of vinculin with ethanol (Table 47.1).

INTERACTION OF VINCULIN WITH LIPOSOMES

Interactions between alpha-actinin and bilayer membranes of liposomes had been studied before (Meyer 1980; Meyer et al. 1982). Addition of Alpha-actinin to suspensions of liposomes resulted in formation of aggregates monitored by an increase in the optical density at 680 nm. For the experiments with vinculin, liposomes were prepared from a lipid mixture extracted from yeast cells. An increase in optical density at 680 nm was recorded by addition of native vinculin (Fig. 47.3A), but no increase was detected after the addition of acid treated (i.e., denatured) vinculin (Fig. 47.3B).

Fig. 47.2. 12.5% $NaDodSO_4$ polyacrylamide gel electrophoresis of vinculin. (A) Native vinculin. (B) Recovered vinculin from the air-water interface. (C) Acid treated vinculin. (D) Acid treated vinculin, renatured by ethanol.

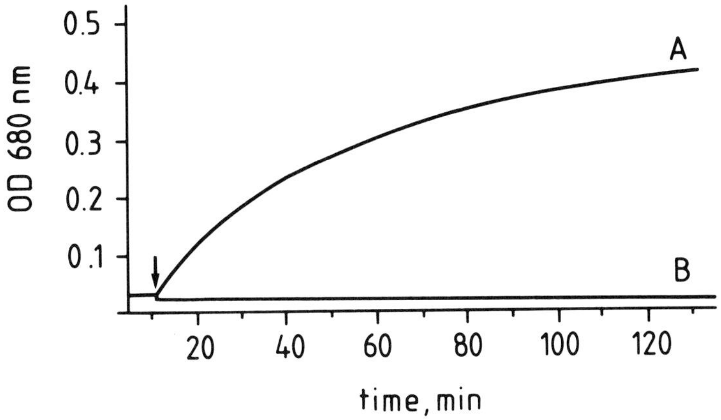

Fig. 47.3. Interaction of native vinculin (A) and acid-treated vinculin (B) with liposomes. Liposomes were prepared from a lipid mixture extracted from yeast cells. 25 μg vinculin were added to the liposome solution (arrow). An increase in optical density was recorded by the addition of native vinculin, but no increase was detected after the addition of acid treated vinculin. (The slight decrease in optical density at the starting point is due to the dilution of the liposome solution by adding the vinculin).

These experiments suggest, that vinculin in its native conformation may interact with biological membranes as well.

SPECTROSCOPIC STUDIES ON THE RIGID VINCULIN LAYER

The molecular structure of the rigid vinculin surface layer formed at the air-water interface is currently under investigation. Infrared attenuated total reflection spectroscopic studies (together with U. P. Fringeli, ETH Zurich) suggest a partially ordered structure within the layer (unpublished data). The acid-treated vinculin changed simultaneously with its surface activity also the CD- and fluorescence-absorption (unpublished data). This indicates a conformation change in the vinculin molecule. The partial renaturation of the acid-treated vinculin by ethanol could also be detected by the CD- and fluorescence-spectra. Both native and denatured spectra were superimposable. This observation confirms the restoration of the insertion rate obtained for the renatured vinculin on the Langmuir trough (Table 47.1).

The monolayer experiments and the interaction of vinculin with liposomes suggest that vinculin may interact with biological membranes. Furthermore, our results argue for a possible vinculin-vinculin interaction and thus an aggregated state for this molecule in or at plasma membranes. (A similar idea was suggested by Avnur, Small and Geiger 1983.) A rigid cluster of vinculin molecules could be formed providing a structural basis for the interaction with other cytoskeletal and membrane proteins.

ACKNOWLEDGMENT

We wish to thank Dr. P. Bruckner and Dr. D. Dimitrov for their help with CD- and fluorescence spectra.

This work is supported by the Swiss National Science Foundation Grant No. 3.269-0.82.

REFERENCES

1. Avnur, Z., J. V. Small, and B. Geiger, 1983. Actin-independent association of vinculin with the cytoplasmic aspect of the plasma membrane in cell-contact areas. J. Cell Biol. 96:1622-30.
2. Birchmeier, W., T. A. Liebermann, B. A. Imhof, and T. E. Kreis, 1982. Intracellular and extracellular components involved in the formation of ventral surfaces of fibroblasts. Cold Spring Harbor Symp. Quant. Biol. 46: 755-67.
3. Burridge, K. and J. R. Feramisco, 1980. Microinjection and localization of a 130K protein in living fibroblasts: a relationship to actin and fibronectin. Cell 19:587-95.
4. Fromherz, P., 1975. Instrumentation for handling monomolecular films at an air-water interface. Rev. Sci. Instrum. 46:1380-85.
5. Geiger, B., 1979. A 130 protein from chicken gizzard: its localization at the termini of microfilament bundles in cultured chicken cells. Cell 18:193-205.
6. Geiger, B., 1981. Transmembrane linkage and cell attachment: the role of vinculin. In International Cell Biology

(ed. H. G. Schweiger), pp. 761-73. Springer Verlag, Berlin.

7. Geiger, B., 1982. Involvement of vinculin in contact-induced cytoskeletal interactions. Cold Spring Harbor Symp. Quant. Biol. 46:671-82.
8. Geiger, B., K. T. Tokuyasu, A. H. Dutton, and S. J. Singer, 1980. Vinculin, an intracellular protein localized at specialized sites where microfilament bundles terminate at cell membranes. Proc. Natl. Acad. Sci. USA 77:4127-31.
9. Jockusch, B. M. and Isenberg, G., 1982. Vinculin and alpha-actinin: interaction with actin and effect on microfilament network formation. Cold Spring Harbor Symp. Quant. Biol. 46:613-23.
10. Meyer, R. K., 1980. Spezifische Lipide ermöglichen eine starke Alpha-Aktinin-Membran Wechselwirkung. Doctoral thesis (University of Basel, Basel, Switzerland).
11. Meyer, R. K., H. Schindler, and M. M. Burger, 1982. Alpha-actinin interacts specifically with model membranes containing glycerides and fatty acids. Proc. Natl. Acad. Sci. USA 79:4280-84.
12. Tokuyasu, K. T., A. H. Dutton, B. Geiger, and S. J. Singer, 1981. Ultrastructure of chicken cardiac muscle as studied by double immunolabelling in electron microscopy. Proc. Natl. Acad. Sci. USA 79:7619-23.

48 On the Presence of Actin Microfilaments in the Cytoplasm of Dark Basal Cells of the Rat Olfactory Epithelium

G. C. Balboni and G. B. Vannelli

Besides the olfactory neurons and the supporting cells, basal cells are the third cellular component of the olfactory epithelium. The basal cells are generally believed to be responsible for the proliferative and regenerative processes in the O.E. (Graziadei et al., 1971, 1978; Mulvaney, 1971, ecc.). Two types of basal cells may be identified: (1) the clear basal cells, able to undergo mitotic division and that are the true stem cells (Balboni and Vannelli, 1982) and (2) the dark basal cells, whose significance is not yet ascertained. Aiming to improve our knowledge of these cells, we have studied them at the light and electron microscope and with the method of indirect immunofluorescence in the O.E. of prepubertal and postpubertal male and female rats.

The dark cells are very irregular in shape and their cytoplasmatic extensions surround the clear basal cells and the inner portion of the other cells. The nucleus too is irregular in shape and dark, and the cytoplasm contains numerous mitochondria and bundles of microfilaments. These cells are connected to each other by desmosomes and to the basal membrane of the epithelium by hemidesmosomes. Some anular gap junctions between the cellular expansions can be observed. In the olfactory epithelium of prepubertal rats and postpubertal females in estrus, in which an evident cellular turnover occurs, the dark basal cells appear extremely irregular in shape with many long and thin cytoplasmatic extensions projecting themselves between other cells up to the outer portions of the epithelium.

The nature of microfilaments was investigated with the method of indirect immunofluorescence according to Weber et al., 1975. Reactions were negative for prekeratin and intensely positive for actin (Balboni and Vannelli, 1982).

Furthermore, applying the method of Falck and Owman for the demonstration of adrenergic innervation, some fluorescent isolated fibers were identified within the O.E. At the E.M. level we have observed the presence in the basal portion of the O.E. of some nervous fibers and varicosities closely related to the branching prolongments of the dark basal cells. In the varicosities a variable number of dense core vesicles, 70-100 nm in diameter, were observed. A cleft of about 20 nm separates the cell membrane of these nervous terminals from the opposite cell membrane of dark basal cell prolongments.

Therefore, in our opinion, these cells must be considered as a particular type of supporting cells, able of some contractile activity, probably under the control of the autonomic nervous system. The close relationship of these cells with the other cells in the O.E. and the possibility that have to modify in an important way their shape, suggests the hypothesis that they can play a role in regulating the cellular arrangement in the epithelium under different conditions and in modulating the activity of the sensory cells.

REFERENCES

Balboni G.C. and Vannelli G.B.: Morphological features of the olfactory peithelium in prepubertal and postpubertal rats. "Olfaction and endocrine regulation." IRL Press, 285-97, 1982.

Balboni G.C. and Vannelli G.B.: Sur les cellules sombres de l'epithelium olfactif du rat. Bull. Ass. Anat., 66:207-14, 1982.

Berkowitz L.R., Fiorello O., Kruger L. and Maxwell D.S.: Selective staining of nervous tissue for light microscopy following preparation for electron microscopy. J. Histochem. Cytochem., 66:808-14, 1968.

Falck B. and Owman C.: A detailed methodological description of the fluorescence method for the cellular demonstration of biogenic monoamines. Acta Univ. Lund., section II, no. 7, 1965.

Graziadei P.P.C. and Metcalf J.F.: Autoradiographic and ultrastructural observation on the frog's olfactory mucosa. Z. Zellforsch., 116:305-18, 1971.

Graziadei P.P.C. and Monti Graziadei G.A.: Continuous nerve cell renewal in the olfactory system. Handbook of Sensory Physiol., Springer Verlag, 9:55-83, 1978.

Mulvaney B.D. and Heist H.E.: Regeneration of rabbit olfactory epithelium. Amer. J. Anat., 131:241-52, 1971.

Weber K., Birring T. and Osborn M.: Specific visualization of tubulin containing structures in tissue culture cells by immunofluorescence. Exp. Cell Res., 95:111-20, 1975a.

PART IV

REGULATORY MECHANISMS OF FUNCTION AND ORGANIZATION OF THE CYTOSKELETAL COMPONENTS

49 Toward an Understanding of the Function of F-Actin Capping and Severing Proteins

G. Isenberg and H. Maruta

GENERAL ASPECTS OF PROTEINS WHICH REGULATE ACTIN POLYMERIZATION

In cells, the regulation of the monomer-polymer transition of actin is a prerequisite for maintaining various coordinated cellular movement phenomena such as cell locomotion, surface mobility, and intracellular motility. Many proteins are known to be involved in organizing the spatial distribution of actin filaments within the cytoplasm. Here, however, we would like to concentrate on those proteins, which from in vitro studies are known to influence the polymerization of actin directly. As shown on Table 49.1 these proteins can be grouped so far into three classes: (1) actin sequestering proteins like profilin and DNAase I, (2) actin filament severing or fragmenting proteins, and (3) actin filament capping proteins. The latter interfere with actin monomer exchange at either the fast growing (plus) end or the slow growing (minus) end of actin filaments. Common to all three groups is probably the ability of the listed proteins to bind to monomeric actin in a 1:1 complex. Since, however, the various proteins bind to different sites on the actin molecule with different affinities, they interfere with the actin monomer-polymer equilibrium in quite different ways.

Concerning the binding affinity it is known that DNAase I forms a very tight complex with actin monomers (Lazarides and Lindberg, 1974) whereas at least the binding of Acanthamoeba profilin to actin is weak (K_D 2-10 μM) (Tobacman and Korn, 1982; Tseng and Pollard, 1982; Tobacman et al., 1983). Most recent results suggest that the major function of profilin is to inhibit the spontaneous nucleation of actin filament growth by binding to actin monomers, but that in addition the profilin-actin complex can bind to actin filament ends (Pollard and

Table 49.1. Actin-Binding Proteins That Regulate Actin Polymerization

Protein	Source	Molecular Weight
	I. G-actin Binding Proteins	
Profilin (a)	Acanthamoeba, Physarum	12,500
Profilin (b)	Spleen, Thymus, Brain, Pancreas	15,200
DNase I	Bovine Pancreas	31,000
	II. F-actin Severing Proteins	
Severin*	Dictyostelium	40,000
Fragmin*	Physarum	42,000
Gelsolin*	Rabbit Macrophage, Human Platelet	90,000-91,000
Villin*	Chicken Intestinal Microvilli	95,000
	III. F-actin Capping Proteins	
	A. Fast-growing End Capping	
Cap(28+31)	Acanthamoeba	28,000 + 31,000
Cap(31+36)	Bovine Brain	31,000 + 36,000
Cap42(a)*	Physarum	42,000
Cap42(b)**	Physarum	42,000
	B. Slow-growing End Capping	
Beta-actinin	Rabbit Skeletal Muscle	34,000 + 37,000
Acumentin	Rabbit macrophage	65,000

*Ca^{++}-dependent.
**Phosphorylatable and Ca^{++}-dependent.

Cooper, 1983; Tilney et al., 1983). Once bound, however, profilin is most likely immediately released from the complex due to a conformational change of the actin monomer which occurs upon polymerization.

The F-actin severing proteins in group II have a much higher affinity for actin monomers, even when those are assembled into a filament. This binding affinity is so high that actin molecules are released from the actin helix during complex formation. In addition, by forming this 1:1 actin monomer complex, which is able to cap the fast growing ends of actin filaments, severing proteins not only fragment actin filaments but also induce a net depolymerization of the capped actin filament fragments at their slow growing ends. Thus, severing proteins will induce a very fast and drastic reduction of actin filament lengths. Although severing proteins share with DNAase I the very high affinity to actin monomers, their binding sites on the actin molecule apparently differ from the DNAase I binding site, since severing proteins seem to be able to bind to the DNAase I-actin complex (Glenney et al., 1981; Wang and Bryan, 1981; Kurth et al., 1983).

The proteins of group III are exclusively capping proteins. Most likely, these proteins also form a 1:1 complex with actin monomers. But due to a lower affinity this complex formation occurs only with unpolymerized actin molecules and not with actin molecules which have assembled into a filament. Like the complex formed with severing proteins and actin monomers, the capping protein-actin monomer complex binds exclusively to one end of the actin filament. Within the group of capping proteins we can distinguish fast growing end (A) and slow growing end capping proteins (B). Comparing the proteins which exclusively cap filaments with the F-actin severing proteins one can assume that only the group A capping proteins share with severing proteins the same binding site on the actin molecule.

THE "CAPPED G-ACTIN" MODEL

In an attempt to explain the effect of capping proteins on actin polymerization in the simplest manner, the "capped G-actin" model has been proposed (for details see Maruta and Isenberg, 1983) (Fig. 49.1). The model should provide an explanation for (1) the inhibition of actin polymerization at the fast growing end, (2) the induction of depolymerization at the slow growing end, and (3) the nucleation of filament growth at an early stage

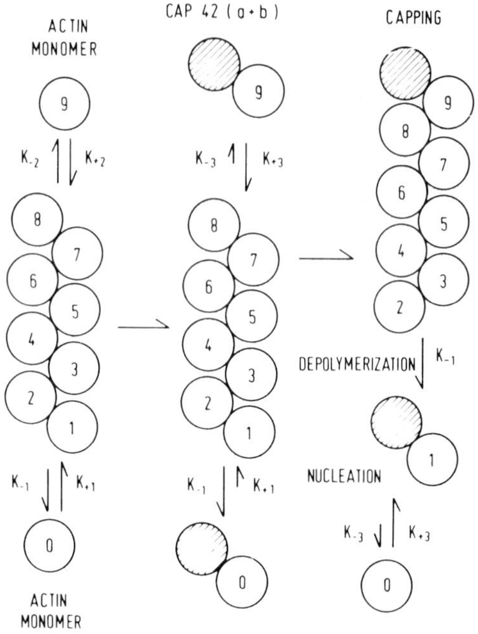

Fig. 49.1. The "capped G-actin" model as a possible mechanism for capping, the induction of filament depolymerization and nucleation of actin filament growth. The open circles represent actin monomers, the shaded circles stand for Cap 42 (a+b), a calcium dependent phosphorylatable capping protein from Physarum. (With permission from Maruta and Isenberg, 1983.)

of actin polymerization. In principle there are two mechanisms by which capping proteins may bind to filament ends: they could bind directly or add to the polymer ends as the above mentioned 1:1 complex with monomeric actin. If capping proteins bind directly to filament ends without forming a complex with G-actin the net loss of actin molecules at the slow growing end should stop when the G-actin concentration reaches the critical concentration for this particular filament end. Furthermore, the final length of actin filaments should be independent of the capping protein concentration if this exceeds the "minimum concentration" necessary to block all available free barbed ends. Thus, it should be impossible for a capped filament to depolymerize completely without assuming a complex formation between capping proteins and G-actin or assuming a repetitive and spontaneous breaking of filaments which subsequently may be capped. However, from the observation that under conditions where spontaneous fragmentation is negligible (Wegner and Savko, 1982) a complete depolymerization of actin filaments could be induced by high concentrations of capping proteins (Maruta and Isenberg, 1983) the following model assumptions could be drawn:

1. Capping proteins bind to G-actin at a 1:1 molar ratio.
2. Under equilibrium conditions this complex binds to the ends of actin filaments. For "barbed" end capping proteins the affinity for the complex is by far higher at the barbed end than at the pointed end. Furthermore, the association rate constant for the complex at the barbed end (k_{+3}) is much higher than that for free G-actin (k_{+2}). The pointed end is unfavorable for binding the complex as is indicated by the high dissociation rate constant (k_{-1}).
3. Once a filament is capped, depolymerization is blocked at the fast growing end, leading to a net depolymerization at the slow growing, pointed end. Shortening at the pointed end until zero length will proceed as long as the free capping protein concentration is high enough to complex all actin molecules which are released at the pointed end, so that the free G-actin concentration is always below the critical concentration for this particular end.
4. A large excess of G-actin over capping protein at the starting point of polymerization will result in the formation of a complex heterodimer which is more stable than the normal actin dimer, binds more readily to another G-actin, and therefore nucleates actin filament formation.

The existence of a 1:1 complex of F-actin capping or severing protein with G-actin is not hypothetical. For villin (Glenney et al., 1981) or gelsolin (Yin et al., 1981) the complex is stable enough to be isolated by gel-filtration. Physarum fragmin is always isolated as a very tight complex (molar ratio 1:1) with the actin monomer (Hasegawa et al., 1980; Hinssen, 1981) and among the capping proteins at least "cap 90," a 90 kd Ca^{++}-dependent F-actin capping protein from vertebrate brain (Isenberg et al., 1983) and a 90 kd protein from human platelets (Kurth et al., 1983) have been shown to form a fairly stable complex with actin.

HOW TO DISTINGUISH F-ACTIN CAPPING AND SEVERING ACTIVITY

As is shown in Table 49.1, F-actin severing proteins, in addition to capping actin filaments as a 1:1 actin complex, have the ability to fragment actin filaments directly. Using the common viscometric assays it is not possible to distinguish between fragmentation (Fig. 49.2) and exclusively capping (Fig. 49.3), since both mechanisms induce a rapid decrease in the viscosity of F-actin. When viewed in the electron microscope, severing proteins will effectively generate short filament fragments of only a few helical repeats within seconds and at a very low stoichiometry. Although by viscometric criteria the exclusively capping proteins may be almost equally effective at similar molar ratios, electron micrographs show that under these conditions the length of the capped filaments is not appreciably reduced. The nevertheless rapid decrease of viscosity may reflect only slight changes of filament length, too small to be detected by electron microscopy and/or the loss of end-to-side interactions between actin filaments, which may considerably contribute to the viscosity of actin networks. Increasing the molar ratio of capping proteins over actin up to 1:50, 1:10, or 1:5 will result in a comparable decrease of filament length as is achieved with severing proteins at a molar ratio of 1:500 or 1:1,000.

In order to discriminate clearly between actin filament capping and actin filament fragmenting activity, we have designed an easy experimental test: severing proteins, which by their high affinity for actin monomers fragment single filaments along their entire length, are still able to act efficiently on Triton-extracted cytoskeletons, leading to the observed destruction of actin meshworks and stress-fibers (Fig. 49.4). "True" capping proteins, however, will not induce any depolymerization, unless actin monomers can be released from the opposite filament end. Since in the Triton-model actin monomers apparently do not dissociate from the pointed ends, capping proteins have no effect on the actin cytoskeleton (Fig. 49.4c) whereas they potently disassemble actin meshworks and stress-fibers when microinjected into living cells (Füchtbauer et al., 1983).

CAPPING PROTEINS AS TOOLS FOR STUDYING ACTIN POLYMERIZATION

By plotting filament growth rates versus actin monomer concentrations, Pollard and Mooseker (1981) were able to

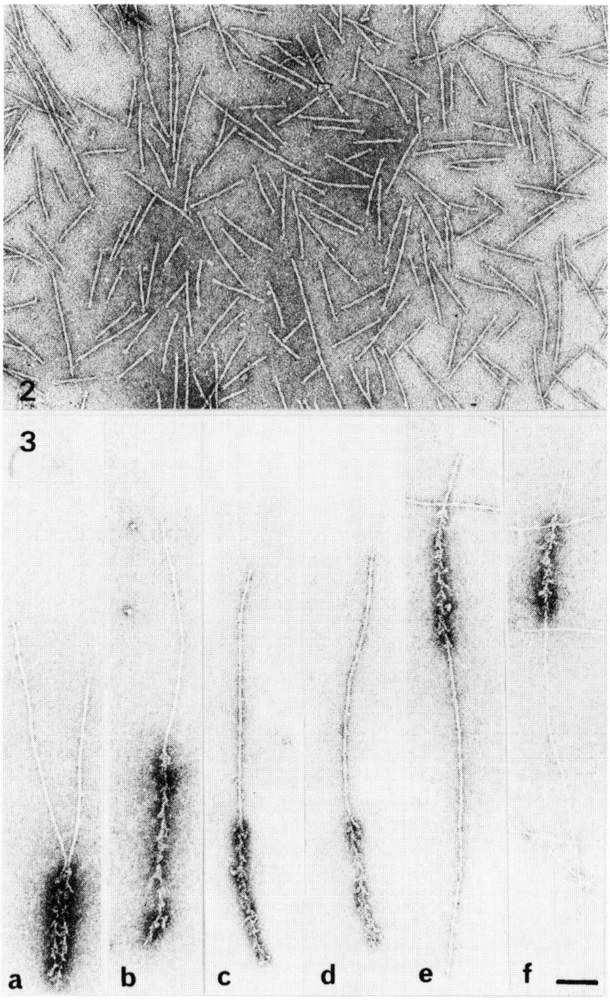

Fig. 49.2. Electron micrograph of actin filaments which have been fragmented by fragmin, an actin filament severing protein from Physarum. Magnification 82,500 ×.

Fig. 49.3. Capping of actin filaments as demonstrated by the inhibition of growth of S-1 decorated, fixed filament fragments at their barbed ends (a-d). In the absence of capping proteins actin filaments grow bidirectionally (e,f). (With permission from Maruta and Isenberg, 1983.)

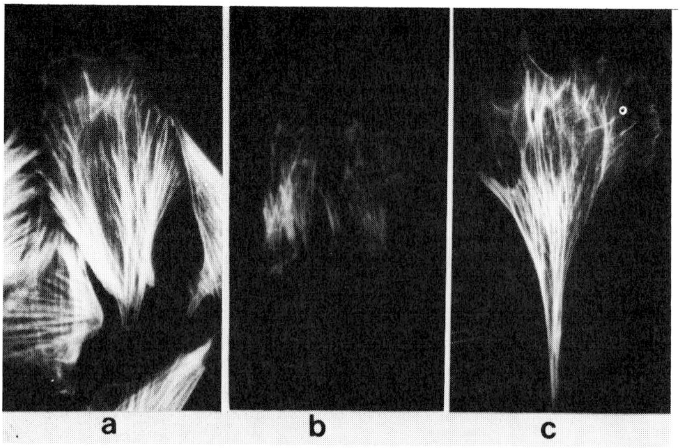

Fig. 49.4. Capping and severing proteins are distinguishable by their different action on the actin cytoskeleton. (a) Triton-extracted cytoskeletons of heart fibroblasts stained specifically with rhodamine-phalloidin for F-actin after treatment with buffer (control), (b) 5 mins incubation with a 90 kd actin filament severing protein from smooth muscle (40 μg/ml) (Hinssen H. et al., in press), and (c) "cap 90," (66 μg/ml) a Ca^{++}-dependent F-actin capping protein from vertebrate brain. Note that all three micrographs have the same magnification: 220 ×.

concentrations, Pollard and Mooseker (1981) were able to extrapolate from kinetic data the critical monomer concentration for each filament end. Since, however, events at the fast growing, barbed end always interfere with the slow growing, pointed end, exact values for the critical concentration at the pointed end were difficult to obtain. This problem was overcome by simply capping the barbed filament end and measuring directly in a fluorescent assay the monomer concentration of free G-actin at which prelabelled and capped actin filaments start to disassemble from their free pointed ends (Wegner and Isenberg, 1983). The critical concentration of the barbed end was reexamined by using the inverse assay: unlabelled actin filaments were mixed with various concentrations of labelled actin monomers and the concentration at which the barbed ends consume monomers was determined by measuring the fluorescent change (Fig. 49.5). In agreement with previous data, the critical concentration for the barbed end was 0.12 μM whereas the critical concentration for the pointed end was much higher than expected. Under physiological salt conditions and at 37°C

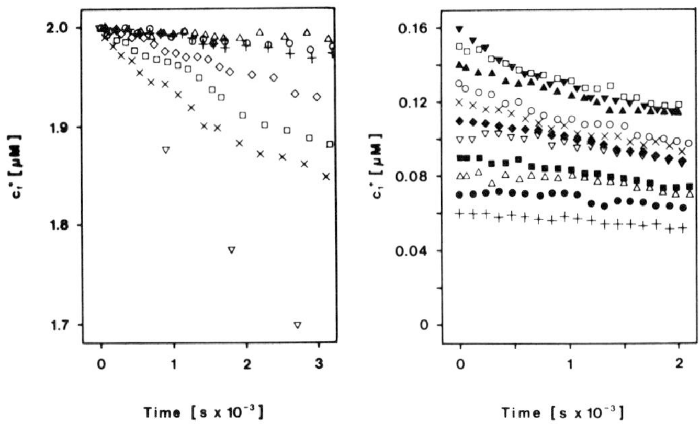

Fig. 49.5. (a) The time course of fluorescent change induced by the disassembly at the pointed end of prelabelled and capped actin filament fragments at various concentrations of actin monomers (see the various symbols). Filaments start to disassemble at a critical concentration for the pointed end, below 1.5 µm. (b) The time course of fluorescent change induced by the consumption of labelled actin monomers at the barbed filament ends. Actin monomers assemble into a filament above the critical concentration of 0.12 µM. (With permission from Wegner and Isenberg, 1983.)

this was 1.5 µM, i.e., 12-15-fold higher than the barbed end critical concentration. In a different approach similar values have been reported by following monomer incorporation at the two ends of the acrosomal process of Thyone sperm using slightly different temperatures and ionic assay conditions (Tilney et al., 1983).

THREE F-ACTIN CAPPING AND SEVERING PROTEINS IN PHYSARUM SEEM TO BE NONPOLYMERIZABLE ACTIN VARIANTS

In Physarum we found that besides the major polymerizable actin there exist. at least three other proteins with the molecular weight of 42 kd (Maruta et al., 1983; Maruta and Isenberg, 1983). Two of them form a heterodimer Cap 42 (a+b). Cap 42 (a+b) is a Ca^{++}-dependent, phosphorylatable F-actin capping protein. Its phosphorylation regulates its actin binding and vice versa. The complex Cap 42 (a+b) as well as its subunits Cap 42 (a) and Cap 42 (b) are able to cap the fast-growing end of actin filaments, to block actin polymerization at this end and to induce a rapid depolymerization of the filaments at the opposite end. These properties are shared with Physarum fragmin, the fourth protein of 42 kd. However, unlike fragmin, neither Cap 42 (a), Cap 42 (b) nor their complex has any detectable F-actin severing activity. In addition, Cap 42 (a), Cap 42 (b), and Cap 42 (a+b) as well as fragmin nucleate the growth of actin filaments at an early stage of actin polymerization. Cap 42 (a) alone requires Ca^{++} for its capping activity. Cap 42 (b) is phosphorylated in vitro at two threonine residues by a specific kinase from Physarum. The same kinase does not phosphorylate either Cap 42 (a), fragmin, or actin. Interestingly, Cap 42 (b) requires Ca^{++} for its capping activity only when it is phosphorylated. Similarly, Cap 42 (a+b), the complex of Cap 42 (a) and Cap 42 (b), requires Ca^{++} for its capping activity only when Cap 42 (b) is phosphorylated. The phosphorylation of Cap 42 (b) is almost completely inhibited by a tertiary complex of Cap 42 (a), actin, and Ca^{++}.

DNAase I which is known to bind actin effectively, also binds very tightly to Cap 42 (b) and almost completely inhibits the phosphorylation of Cap 42 (b) in a Ca^{++}-independent manner. However, neither Cap 42 (a) nor fragmin bind to DNAase I. In common, but unlike actin, Cap 42 (b), Cap 42 (a), and fragmin do not form any filaments by self-assembly. Furthermore, Cap 42 (b), Cap 42 (a), and fragmin in contrast to actin are not phosphorylated by the catalytic subunit of a cAMP-dependent protein kinase which selectively phosphorylates actin in its serine residues. Therefore, the four proteins of 42 kd in Physarum, i.e., actin, fragmin, Cap 42 (a), and Cap 42 (b) are functionally distinguishable from each other (see Table 49.2).

The four proteins of 42 kd in Physarum fall into two groups, a and b. Group a contains Cap 42 (a) and fragmin, whereas group b contains Cap 42 (b) and actin. The tryptic peptide maps of the two proteins in each group obtained by limited proteolysis are almost superimposable, if not identical with each other, indicating that Cap 42 (a) is related to fragmin, and Cap 42 (b) is closely related to actin. In fact, Cap 42 (a) preferentially forms a very tight 1:1 complex with Cap 42 (b), whereas fragmin forms a similar 1:1 complex only with actin. More interestingly, the four proteins of 42 kd share at least a

Table 49.2. Four Members of Physarum Actin Family

	Actin	Cap 42 (b)	Cap 42 (a)	Fragmin
Polymerization	+	-	-	-
Phosphorylation (Ser)[1]	+	-	-	-
Binding to DNase I	+	+	-	-
Phosphorylation (Thr)[2]	-	+	-	-
Capping F-actin	-	+	+	+
Severing F-actin	-	-	-	+
Tryptic digestion[3]	b	b	a	a
Anti-fragmin IgG	-	+	+	+
Anti-Cap 42 (a) IgG	-	+	+	+
Anti-Cap 42 (b) IgG	+	+	+	+

[1]The phosphorylation at the two serine residues by catalytic subunit of a cAMP-dependent protein kinase from bovine heart.

[2]The phosphorylation at the two threonine residues by Physarum Cap 42 (b) kinase.

[3]The tryptic peptide maps of the four proteins indicate that they fall into two groups, a and b.

common antigenic site with each other. For example, rabbit anti-Cap 42 (b) IgG reacts not only with Cap 42 (b) but also with actin, Cap 42 (a), and fragmin. Furthermore, when preincubated with actin, the anti-Cap 42 (b) IgG loses its ability to bind further to actin as well as to the three other proteins, indicating that Cap 42 (b), Cap 42 (a), and fragmin share an antigenic site(s) with actin. Therefore, it is conceivable that actin is structurally related not only with Cap 42 (b) but also with both Cap 42 (a), and fragmin, although the four proteins are functionally distinguishable from each other.

THE PHYSARUM ACTIN GENE FAMILY AND THE POSSIBLE EVOLUTION OF F-ACTIN CAPPING AND SEVERING PROTEINS

It is of great interest to note that Physarum contains at least four and probably five distinct actin genes, whereas Dictyostelium and Drosophila contain at least seventeen and six different actin genes, respectively, which potentially give rise

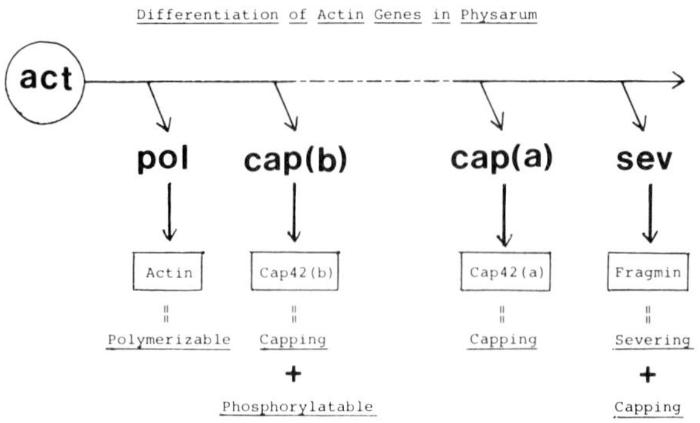

Fig. 49.6. The cap-actin gene family, a hypothesis (for details see text).

to several distinct actin species or variants (Kindle and Firtel, 1978; Firtel et al. 1979; Fyrberg et al., 1981; Schedl and Dove, 1982). If Physarum actin, Cap 42 (b), Cap 42 (a), and fragmin can be proven to be four actin variants which are derived from four distinct actin genes ("pol," "cap (b)," "cap (a)," and "sev" for polymerizable, capping (b), capping (a), and severing, respectively) one may assume that the Physarum actin genes did occur in the following order: "pol"→"cap (b)"→"cap (a)"→ "sev" as shown in Figure 49.6. The duplication and subsequent mutation of an ancestral silent actin gene called "act" might give rise first to the most conserved gene "pol," and then to a second gene "cap (b)" at an early stage of the gene evolution. Its further multiplication and rearrangement, the latter probably involving several steps, might evoke a third gene "cap (a)" and finally a fourth gene "sev." From the evolutionary point of view the existence of several genes coding for actin or its functionally different variants is an intriguing possibility how a regulatory mechanism for actin polymerization could have developed without necessarily creating a complete set of absolutely new proteins. Since at least the Cap 42 (b) seems to be widely structurally homologous to actin, it is conceivable that the Cap 42 (b) monomer during actin polymerization is accepted as an actin analogue and is hence readily incorporated into the filament. The substitution of only a few or even a single amino acid within this actin variant may then, however, enable this protein to block the further addition of actin monomers thus turning the iso-actin into an F-actin capping protein. The further substitution of only a few amino acids may give rise to even more regulatory sites, e.g., phosphorylatable or Ca^{++}-binding sites or high affinity actin binding sites which are involved in severing.

REFERENCES

Firtel RA, Timm R, Kimmel AR, McKeown M (1979). Unusual nucleotide sequences at the 5' end of actin genes in Dictystelium discoideum. Proc Natl Acad Sci (US) 76:6206.

Füchtbauer A, Jockusch BM, Maruta H, Kilimann MW, Isenberg G (1983). Disruption of microfilament organization after injection of F-actin capping proteins into living tissue culture cells. Nature 304:361.

Fyrberg EA, Bond BJ, Hershey ND, Mixter KS, Davidson N (1981). The actin genes of Drosophila: Protein coding regions are highly conserved but intron positions are not. Cell 24:107.

Glenney JR, Kaulfus Ph, Weber K (1981). F-actin assembly modulated by villin: Ca^{++}-dependent nucleation and capping of the barbed end. Cell 24:471.

Hasegawa T, Takahashi S, Hayashi H, Hatano S (1980). Fragmin: a calcium sensitive regulatory factor on the formation of actin filaments. Biochem. 19:2677.

Hinssen H (1981). An actin-modulating protein from Physarum polycephalum. Europ J Cell Biol 23:225.

Isenberg G, Ohnheiser R, Maruta H (1983). "Cap 90," a 90 kd Ca^{++}-dependent F-actin capping protein from vertebrate brain. FEBS Lett, in press.

Kindle K, Firtel RA (1978). Identification and analysis of Dictyostelium actin genes, a family of moderately repeated genes. Cell 15:763.

Kurth MC, Wang LL, Dingus J, Bryan J (1983). Purification and characterization of a gelsolin-actin complex from human platelets. J Biol Chem 258:10895.

Lazarides E, Lindberg U (1974). Actin is the naturally occurring inhibitor of Deoxyribonuclease I. Proc Natl Acad Sci (US) 71:4742.

Maruta H, Isenberg G, Schreckenbach T, Hallmann R, Risse G, Shibayama T, Hesse J (1983). Ca^{++}-dependent actin-binding phosphoprotein in Physarum I. Ca^{++}/actin dependent inhibition of its phosphorylation. J Biol Chem 258:10144.

Maruta H, Isenberg G (1983). Ca^{++}-dependent actin-binding phosphoprotein in Physarum II. Ca^{++}-dependent F-actin capping activity of subunit a and its regulation by phosphorylation of subunit b. J Biol Chem 258:10151.

Pollard TD, Cooper JA (1983). Modification of actin polymerization by Acanthamoeba profilin. J Cell Biol 97:289a.

Pollard TD, Mooseker MS (1981). Direct measurement of actin polymerization rate constants by electron microscopy of actin filaments nucleated by isolated microvillus cores. J Cell Biol 88:654.

Schedl T, Dove WF (1982). Mendelian analysis of the organization of actin gene sequences in Physarum polycephalum. J Mol Biol 160:41.

Tilney LG, Bonder EM, Coluccio LM, Mooseker MS (1983). Actin from Thyone sperm assembles on only one end of an actin filament: A behavior regulated by profilin. J Cell Biol 97:112.

Tobacman LS, Korn ED (1982). The regulation of actin polymerization and the inhibition of monomeric actin ATPase activity by Acanthamoeba profilin. J Biol Chem 257:4166.

Tobacman LS, Brenner StL, Korn ED (1983). Effect of Acanthamoeba profilin on the pre-steady state kinetics of actin polymerization and on the concentration of F-actin at steady state. J Biol Chem 258:8806.

Tseng PCh, Pollard TD (1982). Mechanism of action of Acanthamoeba profilin: Demonstration of actin species specificity and regulation by micromolar concentrations of $MgCl_2$. J Cell Biol 94:213.

Wang LL, Bryan J (1981). Isolation of calcium-dependent platelet proteins that interact with actin. Cell 25:637.

Wegner A, Isenberg G (1983). 12-fold difference between the critical monomer concentrations of the two ends of actin filaments in physiological salt conditions. Proc Natl Acad Sci (US) 80:4922.

Wegner A, Savko P (1982). Fragmentation of actin filaments. Biochem 21:1909.

Yin HL, Hartwig H, Maruyama K, Stossel ThP (1981). Ca^{++} control of actin filament length. J Biol Chem 256:9693.

50 Cytoskeleton Organization and Adrenal Chromaffin Cell Function

J. M. Trifaro, M. F. Bader, A. Côté, R. L. Kenigsberg, T. Hikita, R. W. H. Lee

INTRODUCTION

Adrenal chromaffin cells are among those cells from which contractile proteins have been isolated and characterized (Trifaró, 1978; Trifaró et al., 1982).

Catecholamines are released from chromaffin cells by exocytosis, a mechanism that follows the movement of secretory granules to the plasma membranes and the subsequent extrusion of the soluble granular contents to the cell exterior (Trifaró, 1977; Trifaró and Cubeddu, 1979).

Due to the similarities between stimulus-secretion coupling and excitation contraction coupling in muscle, it has been proposed that the secretory process of the chromaffin cell might be mediated by contractile elements either associated with the chromaffin granule or present elsewhere in the chromaffin cells (Trifaró, 1978).

This paper describes the results obtained in our laboratory on the isolation, characterization, and localization of cytoskeleton proteins in adrenal chromaffin cells. The possible roles for these proteins in chromaffin cell functions are also discussed.

EXPERIMENTAL PROCEDURES

The studies have been performed either on isolated bovine chromaffin cells or on monolayer cultures of bovine chromaffin cells. The cells were isolated and cultured as previously described (Trifaró et al., 1978; Kenigsberg and Trifaró, 1980; Trifaró and Lee, 1980).

The techniques employed in the isolation and characterization of myosin and actin have been described in detail elsewhere

(Trifaró and Ulpian, 1976; Lee et al., 1979). The preparation of affinity chromatography-purified antibodies against contractile proteins and their use in radioimmunoassays and in immunohistochemistry has been previously published (Trifaró et al., 1978; Lee and Trifaró, 1981). Immunocytochemical labelling of cells with protein A-gold complex was performed as described by Roth et al. (1978).

Chromaffin cells were homogenized and their different subcellular fractions were obtained by differential and density gradient centrifugation as described by Trifaró and Duerr (1976).

Chromaffin cell cytoskeletons were prepared as described by Skinert et al. (1982) and two-dimensional electrophoresis of these proteins was carried out as published elsewhere (Bader and Aunis, 1983).

RESULTS

Contractile Proteins

Myosin

Chromaffin cell myosin can be purified by treating a crude actomyosin preparation with KI-ATP buffers and ammonium sulfate precipitation followed by filtration through a 4% agarose column (Trifaró and Ulpian, 1976). The chromaffin cell myosin thus obtained has the following characteristics (Trifaró and Ulpian, 1976): (a) high K^+-EDTA and Ca^{2+} ATPase activities; (b) ATPase, which is inhibited by Mg^{2+}; (c) two 200 KD heavy chains, two 20 KD and two 16.5 KD light chains; (d) forms bipolar filaments upon dialysis against (mM) KCl, 30; DTT, 0.1 and Mg^{2+}, 0.2; (e) two globular heads joined flexibly to a tail with a total length of 1,600 Å; (f) binds to 6 nm actin filaments with a periodicity of 30-35 nm, and (g) is activated by smooth muscle actin in the presence of a chromaffin cell cofactor (MLCK?).

Immunofluorescence studies carried out on isolated chromaffin cells have demonstrated that myosin is mainly localized in the cytosol (Trifaró et al., 1978).

Actin and Alpha-Actinin

Actin can be isolated from a chromaffin cell cytosol preparation by DNAse I-Sepharose affinity chromatography (Lee et al., 1979). Two-dimensional electrophoresis of chromaffin cell actin revealed the presence of two (beta and gamma) isomeric forms (Lee et al., 1979). Affinity chromatography purified antibodies against chicken gizzard actin recognized chromaffin cell actin (Lee and Trifarò, 1981). Immunofluorescence studies carried out on 7-day-old cultured chromaffin cells with this antibody have shown a filamentous and fine granular fluorescence (Fig. 50.1a) in the cytosol of the cell bodies, neurites, and terminal cones (Lee and Trifarò, 1981). This is a quite distinct pattern of fluorescence compared to that observed when the same antibody was used to stain fibroblasts (Fig. 50.1c). The granular fluorescent pattern obtained with antiactin was similar to that observed when chromaffin cells were stained with an antibody against dopamine beta-hydroxylase (Fig. 50.1d). Moreover, when the protein A-gold technique was used to examine actin sites at the electron microscope level, electron opaque gold particles were observed in the proximity of the secretory granules (Trifarò et al., 1983).

Cultured chromaffin cells were also stained with an antibody prepared in our laboratory against chicken pectoral muscle alpha-actinin. Here, again, a granular fluorescent pattern of distribution, similar to that observed with antiactin, was found (Fig. 50.1d).

The aforementioned results seem to suggest that actin and alpha-actinin might be associated with the secretory granules. To further test this possibility, chromaffin granules were isolated by two density gradient procedures differing only in ionic strength. The high ionic strength in one of the procedures was obtained by the addition of 0.15 M KCl to sucrose. The proteins of the granule membranes obtained by the two methods were separated by SDS-polyacrylamide gel electrophoresis. Densitometric scanning of the gels revealed forty-six bands in both membrane preparations (Fig. 50.1e). However, the relative intensity of the bands differed with both preparations were compared. While the amount of dopamine beta-hydroxylase (band 19) was similar in both preparations, the membranes prepared from granules isolated in the presence of KCl showed that bands of lower molecular weight were present in larger amounts (Fig. 50.1e). The protein in band 28, which comigrated with chicken gizzard actin, was found at a relatively higher concentration in membranes prepared from granules isolated by the KCl-sucrose gradients. Furthermore, band 14 of the electrophoretic pattern comigrated with alpha-actinin (Fig. 50.1e). Recently, alpha-actinin has also been identified in electrophoretic patterns of chromaffin granule membrane proteins by immunoblotting techniques (Bader and Aunis, 1983).

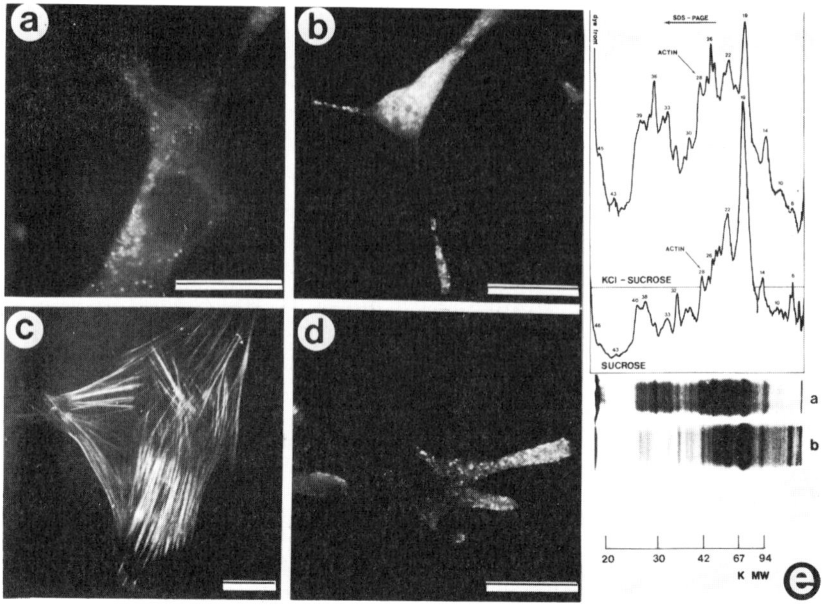

Fig. 50.1a-d. Cultured chromaffin cells (a, b, and d) and fibroblast (c) as they appear after staining with antibodies against actin (a and c), alpha-actinin (b), and dopamine beta-hydroxylase (d). Cells were cultured for 7 days and stained with the antibodies as indicated in Experimental Procedures. The preparations were examined by epifluorescence. A fine granular fluorescence is observed in chromaffin cell bodies and processes. The antiactin staining of the fibroblast (c) shows the characteristic pattern of stress fibers, whereas staining of chromaffin cells with the same antibody shows a granular fluorescent pattern (a). The horizontal bars represent 30 μm.

Fig. 5.1e. Densitometric analysis of SDS-polyacrylamide slab gels after electrophoresis of chromaffin granule membrane proteins. The granules were isolated by KCl-sucrose [gel (a) and top trace] and sucrose [gel (b) and bottom trace] density gradients. Bands 14, 19, and 28 correspond to alpha-actinin, dopamine beta-hydroxylase, and actin, respectively.

Microtubules

The distribution of microtubules in cultured chromaffin cells was studied by one of us, using the indirect immunofluorescence technique (Bader et al., 1981). Staining the chromaffin cells with bovine brain tubulin antibodies showed a network of very thin and highly ramified microtubules at the cell periphery. Microtubules were also visible along the neurites and growth cones. In addition, there was an intense fluorescence around the nucleus, suggesting that microtubules might elongate from the nuclear region toward the cell processes (Bader et al., 1981). Fifteen minutes after colchicine treatment, the cell processes began to decrease in diameter and to retract. This retraction, which was accompanied by the movement of secretory granules from the processes back into the cell body, continued for one hour, and two hours later, the cells were almost round (Bader et al., 1981). At this point, a diffuse antitubulin staining pattern was observed, suggesting a complete depolymerization of the cell microtubules (Bader et al., 1981). Consequently, microtubules are probably involved in the transport of chromaffin granules to the cell periphery. However, they do not seem to be involved in the final stages of secretion since the exocytotic release of catecholamines induced by a depolarizing concentration of K^+ is not blocked by either colchicine or vinblastine (Trifaró et al., 1972).

Neurofilaments

Many cell types contain 10 nm filaments called intermediate filaments. Since intermediate filaments have never been described in chromaffin cells, it was interesting to characterize them and to study their distribution within the cells. The neuronal origin of chromaffin cells suggested the presence of neurofilaments, a class of intermediate filaments which represent a specific marker for neurones and their neural crest relatives. Consequently, we have used an antibody raised against the 68 KD subunit of a rat brain neurofilament preparation and stained chromaffin cells by the indirect immunofluorescence technique. The staining pattern obtained, showed very thin filaments specifically localized around the nucleus (Fig. 50.2). The periphery of the cell body, the neurites and the growth-cone structures were not labelled. A stronger fluorescence was observed on one side of the nucleus and the filaments appeared to grow from this nuclear area, forming a network around the

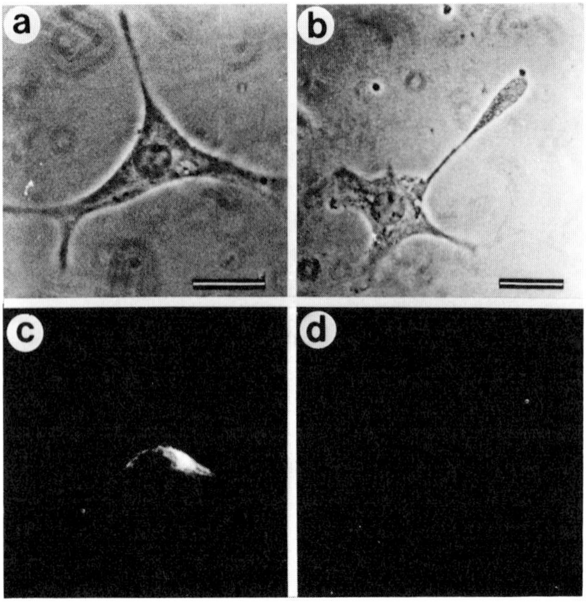

Fig. 50.2. Cultured chromaffin cells as they appear after staining with neurofilament antibodies. Chromaffin cells (a and b) cultured for 7 days and stained with an antibody against the 68 KD neurofilament subunit (c) or with the same antibody after preadsorption (control) with an excess of neurofilament preparation (d). Thin fluorescent filaments are localized around the cell nucleus (c) whereas the cell processes and growth-cone structures seen by phase contrast (a) are not stained. No staining was detected when preadsorbed antiserum was used (d). The horizontal bar represents 30 μm.

nucleus (Fig. 50.2). The same staining pattern was observed when antibodies raised against the 145 KD or the 200 KD neurofilament subunits were used (Bader et al., 1983). As controls, neurofilament-preadsorbed antisera were used, and under these conditions, there was no staining of chromaffin cells (Fig. 50.2). Some fibroblasts were also present in our cultures, but they were never labelled by neurofilament antisera, suggesting that these antisera did not crossreact with vimentin filaments.

Further evidence for the presence of neurofilaments subunits was provided by: (a) the immunoprecipitation of radioactive-labelled neurofilament subunits from homogenates prepared from chromaffin cells previously labelled with [^{35}S]methionine, and

(b) detection in two-dimensional electrophoresis gels of three spots of chromaffin cell cytoskeleton proteins with molecular weights and isoelectric points similar to those of brain bovine neurofilament subunits (Bader et al., 1983, in preparation).

The above results demonstrate the presence of neurofilaments in cultured chromaffin cells. In addition, we have been able to detect neurofilaments in freshly isolated chromaffin cells (Bader et al., 1983), thus suggesting that neurofilaments are expressed in vivo, in the adrenal gland, as well as in the cultured cells.

Calmodulin

Calmodulin is detected in cultured chromaffin cells at a level of 24 ± 5 ng/10^6 cells (n = 6). This value corresponds to 0.04% of the total protein content in chromaffin cell homogenates. The cytosolic fraction contained 69% of the chromaffin cell calmodulin when the cells were homogenized in the presence of 2 mM EGTA. The secretory granule membranes contained 167 ± 28 ng calmodulin/mg protein, and when isolation was carried out in the absence of EGTA, the granule membranes contained 303 ± 9 ng calmodulin/mg protein. The increase in the amount of membrane-associated calmodulin in the absence of EGTA suggested the presence of Ca^{++}-dependent calmodulin binding sites in chromaffin granule membranes. The presence and the properties of these binding sites were examined using ^{125}I-calmodulin. The binding of calmodulin to secretory granule membranes was found to reach a maximum at Ca^{++} concentrations of 5×10^{-5} - 10^{-4} M. In addition, this specific binding was saturable, with an apparent k_d of 16 nM and a B_{max} of 35 pmoles calmodulin/mg protein. The calmodulin-binding sites associated with the chromaffin granule membrane have been identified by use of a photoaffinity cross-linker. Only one protein was found to bind calmodulin specifically in the presence of high levels of free calcium. This calmodulin-binding protein, of molecular weight 85 K as seen on SDS-polyacrylamide gels, is a minor component of the granule membranes since it represents approximately 4% of the total membrane proteins.

It is well known that a large number of calmodulin-dependent processes can be blocked by antipsychotic drugs such as trifluoperazine. The effects of this agent were examined in the ACh- and in the high K^+-induced release of catecholamines from cultured chromaffin cells. Trifluoperazine was found to inhibit the secretory response to the above two secretagogues in a dose-

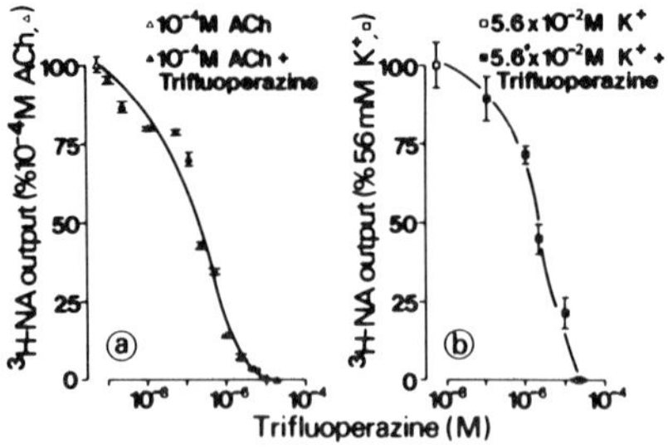

Fig. 50.3. Inhibition of (a) acetylcholine- and (b) K^+-induced release of [^{3}H]noradrenaline from cultured chromaffin cells by trifluoperazine at the concentrations indicated in the figure.

related fashion (Fig. 50.3). Trifluoperazine, at concentrations ranging from 10^{-6} to 10^{-5} M significantly inhibited K^+-induced amine output without modifying ^{45}Ca uptake or efflux. However, only at a high concentration of this agent (2.5×10^{-5} M), which completely blocked K^+-induced amine output, was the ^{45}Ca uptake decreased by 30 ± 14% (n = 6) (Table 50.1). These results show that this agent blocks catecholamine release at a step distal from calcium entry, thus suggesting, indirectly, a role for calmodulin in the secretory process of the chromaffin cells (Kenigsberg et al., 1982).

In order to obtain more direct evidence for the role of calmodulin in secretion, antibodies raised against calmodulin and purified by affinity chromatography were introduced directly into the cytoplasm of viable cultured chromaffin cells by erythrocyte ghost-mediated microinjection (Trifaró and Kenigsberg, 1983). Chromaffin cells in culture have the ability to accumulate exogenous catecholamines (Kenigsberg and Trifaró, 1980). Microinjection of normal IgG (used as a control) or calmodulin antibodies did not affect the cells' ability to accumulate [H]noradrenaline. However, cells which had been injected with anticalmodulin IgG significantly released less catecholamines when stimulated by either ACh or high K^+ than did cells injected with normal IgG (Table 50.2). These results suggest that calmodulin is directly involved in the process of stimulus-secretion coupling in the chromaffin cell.

DISCUSSION

The experiments discussed here have demonstrated the presence of cytoskeleton proteins and their regulatory proteins

TABLE 1. EFFECT OF VARIOUS CONCENTRATIONS OF TRIFLUOPERAZINE ON 56 mM $[K^+]$-INDUCED ^{45}Ca UPTAKE INTO AND $[^3H]$NORADRENALINE OUTPUT FROM CULTURED CHROMAFFIN CELLS.

TRIFLUOPERAZINE	INHIBITION AS % OF CONTROL	
CONCENTRATION	^{45}Ca UPTAKE	$[^3H]$NA OUTPUT
10^{-6}M	0	30 ± 3*
2.5 x 10^{-6}M	0	64 ± 5
10^{-5}M	4 ± 2	79 ± 5
2.5 x 10^{-5}M	30 ± 14	100 ± 1

* MEAN ± S.E.

TABLE 2. Effect of intracellular delivery of antibodies against calmodulin on acetylcholine- and potassium- evoked release of catecholamines from chromaffin cell cultures.

Experiment Number	Globulin Injected*	Stimulus	Catecholamine Release (% of total store)	Inhibition of Catecholamine Release (%)
1	n-IgG	ACh	10.9 ± 2.2**	-
	anti-CM-IgG	"	4.4 ± 1.0	59.6
2	n-IgG	ACh	9.1 ± 1.4	-
	anti-CM-IgG	"	3.4 ± 1.2	62.6
3	n-IgG	ACh	13.1 ± 0.9	-
	anti-CM-IgG	"	2.4 ± 0.9	81.7
4	n-IgG	K^+	11.0 ± 0.7	-
	anti-CM-IgG	"	6.7 ± 0.3	39.1
5	n-IgG	K^+	9.9 ± 0.4	-
	anti-CM-IgG	"	3.5 ± 0.4	64.6

Chromaffin cells were stimulated by either acetylcholine (10^{-4}M ACh) or potassium (56 mM K^+) for 9 minutes. Catecholamine output is expressed as percentage of total cell content.

* Human erythrocyte ghost-mediated microinjection of normal pre-immune IgG (n-IgG) or anti-calmodulin IgG (anti-CM-IgG).

** Mean ± S.E. of four different culture dishes (10^6 cells/dish).

in adrenal chromaffin cells. Microtubules and probably microfilaments are involved in the chromaffin cell with the transport of secretory granules as well as neurite outgrowth and maintenance (Bader et al., 1981). However, unlike tubulin and the contractile proteins, the neurofilament triplet proteins are a minor cytoskeleton component of these cell neurites and therefore, it is also difficult to imagine an active role for the neurofilaments in facilitating the granule transport. Moreover, staining obtained in cultured chromaffin cells using neurofilament antibodies is very similar to the staining obtained in cultured fibroblasts using vimentin antibodies or in cultured epithelial cells using keratin antibodies (Lazarides, 1980). This similar localization might be an indication of similar roles for the different intermediate filaments. It has been proposed that vimentin filaments might provide a mechanical support for the nucleus so to constrain it in a specific place in the cell. A similar function could be suggested for the chromaffin cell neurofilaments in view of their distribution within the cell.

The association of some contractile proteins with chromaffin granules strongly suggests that contractile (myosin, actin, alpha-actinin) and regulatory (calmodulin) proteins might be involved in the secretory process. If exocytosis is a true contractile event (Trifaró, 1978), contractile proteins might be involved with the transport mechanism of secretory granules to release sites on the plasma membrane, with the fusion process, or with the extrusion phenomenon itself. In any of these cases, two molecular mechanisms of action for the contractile proteins might exist.

The first possible mechanism would involve actin, alpha-actinin, myosin, and the regulatory protein calmodulin (Fig. 50.4b). In this case, a sliding filament mechanism similar to that found in muscle, would operate in chromaffin cells. Myosin should be arranged in bipolar filaments. In this regard, we have demonstrated that chromaffin cell myosin can form bipolar filaments (Trifaró and Ulpian, 1976). The alpha-actinin present in granule membranes would provide binding sites for actin. Actin binding sites might also be present in plasma membranes. Stimulation depolarizes the cell. Ca^{++} entry is activated with the subsequent increase in intracellular Ca^{++}. This, in turn, would activate the sliding filament mechanism through a calmodulin-dependent action (Fig. 50.4b). At the molecular level, this would imply the activation of myosin ATPase by actin, a mechanism that requires, in all nonmuscle tissues so far investigated, the phosphorylation of the 20,000 daltons myosin light chain. This phosphorylation step is regulated by

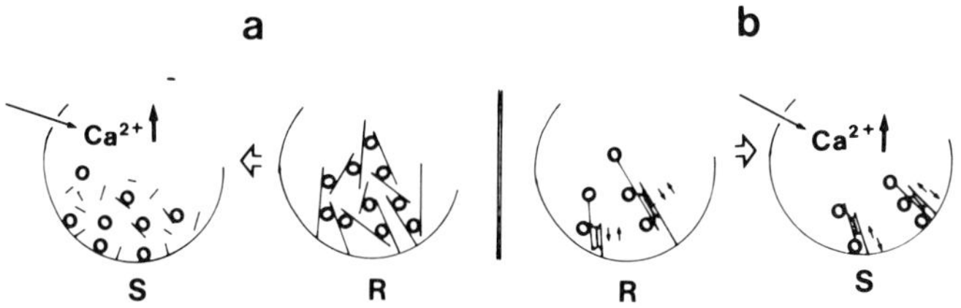

Figure 50.4. Schematic representation of two possible mechanisms (a and b) in which contractile proteins may play a role in chromaffin cell secretion (see the Discussion for explanation of the scheme).

calmodulin (Adelstein et al., 1980). Our experiments with trifluoperazine have shown that this calmodulin antagonist does inhibit catecholamine release in response to either acetylcholine or high potassium stimulation at a step distal from Ca^{++} entry, thus suggesting that the drug is antagonizing calmodulin. Moreover, more direct evidence for the role of calmodulin was obtained from the experiments on the intracellular delivery of calmodulin antibodies. In this case, catecholamine release in response to stimulation was inhibited subsequent to the microinjection of the antibodies. Although the results obtained with the calmodulin antibodies provide us with more direct evidence for calmodulin involvement in the process of secretion of the chromaffin cells, one cannot conclude with certainty at this point that the role of calmodulin in secretion is through the modulation of the myosin light chain phosphorylation. Other calmodulin-dependent processes might be involved in secretion, and these are discussed below.

A second possible theory of a mechanism in which contractile proteins may play a role in secretion is based on the viscosity properties of actin and does not require the intervention of myosin (Fig. 50.4a). In this case, actin would control chromaffin cell cytosol viscosity through the formation of a mesh of microfilaments which are also cross-linked and stabilized by the chromaffin granules. In this regard, it has recently been shown in in vitro experiments that chromaffin granules will induce actin polymerization and gel formation, effects blocked by raising the concentration of Ca^{++} in the medium (Fowler and Pollard, 1982). Therefore, the cytosolic actin network would oppose the movement of secretory granules toward release sites. When the cells are stimulated, Ca^{++} enters the cell. This may produce, directly or through the activation of a specific protein (gelsolin?), a shortening in the length of the microfilaments. This would decrease the viscosity of the cytosol allowing the movement of granules toward release sites (Fig. 50.4a). A protein with these properties has been isolated from macrophages, and it has been named gelsolin (Stendhal and Stossel, 1980). Gelsolin is activated when the concentration of Ca^{++} is increased to 10^{-6} M (Stendhal and Stossel, 1980). Calmodulin might be involved in the modulation of this process; alternatively, it might be involved in the granule-plasma membrane fusion process itself. We have presented here data on the presence, in granule membranes, of high affinity binding sites for calmodulin. In addition, recent morphological observations have suggested that trifluoperazine might inhibit carbamylcholine-induced catecholamine secretion by interfering with the process of fusion between chromaffin granules and plasma membranes (Burgoyne et al., 1982). This could also be the site of inhibition for the calmodulin antibody when delivered intracellularly by erythrocyte ghost-mediated microinjection.

In conclusion, although the contractile proteins of chromaffin cells may be significant to cell functions other than secretion, the study of their possible role in the secretory process is crucial to the understanding of the release process at cellular and molecular levels. Thus, the term stimulus-secretion coupling, coined by Douglas and Rubin (1961) more than twenty years ago, to describe the events involved in the secretory process, seems today perfectly appropriate.

ACKNOWLEDGMENT

This work was supported by grant PG-20 from the MRC of Canada.

REFERENCES

Adelstein, R. S., M. A. Conti and M. D. Pato (1980) Ann N.Y. Acad. Sci. 356:142-50.

Bader, M.-F. and D. Aunis (1983) Neuroscience 8:165-81.

Bader, M.-F., J. Ciesielski-Treska, D. Thierse, J. E. Hesketh and D. Aunis (1981) J. Neurochem. 37:917-33.

Bader, M.-F., E. Georges, W. E. Mushynski and J. M. Trifaró (1983) (in preparation).

Burgoyne, R. D., M. J. Geisow and J. Barrow (1982) Proc. R. Soc. Lond. B:216:111-15.

Douglas, W. W. and R. P. Rubin (1961) J. Physiol. 159:40-57.

Fowler, V. M. and H. B. Pollard (1982) Nature, 295:336-39.

Kenigsberg, R. L., A. Côté, and J. M. Trifaró (1982) Neuroscience, 7:2277-86.

Kenigsberg, R. L. and J. M. Trifaró (1980) Neuroscience, 5: 1547-56.

Lazarides, E. (1980) Nature, 283:249-56.

Lee, R. W. H., W. E. Mushynski, and J. M. Trifaró (1979) Neuroscience, 4:843-52.

Lee, R. W. H. and J. M. Trifaró (1981) Neuroscience, 6:2087-2108.

Poisner, A. M. and J. M. Trifaró (1967) Mol. Pharmacol. 3: 361-71.

Roth, J., M. Bendayan and L. Orci (1978) J. Histochem., 26: 1074-81.

Steinert, P., Zackroff, R., M. Aynardi-Whitman and R. D. Goldman (1982) Meth. Cell Biol. 24 (Ed. L. Wilson), Acad. Press, N.Y., pp. 309-19.

Stendhal, O. I. and T. P. Stossel (1980) Biochem. Biophys. Res. Comm. 92:675-81.

Trifaró, J. M. (1977) A. Rev. Pharmac. Toxicol., 17:27-47.

Trifaró, J. M. (1978) Neuroscience, 3:1-24.

Trifaró, J. M., B. Collier, A. Lastowecka and D. Stern (1972) Mol. Pharmacol., 8:264-67.

Trifaró, J. M. and L. Cubeddu (1979) Trends in Autonom. Pharmacol. (Ed. S. Kalsner) Urban & Schwarzenberg. Baltimore-Munich, pp. 195-249.

Trifaró, J. M. and A. C. Duerr (1976) Bioch. Biophys. Acta 421:153-67.

Trifaró, J. M. and R. L. Kenigsberg (1983) Fed. Proc. 42:457.

Trifaró, J. M., R. L. Kenigsberg, A. Côté, R. W. H. Lee and T. Hikita (1983) Can. J. Phys. Pharm. (in press).

Trifaró, J. M. and R. W. H. Lee (1980) Neuroscience, 5:1533-46.

Trifaró, J. M., R. W. H. Lee, R. L. Kenigsberg and A. Côté (1982) Adv. Bioscien., 36:152-58.

Trifaró, J. M. and C. Ulpian (1975) FEBS Lett. 57:198-202.

Trifaró, J. M. and C. Ulpian (1976) Neuroscience, 1:483-88.

Trifaró, J. M., C. Ulpian and H. Preiksaitis (1978) Experientia, 34:1568-71.

51 Interrelationship of Intermediate Filaments with Microtubules and Microfilaments in Monkey Kidney TC7 Cells

Julio E. Celis, J. Victor Small,
Peter Mose Larsen, Stephen J. Fey,
J. De Mey, Ariana Celis

INTRODUCTION

Intermediate-sized filaments (7-11 Å in diameter) are ubiquitous cytoskeletal elements (for references see 1 and 2) that have been implicated in various intracellular roles including nuclear anchorage (3, 4), organelle interactions (5-12), gene expression (13); however, the molecular mechanisms underlying these functions remain largely unknown. To date, much emphasis has been placed on determining their subunit composition (for references see 1, 2, and 14), and rather less on their structural and functional relationships with other cytoskeletal systems. In this study, we present evidence for the interaction of intermediate filaments (keratins and vimentin) with both microtubules and microfilaments in monkey kidney TC7 cells (15).

MATERIALS AND METHODS

Cells

African green monkey kidney TC7 cells (a gift of Dr. A. Graessmann) were grown as monolayer cultures in Dulbecco's modified Eagle's medium supplemented with 10% fetal calf serum and antibiotics (penicillin, 100 IU/ml; streptomycin, 50 μg/ml).

The procedures for labelling cells with [^{35}S]-methionine (16), two dimensional gel electrophoresis (17, 18), preparation of low and high salt extracted cytoskeletons (4), and indirect immunofluorescence (10) have been described in detail elsewhere. The vimentin antibody was a gift from Dr. S. Blose.

Immunoelectron microscopy

Immunoelectron microscopy was performed on unfixed intermediate-filament enriched cytoskeletons on electron microscope grids (4) using the indirect immunogold staining (IGS) procedure according to De Mey et al. (19). Following antibody treatment cytoskeletons were negatively stained with aqueous uranyl acetate (4).

Intermediate-Sized Filament Proteins of TC7 Cells

Figure 5.1 shows a two dimentsional gel fluorograph (IEF) of proteins from asynchronous TC7 cells labelled for 20 hr with [^{35}S]-methionine. The position of the intermediate-sized filament proteins (keratins IEF 36 [M_r = 48,500] and 46 [M_r = 43,500]; vimentin [IEF 26; HeLa protein catalogue number]) (20-22) as well as of α and β-tubulin and actin are indicated as reference. Keratins IEF 36 and 46 correspond to proteins 8 and 18 in the catalogue of human cytokeratins (14). The authenticity of the intermediate filament proteins has been confirmed by two dimensional gel electrophoretic analysis of [^{35}S]-methionine labelled cytoskeletons enriched in intermediate filaments (4,5; not shown), and of immunoprecipitates of total TC7 proteins reacted with a broad specificity keratin antibody (15, not shown). Similar results have been obtained when total TC7 proteins are immunoprecipitated with a polyclonal antibody raised against human keratin IEF 46 (24).

Intermediate Filaments of TC7 Cells: Perinuclear Bodies (IFFC) and Their Relationship with MTOC

Immunofluorescence staining of methanol:acetone fixed TC7 cells with a mouse polyclonal antibody raised against HeLa keratin IEF 46 (24) (Figs. 51.2a and b) or the broad specificity keratin antibody (not shown) revealed a discontinuous staining of filaments (Figs. 51.2a,b) as well as the presence of strongly fluorescent perinuclear bodies (arrows in Fig. 51.2a; such bodies were seen in about 58% of the cells in interphase) similar to those described by Borefreund et al. (25) in cultured Hepatoma cells and by Hormia et al. (26) in endothelial cells. Immunoelectron microscopy (keratin IEF 46 antibody) using the immunogold staining (IGS) procedure (19) revealed that the discontinuous staining was not due to the fragmentation of the

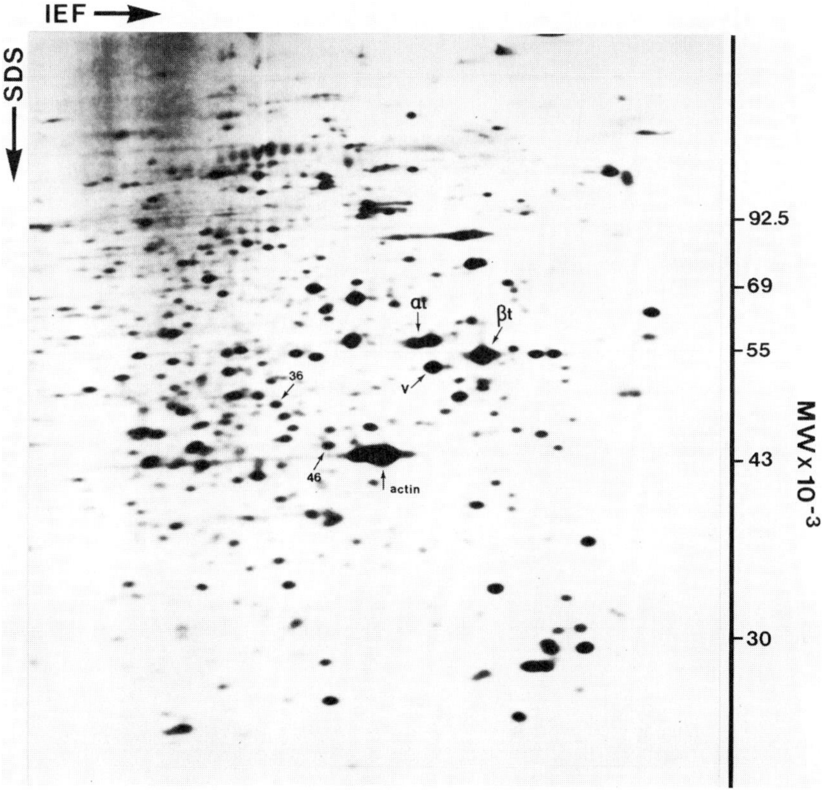

Fig. 51.1. Intermediate-sized filament proteins of TC7 cells. Two dimensional gel electrophoresis (IEF) of [^{35}S]-methionine labelled proteins from TC7 cells labelled for 20 hr.

filament system but to a real discontinuous distribution of label along segments of the filament net (Fig. 51.3a). The same antibody stained filaments in a continuous fashion in transformed human amnion cells (AMA) (24) as determined both by immunoelectron microscopy (small particles in Fig. 51.3b) and immunofluorescence microscopy (not shown). A discontinuous staining of TC7 cells was also observed with the broad specificity keratin antibody (not shown). Staining of TC7 cells with vimentin antibodies revealed a continuous staining of subset of filaments, as judged by immunoelectron microscopy (not shown).

The perinuclear bodies (IFFC) could be shown, by analysis of enucleated cells (cytoplasts; antikeratin staining; Fig. 51.4a) to be cytoplasmic structures. In many cases, filaments were

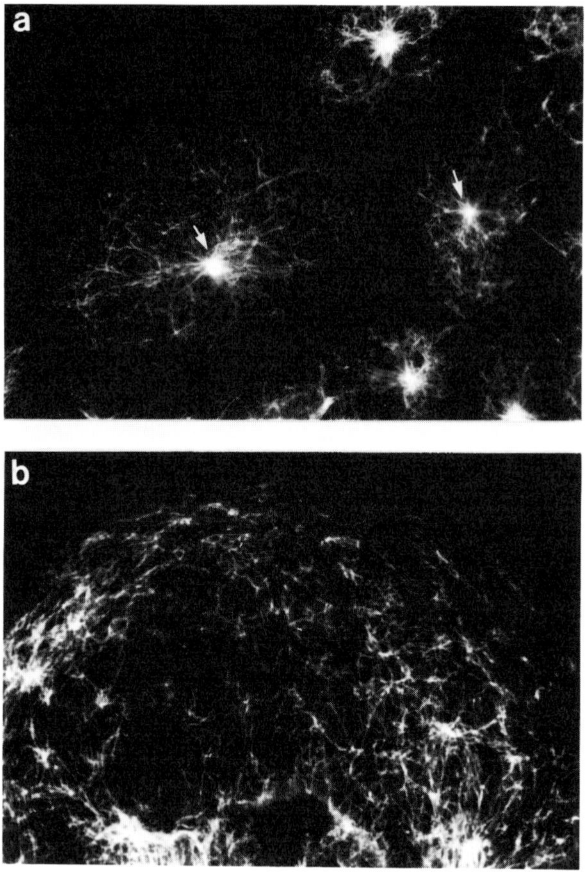

Fig. 51.2. Discontinuous staining of intermediate-sized filaments with keratin antibodies. For immunofluorescence the cells were treated first with methanol (4 min. at -20°C followed by acetone (as for methanol) previous to incubation with the antibody. (a,b) TC7 cells reacted with HeLa keratin IEF 46 antibody. (a) ×300; (b) ×600.

seen radiating from the bodies (Fig. 51.4b), (25,27,28). Normally, one body per cell was observed although in some cases two or three focal centers were seen and these were often interconnected by filaments. The perinuclear bodies also stained brightly with vimentin antibodies (arrows in Fig. 51.4c).

Methanol:acetone treated TC7 cells reacted with tubulin antibodies also revealed strong perinuclear staining sites in

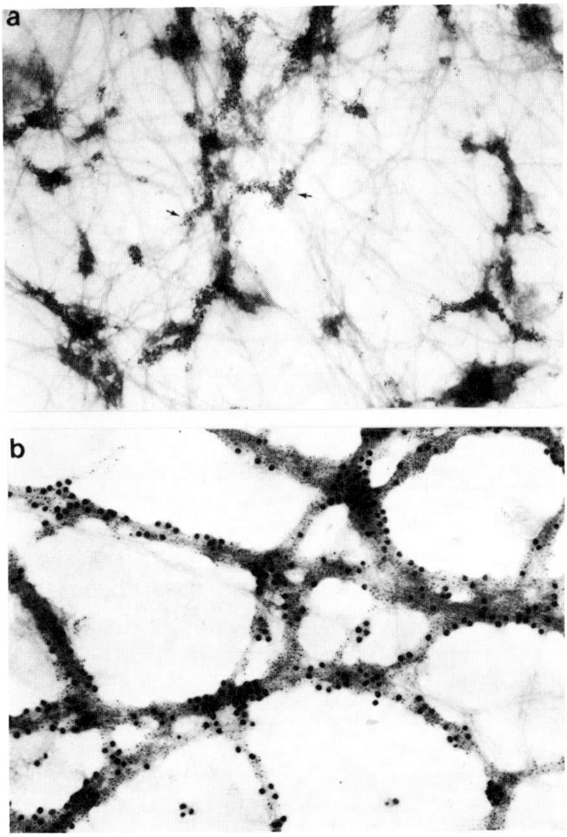

Fig. 51.3. Immunoelectron microscopy of intermediate filaments enriched TC7 cytoskeletons. The cytoskeletons were reacted with HeLa keratin IEF 46 antibody (small gold particles) as described in reference 19. (a) TC7 cells, (b) transformed amnion cells (AMA). The large gold particles in (b) arise from a second label for vimentin (suboptimal concentration of second antibody) and should be ignored. (a) ×40,000 (b) ×55,000.

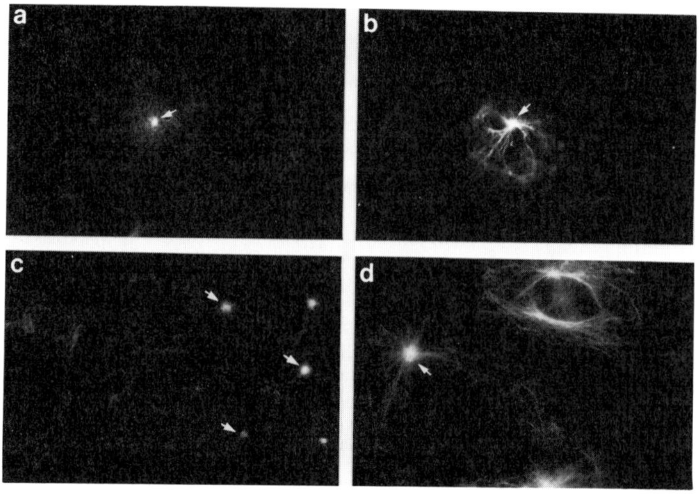

Fig. 51.4. Indirect immunofluorescence staining of TC7 cells with various antibodies. (a) TC7 cytoplast reacted with keratin IEF 46 antibody, (b) TC7 cell reacted with keratin IEF 46 antibody, (c) TC7 cells reacted with vimentin antibodies, and (d) as (c) but reacted with tubulin antibodies. (a,b) ×380 (c) ×280 (d) ×240.

some of the interphase cells (arrow in Fig. 51.4d) and in most cases, these codistributed with the IFFC as judged by double label immunofluorescence (Fig. 51.5a; tubulin staining; Fig. 51.5b keratin staining). Similar analysis of mitotic cells showed, however, that the centrioles (Fig. 51.5c) do not codistribute with the IFFC (Fig. 51.5d; seen in only 20% of the mitotic cells). Also, following nocodazole treatment (10 μg/ml, 20 hr, 29-30) and recovery (3 min) the MTOC (31) and the IFFC usually had different cellular locations (Figs. 51.5e and f; arrows indicate the position of the intermediate filaments focal centers) although in some cases, they were in close proximity or were coincident (Figs. 51.5g and h).

By double immunofluorescence microscopy using Rhodamine-labelled phaloidin and keratin IEF 31 antibody (24) the perinuclear bodies were found to be negative for actin (not shown). The IEF 31 antibody was used so as to obtain a weak staining of the IFFC. This antibody cross-reacted weakly with both keratins present in TC7 cells (see also reference 24 for cross-reactivity of this antibody).

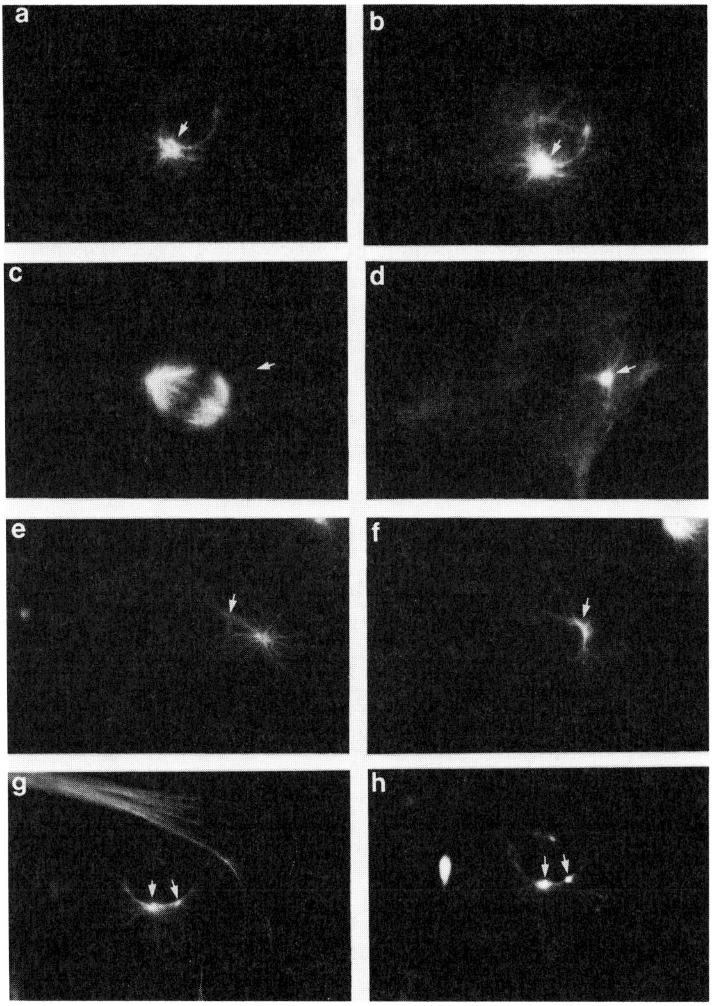

Fig. 51.5. Distribution of IFFC and MTOC in TC7 cells. (a,b) double immunofluorescence of methanol:acetone fixed interphase TC7 cells reacted with tubulin (a) and broad specificity keratin antibody (b). (c,d) double immunofluorescence of mitotic TC7 cells reacted with tubulin (c) and broad specificity keratin antibody (d). (e,f,g,h) double immunofluorescence of nocodazole treated cells (10 μg/ml; 20 hr, 3 min. recovery), fixed and reacted with tubulin (e,g) and the broad specificity keratin antibody (f,h).

Demecolcine Induces a Redistricution of the Keratin and Vimentin Filaments in TC7 Cells

Treatment of TC7 cells with demecolcine (20 hr at 37°C) resulted in the complete disassembly of microtubules (not shown) and a drastic redistribution of the keratin (Fig. 51.6a) and vimentin (not shown) intermediate filaments. Fewer discontinuously stained filaments were also observed toward the cell periphery. The demecolcine treatment also affected the focal centers (arrow in Fig. 51.6b) which were then observed in only about 19% of the cells. A partial although less dramatic effect of antimitotic drugs on keratin containing filaments has also been reported in PTK2 cells (32, and our own observations) (Fig. 51.6b).

Cytochalasin B Affects the Organization of Intermediate-Sized Filaments and Microfilaments

Treatment of TC7 cells with cytochalasin B (10 μg/ml, 1 hr at 37°C) followed by methanol:acetone fixation revealed a starlike arrangement of the keratin (arrows; Fig. 51.6c) and vimentin filaments (not shown) similar to that recently reported in mouse KLN 205 carcinoma cells treated simultaneously with colchicine and cytochalasin D (33). The effect of cytochalasin B on intermediate filaments in TC7 cells could be observed already 20 min after incubation; the experiments reported here have been carried out after 60 min treatment. In parallel experiments no significant effect of cytochalasin B was observed with PTK2 cells (Fig. 51.6d; keratin staining). Doublelabel immunofluorescence of cytochalasin B treated TC7 cells (acetone fixation) with the keratin IEF 31 antibody (Fig. 51.6e) and Rhodamine-labelled phalloidin (Fig. 51.6f) showed that most of the keratin starlike aggregated codistributed with the actin aggregates (arrows in Figs. 51.6e and f). In many cells, we observed a diffuse fibrillar region surrounding the focal centers that reacted with both the keratin antibody and Rhodamine phalloidin (not shown) (15). These did not stain with Rhodamine phalloidin even after cytochalasin B treatment (15). Similar studies using tubulin antibodies did not reveal any dramatic effect of cytochalasin B on the distribution of these filaments (results not shown. shown.

DISCUSSION

A characteristic difference between the intermediate filament and microtubule networks is the absence in the former of freely

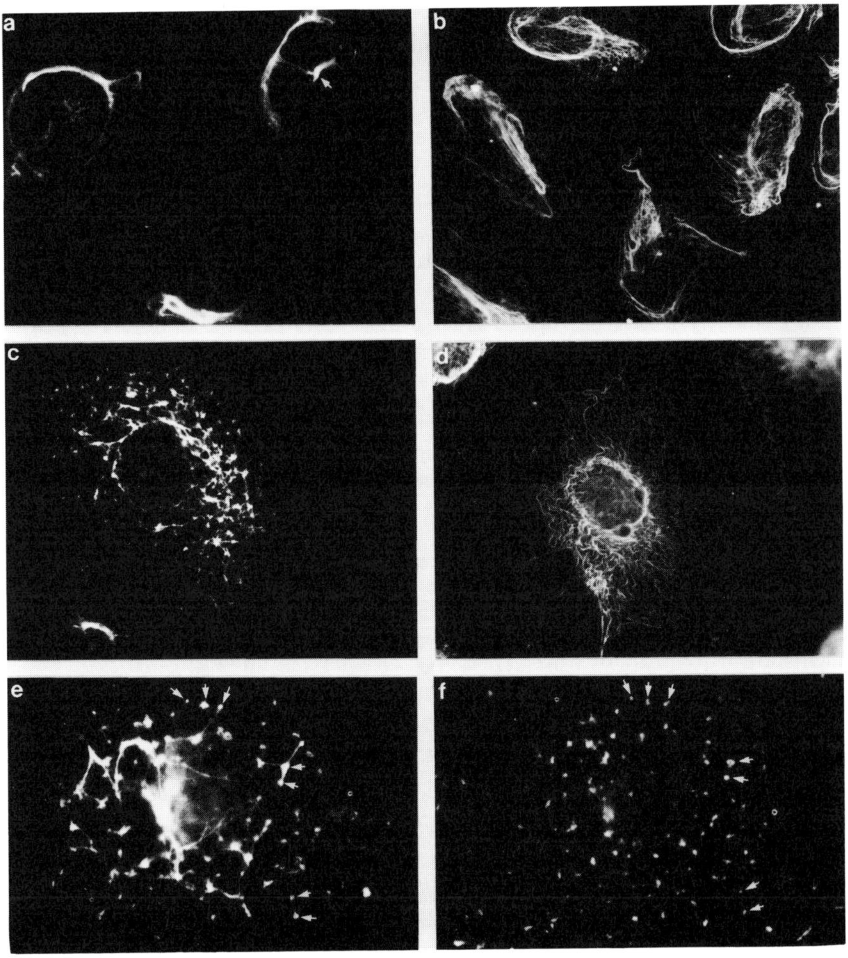

Fig. 51.6. Fluorescence microscopy of TC7 cells treated with various drugs. (a) TC7 cells treated with demecolcine (10 μg/ml; 20 hr at 37°C) and reacted with the keratin IEF 46 antibody, (b) PTK2 cells treated as above but reacted with the keratin IEF 31 antibody, (c,d) TC7 (c) and PTK2 cells (d) treated with cytochalasin B (60 min. at 37°) and reacted with a broad specificity keratin antibody. (e,f) double immunofluorescence of acetone fixed TC7 cells reacted with keratin IEF 31 antibody (e) and Rhodamine phalloidin (f). Arrows indicate equivalent points in the micrographs. (a,b) ×380.

exposed filaments ends. This is most readily apparent in whole mount cytoskeletons viewed by electron microscopy. The discontinuous staining of the intermediate filaments in TC7 cells with keratin antibodies indicates a segmental distribution of antigenic sites along parts of the filament networks. Whether or not this results from an absence of certain keratin polypeptides in the unstained sections or a structural conformation leading to unexposed sites is at present unknown. It is noteworthy, however, that the anti-IEF 46 antibody showed a continuous staining in AMA cells. A granular staining of keratin filaments has recently been reported in a clonal cell line (BMGE + HM) selected from bovine mammary gland epithelial cell cultures that lacks both vimentin and desmin filaments (34).

The presence of perinuclear bodies (IFFC) that may contain both keratins and vimentin is of interest as these cytoplasmic bodies may correspond to intermediate filament organizing centers (25,27,28). These centers have not been observed in many other cultured cell lines studied and therefore may not represent typical organizing centers as in the case of microtubules (31). In interphase, the IFFC codistributed with focal arrays of microtubules (most likely centrosomes) but were separate from the centrioles in mitotis and (in many cases) the MTOC induced after brief recovery from nocodazole treatment. These results suggest that MTOC and IFFC are independent structures (see also refs. 26 and 27) although there may be some degree of association. Indeed, an association between centrioles and vimentin filaments has been proposed in various cell types (6,25,35-37).

Contrary to what is observed in many epithelial cells treated with antimitotic drugs the keratin filaments in TC7 cells are drastically rearranged following treatment with demecolcine or nocodazole. Even though the molecular mechanism(s) underlying this redistribution is at present unknown the results clearly suggest that the organization of the intermediate filaments (keratins and vimentin) is dependent of microtubules at least in this cell line. This is consistent with the close association between vimentin filaments and microtubules that has been reported in various cell lines (see 38 and references therein).

The effect of cytochalasin B on the organization of intermediate filaments (keratins and vimentin) in TC7 cells is a novel observation. Recently, Knapp et al. (33) showed that treatment of KLN 205 carcinoma cells with colchicine together with cytochalasin D converted the keratin cytoskeleton into a series of starlike structures which they suggested are maintained by multiple membrane attachment sites. In our studies (using only

Cytochalasin B) we have further shown that most of the starlike structures codistribute with actin patches, a fact that would strongly suggest that intermediate filaments may interact with microfilaments at least in certain regions of the cell.

In summary the foregoing studies point toward a close relationship (structural and/or functional) between the various cytoskeletal systems in monkey kidney TC7 cells and add support to the notion that cytoskeletal interactions may be different in various cell types.

ACKNOWLEDGMENTS

We thank O. Jensen for photography and K. Dejgaard and M. Hattenberger for assistance. We also thank Prof. Th. Wieland for a generous gift of rhodamine-labelled phalloidin. P.M.L. is a recipient of a fellowship from the Aarhus University. S.J.F. is a recipient of a fellowship from the Danish Cancer Foundation. This work was supported by grants from the Danish Natural Science and Medical Research Councils, the Danish Cancer Foundation, the Carlsberg Foundation, and NOVO (to J.E.C.) and the Austrian Science Research Council (to J.V.S.).

REFERENCES

1. Lazarides, E. (1980) Nature 283, 249-56.
2. Cold Spring Harb. Symp. Quant. Biol., 46, on "Organisation of the cytoplasm," Cold Spring Harbor, 1982.
3. Letho, V. P., Virtanen, J. & Kurki, P. (1978) Nature 272, 175-77.
4. Small, J.V. & Celis, J.E. (1978). J. Cell Sci. 31, 393-409.
5. Lee, C. S., Morgan, G. & Wooding, F. B. P. (1979) J. Cell Sci. 38, 125-35.
6. Goldman, R. D., Zackroff, R. V., Starger, J. M. & Whitman, M. (1979) J. Cell Biol. 83 (2 part 2), 343a abstr.
7. David-Ferreira, K. L. & David-Ferreira, J. F. (1980) Int. Cell Biol. Rep. 4, 655-62.
8. Pharie-Washington, L., Silverstein, S. C. & Wang, E. (1980) J. Cell Biol. 92, 575-78.
9. Chen, L. B., Summerhayes, I. C., Johnson, L. V., Walsh, M. L., Bernal, S. D. & Lampidis, T. J. (1982). Cold Spring Harbor Symp. Quant. Biol., 46, 141-51.
10. Mose Larsen, P., Bravo, R., Fey, S. J., Small, J. V. & Celis, J. E. (1982) Cell, 31, 681-92.

11. Mose Larsen, P., Fey, S. J., Bravo, R. & Celis, J. E. (1983) Electrophoresis, 4, 247-56.
12. Tokuyasu, K. T., Dutton, A. H. & Singer, S. J. (1983) J. Cell Biol., 96, 1727-35.
13. Traub, P., Nelson, J. W., Kühn, S. & Vorgias, C. (1983). J. Biol. Chem. 258, 1456-66.
14. Moll, R., Franke, W. W., Schiller, D. L., Geiger, B. & Knepler, R. (1982) Cell, 31, 11-24.
15. Celis, J. E., Small, J. U., Mose Larsen, P., Fey, S. J., De May, J., and Celis, A., Proc. Natl. Acad. Sci., in press.
16. Celis, J. E. & Bravo, R. (1981) Trends in Biochem. Sci. 6, 197-202.
17. Bravo, R., Small, J. V., Fey, S. J., Mose Larsen, P. & Celis, J. E. (1982) J. Mol. Biol. 154, 121-42.
18. Bravo, R. (1983). In: Two dimensional gel electrophoresis of proteins, Methods and Applications. (J. E. Celis and R. Bravo, eds.), Academic Press, New York, in press.
19. De Mey, J., Moerermans, M., Geuens, G., Nuydens, R. & De Brabander, M. (1981) Cell Biol. Int. Rep., 5, 889-99.
20. Bravo, R., Bellatin, J. & Celis, J. E. (1981) Cell Biol. Int. Rep., 5, 93-96.
21. Bravo, R. & Celis, J. E. (1982). Clin. Chem. 28, 766-81.
22. Bravo, R. & Celis, J. E. (1983) In: Two dimensionsal gel electrophoresis of proteins, Methods and Applications. (J. E. Celis and R. Bravo, eds), Academic Press, New York, in press.
23. Fey, S. J. Mose Larsen, P. & Celis, J. E. (1983) FEBS Lett., 157, 165-69.
24. Bravo, R., Fey, S. J., Mose Larsen, P., Coppard, N. & Celis, J. E. (1983) J. Cell Biol., 96, 416-23.
25. Borenfreund, E., Schmid, E., Bendich, A. & Franke, W. W. (1980) Exp. Cell Res., 177, 215-35.
26. Hormia, M., Linder, E., Letho, V.-P., Vartio, T., Badley, R. A., & Virtanen, I. (1982). Exp. Cell Res. 138, 159-66.
27. Eckert, B. S., Daley, R. A. & Parysek, L. M. (1982). J. Cell Biol. 192, 575-78.
28. Fey, S. J., Mose Larsen, P., Bravo, R., Celis, A. and Celis, J. E. (1983) Proc. Natl. Acad. Sci. 89, 1905-09.
29. De Brabander, M. J., Van de Veire, R. M. L., Aerts, F. E. M., Borges, M. & Janssen, P. A. J. (1976). Cancer Res. 36, 905-16.
30. Hoebeke, J., Van Nijen, G., & De Brabrander, M. (1976) Biochem. Biophys. Res. Commun. 69, 319-24.

31. Pickett-Heaps, J. D. (1969). Cytobios, 1, 257-80.
32. Klymkowsky, M. W. (1982) EMBO J. 1, 161-65.
33. Knapp, L. W., O'Guin, W. M. & Sawyer, R. H. (1983). Science, 219, 501-03.
34. Schmid, E., Franke, W. W., Grund C., Schiller, D., Kolb, H., & Paweletz, N. (1983) Exp. Cell Res., 146, 309-28.
35. Wang, E., Connolly, J. A., Kalnins, U. K. & Choping, P. W. (1979) Proc. Natl. Acad. Sci. 76, 5719.
36. Aubin, J. E., Osborn, M., Franke, W. W. & Weber, K. (1980) Exp. Cell Res., 129, 149-65.
37. Blose, S. H. (1981) Cell Motility, 1, 417-31.
38. Geuens, G., de Brabander, M., Nuydens, R. & De Mey, J. (1983) Cell Biol. Int. Rep., 1, 35-47.

52 Regulation of the Actin-activated ATPase Activity of Acanthamoeba Myosin II

Edward D. Korn, Joseph P. Albanesi,
Mark A. L. Atkinson, Graham Cote,
John A. Hammer III, Jacek Kuznicki

Previously, we have described the isolation and general properties of three myosin isoenzymes of the soil amoeba Acanthamoeba castellanii. Their general physical and enzymatic features are as follows. Myosins IA and IB are small, globular molecules with no detectable tail portion and they seem to be unable to form filaments or any other type of aggregates (Maruta et al., 1979; Gadasi et al., 1979; Gadasi and Korn, 1979). Myosin IA consists of two polypeptide chains of molecular weights 130,000 and 17,000 in 1:1 molar ratio and has a native molecular weight of 159,000 by sedimentation equilibrium (Albanesi and Korn, unpublished). Myosin IB contains two polypeptides of molecular weights 125,000 and 27,000 in a 1:1 molar ratio and has a native molecular weight of 150,000 by sedimentation equilibrium (Albanesi and Korn, unpublished). Sometimes less than stoichiometric amounts of a third polypeptide of molecular weight 14,000 is isolated with both myosins IA and IB. Both myosins IA and IB are phosphorylated at a single serine residue on their heavy chains (Maruta and Korn, 1977b; Hammer et al., 1983). No physical changes in the molecules have been observed as a consequence of phosphorylation but their enzymatic activities are profoundly affected. In particular, only the phosphorylated forms of myosins IA and IB have actin-activated ATPase activity although the Ca^{2+}- and K^+/EDTA-ATPase activities of these enzymes are the same whether or not they are phosphorylated.

Phosphorylation of myosins IA and IB appears to reverse an inhibition caused by the unphosphorylated serine because actin-activated ATPase activity can also be obtained by proteolytically removing the phosphorylatable domain from the catalytic domain (Maruta and Korn, 1981). With the availability of a homogeneous preparation of Acanthamoeba myosin I heavy chain kinase (Hammer et al., 1983), it has been possible to explore further the properties of the phosphorylated myosins IA and IB. In summary, we find (Albanesi et al., 1983) that both the phosphorylated and unphosphorylated myosins bind equally tightly to F-actin in the presence of Mg^{2+}-ATP, i.e., under conditions identical to those of the enzymatic assay, and that the binding is very tight (K_D of about 0.1 μM for myosin IA and 0.2 μM for myosin IB). These dissociation constants are also very similar to the K_{ATPase} values for actin concentration required for half-maximal ATPase activity) which are about 0.2 and 0.3 μM for myosins IA and IB, respectively. From these and other data it has been concluded that phosphorylation regulates the actin-activated ATPase activity of myosins IA and IB through an effect on one or more steps of the catalytic cycle rather than by affecting the binding of the myosins to actin.

Myosin II, the third isoenzyme in the Acanthamoeba, is of a very different, and more conventional, structure. The molecule has a native molecular weight of about 450,000 by sedimentation equilibrium analysis (Kuznicki, Albanesi, and Korn, unpublished) and contains six polypeptides: two of molecular weight 185,000, two of about 17,500, and two of about 17,000 (Maruta and Korn, 1977a). In the electron microscope, myosin II is visualized as two globular heads attached to a coiled-coiled tail and, under appropriate ionic conditions, it forms the bipolar filaments (Collins et al., 1982b) characteristic of all known myosins, except for myosins IA and IB from Acanthamoeba. Myosin II is also phosphorylated on its heavy chains but at three serine residues on each heavy chain (Collins and Korn, 1981; Cote et al., 1981). In contrast to myosins IA and IB, however, it is the dephosphorylated form of myosin II that exhibits actin-activated ATPase activity while the fully phosphorylated molecule is inactive under all conditions we have tested. It is not possible to determine a true K_D for the binding of myosin II to F-actin under assay conditions because the myosin is filamentous, but available evidence (Collins et al., 1982b) suggests that the effects of phosphorylation on the K_D and the K_{ATPase} do not explain the effects of phosphorylation on the actin-activated ATPase activity of myosin II. Therefore, as for myosins IA and IB, it must be assumed that phosphorylation regulates the actin-activated ATPase activity of Acanthamoeba myosin II by affecting one or more steps in the catalytic cycle.

In the remainder of this paper, we will describe some of our recent studies that suggest that the regulation of myosin II is complex and cannot be explained by a simple intramolecular process, i.e., the catalytic site on a given heavy chain is not directly regulated by the state of phosphorylation of the serine residues on the same chain, but rather, at least in part, by a supramolecular effect of the state of phosphorylation on the conformation of the myosin filament as a unit.

We previously showed that the three phosphorylation sites on a single heavy chain are grouped very close together, within a region of 3,000-9,000 daltons, very near the carboxyl end of the heavy chain and very far from the catalytic site in the globular head (Collins et al., 1982a). A 9,000-Dalton peptide that contains all three phosphorylation sites is now being sequenced (Cote and Korn, unpublished). The first 35 residues at the N-terminal end of the peptide (i.e., from the site of chymotryptic cleavage) contain hydrophobic amino acids at alternate third and fourth positions indicative of a coiled-coil structure. This region is followed by proline and glycine residues and the characteristic pattern of hydrophobic amino acids stops. The three phosphorylatable serines are probably contained in the subsequent domain (perhaps as small as 10-15 residues) about 30 residues from the C-terminus. We are faced then with the apparent paradox that the regulatory sites are very far removed from the catalytic and actin-binding sites and that the two domains are separated by a relatively rigid coiled-coil region.

The first clue to understanding the mechanism of regulation came with the observation that the specific activity of the actin-activated ATPase activity of myosin II increased as a function of the myosin concentration (Collins and Korn, 1981), suggesting that myosin-myosin interactions might be involved, either directly or indirectly. Also, early data showed that the actin-activated ATPase activity was very responsive to the Mg^{2+} concentration (Collins and Korn, 1981) indicative of a cooperative response such as filament formation. The optimal Mg^{2+} concentration increases with the ionic strength and pH as well as with the extent of phosphorylation of the myosin preparation. Finally, we determined that myosin II was, in fact, filamentous under all conditions in which actin-activated ATPase activity could be demonstrated (Kuznicki et al., 1983). This seems to eliminate the possibility that the regulatory sites might come into close proximity with the catalytic and actin-binding sites by the myosin molecule curling back on itself, as smooth muscle and thymus myosins seem to do under certain conditions.

Filaments made from dephosphorylated myosin II (the actin-activatable form) are more stable to the dissociating effects of high concentrations of salt and ATP than are filaments of phosphorylated myosin II (Collins et al., 1982b). But this difference is apparent only at concentrations of the effectors which are much higher than physiological and which inhibit the actin-activated ATPase activity of dephosphorylated myosin II. Under optimal assay conditions in vitro, however, the filaments formed by dephosphorylated myosin II are generally appreciably larger than those formed by phosphorylated myosin II (Kuznicki et al., 1983). For example, in 7 mM $MgCl_2$, dephosphorylated myosin II has a sedimentation coefficient of about 125 s while phosphorylated myosin II has a sedimentation coefficient of about 25 s (monomeric myosin II has a sedimentation coefficient of about 4.8 s). This difference in filament size does not, however, appear to account for the difference in the actin-activated ATPase activity of the two forms of myosin II because conditions can be found (3-4 mM $MgCl_2$) where both dephosphorylated and phosphorylated myosin filaments have the same sedimentation coefficient (about 20 s) yet differ greatly in enzymatic activity. Moreover, the enzymatic activities of small and large filaments of dephosphorylated myosin II are essentially the same.

The exciting result was obtained when the actin-activated ATPase activities of copolymers of dephosphorylated and phosphorylated myosin II were compared to the enzymatic activity of equivalent mixtures of their homopolymers (Kuznicki et al., 1983). First, sedimentation velocity experiments demonstrated that copolymers were, in fact, formed because only one species (with a sedimentation coefficient similar to that of phosphorylated myosin) was observed for 1:1 copolymers of dephosphorylated and phosphorylated myosin II under conditions in which the homopolymers had different sedimentation coefficients. Most interesting, the actin-activated ATPase activity of the 1:1 copolymer was only about 50% of the activity of a 1:1 mixture of the homopolymers. Consistent results were obtained with copolymers varying from 90% dephosphorylated myosin II to 90% phosphorylated myosin II. Thus, the conclusion seems inevitable that the presence of phosphorylated myosin II in the same filament with dephosphorylated myosin II decreases the enzymatic activity of the dephosphorylated molecules. If all the activity in the 1:1 copolymer, for example, were attributable to the dephosphorylated molecules they would be inhibited about 60% by the presence of phosphorylated molecules in the same filament. Alternatively, it might be that within the copolymer the heads of dephosphorylated and phosphorylated myosins are both active, in which case their average specific activity in a 1:1 copolymer would be about 20% of the specific activity of the molecules in the homopolymer of dephosphorylated myosin II.

Very recently, we have carried out similar studies with copolymers of dephosphorylated myosin II and trypsin-digested dephosphorylated myosin II (Kuznicki, Atkinson, and Korn, unpublished). Limited trypsin digestion produces a single cleavage in the heavy chain of myosin II forming a head-peptide of about 75,000 daltons and a tail-peptide of about 110,000 daltons (Collins et al., 1982a). Neither of the two myosin light chains is affected, as demonstrated by their unchanged electrophoretic mobilities both in SDS-gels and in urea-gels. Under nondenaturing conditions the trypsin-cleaved molecule remains intact and forms filaments that are similar to, if not indistinguishable from, filaments of intact myosin. The trypsin-cleaved dephosphorylated myosin retains full Ca^{2+}- and K^+/EDTA-ATPase activities but has no actin-activated ATPase activity at pH 7 at any concentration of $MgCl_2$.

The sedimentation coefficient of trypsin-cleaved myosin II under enzymatic assay conditions is more like that of phosphorylated myosin than like that of the dephosphorylated myosin from which the trypsin-cleaved enzyme was prepared. And, also similarly to the effect of phosphorylated myosin, copolymers of trypsin-cleaved dephosphorylated myosin and native dephosphorylated myosin have lower actin-activated ATPase activity than the sum of their homopolymers assayed separately. In fact, the effect of the trypsin-cleaved myosin is even more pronounced than that of phosphorylated myosin II. A 1:1 copolymer of trypsin-cleaved and native dephosphorylated myosin II has only about 20% of the actin-activated ATPase activity of dephosphorylated myosin alone at pH 7, 4 mM $MgCl_2$.

The results with trypsin-cleaved myosin II, then, fit into the pattern that suggests that for actin-activated ATPase activity to be expressed, the myosin II filament must have an appropriate conformation. This active conformation can apparently be disrupted by the presence of molecules phosphorylated at one or more of three closely grouped sites near the tail of the heavy chain or of molecules in which the heavy chains have been cleaved by trypsin into two still-associated domains.

Experiments in progress with chymotrypsin-cleaved myosin (Cote, Kuznicki, and Korn, unpublished) add more complexity to the story. Limited chymotrypsin cleavage removes, as mentioned above (Collins et al., 1981a), only a 9,000-Dalton peptide (containing the three phosphorylatable serines) from the tail of the heavy chains leaving the rest of the molecule, including the two light chains, unaffected. Filaments of chymotrypsin-cleaved myosin II have full Ca^{2+}- and K^+/EDTA-ATPase activities but no actin-activated ATPase activity. In this case, however, the chymotrypsin-cleaved myosin does not seem to affect appreciably the actin-activated ATPase activity of dephosphorylated myosin in copolymers. Thus, the small C-terminal portion of the heavy chain that can be removed by chymotrypsin inhibits actin-activated ATPase activity when it is phosphorylated but seems to be necessary in its dephosphorylated form to produce enzymatically active filaments.

Finally, the ability of phosphorylated myosin and trypsin-cleaved myosin to inhibit the actin-activated ATPase activity of native, dephosphorylated myosin in copolymers can be overcome by raising the Mg^{2+} concentration even though the ATPase activities of the homopolymers are not enhanced at higher concentrations of Mg^{2+}. Also, the actin-activated ATPase activity of homopolymers of trypsin-cleaved myosin II, which has no enzymatic activity at pH 7, is the same as that of native, dephosphorylated myosin II at pH 6.1 and 1 mM Mg^{2+} (and 1 mM ATP).

In summary, actin-activated ATPase activity of *Acanthamoeba* myosin II is regulated by phosphorylation of three serine residues closely grouped near the carboxyl terminus of the heavy chain. The totally dephosphorylated molecule has maximal actin-activated ATPase activity and the totally phosphorylated molecule has no actin-activated ATPase activity under the conditions tested thus far. The ATPase activity of filaments of partially phosphorylated myosin II or copolymers of dephosphorylated and phosphorylated myosin II is optimal at a higher Mg^{2+} concentration than is required for optimal actin-activated ATPase activity of the homopolymer of dephosphorylated myosin. The optimal Mg^{2+} concentration is higher at higher ionic strengths and lower at lower pH. Under identical conditions, filaments of dephosphorylated myosin are larger than filaments of phosphorylated myosin but there is no correlation between filament size and actin-activated ATPase activity. The actin-activated ATPase activity of dephosphorylated molecules of myosin II is inhibited when they are present in copolymers with either phosphorylated or trypsin-cleaved molecules. This inhibition can be at least partially overcome by raising the Mg^{2+} concentration. These, and other observations, suggest that phosphorylation inhibits the actin-activated ATPase activity of *Acanthamoeba* myosin II by affecting the conformation of the myosin filaments. In addition, there are pronounced, but little understood, effects of pH and Mg^{2+}. It remains to be established whether myosin heads attached to phosphorylated tails can have actin-activated ATPase activity.

REFERENCES

Albanesi, J. P., Hammer, J. A. III and Korn, E. D.: The interaction of F-actin with phosphorylated and unphosphorylated myosins IA and IB from Acanthamoeba castellanii. J. Biol. Chem. 258:10176-81, 1983.

Collins, J. H., Cote, G. P. and Korn, E. D.: Localization of the three phosphorylation sites on each Acanthamoeba myosin II heavy chain to a segment at the end of the tail. J. Biol. Chem. 257:4529-34, 1982a.

Collins, J. H. and Korn, E. D.: Actin-activation of Ca^{2+}-sensitive Mg^{2+}-ATPase activity of Acanthamoeba myosin II is enhanced by dephosphorylation of its heavy chains. J. Biol. Chem. 255:8011-14, 1980.

Collins, J. H. and Korn, E. D.: Purification and characterization of actin-activatable, Ca^{2+}-sensitive myosin II from Acanthamoeba. J. Biol. Chem. 256:2586-95, 1981.

Collins, J. H., Kuznicki, J., Bowers, B. and Korn, E. D.: Comparison of the actin binding and filament formation properties of phosphorylated and dephosphorylated Acanthamoeba myosin II. Biochemistry 21:6910-15, 1982b.

Cote, G. P., Collins, J. H. and Korn, E. D.: Identification of three phosphorylation sites on each heavy chain of Acanthamoeba myosin II. J. Biol. Chem. 256:12811-16, 1981.

Gadasi, H. and Korn, E. D.: Immunochemical analysis of Acanthamoeba myosins IA, IB, and II. J. Biol. Chem. 254: 8095-98, 1979.

Gadasi, H., Maruta, H., Collins, J. H. and Korn, E. D.: Peptide maps of myosin isoenzymes of Acanthamoeba castellanii. J. Biol. Chem. 254:3631-36, 1979.

Hammer, J. A. III, Albanesi, J. P. and Korn, E. D.: Purification and characterization of a myosin I heavy chain kinase from Acanthamoeba castellanii. J. Biol. Chem. 258:10168-75, 1983.

Kuznicki, J., Albaneis, J. P., Cote, G. P. and Korn, E. D.: Supramolecular regulation of the actin-activated ATPase activity of filaments of Acanthamoeba myosin II. J. Biol. Chem. 258:6011-14, 1983.

Maruta, H., Gadasi, H., Collins, J. H. and Korn, E. D.: Multiple forms of Acanthamoeba myosin I. J. Biol. Chem. 254:3624-30, 1979.

Maruta, H. and Korn, E. D.: Acanthamoeba myosin II. J. Biol. Chem. 252:6501-09, 1977a.

Maruta, H. and Korn, E. D.: Acanthamoeba cofactor protein is a heavy chain kinase required for actin-activation of the Mg^{2+}-ATPase activity of Acanthamoeba myosin I. J. Biol. Chem. 252:8329-32, 1977b.

Maruta, H. and Korn, E.: Proteolytic separation of the actin-activatable ATPase from the phosphorylation site on the heavy chain of Acanthamoeba myosin IA. J. Biol. Chem. 256:503-06, 1981.

53 Modification of the Ca++ Requirement for Cardiac Myofibrillar Activation: A Cardiotonic Principle?

J. W. Herzig, L. H. Botelho, R. J. Solaro

INTRODUCTION

During the last years, increasing evidence has accumulated that the relationship between sarcoplasmic Ca^{++} concentration and myocardial activation is not invariant but is subject to intracellular regulation. The still often used simplification of considering myocardial force as a monitor of the intracellular free Ca^{++} concentration appears therefore increasingly unjustified. Alterations in the Ca^{++} sensitivities of myocardial myofibrillar ATPase (Ray and England, 1976) or force development of functionally isolated contractile structures from heart muscle (McClellan and Winegrad, 1978; Herzig et al., 1981b) have been attributed to corresponding differences in the phosphorylation degree of troponin I (Tn I). This phosphorylation reaction is governed by the opposing actions of a cAMP dependent protein kinase and a phosphatase, cAMP forming a link with adrenoceptors and the sympathetic nervous system. As cAMP alone, after application to cardiac myofibrils or "skinned," demembranated myocardial cells, induces Tn I phosphorylation and decreases Ca^{++} sensitivity, the cAMP dependent protein kinase seems to be an intrinsic constituent of the contractile structures (McClellan and Winegrad, 1978; Herzig et al., 1981b). The negative inotropic effect resulting from Tn I phosphorylation is, during adrenergic stimulation of the heart, superimposed by an increase in Ca^{++} influx due to cAMP dependent phosphorylation of membrane proteins (Reuter and Scholz, 1977). At the same time, phospholamban phosphorylation induces an increase in the Ca^{++} sequestration rate of the sarcoplasmic reticulum (SR) which is assumed to be responsible for the observed acceleration of relaxation due to catecholamine stimulation (Tada et al., 1974). Phosphorylation of both Tn I and phospholamban appears as a kind of defense mechanism limiting energy consumption in the course of adrenergic activation. The relaxant effect of Tn I phosphorylation and concomitant Ca^{++} desensitization is most directly visualized by the ability of catecholamines to relax cardiac K^+ contracture after exclusion of SR pumping activity at constant sarcoplasmic Ca^{++} concentration (Marban et al., 1980).

In the light of this knowledge, it may be hypothesized that at least certain forms of cardiac insufficiency may be the result of a decrease in Ca^{++} sensitivity of the contractile structures rather than of a lack of activator Ca^{++}. Evidence for this hypothesis is derived from the observation that experimental cardiac insufficiency induced by acute administration of isoprenaline in μM concentrations coincides with Tn I phosphorylation (Solaro et al., 1980). Furthermore, Allen et al. showed in 1982 that heart failure due to hypoxia occurs at unaltered sarcoplasmic Ca^{++} concentration. Consequently, it appears justified to consider mechanisms by which the Ca^{++} sensitivity of the contractile structures is increased (or brought back to normal) as a more causal approach to the treatment of at least certain forms of cardiac failure. APP 201-533 (3-Amino-6-methyl-5-phenyl-2(1H)-pyridinone) is a novel cardiotonic agent which exerts its positive inotropic action (Salzmann et al., 1983) neither by beta adrenoceptor stimulation nor via inhibition of the Na^+/K^+ ATPase (Herzig et al., 1983). We therefore investigated the influence of APP 201-533 upon the responsiveness of functionally isolated contractile structures and myofibrils from heart muscle to Ca^{++} ions at well defined, buffered concentrations.

MATERIALS AND METHODS

1. Detergent treated myocardial strips

Subendocardial muscle fiber bundles (∿4 mm in length, ∿0.2 mm in diameter) were prepared from trabecula septo marginalis from freshly slaughtered pigs. The preparations were then shaken at 4°C for 24 hours in a solution consisting of 50% (v/v) glycerol, 20 mM histidine HCl, 10 mM NaN_3, 0.5% (w/v) Lubrol WX, pH 7.3. The preparations were then stored until use at -18°C in 50% (v/v) glycerol, 20 mM histidine HCl, 10 mM NaN_3, pH 7.0. For force measurements, the preparations were attached to an isometric force transducer (AME AE 801 OEM) with a stiffness better than 1 mN/μm, by means of a fast setting glue (cellulose nitrate dissolved in acetone). The preparations were relaxed in a solution containing 10 mM $ATPNa_2$, 12 mM $MgCl_2$, 5 mM EGTA, 20 mM imidazole, 5 mM NaN_3, 60 mM KCl, 5 mM phosphoenole pyruvate, 4 U/ml pyruvate kinase from rabbit muscle, pH 6.7 at 22°C, free Ca^{++} concentration 1.66×10^{-10}M. Contraction was induced by replacing EGTA with Ca-EGTA or by mixing both in distinct ratios. The resulting free Ca^{++} concentrations (up to 4.15×10^{-5}M) were calculated using a computer program which is a modified version of a program used by D. C. S. White, York, based on the one published by Perrin and Sayce in 1967.

2. Preparation of cardiac myofibrils

Myofibrils were prepared from dog hearts using Triton X-100 according to Solaro et al. (1971).

3. Measurement of myofibrillar ATPase

The Ca^{++} activated, Mg dependent ATPase of the myofibrils was measured by determination of liberated inorganic phosphate (cf. also the following paragraph).

4. Myofibrillar Ca^{++} binding

Myofibrillar Ca^{++} binding was determined by a centrifugation method using ^3H glucose as solute space marker. Solutions for ATPase measurements and Ca^{++} binding studies were identical with the exception that for binding studies 0.3 MCi/1 $^{45}Ca^{++}$ and 0.3 mCi/1 ^3H glucose were added. Other conditions were: 0.5-1.0 mg myofibrillar protein/ml (determined according to Lowry et al., 1951), 65 mM KCl, 40 mM imidazole, 3 mM $MgCl_2$, 2 mM $ATPNa_2$, 0-1 mM $CaCl_2$, 0-1 mM EGTA, pH 7.0 at 30°C. Details of the procedures for ATPase measurement and Ca^{++} binding were as described by Solaro and Shiner (1976).

5. cAMP dependent protein kinase assay

The cAMP dependent protein kinase activity was determined in a 48,000 × g supernatant from homogenized rat liver, by a slightly modified version (MES instead of MOPS buffer) of the phosphocellulose absorption method described by Witt and Roskoski in 1975, using Histone f_2b as substrate.

RESULTS

Influence of APP 201-533 on the Ca^{++} Activated Force in Detergent Treated Myocardial Strips

Fiber bundles from porcine trabecula septo marginalis were extracted and Ca^{++} activated as described in the Methods. According to the free Ca^{++} concentration in the incubation bath which, after a brief equilibration period, equalled the "intracellular" Ca^{++} concentration in the immediate surroundings of the contractile structures, sustained contractions were elicited (Fig. 53.1a, inset), resulting in the well-known sigmoidal relationship between force and pCa, i.e., the negative logarithm of the molar Ca^{++} concentration (Fig. 53.1b). Half maximal activation (EC_{50}) was reached at approximately pCa 6.075, full relaxation near pCa 7.5, and full activation at pCa 4.8. Maximum Ca^{++} activated force ranged from 250 to 1,000 µN, depending on preparation diameter (100 to 300 µm). In all experiments, 0.5% (v/v) of 1-Methyl-2-pyrrolidone were present. This organic solvent is necessary to keep APP 201-533 in solution at near neutral pH. Application of APP 201-533 (but not Amrinone, cf. Fig. 53.1a, inset) in a concentration of 5×10^{-4}M led to a leftward shift of the Ca^{++} activation curve, displacing the EC_{50} to lower concentrations by 0.225 pCa units (Fig. 53.1b), without significantly affecting relaxation or maximal Ca^{++} activation. This leftward shift was reversible and concentration dependent in concentrations of APP 201-533 exceeding 10^{-4}M (Fig. 53.1a) and did not seem to be saturated at a concentration of 2×10^{-3}M where the upper limit of solubility is reached. Such a Ca^{++} sensitization may be the result of an increase in the Ca^{++} affinity of the regulator protein troponin C. We therefore investigated the influence of APP 201-533 on myofibrillar Ca^{++} binding.

Influence of APP 201-533 on Myofibrillar Ca^{++} Binding

Canine cardiac myofibrils were isolated as described in the Methods. Using a centrifugation technique, binding of $^{45}Ca^{++}$ to the myofibrils was measured as a function of pCa (Fig. 53.2a). In parallel, the Ca^{++} activated, Mg dependent ATPase of the myofibrils was measured in absence and presence of 5×10^{-4}M APP 201-533, as a monitor for drug action. Typical values at pCa 7.5 and pCa 5.25 were 12-15 and 90-100 nmoles P_i per mg protein per minute at 30°C. It is shown in Figure 53.2b that,

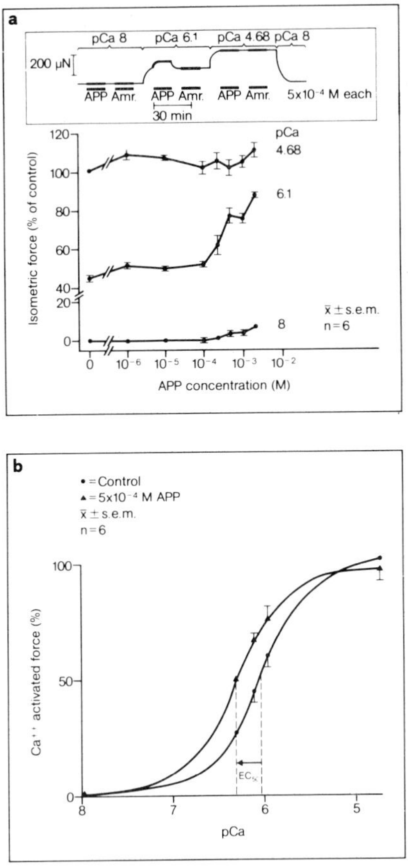

Fig. 53.1. Influence of APP 201-533 on the Ca^{++} sensitivity of force in detergent treated cardiac muscle. (a) Reversible "extra activation" at constant Ca^{++} concentration by APP 201-533. Note: Relaxation (pCa 8) and maximum force (pCa 4.68) are not influenced. Amrinone is ineffective in this model (inset). (b) Left shift of the Ca^{++} activation curve by 5×10^{-4}M APP 201-533. For experimental conditions see Methods.

similar to the results obtained in detergent treated cardiac fiber bundles (cf. Fig. 53.1b), there was a marked leftward shift of the Ca^{++} activation curve for myofibrillar ATPase, by about 0.25 pCa units. This coincides with a significant leftward displacement of the Ca^{++} binding curve (Fig. 53.2a), showing an increase in the amount of myofibrillar bound $^{45}Ca^{++}$ at a given Ca^{++} concentration. A similar observation has been reported by Holroyde et al. in 1979 as the result of treatment of cardiac myofibrils with alkaline phosphoprotein phosphatase. As the phosphorylation degree of Tn I which determines the Ca^{++} affinity of troponin C is governed by the opposing actions of cAMP dependent protein kinase and phosphoprotein phosphatase and as furthermore the Ca^{++} sensitizing effect of APP 201-533 resembles the action of phosphatase, we investigated the influence of the compound on the activity of cAMP dependent protein kinase.

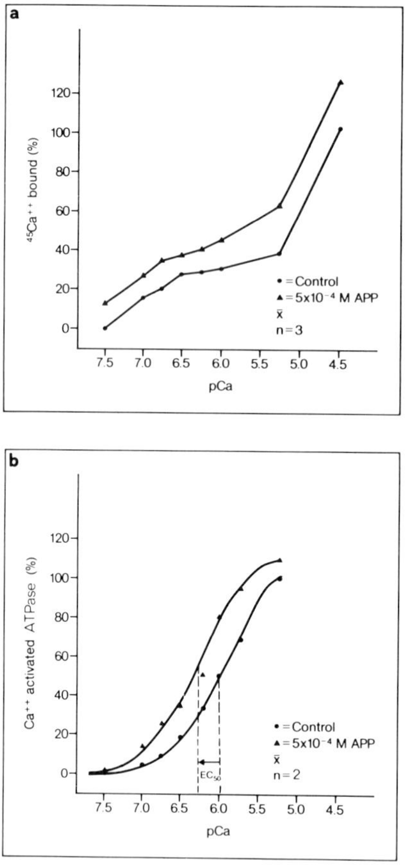

Fig. 53.2. Influence of APP 201-533 on the Ca^{++} sensitivities of $^{45}Ca^{++}$ binding (a) and Ca^{++} activated ATPase (b) in cardiac myofibrils. For experimental conditions see Methods.

Influence of APP 201-533 on the Activity of cAMP Dependent Protein Kinase from Rat Liver

Using Histone f_2b as substrate for phosphorylation, the activity of cAMP dependent protein kinase from rat liver supernatant was measured by means of a filter assay, in absence and presence of various concentrations of APP 201-533. In Figure 53.3, it is shown that the compound concentration dependently reduced the activity of the cAMP dependent protein kinase, although the inhibition was incomplete and amounted to maximally 60% decrease in activity at 10^{-3}M. Half maximal inhibition was reached at a drug concentration smaller than 10^{-4}M.

DISCUSSION

As shown in the present study, APP 201-533 increases the Ca^{++} sensitivities of force and myofibrillar ATPase. This is paralleled by enhanced Ca^{++} binding to the myofibrils. A similar observation has been reported by Solaro and Rüegg in 1982 for Sulmazole (AR-L 115 BS), another compound with Ca^{++} sensitizing properties (Herzig et al., 1981a). It may be stated here that the cardiotonic agents Amrinone and Milrinone which are structurally related to APP 201-533, show no Ca^{++} sensitizing activity in our model (unpublished observations, cf. also Fig. 53.1a, inset). Furthermore, APP 201-533 has an inhibitory action on cAMP dependent protein kinase from rat liber. A cAMP dependent protein kinase is an intrinsic constituent of myocardial contractile structures and is not removed from the contractile system by the detergent treatment used in this study (cf. also Herzig et al., 1981b).

As cAMP dependent phosphorylation of Tn I is known to reduce the Ca^{++} affinity of troponin C and vice versa (Holroyde et al., 1979), we propose that the Ca^{++} sensitizing action of APP 201-533 may be based on a shift in the relative activities of the enzyme couple of cAMP dependent protein kinase and phosphoprotein phosphatase, thereby favoring rroponin I to be dephosphorylated, i.e., to be in a state where the Ca^{++} affinity of troponin C is enhanced. Assuming that at least certain forms of cardiac insufficiency (e.g., hypoxia or isoprenaline induced cardiac failure) may be due to a decrease in Ca^{++} sensitivity of the contractile structures rather than a lack of Ca^{++} available for activation, the concept of Ca^{++} sensitization of the contractile proteins, possibly via inhibition of Tn I phosphorylation, with compounds of the type of APP 201-533 appears to provide a more causal approach to the treatment of at least certain forms of heart failure. Yet, as the described Ca^{++} sensitizing properties of APP 201-533 are only observed in concentrations presumably exceeding therapeutically obtainable plasma levels and as

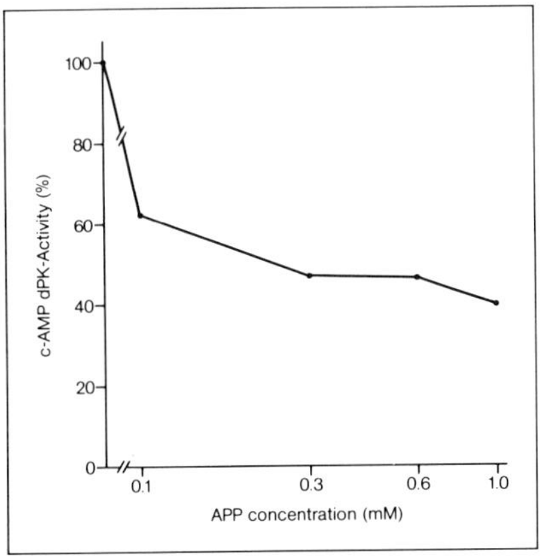

Fig. 53.3. Inhibitory effect of APP 201-533 on the activity of cAMP dependent protein kinase (cAMP dPK) from rat liver supernatant. For experimental details see Methods.

the local intracellular concentrations in intact cardiocytes governed by possible accumulation and compartmentalization are not known, it must remain an open question whether the cardiotonic action of the compound in vivo (cf. Salzmann et al., 1983) is based on Ca^{++} sensitization. The search for more potent compounds with comparable mechanism of action must progress.

ACKNOWLEDGMENT

The authors would like to thank Dr. D. C. S. White, Department of Biology, University of York, Heslington, UK, and Dr. R. Hummel, SANDOZ Ltd., Basel, Switzerland, for their help in establishing the computer program used in this study to calculate concentrations of free Ca^{++} and various complex species in the experimental solutions.

REFERENCES

Allen DG, Kurihara S and Orchard CH (1982): The effects of hypoxia on intracellular calcium transients in mammalian cardiac muscle. J. Physiol. (Lond.) 328, 22-23P.

Herzig JW, Feile K and Rüegg JC (1981a): Activating effects of AR-L 115 BS on the Ca^{++} sensitive force, stiffness and unloaded shortening velocity (V_{max}) in isolated contractile structures from mammalian heart muscle. Arzneim. Forsch./ Drug res. 31, 188-91.

Herzig JW, Köhler G, Pfitzer G, Rüegg JC and Wölffle G (1981b): Cyclic AMP inhibits contractility of detergent treated glycerol extracted cardiac muscle. Pflüger's Arch. Eur. J. Physiol. 391, 208-212.

Herzig JW, Bormann G, Botelho L, Erdmann E, Salzmann R and Solaro RJ (1983): Intracellular actions of APP 201-533, a novel cardiotonic agent: increase in Ca^{++} sensitivity and economization of the myocardial contractile process. J. Molec. Cell. Cardiol. 15 (Suppl. 1), 244.

Holroyde MJ, Howe E and Solaro RJ (1979): Modification of the Ca^{++} requirements for activation of cardiac myofibrillar ATPase by cyclic AMP dependent phosphorylation. Biochim. Biophys. Acta 586, 63-69.

Lowry OH, Rosebrough ND, Farr AC and Randall RJ (1951): Protein measurements with the Folin phenol reagent. J. Biol. Chem. 193, 265-71.

Marban E, Rink TJ, Tsien RW and Tsien RY (1980): Free calcium in heart muscle at rest and during contraction measured with Ca^{++} sensitive microelectrodes. Nature 286, 845-50.

McClellan G and Winegrad S (1978): The regulation of calcium sensitivity of the contractile system in mammalian cardiac muscle. J. Gen. Physiol. 72, 737-67.

Perrin DD and Sayce IG (1967): Computer calculation of equilibrium concentrations in mixtures of metal ions and complexing species. Talanta 14, 833-42.

Ray KP and England PJ (1976): Phosphorylation of the inhibitory subunit of troponin and its effects on the calcium dependence of cardiac myofibrillar ATPase. FEBS Lett. 70, 11-16.

Reuter H and Scholz H (1977): The regulation of the calcium conductance of cardiac muscle by adrenaline. J. Physiol. (Lond.) 264, 49-62.

Salzmann R, Bormann G and Scholtysik G (1983): APP 201-533, a new positive inotropic drug. Naunyn Schmiedeberg's Arch. Pharmacol. 324 (Suppl.), R 35.

Solaro RJ, Pang DC and Briggs FN (1971): The purification of cardiac myofibrils with Triton X-100. Biochim. Biophys. Acta 245, 259-62.

Solaro RJ and Shiner JS (1976): Modulation of the Ca^{++} control of dog and rabbit cardiac myofibrils by Mg^{++}: comparison with rabbit skeletal myofibrils. Circ. Res. 39, 8-14.

Solaro RJ, Holroyde MJ, Herzig JW and Peterson JW (1980): Cardiac relaxation and myofibrillar interactions with phosphate and vanadate. Europ. Heart J. 1 (Suppl. A), 21-27.

Solaro RJ and Rüegg JC (1982): Stimulation of Ca^{++} binding and ATPase activity of dog cardiac myofibrils by AR-L 115 BS, a novel cardiotonic agent. Circ. Res. 51, 290-94.

Tada M, Kirchberger MA and Katz AM (1974): Phosphorylation of a 22,000 dalton component of the cardiac sarcoplasmic reticulum by adenosine 3':5'-monophosphate-dependent protein kinase. J. Biol. Chem. 250, 2640-47.

Witt JJ and Roskoski RJr (1975): Rapid protein kinase assay using phosphocellulose paper absorption. Anal. Biochem. 66, 253-58.

54 Effect of Light Chain Phosphorylation on the Monomer–Polymer Equilibrium of Thymus Myosin

Robin Smith and John Kendrick-Jones

INTRODUCTION

The level of phosphorylation of the 20,000 M_r (regulatory) light chains of bovine thymus controls the state of activation of this myosin. Phosphorylation of the light chain by a Ca^{2+} calmodulin-light chain kinase complex switches on the myosin and allows it to interact with actin and thus generate force. Dephosphorylation of the regulatory light chain by a light chain phosphatase induces the disassociation of the myosin from the actin and hence causes relaxation. Nonphosphorylated myosin in physiological conditions (150 mM NaCl, 5 mM $MgCl_2$, 25 mM imidazole pH 7.0) exists in a polymeric form or filament and MgATP in stoichiometric amounts can disassemble these filaments into their constituent monomers. Phosphorylation of the regulatory light chain induces these disassembled myosins to reassemble into filaments (Scholey et al. 1982). Thus, light chain phosphorylation-dephosphorylation not only affects the interaction of the myosin with actin but also the interaction between the myosin molecules themselves.

Electron microscopical examination of the phosphorylated and the dephosphorylated myosin molecules in the presence of MgATP after rotary shadowing has shown that the myosin monomer exists in two conformations. In phosphorylated myosin the rod portion of the molecule has an extended conformation and the molecule has an $S_{20,w}$ = 6S; whereas the nonphosphorylated myosin monomer in MgATP has its rod section folded at two hinge regions into a compact structure with an $S_{20,w}$ = 10S (Craig et al. 1983) (Fig. 54.1). These extended and folded conformations have also been observed in smooth muscle myosin (Trybus et al. 1982; Onishi et al. 1982; Craig et al. 1983).

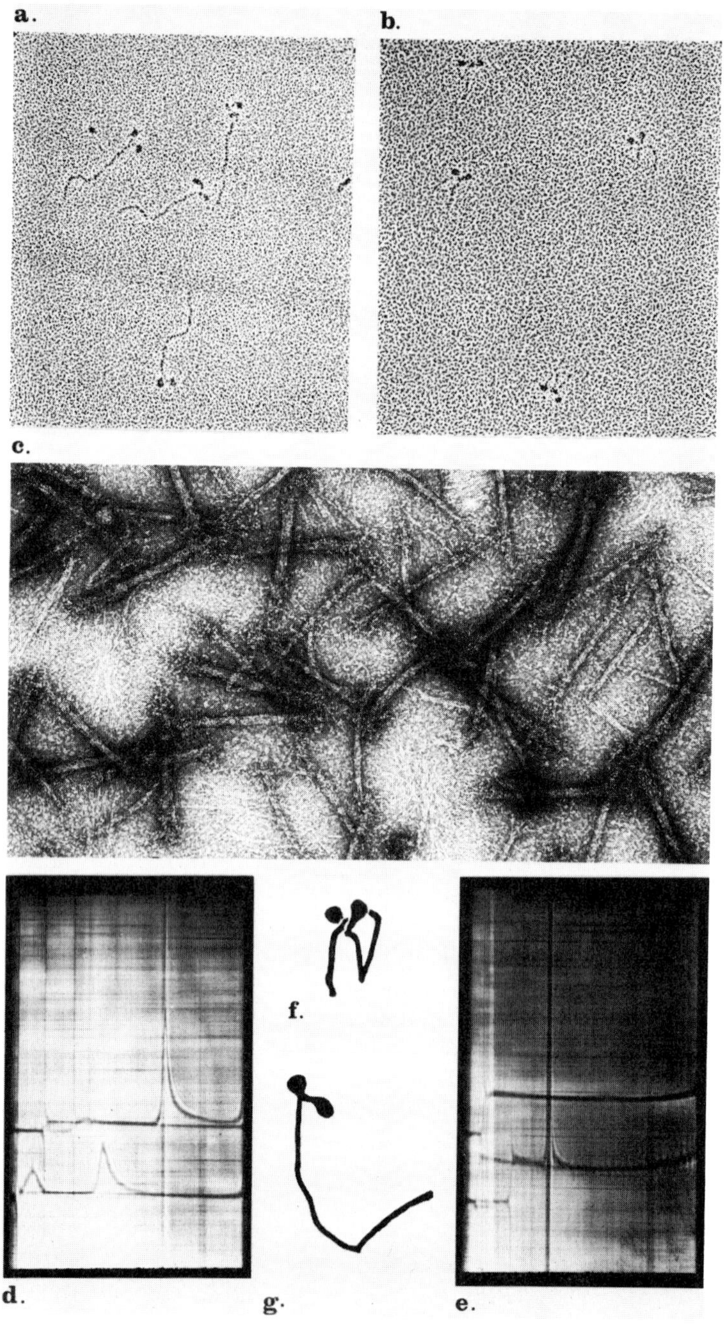
a.
b.
c.
f.
d.
g.
e.

In skeletal muscle myosin it has been shown (Josephs and Harrington 1966, 1968) that the process of filament polymerization is consistent with a monomer ↔ polymer equilibrium obeying the conditions described by Gilbert (1955, 1958, 1959) in which the rate of attainment of equilibrium is rapid. It was further shown that the equilibrium system could be considered to be composed of two basically homogenous groups such that:

$$n. \text{ monomers} \rightarrow \text{polymer} \tag{1}$$

and

$$K_{eq} = \frac{[\text{polymer}]}{[\text{monomers}]^n} = \frac{[P]}{[M]^n} \tag{2}$$

A characteristic feature of this system is that as the total protein concentration (C_t) increases, the concentration of free monomer in solution (C_m) increases up to a fixed "critical concentration" after which it remains constant, whereas the polymer concentration (C_p) then continues to increase in parallel with C_t.

Since thymus myosin has two different monomer species (6S and 10S), interconvertible by phosphorylation, we have investigated the equilibrium characteristics of both phosphory-

Fig. 54.1. Monomers and Polymers of Thymus Myosin. (a) Rotary shadowed electron micrograph of extended, 6S, thymus myosin molecules in 150 mM NaCl that have been phosphorylated by myosin light chain kinase in the presence of MgATP. Sample taken from an assembly assay before the filaments have formed. (b) Rotary shadowed electron micrograph of folded, nonphosphorylated 10S thymus myosin molecules in 150 mM NaCl in the presence of MgATP. (c) Nonphosphorylated thymus myosin filaments (polymers) in 150 mM NaCl, pH 7.0. Length ∿ 0.43 μm. $S_{20,w}$ = 50-60S. (d) Schlieren pattern of thymus myosin in 150 mM NaCl pH 7.0. 20,000 r.p.m. Top pattern: no ATP present. Polymer peak 50-60S. Bottom pattern: +2.5 mM ATP. Polymer peak (right) 50-60S; folded monomer peak (left) 10S. Ultracentrifugation was performed in a Beckman Model E analytical ultracentrifuge operated at 7-12°C. Protein concentration = 2.5 mg/ml^{-1}. (e) Schlieren pattern of thymus myosin in 0.6 M NaCl pH 7.0. 56,000 r.p.m. Bottom pattern only: 6S monomer peak. (f) Tracing of a folded 10S monomer. (g) Tracing of an extended 6S monomer.

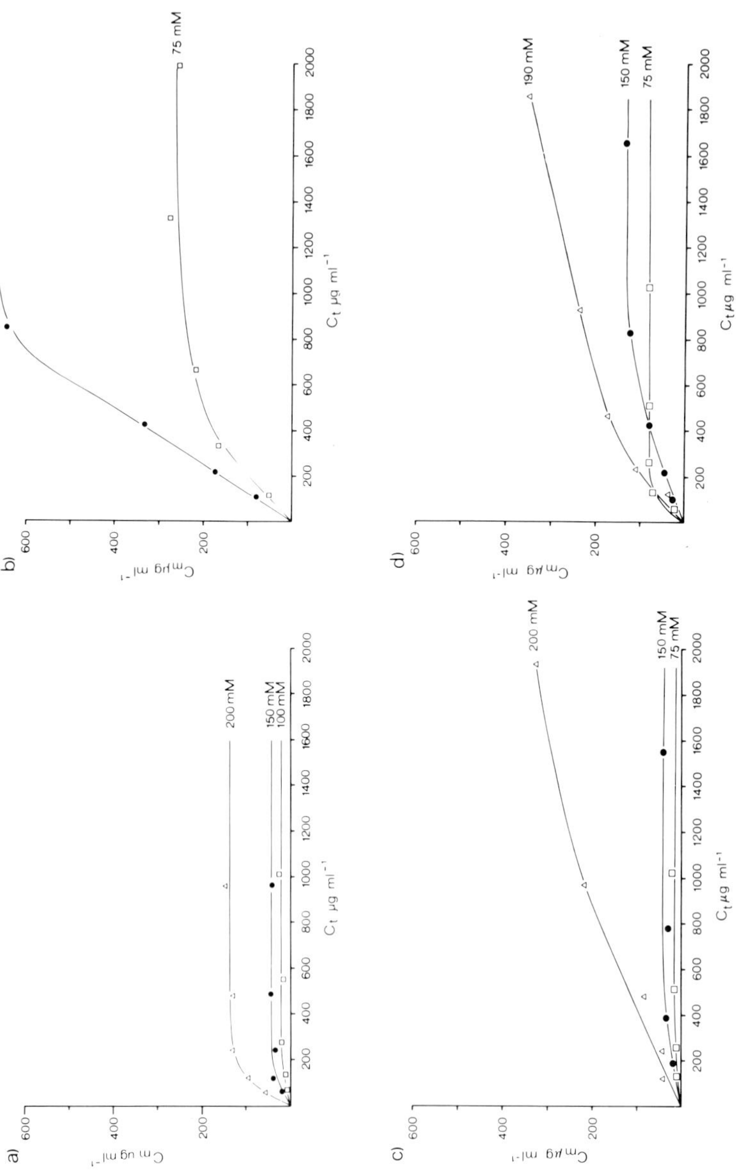

a)
C_m μg ml^{-1}
C_t μg ml^{-1}
200 mM
150 mM
100 mM
b)
C_m μg ml^{-1}
C_t μg ml^{-1}
150 mM
75 mM
c)
C_m μg ml^{-1}
C_t μg ml^{-1}
200 mM
150 mM
75 mM
d)
C_m μg ml^{-1}
C_t μg ml^{-1}
190 mM
150 mM
75 mM

lated and nonphosphorylated myosin under differing conditions in order to help shed light on the mechanism by which phosphorylation exerts its effect on filament stability.

METHODS

Purified bovine thymus myosin was used for all these experiments and was prepared by the ammonium sulphate procedure described in Scholey et al. (1982). Gizzard myosin light chain kinase and thymus calmodulin were purified according to the methods described in Kendrick-Jones et al. (1982) and Adelstein and Klee (1981), respectively. The myosin was thiophosphorylated rather than phosphorylated to prevent dephosphorylation occurring during the experiments. To obtain thiophosphorylated myosin, the myosin was incubated at a concentration of 7 mg/ml^{-1} with 4 mM ATP-γ-S, 0.6 mM $CaCl_2$, 5 mM $MgCl_2$, 15 $\mu g/ml^{-1}$ calmodulin and 10 $\mu g/ml^{-1}$ light chain kinase in 150 mM NaCl, 25 mM imidazole pH 7.0 for 30 minutes at room temperature. The thiophosphorylation state of the myosin was checked by using urea-glycerol polyacrylamide gel electrophoresis (Perrie and Perry 1970).

A dilution series of myosin in high salt buffer (0.6 M NaCl, 25 mM imidazole pH 7.0, 5 mM $MgCl_2$, 0.2 mM EGTA, 0.2 DTT) was made and the dilutions were dialyzed separately with or without 2.5 mM ATP against a buffered solution of known ionic strength (75-225 mM NaCl, 5 mM $MgCl_2$, 25 mM imidazole pH 7.0. 0.2 mM EGTA, 0.2 mM DTT) for 2 hours. After dialysis a sample of each solution was taken for total protein determination (C_t). The remainder was centrifuged at 24 p.s.i. in a Beckman Airfuge bench centrifuge for 20 minutes ($\sim$110,000 $\times$ g). This centrifugation sediments any polymers formed during dialysis and a sample of the supernatant, which now contains only soluble monomer, was taken for protein determination (C_m). This procedure is based on that proposed by Plllard (1982).

Fig. 54.2. Graphs of the value of C_m versus C_t for nonphosphorylated and thiophosphorylated thymus myosin at differing ionic strengths. The values of C_p have not been drawn on for clarity but can easily be obtained since $C_p = C_t - C_m$. The value where the C_m forms a plateau is the C_c for polymer formation. (a) Nonphosphorylated myosin. (b) Nonphosphorylated myosin + 2.5 mM ATP. (c) Thiophosphorylated myosin. (d) Thiophosphorylated myosin + 2.5 mM ATP.

Protein determinations were performed by the Bradford (1976) procedure and a bovine serum albumin standard curve was constructed for each set of readings to ensure accuracy. The protein concentrations are related by the following simple equation:

$$C_t = C_m + C_p \qquad (3)$$

where C_t is the total protein concentration, C_m is the monomer concentration (supernatant concentration after centrifugation), and C_p is the polymer (filament) concentration.

RESULTS AND DISCUSSION

Nonphosphorylated thymus myosin at 150 mM NaCl pH 7.0 assembled into filaments of length 0.43 μm + 0.05 μm (N = 100) (Fig. 54.1c). The filaments appear to be a mixture of bipolar (Niederman et al. 1975) and sidepolar filaments (Craig et al. 1977). When stoichiometric levels of MgATP are added, these nonphosphorylated filaments disassemble and in the electron microscope after rotary shadowing the molecules are seen to have a folded conformation (Fig. 54.1b,f). Phosphorylation of the light chains induces these molecules to unfold (Craig et al. 1983) (Fig. 54.1a) and reassemble into filaments.

The sedimentation profiles of these myosins in the analytical ultracentrifuge reveal that for a given set of conditions, basically only one polymeric, i.e., filament, and one monomeric species is seen. For example, at 150 mM NaCl pH 7.0 the polymeric species for nonphosphorylated myosin has an $S_{20,w}$ of 50-60S; whereas at 170 mM NaCl the polymer has an $S_{20,w}$ of about 40S. In the presence of ATP at an ionic strength from 75 mM to 170 mM nonphosphorylated myosin shows, apart from a polymer peak, a symmetrical peak with an $S_{20,w}$ of 9-12S (the folded monomer conformation) (Fig. 54.1d). Above 170 mM NaCl the folded form extends until, by 200 mM, it has the normal extended conformation with a sedimentation coefficient of $S_{20,w}$ = 6S. Phosphorylated thymus myosin is primarily filamentous so that the small concentration of monomer present cannot be detected by the schlieren optical system.

Using the sedimentation assay it can be shown that at physiological ionic strength in the presence and absence of ATP the myosin has a critical concentration. This observation, together with the information provided by the analytical ultracentrifuge that only one monomeric and polymeric species are present in the solution under these conditions, suggests that the polymerization of thymus myosin is a condensation process very similar to the skeletal myosin system detailed by Josephs and Harrington (1966).

The C_c for skeletal muscle myosin at 170 mM pH 8.3 is about 1 mg/ml^{-1} (Josephs and Harrington 1966) whereas for nonphosphorylated thymus myosin at the same ionic strength but at pH 7.0 it is 50 g/ml^{-1}. This large difference is probably due, at least in part, to the fact that the value of n in equation (1) is very much smaller for thymus myosin than for skeletal myosin since the filaments are smaller.

The striking effect that ATP has on nonphosphorylated myosin at 150 mM is shown by the increase in the C_c from 40 μg/ml^{-1} without ATP to 675 μg/ml^{-1} with ATP. This increase in C_c explains the previous observations that MgATP disassembles thymus myosin filaments (Scholey et al. 1980). We speculate that MgATP, by binding to the MgATPase site induces the myosin to fold into the 10S form (Scholey et al. 1980, Kendrick-Jones et al. 1981) which reduces the equilibrium constant K_{eq} and hence induces disassembly of the filaments. The 6S to 10S transition induced by MgATP presumably accounts for a large part of the reduction in K_{eq} since a folded 10S molecule would be unable to form a normal filament. Light chain phosphorylation, by promoting the unfolding of the 10S molecule, is able to increase the equilibrium constant so that filaments predominate in solution. Increasing ionic strength solubilizes the filaments, and this is reflected, as expected, by an increasing C_c as the salt concentration rises.

If phosphate buffer (KP_i pH 7.0) is used instead of imidazole it is found that at 150 mM KCl, without ATP, nonphosphorylated myosin is mainly filamentous, but that some monomer can be seen. The monomer peak sediments at 10S. So phosphate is able to promote the folding of the myosin monomer, but it is very much less effective than ATP at doing so.

All these studies have been performed in vitro and the precise relevance of these results to the living cell remains to be established. In particular, knowledge of the internal conditions of the cells would be very helpful in attempting to understand the position of the various equilibria for myosin filament formation in vivo. A hypothetical scheme for nonmuscle motility is shown in Figure 54.3. It would be interesting to establish the localized myosin concentrations within the cell, since, if the concentration anywhere was greater than about 675 μg/ml^{-1} (150 mM salt) then filaments would be present, whatever the phosphorylation state of the light chains. In the cell these

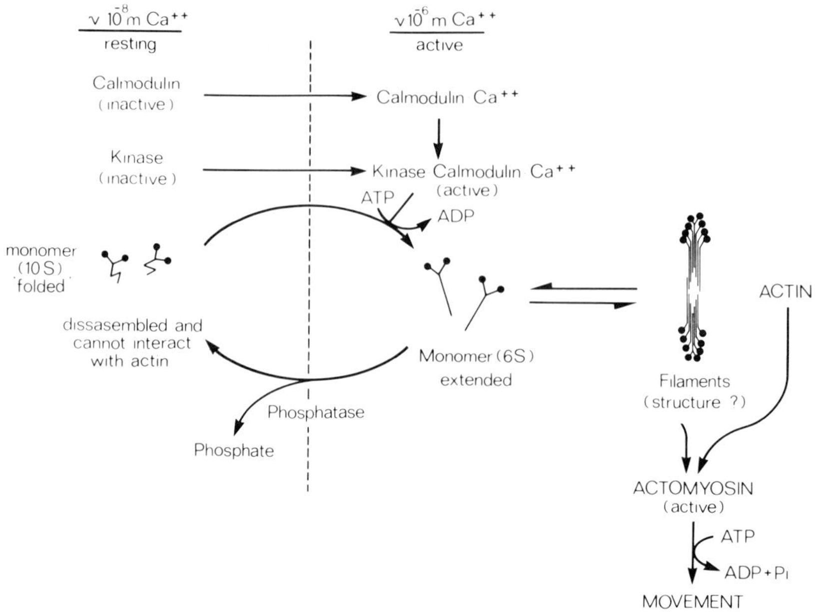

Fig. 54.3. A hypothetical scheme for nonmuscle motility. Incorporating the equilibria talked about in the text. Note this scheme assumes that assembled filaments are required for movement.

equilibria between the two monomeric myosins and the polymer could provide a sensitive method of regulation which might have an important role in modulating the force of motile events.

REFERENCES

Adelstein, R. S., Klee, C. B., 1981. Purification and characterisation of smooth muscle myosin light chain kinase. J. Biol. Chem. 256, 7501-09.

Bradford, M., 1976. A rapid and sensitive method for the quantitation of microgram quantities of protein utilising the principle of protein-dye binding. Anal. Biochem. 72, 248-54.

Craig, R., Megerman, J. 1977. Assembly of smooth muscle myosin into side-polar filaments. J. Cell. Biol. 75, 990-96.

Craig, R., Smith, R. C., Kendrick-Jones, J., 1983. Light chain phosphorylation controls the conformation of vertebrate non-muscle and smooth muscle myosin molecules. Nature. Lond. 302, 436-39.

Gilbert, G. A., 1955. In a general discussion (no title). Discussions Faraday. Soc. 20, 68-71.

Gilbert, G. A., 1959. Sedimentation and electrophoresis of interacting substances I. Idealised shape for a single substance aggregating reversibly. Proc. Roy. Soc. Lond. Ser A. 250, 354-66.

Gilbert, G. A., 1963. Sedimentation and electrophoresis of interacting substances III. Sedimentation of a reversibly aggregating substance with concentration dependent sedimentation coefficients. Proc. Roy. Soc. Lond. Ser A. 276, 354-66.

Josephs, R., Harrington, F. W., 1966. Studies on the formation and physical chemical properties of synthetic myosin filaments. Biochemistry. 5, 3474-87.

Josephs, R., Harrington, F. W., 1968. On the stability of myosin filaments. Biochemistry. 8, 2834-47.

Kendrick-Jones, J., Taylor, K. A., Scholey, J. M., 1982a. Phosphorylation of non-muscle myosin and stabilization of thick filament structure. Methods. Enzymol. 85, 364-73.

Kendrick-Jones, J., Jakes, R., Tooth, P. J., Craig, R., Scholey, J. M., 1982b. Role of myosin light chains in the regulation of contractile activity. In: Basic Biology of Muscles: a comparative approach (ed. B. M. Twarog, R. J. C. Levine and M. M. Bewey), pp 255-72. New York: Raven Press.

Niederman, R., Pollard, T. D., 1975. Human platelet myosin II. In vitro assembly and structure of myosin filaments. J. Cell. Biol. 67, 72-92.

Onishi, H., Wakabayashi, T., 1982. Electron microscopic studies of molecules from chicken gizzard muscle. J. Biochem. Tokyo, 92, 8718-879.

Perrie, N. T., Perry, S. V., 1970. An electrophoretic study of the low molecular weight components of myosin. Biochem. J. 119, 31-38.

Pollard, T. D., 1982. Structure and polymerisation of Acanthamoeba myosin II filaments. J. Cell. Biol. 95, 816-25.

Scholey, J. M., Taylor, K. A., Kendrick-Jones, J., 1980. Regulation of non-muscle myosin assembly by calmodulin-dependent light chain kinase. Nature. Lond. 287, 233-35.

Scholey, J. M., Smith, R. C., Drenckahn, D., Groschel-Stewart, U., Kendrick-Jones, J., 1982. Thymus myosin. J. Biol. Chem. 257, 7737-45.

Smith, R. C., Cande, W. Z., Craig, R., Tooth, P. J., Scholey, J. M., Kendrick-Jones, J., 1983. Regulation of myosin filament assembly by light chain phosphorylation. Phil. Trans. Roy. Soc. Lond. Ser B. 302, 73-82.

Trybus, K. M., Huiatt, T. W., Lowey, S., 1982. A bent monomeric conformation of myosin from smooth muscle. Proc. Nat. Acad. Sci. U.S.A. 79, 6151-55.

55 Protein Factor(s) from Chicken Gizzard Muscle Affecting the Mg^{2+}- ATPase Activity of Skeletal Muscle Actomyosin

Jan Sosiński, Adam Szpacenko, Renata Dabrowska

INTRODUCTION

Although smooth muscle regulation by phosphorylation of myosin light chains is widely accepted (Adelstein and Eisenberg, 1980; Walsh and Hartshorne, 1982) there are some indications for the existence of an alternative or complementary mechanism of regulation through the protein(s) associated with thin filaments. The alternative mechanism was proposed by Ebashi and his coworkers (1982), who think that the protein called leiotonin (composed of two subunits: A and C) is a Ca^{2+}-dependent activator of actomyosin ATPase. Leiotonin has no myosin light chain kinase activity. On the other hand, Marston (1982) claimed that a dual mechanism of regulation exists in smooth muscle through phosphorylation of myosin light chains (primary) and troponinlike proteins (secondary).

It has been shown previously that thin filaments isolated from pig aorta have the ability to activate skeletal muscle myosin Mg^{2+}-ATPase in a Ca^{2+}-sensitive way (Marston et al., 1980). We have noticed the same behavior in the thin filaments of chicken gizzard. The identification of the factor(s) responsible for this phenomenon is described in this paper.

RESULTS AND DISCUSSION

Gizzard thin filaments activate the Mg^{2+}-ATPase of rabbit skeletal muscle myosin with a Ca^{2+}-sensitivity of up to 70%. As shown in Figure 55.1, the ATPase activity of the gizzard thin filament and rabbit skeletal muscle myosin complex, measured in the presence of Ca^{2+}, significantly exceeded that of the complexes gizzard actin-skeletal muscle myosin or gizzard actin-

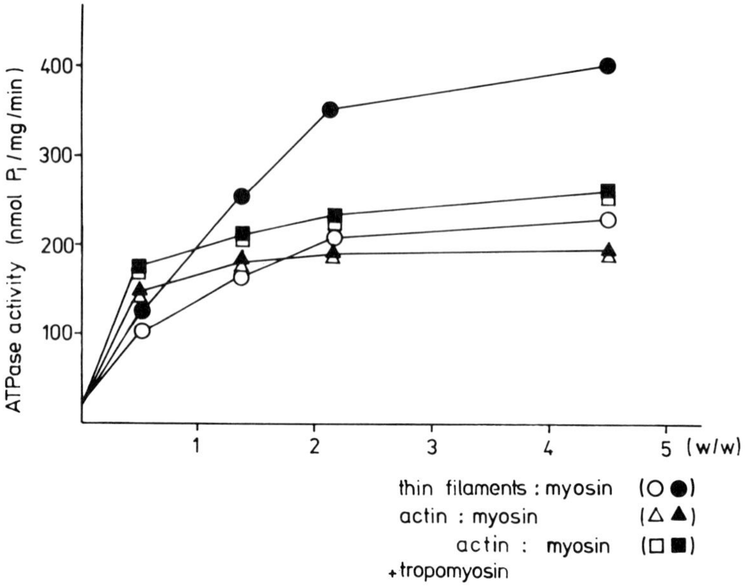

Fig. 55.1. The effect of chicken gizzard thin filaments (○, ●), gizzard actin (△, ▲) and gizzard actin-gizzard tropomyosin complex (□, ■) on Mg^{2+}-ATPase activity of rabbit skeletal muscle myosin. Assay conditions: 50 mM KCl, 2 mM ATP, 10 mM $MgCl_2$, 20 mM Tris-acetate pH 7.0, 30°C, rabbit skeletal muscle myosin 120 μg/ml, gizzard actin 80 μg/ml, gizzard tropomyosin:actin 1:7 (molar ratio). Open symbols indicate the inclusion of 2 mM EGTA.

gizzard tropomyosin-skeletal muscle myosin at the weight ratios of actin to myosin over 1:1. Since the ATPase activity, estimated in the absence of Ca^{2+}, for the thin filament-myosin complex and for the gizzard actin-gizzard tropomyosin-myosin complex was nearly the same, we have assumed that the thin filaments contained Ca^{2+}-sensitive activator.

Taking into account that myosin light chain kinase might be associated with thin filaments (Dabrowska et al., 1982) and that affects the interaction of thin filaments with myosin in the presence of Ca^{2+}, through phosphorylation of myosin (Pemrick, 1980), we checked the amount phosphorylated myosin light chain before and after the ATPase assay by urea gel electrophoresis. The stable level of myosin light chain phosphorylation (data not shown) indicated that the action of activator is not related to phosphorylation of myosin.

Thin filaments isolated from chicken gizzard muscle consist mainly of actin and tropomyosin, and in addition they contain a number of unidentified minor proteins with molecular weights lower and higher than actin. Digestion of thin filaments with trypsin at the ratio of enzyme to substrate 1:1,000 (w/w) produced a decrease in the activation of myosin Mg^{2+}-ATPase in the presence of Ca^{2+} to the level of ATPase produced by gizzard tropomyosin and actin; lower concentrations of enzyme (1:10,000) caused only partial degradation of activator. These experiments indicated the protein nature of the activator. However, from the SDS-gels of digested samples it was difficult to point out definitively which protein of the thin filaments is the activator, although it seems to be a protein with molecular weight higher than actin (data not shown).

When the proteins of gizzard thin filaments were fractionated with ammonium sulphate, the active fraction was found in the 0-40% saturated ammonium sulphate precipitate. Activation of the Mg^{2+}-ATPase of skeletal muscle actomyosin by this fraction was, however, Ca^{2+}-independent (Fig. 55.2). The 0-40% saturated ammonium sulphate fraction did not affect the Mg^{2+}-ATPase activity of myosin. The active fraction contained mainly denatured actin and a few minor proteins with molecular weights higher than actin. The fraction salted out between 40-70% saturated ammonium sulphate affected actomyosin ATPase only slightly. This was probably caused by tropomyosin—the main component of this fraction. Combination of both fractions, i.e., 0-40% and 40-70% saturated ammonium sulphate fractions, did not restore the Ca^{2+}-sensitivity of the activator effect on actomyosin ATPase. Addition of calmodulin to the fraction salted out at 0-40% saturated ammonium sulphate was also without effect (Fig. 55.2).

It was noticed that the supernatant obtained at the last step of the thin filaments preparation (according to Driska and Hartshorne, 1975), i.e., during their ultracentrifugation, has a very strong activity as a Ca^{2+}-independent activator of skeletal muscle actomyosin ATPase (Fig. 55.2). It enhanced the ATPase activity 2-3-fold. It is possible that the protein activator which is present in thin filaments is partially released (similar to tropomyosin) in the supernatant during ultracentrifugation of thin filaments. This idea was supported by the observation that during ammonium sulphate fractionation of the supernatant, the activator was salted out between 0-40% saturated ammonium sulphate, as well as by the fact that the same concentration of trypsin that caused degradation of activator present in thin filaments also destroys its activity.

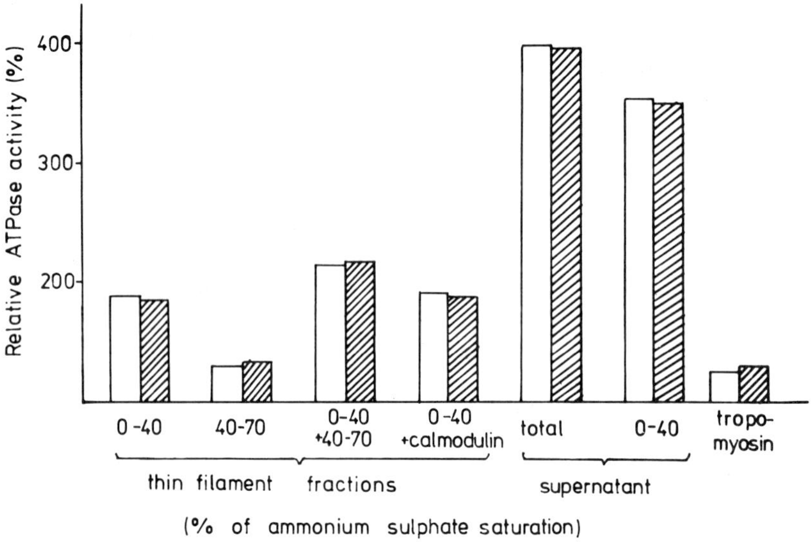

Fig. 55.2. The distribution of the activator of Mg^{2+}-ATPase of skeletal muscle actomyosin in various protein fractions obtained from chicken gizzard thin filaments and the supernatant after ultracentrifugation of the thin filaments. Assay conditions as in Fig. 55.1. Proteins were used at the following concentrations: rabbit skeletal muscle myosin 120 μg/ml, rabbit skeletal muscle actin 80 μg/ml, gizzard tropomyosin 20 μg/ml, brain calmodulin 5 μg/ml, fractions of gizzard thin filaments and supernatant as well as total supernatant 30 μg/ml each. Empty bars indicate the inclusion of 2 mM EGTA.

The fraction of supernatant salted out between 0-40% saturated ammonium sulphate was applied to a DE-52 column and the proteins were eluted with linear gradient of 0.06-0.3M NaCl. The active fraction was eluted at about 0.1M NaCl and, as shown by SDS-gels, it contained mainly proteins with molecular weights 130,000, 96,000, and 42,000 (Fig. 55.3). It might be supposed that α-actinin (mol. wgt ∿ 95,000) is the protein responsible for the activation of the actomyosin ATPase by this fraction; however, some properties of the activator distinguish it from α-actinin. For example, α-actinin is eluted from a DE-52 column under the conditions used at approximately 0.28M NaCl which suggests that it is more acidic than the activator. Also, tropomyosin did not abolish the activation of the actomyosin ATPase by the activator as it does in the case of activation of actomyosin ATPase by α-actinin (Endo and Masaki, 1982).

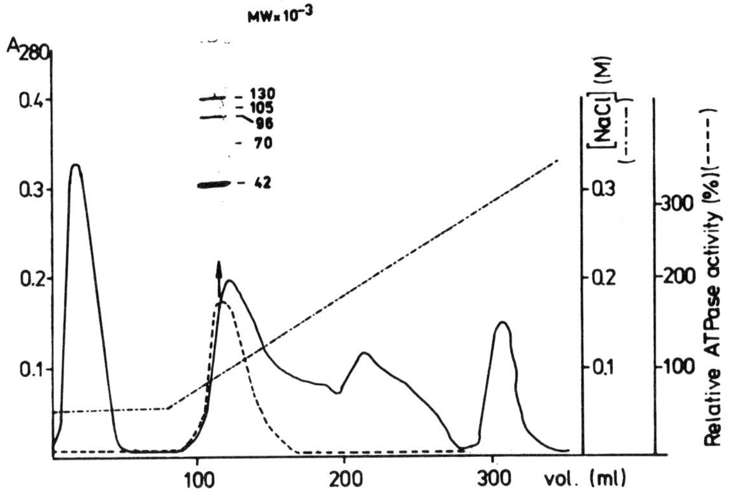

Fig. 55.3. DEAE-cellulose chromatography of the fraction of supernatant salted out at 0-40% saturated ammonium sulphate. Approx. 30 mg protein was applied to a 1 × 20 cm column of DE-52 equilibrated with 60 mM KCl, 20 mM Tris-HCl, pH 7.5 and 0.2 mM DTT. A linear gradient (0.06-0.3 M) NaCl was applied as indicated. Flow rate 10 ml/hour. ATPase activity was assayed as described in Fig. 55.1, with skeletal muscle myosin 120 µg/ml, skeletal muscle actin 80 µg/ml, and 200 µl of each (3 ml) fraction from column. Insert shows SDS-gel electrophoresis of active fraction on 8% acrylamide according to Laemmli (1970).

The disappearance of the Ca^{2+}-sensitivity of the activator after fractionation of the thin filaments indicates its complexity. It is possible that, like leiotonin, it is composed of two subunits and removal or destruction of one of them makes its effect on actomyosin ATPase Ca^{2+}-insensitive. Also, it cannot be excluded that activator is Ca^{2+}-independent and thin filaments contain, in addition, a Ca^{2+}-dependent inhibitor of actomyosin ATPase. Further investigations are necessary to solve this problem.

REFERENCES

Adelstein R. S. and Eisenberg E., Regulation and kinetics of the actin-myosin-ATP interaction. Ann. Rev. Biochem. 49, 921-56, 1980.

Dabrowska R., Hinkins S., Walsh M. P. and Hartshorne D. J., The binding of smooth muscle myosin light chain kinase to actin. Biochem. Biophys. Res. Commun. 107, 1524-31, 1982.

Driska S., and Hartshorne D. J., The contractile proteins of smooth muscle. Properties and components of a Ca^{2+}-sensitive actomyosin from chicken gizzard. Arch. Biochem. Biophys. 167, 203-212, 1975.

Ebashi S., Nonomura Y., Nakamura S., Nakasone H. and Kohama K., Regulatory mechanism in smooth muscle: actin-linked regulation. Fed. Proc. 41, 2863-67, 1982.

Endo T. and Masaki T., Molecular properties and functions in vitro of chicken smooth-muscle α-actinin in comparison with those of striated muscle α-actinins. J. Biochem. (Tokyo) 92, 1457-68, 1982.

Laemmli K. U., Cleavage of structural proteins during the assembly of the head of bacteriophage T-4. Nature 227, 680-85, 1970.

Marston S. B., The regulation of smooth muscle contractile proteins. Prog. Biophys. Molec. Biol. 41, 1-41, 1982.

Marston S. B., Trevett R. M. and Walters M., Calcium ion-regulated thin filaments from vascular smooth muscle. Biochem. J. 185, 355-65, 1980.

Pemrick S. M., The phosphorylated L_2 light chain of skeletal myosin is a modifier of the actomyosin ATPase. J. Biol. Chem. 255, 8836-41, 1980.

Walsh M. P. and Hartshorne D. J., Actomyosin of smooth muscle. Calcium and Cell Function 3, 223-69, 1982.

56 Structural Organization of Rabbit Fast Skeletal Myosin as a Modulator of Light Chain Proteolytic Susceptibility

R. Cardinaud

INTRODUCTION

In all types of muscle cells studied up to the present time, Ca^{2+} is reputed to trigger the contraction but the mechanisms vary in a very interesting fashion according to the type of muscle (for a review see e.g., 1,2 and ref. therein). In vertebrate fast skeletal muscle the essential regulatory system is known to be located on the complex actintroponin-tropomyosin system. In this type of muscle there is no clear evidence for a specific role for the LC2 light chain which nevertheless binds Ca^{2+} and can be phosphorylated on Ser-15. Besides the phosphorylatable light chain, each head contains either one of two different types of light chains: LC1 (often called A1, M_r app:25 kDa) and LC3 (A2, M_r app:17 kDa). In spite of observed differences in behavior of S1-(A1) and S1-(A2) under various conditions, the respective roles of these two types of light chains remain unknown.

For technical reasons (myosin is soluble only at high ionic strength), most studies are currently carried out with the "active" fragments HMM and S1. However, a suspension of reconstituted filaments is a much closer model of myosin in its "physiological" physical state. The present study shows that the light chains exhibit characteristic differences in behavior when myosin is in the monomeric state as compared to filaments. It is also reported that the filament structure responds differently to ligands such as Mg^{2+}. A comparison of proteolytic susceptibilities constitutes a very sensitive probe of conformational states. It was observed: (1) in the presence of Ca^{2+} there is an increased susceptibility of the HMN-LMN junction to chymotrypsin at high ionic strength while the opposite is observed for the clevage at Phe-18 of the LC2 light chain. The absence of S1 in the digest is also an indication that the N-terminal segment of LC2 plays no essential role in the protection of the S1-rod junction. (2) The specific cleavage of LC1 at Lys-17 obtained with trypsin is faster at low ionic strength. (3) The specific and very fast cleavage of LC_2 at Arg-7 by trypsin is also faster at low ionic strength but only in the presence of Mg^2. In the presence of EDTA the proteolytic susceptibility is identical at low and high ionic strength. (4) Further cleavage of LC2 by trypsin is very slow except at low ionic strength in the presence of EDTA.

MATERIAL AND METHODS

Myosin preparation, sodium dodecyl, sulfate-polyacrylamide gel electrophoresis, densitometry, and radioactive distribution analyses were performed as described in (6,7). All comparisons were made under strictly identical conditions using the same myosin preparation and the same protease solutions. The compared assays were run in parallel with a negligible 0-time shift (usually less than 10 min). Cylindrical gels were used since they were found to give more reproducible quantitative results. Staining with Coomassie Brilliant Blue R250 and destaining of the gels were performed with a highly standardized system to minimize differences in staining efficiency. All reported values are the average of 3 to 5 independent measurements.

RESULTS

(1) Chymotryptic cleavage rate at Phe-18 of LC2. When carried out in the presence of Ca^{2+} this cleavage produces a relatively stable LC2" species deprived both of a basic N-terminal segment ($\alpha N^+(CH_3)_3$-Ala-Pro-Lys-Lys-Ala-Lys-Arg-Arg- . . .) and the phosphorylatable serine (Ser-15) (8). We observed a pseudo-first order kinetics with $k_H = 1.5\ 10^{-3}s^{-1}$ (high ionic strength) and $k_L = 5.2\ 10^{-3}s^{-1}$ (low ionic strength) (Fig. 56.1a). A characteristic feature of these kinetics is that paradoxically the more compact filament structure seems to expose the LC2-Phe-18 region. It was interesting to study in parallel possible breaks in the heavy chain (Fig. 56.1b). At high ionic strength we observed a characteristic dual cleavage at two points in the HMM-LMM junction, $k_H = 2.7\ 10^{-3}s^{-1}$. At low ionic strength ($+Ca^{2+}$) the heavy chain cleavage is much slower: $k_L = 1.3\ 10^{-2}s^{-1}$ and after 8 min no significant formation of S1 could be detected; moreover, the electrophoretic analysis of the total digest showed that very little LMM was formed.

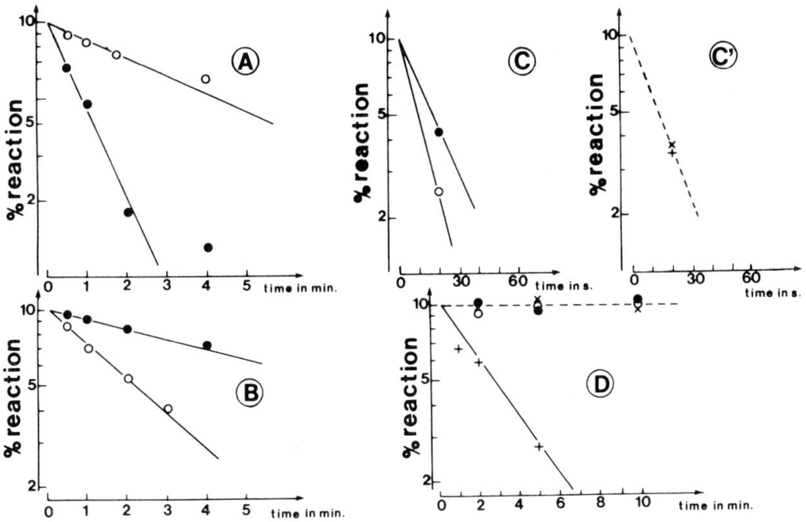

Fig. 56.1. Kinetic study of the proteolytic susceptibility of various peptide bonds in LC2. Myosine suspension or solutions (10 mg/ml^{-1}) were digested at 24°. The total digest were analyzed by SDS-PAGE and product formation estimated by densitometric measurements with a Vernon PH6 densitometer equipped with an integrator. (a) Fragmentation of the heavy chain (see text). (1) ○ = digestion by chymotrypsin (0.03 mg/ml^{-1} in 0.06 M NaCl, 0.01 M PO_4, pH 7.0, 0.3 mM $CaCl_2$). (2) ● = as in (1) in 0.6 M NaCl, 0.01 M PO_4, pH 7.0, 0.3 mM $CaCl_2$. (b) Fragmentation of LC_2. Kinetics of the formation of LC_2'' by cleavage at Phe-18. (1) ○ = as in (a) (1); (2) ● = as in (a) (2). (c) Tryptic fragmentation of LC_2 by cleavage at Arg-7 (formation of LC_2'); enzyme:0.03 mg/ml^{-1}. (1) ● = in 0,5 M NaCl, 0,02 M PO4, pH 7.3, 0.01 M $MgCl_2$; (2) ○ = in 0,1 M NaCl, 0,02 M PO_4, PH 7.3, 0.01 M $MgCl_2$. (c') (1) x = as in (c)(1), 0.01 M EDTA replacing $MgCl_2$. (2) + = as in (c)(2), 0.01 M EDTA replacing $MgCl_2$. (d) Tryptic degradation of LC2' to LC_2''' (14 KDa, a transient species). Symbols cover the same conditions as in c and c').

(2) Tryptic cleavage rate at Arg-7 of LC2. The characteristics of this cleavage (LC2 → LC2') have been given in detail in (8). It occurs extremely rapidly (Fig. 56.1cc'). In the presence of EDTA there is no effect of ionic strength whereas a significant effect can be observed in the presence of Mg^{2+} with pseudo-first order rate constants of $k_H = 4.06\ 10^{-2}s^{-1}$ and $k_L = 7.7\ 10^{-2}s^{-1}$. Trypsin degrades this light chain further through a transient intermediate LC2''' of about 14 kDa. This degradation was very slow at high ionic strength in the presence as well as in the absence of Mg^{2+}, but much faster at low ionic strength in the presence of EDTA.

(3) Tryptic cleavage rate at Lys-17 of LC1. This cleavage was described earlier and compared to the same type of cleavage obtained with papain (8). In the presence of Mg^{2+} pseudo-first order rate constant were: $k_H = 3.03\ 10^{-3}\ s^{-1}$ and $k_L = 4.42\ 10^{-3}s^{-1}$. Here again the reaction is slower at high ionic strength. In the presence of EDTA the rate constants were not significantly affected with the same difference between high and low ionic strength ($k_H = 2.87\ 10^{-3}\ s^{-1}$ and $k_L = 4.26\ 10^{-3}\ s^{-1}$). Thus the cleavage of the LC1 N-terminal segment is affected by ionic strength in the same manner as LC2 was in the presence of Mg^{2+}.

DISCUSSION

In this study two observations of importance are: (1) the N-terminal portion of LC2 (about 20 residues) is not implicated in the LC2 protective role of the S1-rod junction observed in the presence of divalent cations; (2) the filament structure imparts some new or enhanced characteristics to the myosin molecule entering into this structure, particularly with regards to the proteolytic susceptibility of the N-terminus portion of its LC1 and LC2 light chains. These two observations and a number of others not reported here suggest intuitively that both these light chains have a highly mobile N-terminal segment or "antenna" which can be found under either one of two extreme conformations: one with the antenna folded back on the surface of the molecule (this would be the case in the monomeric state) and another in which the antenna extends toward the shaft of the filament where it could possibly find some anchoring point. It must be noted that a regulation of these configurations can easily be imagined to be under the control of phosphorylation and/or cation divalent binding in the case of LC2, but on the other hand no agent other than ionic strength and the filament assembly-disassembly can be presently put forward in the case

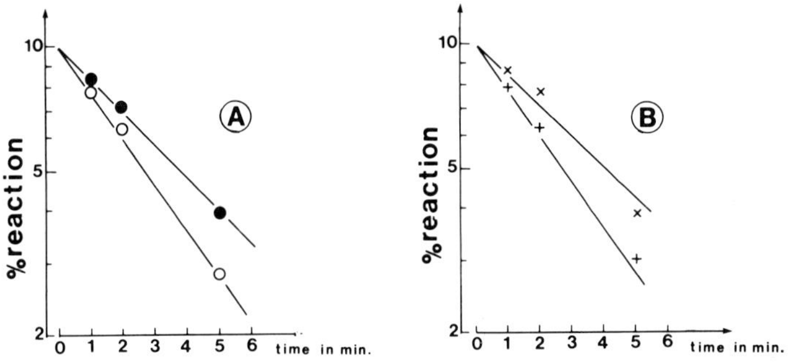

Fig. 56.2. Kinetic study of the proteolytic susceptibility of the Lys-17-Ala-18 bond in LC_1. General conditions as described in Fig. 56.1. Tryptic cleavage under conditions described in c an c' of Fig. 56.1, symbols covering the same conditions. Values in c, c', d and this figure were obtained from the same series of experiments.

of LC1. The present discussion aims at showing that the observations reported here are reasonably accounted for by this hypothesis.

The observation that SDS-polyacrylamide gels show no S1 even though the LC2 → LC2" conversion is nearly 100% suggests that the N-terminal antenna is "available" for a possible interaction with the shaft. At high ionic strength (monomeric myosin) the lower proteolytic susceptibility observed argues in favor of a compact conformation, the antenna being folded back on the bulk of the peptide. In filaments where the structure could be expected to hinder proteolytic attack the reverse is observed and interpreted as evidence that the unfolded antenna is freely floating or extends toward some anchoring point very likely to be located in the shaft, and under this configuration exposes a zone sensitive to the protease. The tryptic conversion of LC2' to the transient LC'" (14 kDa) is also very informative. In the presence of Mg^{2+} the reaction is very slow, exhibiting the very clear protective effect expected from a divalent cation (9,10) with respect to a break somewhere in or beyond the metal binding site. This break seems to be a key to further degradation. A striking observation is that the monomeric structure by itself provides practically the same protection as Mg^{2+} whereas at low ionic strength without Mg^{2+} the degradative proteolytic attack is considerably faster, evidence that in the filament structure the extended antenna exposes also some sensitive site situated on the peptide chain at least as far as the metal binding cleft.

The specific cleavage of LC1 responds to ionic strength in a similar manner but is insensitive to Mg^{2+}. Interaction of this LC1 N-terminal residue with actin postulated by Prince et al. (11) is not at variance with the present observation. It may be imagined that in filaments released LC1 have their N-terminus available for interaction with either the thick filament shaft or the thin filament.

SUMMARY

Proteolytic susceptibilities of various specified bonds were compared in monomeric and aggregated myosin (reconstituted filaments) in the presence and absence of divalent cations. LC2 was cleaved by chymotrypsin $+Mg^{2+}$ at Phe-18 more rapidly in filaments. The tryptic cleavage at Arg-7 was comparatively very fast and faster still at low ionic strength. Another cleavage produced by trypsin and resulting in loss of LC2 is very slow in the presence of divalent cations (both in monomeric myosin and filaments).

It is also slow in the presence of EDTA at high ionic strength but considerably faster in filaments. A break at Lys-17 in LC1 is sensitive only to ionic strength but not to divalent cations. These results suggest: (1) the N-terminal portion of LC2 (about 20 residues) is not implicated in the LC2 protective role of the S1-rod junction observed when divalent cations are present; (2) both LC1 and LC2 N-terminal portions act as a mobile "antenna" which can be found in either of the following two conformations: (a) folded back and somewhat buried in the monomer; (b) extended toward the shaft of the filament at low ionic strength.

REFERENCES

1. Kendrick-Jones, J. and Scholey, J. M. (1981), J. Muscle Res. and Cell Motil. 2, 347-72.
2. Perry, S. V. (1979), Biochem. Soc. Trans. London, 7, 593-617.
3. Wagner, P. D., Slater, S. C., Pope, B. and Weeds, A. G., (1979), Eur. J. Biochem. 99, 385-94.
4. Wagner, P. D. and Weeds, A. G. (1977), J. Mol. Biol. 109, 455-73.
5. Winstanley, M., Trayer, H. R., and Trayer, I. P. (1977), FEBS Lett. 77, 239-42.

6. Cardinaud, R. (1979), Biochimie, 61, 807-21.
7. Cardinaud, R. and Drifford, M. (1982), J. Muscle Res. and Cell Motil. 3, 313-22.
8. Cardinaud, R. (1982), Eur. J. Biochem. 122, 527-33.
9. Weeds, A. G. and Pope, B. (1977), J. Mol. Biol. III, 129-57.
10. Bagshaw, C. R. (1977), Biochemistry, 16, 59-67.
11. Prince, H. P., Trayer, H. R., Henry, G. D., Trayer, I. P., Dalgarno, D. C., Levine, B. A., Cary, P. D., and Turner, C. (1981), Euro. J. Biochem. 121, 213-19.

57 Structural and Functional Specialization in Spectrin: A Comparison of Human Erythrocyte and Brain Proteins

Jon S. Morrow, Alan S. Harris, Peter Shile,
L. A. David Green, Kevin J. Ainger

Nine unique proteolytically resistant peptide domains have been previously identified in human erythrocyte spectrin (1). Associated with many of these domains are specific binding sites responsible for several of the known properties of this molecule (2). The unique ability of this protein to both self associate to very high-molecular weight oligomers, as well as bind F-actin is believed to account for its specialized function in the erythrocyte membrane skeleton. The recent demonstration of similar proteins in most other cells suggests that all cells may share the requirement for a submembranous protein skeletal matrix which acts to link the transcellular cytoskeleton to membrane bound receptors (3-16). However, while the various spectrins have been reported to share antigenic, morphologic, and limited functional similarities, there are also differences. It is likely that the spectrins as a class of proteins display functional and structural specializations necessary to serve a variety of roles in different cells or at different stages of growth and differentiation in the same cell. To explore these differences, as well as similarities, the structural, functional, and antigenic domain structure of human erythrocyte and brain spectrin has been compared. Brain spectrin has also been called fodrin (10) or calspectrin (17). The results indicate that human brain spectrin shares a proteolytic resistant domain structure reminiscent of but not identical to erythrocyte spectrin. Both proteins share identical antigenic determinants, as detected by quantitative ELISA assay, but the similarity is not uniform throughout both subunits. Specifically, the proteins are disparate most significantly at the terminal domains of both subunits. The functions associated with these dissimilar domains also appear to differ between these two spectrins. It is anticipated that understanding the structural and functional specializations of spectrin characteristic of different tissues will allow a better understanding of their functional significance in health and disease.

METHODS

Extraction and Purification of Spectrin

Human erythrocyte spectrin was prepared from whole blood within 24 hours of collection by low ionic strength extraction of white ghosts, as previously described (2).

Human brain spectrin was prepared from cadaver brains removed at autopsy 2-4 hours postmortem, with permission of the family. The brain tissue was cleaned of all extraneous membranes and blood vessels, the cerebral cortex removed, diced, and quick frozen in 2-methylbutane cooled in liquid nitrogen. Samples so prepared were stored at -60°C until use. Aliquots of frozen brain (25-35g) were thawed and homogenized in 180 ml of buffer A (200 mM sucrose, 10 mM glucose, 10 mM N-2-hydroxyethyl piperazine-N'-2 ethane sulfonic acid (HEPES), 1 mM EGTA, 1 mM ATP, 1 mM MgC12, 1 mM dithiothreitol (DTT), 0.5 mM diisopropyl fluorophosphate (DFP), 0.09 mM phenylmethylsulfonyl fluoride (PMSF), 1 microgram (μg)/ml bestatin (Sigma), 1 μg/ml antipain (Sigma), 5 μg/ml leupeptin (Sigma), pH 7.4 at 0°C) with a Dounce glass on glass homogenizer. All procedures were done at 0-4°C. This homogenate was centrifuged 10 min at 3,700 × g to remove nuclei and large debris. The floating membrane fractions were then rehomogenized by a Tissue-Tek (Tekmar) steel homogenizer for 30 sec. and sedimented 35 min. at 300,000 × g. The resulting membrane pellet was washed twice in buffer A; the spectrin was extracted in 250 ml of buffer B (0.1 mM HEPES, 0.1 mM EGTA, 1 mM DTT, 0.5 mM DFP, 0.09 mM PMSF, 1 μg/ml antipain, 1 μg/ml bestatin, 5 μg/ml leupeptin, pH 9.0 c. 37°C) by incubation at 37°C for 60 min. The extracted protein was concentrated by precipitation with 50% saturated ammonium sulfate, and subsequently further purified by gel filtration on a 2.5 × 90 cm column on CL-4B (Pharmacia) in buffer C (260 mM KCl, 40 mM NaCl, 20 mM tris-HCl, 2.5 mM EDTA, 1 mM DTT, .09 mM PMSF, pH 7.6 at 0°C). Samples containing spectrin were pooled and concentrated by dialysis against 10% polyethyleneglycol in the appropriate buffer. Further purification when necessary was done by sedimentation in 5-20% sucrose gradients.

Preparation of Immune Sera

Spectrin (brain or erythrocyte) was further purified by preparative SDS-PAGE using 3.5-6.0% acrylamide slab gels. Fluorescamine labeling prior to electrophoresis allowed the appropriate spectrin peptides to be visualized and subsequently electroeluted from the gels by published metholds (18). The eluted peptides were elusified with Freund's complete adjuvant, and injected intradermally at multiple sites into New Zealand white rabbits. Two boosting injections were given at biweekly intervals, after which the antibody was harvested by arterial ear puncture. Antibody titers were monitored by ELISA assay (19). Affinity purified antibodies were prepared using spectrin coupled by CNBr activation to sepharose CL-4B columns (Pharmacia (20).

Western Blot Analysis and Calmodulin Overlay

Peptides separated by either one or two-dimensional PAGE were analyzed for immunoreactivity by electrophoretic transfer of the peptides to nitrocellulose sheets, followed by incubation with antibody (21,22). Visualization of the reactive peptides was accomplished by autoradiography after incubation with ^{125}I-labeled Staph Protein A (Pharmacia).

Calmodulin binding peptides were detected in the SDS gels by incubating the gel directly with ^{125}I-labeled bovine or human brain calmodulin, following the method of Carlin (23).

Quantitative ELISA Inhibition Assays

The reactivity of the antisera and purified antibodies was measured by an ELISA assay using an avidin-biotin-peroxidase system in which biotinylated goat-anti-rabbit antibodies (Vector) were used to detect rabbit antispectrin antibodies retained on the spectrin coated ELISA plates. For both routine screening and the quantitative inhibition assays, the plates were coated with 50 ng of spectrin per well. The inhibition experiments followed procedures published for keratin (24), in which antibody at high dilution (1:40,000) was preincubated with the inhibitory peptides for 30 min. at 25°C prior to its introduction into the ELISA plate containing the bound spectrin. All inhibition experiments were performed in quadruplicate. Protein determinations were performed by the method of Lowry (25).

Polyacrylamide Gel Electrophoresis (PAGE)

Sodium dodecylsulfate electrophoresis was performed by the method of Laemmli (26) under the conditions previously described (1). Nondenaturing polyacrylamide gels used to detect oligomeric complexes of spectrin and the binding of smaller peptides were performed in 2-4% acrylamide gradients in 40 mM tris, 20 mM sodium acetate, 2 mM EDTA, pH 7.4 at 4°C (2). Two-dimensional isoelectric focusing-SDS PAGE analysis was performed by the method of O'Farrell (27).

RESULTS

Human Erythrocyte and Brain Spectrin Are Characterized by Unique Protease Resistant Domains

Spectrin purified from either brain or erythrocyte displays a number of unique protease resistant domains when subjected to limited trypsin digestion at 0°C. In the case of erythrocyte spectrin, nine major domains have been operationally identified, which in aggregate account for nearly the total peptide mass of the protein (1). However, despite the relative resistance to proteolysis characteristic of these domains, fragmentation within each domain does occur, leading to complex peptide digestion patterns after analysis by two-dimensional isoelectric focusing-SDS PAGE (Fig. 57.1, top). This pattern of cleavage, while complex, is highly reproducible, and represents a sensitive "map" related to the primary, secondary, and tertiary structure of the molecule. The detection of minor alterations in this pattern of digestion has allowed a determination of the site of abnormality in several human spectrin variants (28), and in the case of hereditary pyropoikilocytosis an explanation of the disease process at the molecular level (29).

The digestion of human brain spectrin under identical conditions as erythrocyte spectrin also produces a cascade of intermediate sized peptides displaying relative proteolytic resistance. The conditions of the digestion shown in Figure 57.1 were 20 mM tris-HCl, 1 mM EDTA, pH 8.0, 0°C, for one hour at an enzyme:substrate of 1:20. Prominent and relatively stable peptides in this digest appear at Mr of 118; 89; 77; 71; 46; 41; 35; 27; 21; and 12 kilodaltons (kd). In separate experiments studying the time course of digestion there was progressive loss of the largest of these (118 kd), while the other peptides remained prominent even after eight hours of digestion (data not shown).

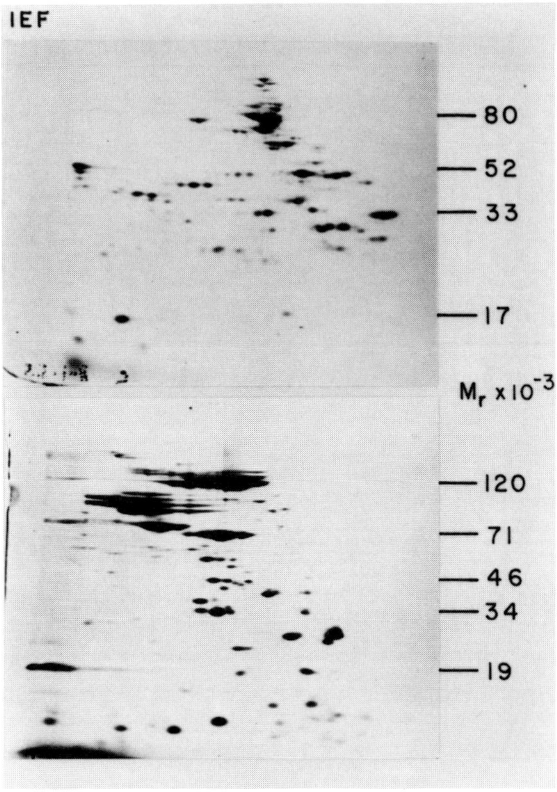

Fig. 57.1. Two-dimensional IEF/SDS-PAGE Analysis of Tryptic Peptides. (Top) Human erythrocyte spectrin was digested with trypsin (20 mM tris-HCl, 0.1 mM EDTA, 0.5 mM 2-mercaptoethanol, pH 8.0, 0°C, enzyme/substrate, 1/20, (w/w)) for 60 minutes. (Bottom) Human brain spectrin digested and analyzed under the same conditions.

Although the size of the peptides generated by trypsin digestion in both brain and erythrocyte spectrin are similar (cf. 80; 74; 52; 46; 41; 35; 28; 17; and 12 kd for the erythrocyte material), they differ in many other respects. The peptides from brain spectrin have isoelectric points more basic than the erythrocyte protein. The two analyses shown in Figure 57.1 were conducted over the same pH range in the focusing dimension. The 0.3 pH unit basic shift of the peptides in the brain material (Fig. 57.1, bottom) reflects their altered charge character. In addition, ^{125}I-labeled peptide nitrocellulose maps prepared from similar sized peptides of the erythrocyte and brain material after exhaustive chymotryptic digestion (30) fail to show a high degree of similarity between the two proteins (data not shown). Finally, monoclonal antibodies prepared against the alpha-I domain of erythrocyte spectrin (22) fail to react with the brain material.

Brain Spectrin Tetramers Do Not Oligomerize Nor Do They Bind the Erythrocyte Alpha-I Domain

A characteristic and important property of erythrocyte spectrin is its ability to undergo self-association under physiologic conditions to form high-molecular weight noncovalent complexes (31,32). This process is driven primarily by mass-action due to the high local concentrations of spectrin maintained at the membrane surface. The process of oligomer formation involves paired interactions between the amino-terminal alpha-I domain (80 kd Mr) and the carboxy terminal beta-I domain (28 kd Mr) (31,32). This interaction may be modeled in vitro by the binding of the isolated and purified alpha-I spectrin domain to either the intact spectrin dimer or to purified beta-I domain (32, Ainger and Morrow, unpublished observations). The conditions which favor the interconversion of erythrocyte spectrin oligomeric species are the same that allow the binding of the isolated active alpha-I fragment: moderate ionic strength (>20 mM) and temperature (>10°C). Thus, at 30-37°C and isotonic salt conditions, pH 7.4, the binding of the alpha-I (80 kd) peptide or the concentration dependence of spectrin oligomers can be easily demonstrated by nondenaturing PAGE (Fig. 57.2).

Human brain spectrin under identical conditions does not undergo the same interconversions (Fig. 57.2). As isolated, this spectrin migrates as a tetramer, consistent with the appearance after rotary shadowing and hydrodynamic properties of a similar protein isolated from pig brains (5). In addition, even after incubation at a variety of concentrations, or in the presence of active alpha-I domain peptide from erythrocyte spectrin, there is no change in the tetrameric state of the brain material. Thus, it appears that the brain and erythrocyte spectrins also differ enormously in the facility with which they undergo changes in their state of self-association; brain spectrin remains steadfastly in the tetrameric state under conditions in which erythrocyte spectrin undergoes facile interconversions.

Brain and Erythrocyte Spectrins Bind Calmodulin at Different Sites

Another property of the spectrins is their ability to bind the calcium regulatory protein calmodulin (14,33). The func-

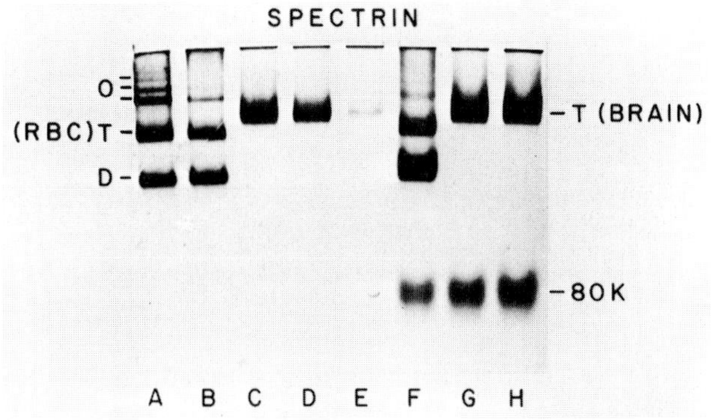

Fig. 57.2. Nondenaturing Polyacrylamide Gel Analysis of Erythrocyte and Brain Spectrin. (A,B) Erythrocyte spectrin at 5 mg/ml before and after incubation at 37°C. Note the fascile change in the equilibrium distribution of oligomeric species. (C,D,E) Brain spectrin at 3.8 mg/ml before and after incubation (C,D); and at 0.2 mg/ml after incubation (D). Note the presence of only tetramer. (F,G,H) Incubation of the alpha-I domain (80 kd) of erythrocyte spectrin with either erythrocyte of brain spectrin. Note binding to only the red cell protein.

tional significance of this binding is unknown. The spectrins from mammalian erythrocytes are unusual in that they have so far only been demonstrated to bind calmodulin in the presence of 3 molar urea (34,35). However, despite this, the site of calmodulin binding to human erythrocyte spectrin has been localized to a specific site within the beta-IV domain (35). By comparison, brain spectrin from cow and pig has been shown by gel overlay techniques to bind calmodulin to the alpha subunit (240 kd) (7). We have confirmed this observation for human brain spectrin, using bovine brain calmodulin (data not shown). Thus, it appears that the functional domain of erythrocyte spectrin beta-chain responsible for spectrin's unusual calmodulin binding behavior is not preserved in the brain protein.

Erythrocyte and Brain Spectrin Possess Many Antigenic Similarities

Polyclonal antibodies prepared to either brain or erythrocyte spectrin cross-react with the other protein. However, the extent and nature of the cross-reactivity is limited. Of three rabbits immunized with erythrocyte spectrin, one responded with antibodies directed exclusively against the alpha (240 kd) subunit, one responded with antibodies primarily against the alpha subunit with only weak activity for the beta subunit, and one rabbit produced high-titer antisera directed against both spectrin subunits. The two rabbits immunized with brain spectrin both produced high-titer antisera to the alpha subunit exclusively. A comparison of the cross-reactivity of the whole proteins, as detected by Western immunoblotting, confirmed the observations made for the nonhuman spectrins that the 240 kd subunits of both proteins (alpha subunits) shared antigenic cross-reactivity. To further explore the nature and extent of antigenic similarity, quantitative ELISA inhibition assays were performed using the antierythrocyte spectrin antisera which reacted with both subunits. Such assays are useful for detecting not only the extent of shared antigenic determinants between two proteins, but also for measuring the relative similarity of cross-reacting epitopes and their relative abundance (24).

Briefly, in such studies, parallel inhibition plots result from identity of antigenic determinants. The extent of their relative displacement is related to the abundance of shared determinants. The limiting degree of inhibition attainable is proportional to the extent of antigenic uniqueness. Inhibition curves for erythrocyte spectrin, alpha-I domain of erythrocyte spectrin, and for brain spectrin are shown in Figure 57.3. The similar slope of all three inhibition plots indicates that both erythrocyte and brain spectrin share nearly identical antigenic determinants. The displacement of the alpha-I (80 kd) and brain spectrin curves to higher amounts of inhibitor indicates that these peptides/proteins contain fewer of the cross-reacting determinants compared to the complete erythrocyte protein. Finally, while a limiting inhibition has not been achieved for either the alpha-I or the brain peptides, at least 80% of the total epitopes present in erythrocyte spectrin are represented in the alpha-I domain (1/6 of the molecule), while perhaps more than 50% of the erythrocyte determinants are present in brain spectrin. To further explore the precise disposition of the cross-reacting determinants within brain and erythrocyte spectrin, Western immunoblots of erythrocyte spectrin tryptic digests (Fig. 57.1) were reacted with rabbit antisera affinity purified on either an erythrocyte or brain spectrin affinity column. Affinity purified antisera to the erythrocyte spectrin reacted strongly with every major spectrin domain; i.e., domains alpha I to V and beta I to IV. Conversely, when the same experiment was carried out with the same antisera affinity purified on a brain spectrin affinity column, there was no reactivity to peptides arising from the amino-terminal portions of the beta IV domain or from the beta I domain. A similar experiment carried

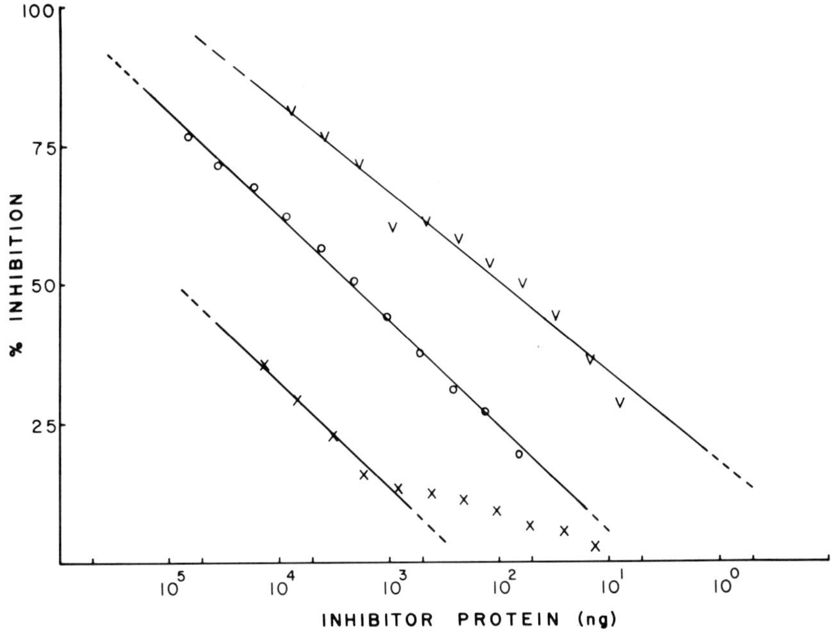

Fig. 57.3. Quantitative ELISA Inhibition Assays of Antierythrocyte Spectrin. (V) erythrocyte spectrin, (O) alpha-I domain of erythrocyte spectrin, (X) human brain spectrin.

out with antibrain spectrin antisera demonstrated strong activity against all of the erythrocyte alpha subunit domains except the alpha-I. The conclusion from these experiments (Fig. 54.4) is that there is selective preservation of antigenic similarity between human brain and erythrocyte spectrins; the greatest degrees of dissimilarity arise at the end of the molecules. It is noteworthy that the areas of greatest dissimilarity also correspond with the functions which are not shared between the two molecules.

DISCUSSION

Recent reports focus on the similarities of spectrin isolated from brain or intestinal microvilli with erythrocyte spectrin (7,36). These studies support the notion that to be "spectrinesque," a protein must: (i) consist of a heterodimer of two high molecular weight polypeptide subunits, of approximately

220 kd to 260 kd Mr; (ii) exhibit a highly asymmetric shape when visualized by electron microscopy after rotary shadowing; (iii) bind the proteins F-actin and ankyrin; (iv) bind calmodulin (calcium dependent); (v) self-associate to at least the tetrameric state; and (vi) share antigenic cross reactivity (3-16). More recently, it has been demonstrated that all "spectrins" appear to have a subunit of 240,000 Mr ("alpha"), while the other subunit appears to vary in molecular weight (220,000 for mammalian erythrocyte, 235,000 for mammalian brain, 260,000 for avian intestinal brush border). During embryogenesis in avian muscle and other tissues several putative spectrin variants have also been observed (37). On the basis of antigenic and molecular weight similarities it has been postulated that the spectrins share a common alpha subunit joined to a tissue or function specific beta subunit. This diversity implies that spectrin is a molecule which has evolved to fulfill a multitude of specialized roles in different cells. The studies reported here suggest that this diversity of function is achieved not only by the combination of different subunits, but also by the preservation or alteration of specific structural or functional domains within each subunit.

Several observations described above support this conclusion. Human erythrocyte spectrin is now well understood in terms of its overall structural organization, and primary sequence data for the alpha-I domain has been reported (38). By isolating and characterizing the spectrin from human brain tissue, variability due to interspecies differences are eliminated. Thus, direct comparisons of peptide maps, sequence data, and functional specialization become both possible and meaningful. As described above, both subunits of human brain spectrin differ significantly from the erythrocyte material when compared by the criteria of high-resolution peptide mapping and quantitative antigenic cross-reactivity. However, these differences do not extend uniformly throughout the molecule. Using antigenic cross-reactivity as a probe, one finds extensive preservation of structure within those portions of the molecule which are remote from the ends, with the sole exception of the erythrocyte alpha-V domain. Since it is at the ends of the molecule that erythrocyte spectrin either binds F-actin or undergoes self-association, it seems likely that the two spectrins should differ in these activities. As described above, brain spectrin does indeed display markedly different properties of self-association, and does not participate in the formation of extended oligomers so characteristic of the erythrocyte material. Actin also appears to bind more strongly to the brain material. The calmodulin binding domain of erythro-

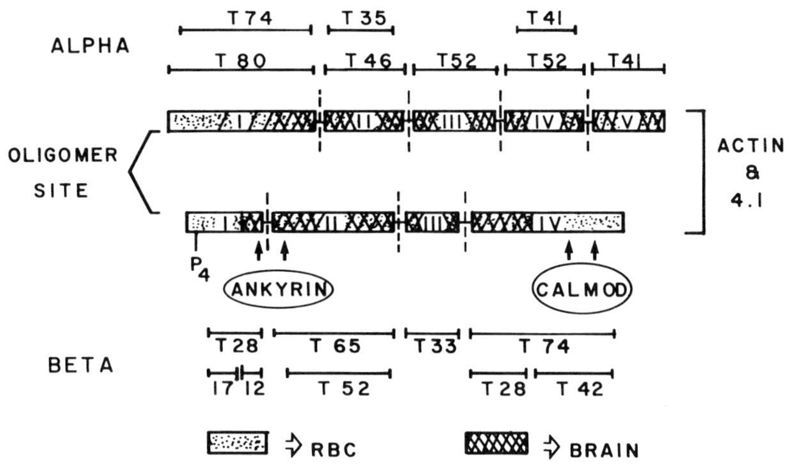

Fig. 57.4. Proposed Structural and Functional Domain Structure of Spectrin. Areas of cross-hatch are those regions of spectrin closely similar in the erythrocyte and brain peptides.

cyte spectrin also does not cross-react with brain spectrin, and correspondingly neither does the brain material bind calmodulin on the beta subunit, rather on the alpha subunit. Conversely, both spectrins appear to retain a highly preserved ankyrin binding capability (5,9), consonant with the high degree of antigenic similarity observed between the spectrins for the ankyrin binding domain.

It is pertinent to consider what portions of the spectrin structure the cross-reacting antibodies are sensing. While our polyclonal rabbit antisera were raised against protein removed from SDS-acrylamide gels, it is likely that both conformational and sequence determinants were responded to. This contention is based on the observation that these antisera display strong reactivity to either native spectrin in immunoassays or to denatured spectrin or spectrin peptides after SDS-PAGE and electroblotting. However, our results cannot be interpreted as evidence that strong sequence homologies do not exist between the spectrins, even in those domains showing little cross-reactivity. The primary sequence of the alpha-I spectrin domain of erythrocytes shows a repeating internal sequence of 106 residues (38, D. Speicher, personal communication). This sequence has also been identified in all of the other major spectrin domains so far examined, including those from the beta subunit. Yet, all of our monoclonal antibodies (22), as well as most of our antisera display absolute specificity for either a specific domain or a specific subunit. Thus, it appears that the antibodies are responding most reliably to epitopes on the protein which represent specific areas of specialization or nonhelical structure. This phenomena is well documented for many smaller proteins (39,40). Because areas of functional specialization are likely to be "hot-spots" for antigenic activity, cross-reactivity may be an ideal way of screening for regions of functional similarity.

Finally, one must consider the implications of these findings for the role of the different spectrins. Our results indicate that spectrinlike proteins display considerable structural and functional heterogeneity within both subunits. It seems likely that specialized functions have evolved to serve specific roles in different tissues. Despite the similarities in molecular weight or overall antigenic cross-reactivity cited by many studies it is likely that specific changes in the functional domains of these proteins will be observed. Erythrocyte spectrin probably represents the most specialized of the spectrin molecules.

ACKNOWLEDGMENTS

J.S.M. is a John A. and George L. Hartford fellow. This work was supported in part by grants from the NIH-PHS; the Will's Foundation; and the March of Dimes.

REFERENCES

1. Speicher, D. W., Morrow, J. S., Knowles, W. K., and Marchesi, V. T. (1980) Proc. Natl. Acad. Sci. U.S.A. 77:5673-77.
2. Morrow, J.S., Speicher, D. W., Knowles, W. K., Hsu, C. J., and Marchesi, V. T. (1980) Proc. Natl. Acad. Sci. U.S.A. 77:6592-96.
3. Goodman, S. R., Zagon, I. S., and Kulikowski, R. R. (1981) Proc. Natl. Acad. Sci. U.S.A. 78:7570-74.
4. Davis, J. and Bennett V. (1982) J. Biol. Chem. 257: 5816-20.
5. Bennett, V., Davis, J. and Fowler, W. E. (1982) Nat 299:126-31.
6. Glenney, J. R., Glenney, P., Osborn, M., and Weber, K. (1982) Cell 28:843-54.
7. Glenney, J. R., Glenney, P., and Weber, K. (1982) Proc. Natl. Acad. Sci. U.S.A. 79:4002-05.

8. Glenney, J. R., Glenney, P., and Weber, K. (1982) J. Biol. Chem. 257:9781-87.
9. Burridge, K., Kelly, T., and Mangeat, P. (1982) J. Cell. Biol. 95:478-86.
10. Levine, J. and Willard, M. (1981) J Cell Biol. 90:631-43.
11. Repasky, E. A., Granger, B. L., and Lazarides, E. (1982) Cell 29:821-33.
12. Palfrey, H. C., Schiebler, W., and Greengard, P. (1982) Proc. Natl. Acad. Sci. U.S.A. 79:3780-84.
13. Nelson, W. J., and Lazarides, E. (1983) Proc. Natl. Acad. Sci. U.S.A. 80:363-67.
14. Kakuichi, S., Sobue, K., and Fujita, M. (1981) FEBS Lett 132:144-48.
15. Shimo-Oka, T., and Watanabe, Y. (1981) J. Biochem. 90: 1297-1307.
16. Davies, P. J. A., and Klee, C. B. (1981) Biochem. Int. 3:203-212.
17. Kakiuchi, S., Sobue, K., Kauda, K., Morimoto, K., Tsukita, S., Tsukita, S., Ishikawa, H. and Kurokawa, M. (1982) Biomed. Res. 3:400-410.
18. Knowles, W. J. and Bolonga, M. (1982) Meth. in Enzymol. 96:305-13.
19. Voller, A., Bidwell, D. E., and Bartlett, A. (1979) The Enzyme Linked Immunoabsorbant Assay (ELISA), Dynatech Laboratories, Alexandria, VA.
20. March, S. C., Parikh, I., and Cautrecasas, P. (1974) Anal. Biochem. 60:149-52.
21. Towbin, H., Staehelin, T., and Gordon, J. (1979) Proc. Natl. Acad. Sci. U.S.A. 76:4350-54.
22. Yurchenco, P. D., Speicher, D. W., Morrow, J. S., Knowles, W. J., and Marchesi, V. T. (1982) J. Biol. Chem. 257:9102-07.
23. Carlin, R. K., Grab, D. J., and Seikevitz, P. (1980) Ann. N.Y. Acad. Sci. 356:73-74.
24. Madri, J. A., and Barwick, K. W. (1983) Lab. Invest. 48:98-107.
25. Lowry, O. H., Rosebrough, N. J., Farr, A. L., and Randall, R. J. (1951) J. Biol. Chem. 193:265-75.
26. Laemmli, U. K. (1970) Nature 227:680-85.
27. O'Farrell, P. H. (1975) J. Biol. Chem. 250:4007-4021.
28. Knowles, W. J., Marchesi, S. L., and Marchesi, V. T. (1983) Sem. Hemat. 20:159-74.
29. Knowles, W. J., Morrow, J. S., Speicher, D. W., Zarkowsky, H. S., Mohandas, N., Mentzer, W. C., Shohet, S. B., and Marchesi, V. T. (1983) J. Clin. Invest. 71: 1867-77.

30. Elder, J. H., Pickett R. A., II, Hampton, J., and Lerner, R. A. (1977) J. Biol. Chem. 252:6510-15.
31. Morrow, J. S. and Marchesi, V. T. (1981) J. Cell Biol. 88:463-68.
32. Morrow, J. S., Haigh, W. B., Jr., and Marchesi, V. T. (1981) J. Supramol. Struct. 17:275-87.
33. Grab, D. J., Carlin, R. K., and Siekevitz, P. (1980) Ann. N.Y. Acad. Sci. 356:55-72.
34. Sobue, K., Fujita, M., Muramoto, Y., and Kakiuchi, S. (1980) Biochem. Internat. 1:561-67.
35. Sears, D. E., Morrow, J. S., and Marchesi, V. T. (1982) J. Cell. Biol. 95:251a.
36. Pearl, M., Fishkind, D., Mooseker, M., Keene, D., and Keller, T. (1983) J. Cell Biol. in press.
37. Nelson, W. J., and Lazarides, E. (1983) Nat. 304:364-68.
38. Speicher, D. W., Davis, G., and Marchesi, V. T. (1983) J. Biol. Chem. in press.
39. Atassi, M. Z. (1975) Immunochemistry 12:423-38.
40. Atassi, M. Z. (1978) Immunochemistry 15:909-36.

58 Effect of S-100 Protein on Assembly of Brain Microtubule Proteins: Direct Effect of S-100 on Tubulin

Rosario Donato

INTRODUCTION

The acidic Ca^{2+}-binding S-100 protein (Moore, 1965; Calissano et al., 1969) is structurally related to calmodulin and other Ca^{2+}-binding proteins (Isobe and Okuyama, 1978). S-100 has been localized, inter alia, at the level of the axonemes of the cilia of: (i) ependymal cells in the mammalian brain (Cocchia, 1981); (ii) epidermal cells of a planarian (Michetti and Cochia, 1982); and (iii) the marine protozoan Euplotes crassus (Cocchia et al., this symposium). These data suggest that S-100 may be structurally and, possibly, functionally related to microtubules (MTs).

We have previously shown that S-100 inhibits the assembly of brain MT proteins in a dose-dependent way and potentiates the disassembling effect of 0.1 to 1 mM Ca^{2+}, in vitro (Donato, 1983). We show here that S-100 primarily affects the nucleation of MTs and has a distinct effect on the elongation reaction, and that the protein brings about its effect by interacting with tubulin.

EXPERIMENTAL PROCEDURES

S-100 was purified from ox brain by the method of Moore (1965) slightly modified as reported (Donato, 1978).

MT proteins were obtained from adult rat brain by the method of Shelanski et al. (1973). Tubulin represented 80-85% of thrice cycled MT proteins. Tubulin was deparated from MT-associated proteins (MAPs) by phosphocellulose chromatography (Weingarten et al., 1975). Assembly of MT proteins was followed spectrophotometrically as the increase in absorbance at 350 nm

(A_{350}). Assembly was initiated by adding GTP (1 mM final concentration). The assembly buffer contained 20 mM MES, pH 6.7, 1 mM EGTA, 1 mM $MgCl_2$, 0.12 M KCl and 10 μM free Ca^{2+}. Assembly of purified tubulin (PC-tubulin) was followed in assembly buffer containing 1 mM GTP and initiated by adding DMSO (10% final concentration) (Himes et al., 1976).

Protein was measured by the method of Lowry et al. (1951).

RESULTS AND DISCUSSION

MT protein assembly has been described in terms of two consecutive reactions, a nucleation step preceding the elongation

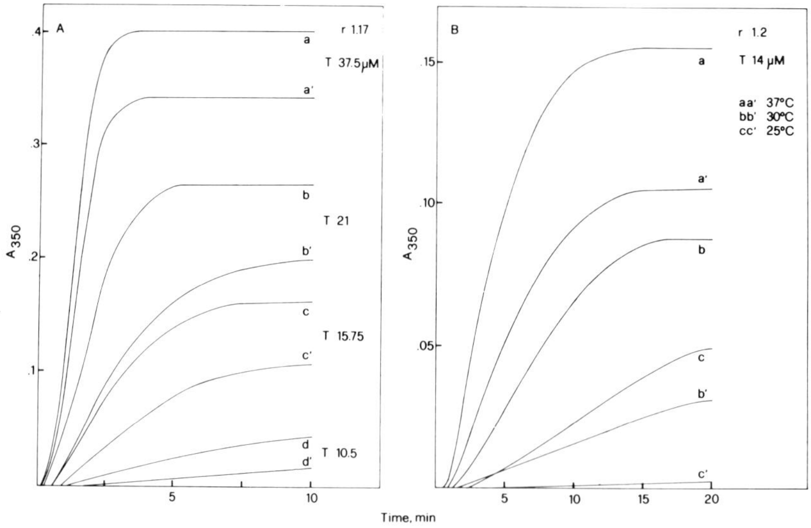

Fig. 58.1. Dependence of the S-100 effect on the MT protein concentration and the temperature of assembly. A. Increasing MT protein concentrations corresponding to the indicated concentrations of tubulin dimer (T) were incubated at 37°C in assembly buffer in the absence (a,b,c) and in the presence (a',b',c') of S-100 at a fixed S-100/tubulin dimer molar ratio (r). After 1 min (zero time), GTP was added and the increase in A_{350} followed for 10 min at 37°C. B. MT proteins corresponding to 14 μM tubulin dimer (T) were incubated at 37°C, 30°C, and 25°C in the absence (a,b,c) and in the presence (a',b',c') of S-100 (r 1.2). After 1 min (zero time), GTP was added and the reaction followed at the same temperatures.

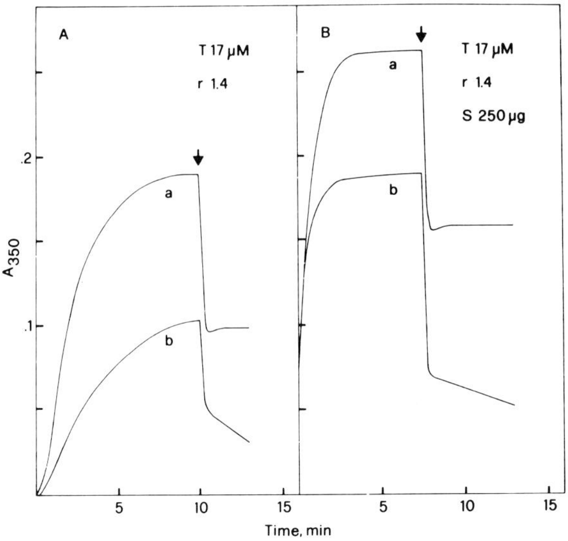

Fig. 58.2. Effect of MT seeds (S) on assembly of MT proteins in the absence and presence of S-100. A. MT proteins corresponding to 17 μM tubulin dimer (T) were incubated at 37°C without (a) and with (b) S-100 at the indicated molar ratio (r). After 1 min (zero time), GTP was added. At the arrow Ca^{2+} was added to a final concentration of 1 mM. B. Conditions were as in A, except that at zero time 20 μl of MT seeds (S) (11 mg protein/ml) were added along with GTP.

reaction (Johnson and Borisy, 1977; Engelborghs et al., 1977). If the S-100 inhibitory effect on MT protein assembly is due to interference with the nucleation step, the assembly in the presence of S-100 should be characterized by an increase in the lag between GTP addition and the onset of the turbidity increase, and in a decrease in the rate of assembly (Donato, 1983). Moreover, the S-100 effect should increase, at a given S-100/tubulin dimer molar ratio, as the MT protein concentration and/or the temperature decrease. Figure 58.1 documents that this actually occurs. The S-100 effect on the lag and on the rate of assembly can be prevented if the MT protein solution is supplemented with seeds of MTs obtained by mechanical disruption of preformed MTs (Fig. 58.2), as expected for a factor interfering with nucleation. However, the effect on the extent is still observed under these conditions, suggesting a distinct interference with the elongation reaction. This conclusion is

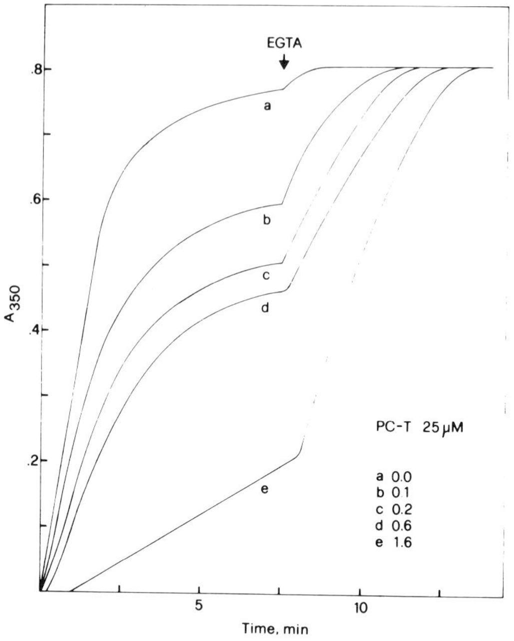

Fig. 58.3. Effect of S-100 on assembly of PC-tubulin. PC-tubulin (PC-T) (25 μM) was incubated at 37°C for 7.5 min in assembly buffer containing 1 mM GTP in the absence and in the presence of increasing S-100 concentrations (the S-100/tubulin dimer molar ratio, r, in single tests is indicated). At zero time, DMSO was added to a final concentration of 10%. After 7.5 min, EGTA was added to a final concentration of 4 mM.

further supported by the effect of 1 mM Ca^{2+} on MT proteins assembled under the conditions depicted in Figure 58.2.

In principle, S-100 could affect MT protein assembly by interacting with either tubulin or one or more MAPs. Figure 58.3 shows that S-100 also inhibits the assembly of PC-tubulin in a dose-dependent way. The S-100 effect is greater on assembly of PC-tubulin than on assembly of whole MT proteins, particularly at the low S-100/tubulin dimer molar ratios. This is probably because in the absence of MAPs, S-100, when present, is the only factor controlling the assembly of tubulin, besides tubulin itself. It could be that in the presence of S-100, tubulin has a reduced sensitivity to MAPs because of either the steric conformation of the tubulin-S-100 complex or of competition between S-100 and MAPs for binding sites on tubulin. Radiolabeled S-100 binds to PC-tubulin but not to MAPs (not shown).

Calmodulin was the first acidic Ca^{2+}-binding protein shown to affect the assembly of MT proteins in vitro (Marcum et al., 1978; Nishida et al., 1979). Calmodulin brings about its inhibitory effect by interacting with the tau factor and MAP2, and has no effect on assembly of PC-tubulin (Sobue et al., 1981; Lee and Wolff, 1982). S-100 interferes with the process of nucleation and with the elongation reaction of MTs by binding to tubulin.

In conclusion, on the basis of these and previously presented data (Donato, 1983), we propose that S-100 is involved in the control of the state of assembly of MTs in a Ca $^{+}$-mediated way.

REFERENCES

Calissano P., Moore B.W. and Friesen A. Effect of calcium ion on S-100, a protein of the nervous system. Biochemistry 8: 4318-26 (1969).

Cocchia D. Immunocytochemical localization of S-100 protein in the brain of adult rat. An ultrastructural study. Cell Tissue Res. 214:529-40 (1981).

Cocchia D., Michetti F., Ruffioni S. and Donato R. Immunochemical and immunocytochemical localization of S-100 protein in the cilia of cell types of different species (this symposium).

Donato R. The specific interaction of S-100 protein with synaptosomal particulate fractions. Site-site interactions among S-100 binding sites. J. Neurochem. 30:1105-1111 (1978).

Donato R. Effect of S-100 protein on assembly of brain microtubule proteins in vitro. FEBS Lett., in press.

Himes R.H., Burton P.R., Kersey R.N. and Pierson G.B. Brain tubulin polymerization in the absence of "microtubule-associated proteins." Proc. Natl. Acad. Sci. USA 73:4397-99 (1976).

Isobe T. and Okuyama T. The amino-acid sequence of S-100 protein (PAP I-b protein) and its relationship to calcium-binding proteins. Eur. J. Biochem. 89:379-88 (1978).

Lee Y.C. and Wolff J. Two opposing effects of calmodulin on microtubule assembly depend on the presence of microtubule-associated proteins. J. Biol. Chem. 257:6306-10 (1982).

Lowry O.H., Rosebrough N.J., Farr A.L. and Randall R.J. Protein measurement with the Folin phenol reagent. J. Biol. Chem. 193:265-75 (1951).

Marcum J.M., Dedman J.R., Brinkley B.R. and Means A.R. Control of microtubule assembly-disassembly by calcium-dependent regulator protein. Proc. Natl. Acad. Sci. USA 75:3771-75 (1978).

Michetti F. and Cocchia D. S-100-like immunoreactivity in a planarian. Immunochemical and immunocytochemical study. Cell Tissue Res. 223:575-82 (1982).

Moore B.W. A soluble protein characteristic of the nervous system. Biochem. Biophys. Res. Commun. 19:739-44 (1965).

Nishida E., Kumagai H., Ohtsuki I. and Sakai H. The interactions between calcium-dependent regulator protein of cyclic nucleotide phosphodiesterase and microtubule protein. I. Effect of calcium-dependent regulator protein on the calcium sensitivity of microtubule assembly. J. Biochem. 85:1257-66 (1979).

Shelanski M.L., Gaskin F. and Cantor C.R. Microtubule assembly in the absence of added nucleotides. Proc. Natl. Acad. Sci. USA 70:765-68 (1973).

Sobue K., Fujita M., Muramoto Y. and Kakiuchi S. The calmodulin-binding protein in microtubules is tau factor. FEBS Lett. 132:137-40 (1981).

Weingarten M.D., Lockwood A.H., Hwo S. and Kirschner M.W. A protein factor essential for microtubule assembly. Proc. Natl. Acad. Sci. USA 72:1858-62 (1975).

59 Do the Myosin Filaments in the Crinoid Arm Muscles Vary in Thickness because of a Fusion Mechanism? : Electron Microscopic and Computer Analysis

M. Daniela Candia Carnevali,
Abele Saita, Guido Pacchetti

INTRODUCTION

According to recent reports (Smith et al., 1981; Saita and Candia Carnevali, 1982; Saita et al., 1982) the echinoderms offer stimulating cues for physiological and morphological studies of muscle tissue, since the various groups (and the same animal too) have functionally and structurally very different muscle systems going from Smooth to a typical obliquely striated model. The arm muscles of the crinoid Antedon consist of fibers all belonging to the latter category, but with striking differences as regards functional requirements and ultrastructural features. These fibers can be distinguished mainly on the basis of different myofilament arrangements, which we previously called A-type and B-type patterns and are variously combined in the arm, according to the different levels (proximal, intermediate, and distal portion) (Candia Carnevali and Saita, in press). The characterizing feature of these obliquely striated fibers is the very heterogeneous arrangement of myosin filaments (Candia Carnevali and Saita, 1982, 1983a,b in press) varying widely in size, number, and distribution from section to section.

The surprising variability of the myosin filament thickness and order can be interpreted on the basis of possible fusion between filaments. This hypothesis is supported in the present paper by direct electron microscope observations and by related computer-aided analysis of quantitative data. The results and their interpretation seem compatibly to fit with the different functional properties, and most of all with the various fiber's differing ability to maintain tension (i.e., active or catch contraction).

MATERIAL AND METHODS

Arms of adult specimens of Antedon rosacea were carefully detached from the animal's central body and dissected into proximal, intermediate, and distal portions. The pieces were fixed (Karnowsky's paraformaldehyde-glutaraldehyde mixture diluted with 0.1 M cacodylate buffer + sea water, 1:1 ratio) at room temperature for five hours, then overnight washed in buffer, postfixed in cacodylate buffered osmium tetroxide, prestained in 1% aqueous uranyl acetate, dehydrated in a graded ethanol series, and embedded in Epon- Araldite mixture. The sections, cut with an LKB V Ultrotome (diamond knife) and stained with lead citrate were observed under Hitachi HS8 and JEOL 100B electron microscopes. The frequencies of thick filaments per 1/4 μm^2 unit area of fiber cross-section were obtained by randomly superimposing a grid on the micrographs and counting chosen areas which fell entirely into fiber contours. The resulting values (means, standard errors, percentage, surface coverage, etc.) were obtained by the SPSS package of programs (Nie et al., 1979) on a UNIVAC 1100/80 computer.

OBSERVATIONS

As previously observed, the flexor muscle bundles of Antedon arm are made up of different fiber types belonging to two main groups (other types of fibers are detected more rarely): the A-type fibers run lengthwise down the whole arm and constitute the only muscle mass in the distal arm; the B-type fibers develop from the intermediate tract and run upwards to the proximal end, where they form a considerable mass surrounding the central A-type core.

Both A- and B-type fibers are obliquely striated, with small oblong cross-sections, completely filled by rectangular fields of thick and thin filaments, spaced by finely granular Z-lines (Figs. 59.1, 59.2 and 59.3). The thick filament pattern is the parameter that varies most in the two types of fiber. The A-type contractile apparatus consists of hexagonally arranged thick filaments, showing slowly increasing diameters (from 20 to 40 nm), so gradually distributed that they must surely represent various levels of the same tapered filament (Fig. 59.2).

The B-type fibers, in contrast, in both cross and longitudinal section, show filaments of exceptional caliber (some more than 110 nm thick!) apparently randomly mixed among normal thick and thin filaments, but regularly arranged in the sarcomere like normal myosin filaments (Figs. 59.3, 59.4). Since they too are hexagonally ordered, surrounded by crowns of actin filaments and even structured like the myosin filaments, they cannot be considered a third type of filament but rather oversized myosin filaments. No periodic structure is recognizable in magnified longitudinal sections (Figs. 59.5, 59.6), but they can be expected to have a paramyosin core inside (biochemical analysis is in progress).

Because of the remarkable variety in their thickness and the frequency distribution from fiber to fiber they cannot be taken as constituting a homogeneous population of myosin filaments; this also seems impossible in view of the measured filament length (3.5-4.0 μm). On the hypothesis of a gradual increase in caliber along the same tapered filament, going from 20 to 110 nm in cross-section diameter, we should obtain a theoretical length of roughly 18 μm.

The number of aberrant filaments seems not to be casual, but strictly related to the height of the arm section (Figs. 59.2, 59.3, 59.4). They are obviously completely lacking in all the distal sections and in the central core of the bundle, i.e., where only A-type fibers are developed; in the B-type fibers their number varies from quite small in the intermediate sections (B II-type fibers) (Fig. 59.3), to progressively more in proximal sections, at which height (B I-type fibers) (Fig. 59.4) almost all the filaments are the oversized type.

Magnified cross and longitudinal sections suggest that these extra-thick filaments originate from some sort of fusion mechanism between thick filaments: often the cross-sections show aberrant filaments with very irregular outlines resembling two or more than two, very close, semifused thick filaments (Fig. 59.8a,b,c,d,e,f,g,h), while in longitudinal sections some Y-shaped filament forked tracts can be identified, where normally thick filaments seem to merge (Fig. 59.7).

DISCUSSION

It is difficult at the moment to draw any firm conclusion, because of the different facets of the oversized filament question. Experimental and theoretical arguments can be debated and completely verifiable hypothesis proposed only when conclusive biochemical and physiological evidence becomes available (research is in progress).

Exceptionally large, thick filaments have often been observed in other invertebrate obliquely striated and smooth fibers

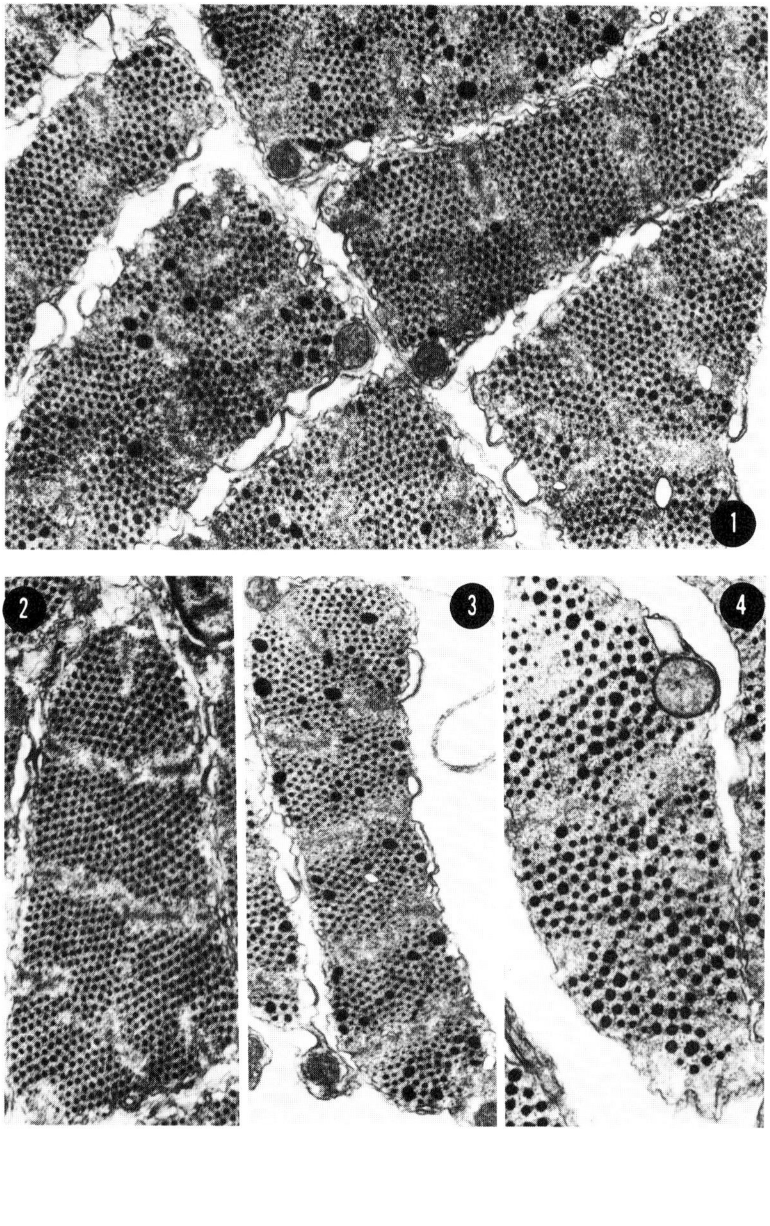
1
2
3
4

(annelids, molluscs), always related to conditions requirring the development and maintenance of strong tension (Sobieszek, 1973; Morrison and Odense, 1974; Lanzavecchia, 1971, 1977; Nonomura, 1974; Gilloteaux and Baguet, 1977; Achazi, 1982; Cohen, 1982). Most of these muscles, in fact, exhibit a typical catch contraction (for the others this function is not yet verified); their fibers, in any case, show a homogeneous thick filament population, i.e., the thick filaments tend to be exceptionally large, often both in diameter and length, and have a large paramyosin core surrounded by a myosin surface layer.

The peculiar feature of the Antedon B-type fibers is the marked polymorphism of the thick filaments, whose thickness and frequency seem to vary in the different fibers and in the same fiber too, presumably with precise meanings. It can be hypothesized that during fiber differentiation the contractile apparatus pattern, namely the thick filament molecular and spatial assembly, is determined and modulated in response to the momentum. In other words, according to the stress it has to bear, a single fiber may be able to rearrange its thick filaments to some degree, from a pattern designed for quick active contractions to another, more suited to isometric maintenance of contraction (catch function).

Through a possible fusion mechanism between myosin filaments, compatible with the main different theories put forward

Fig. 59.1. Interbrachial flexor muscles of Antedon: cross-section of B-type fibers. These are obliquely striated, small, internally organized in regular fields of thick and thin filaments, spaced by granular Z-lines. Surprisingly, thick filaments can be detected, apparently distributed at random among normal actin and myosin filaments. They seem to vary in number from fiber to fiber but are arranged in the sarcomere just like normal myosin filaments. ×28,000

Fig. 59.2. Cross-section of A-type fiber (distal portion of the arm), showing no oversized myosin filaments in its contractile apparatus. ×30,000

Fig. 59.3. B II-type fiber (intermediate portion of the arm) in cross-section. The very thick myosin filaments are not very numerous and they vary in caliber. ×30,000

Fig. 59.4. B I-type fiber (proximal portion of the arm) in cross-section. Almost all the myosin filaments seem extra-thick. ×30,000

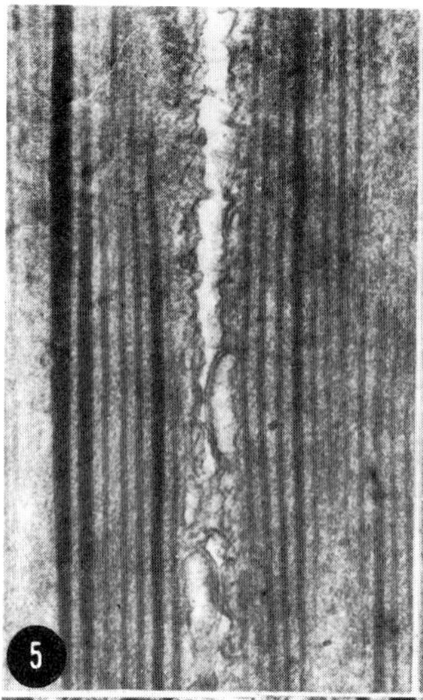

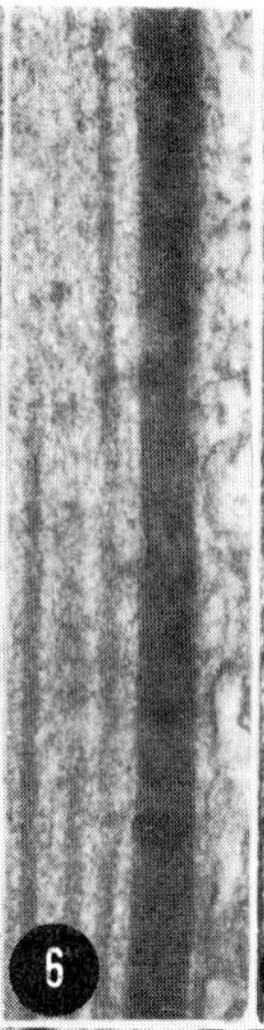

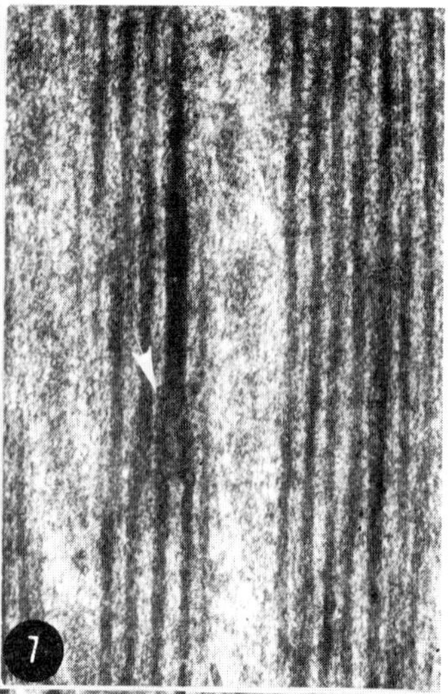

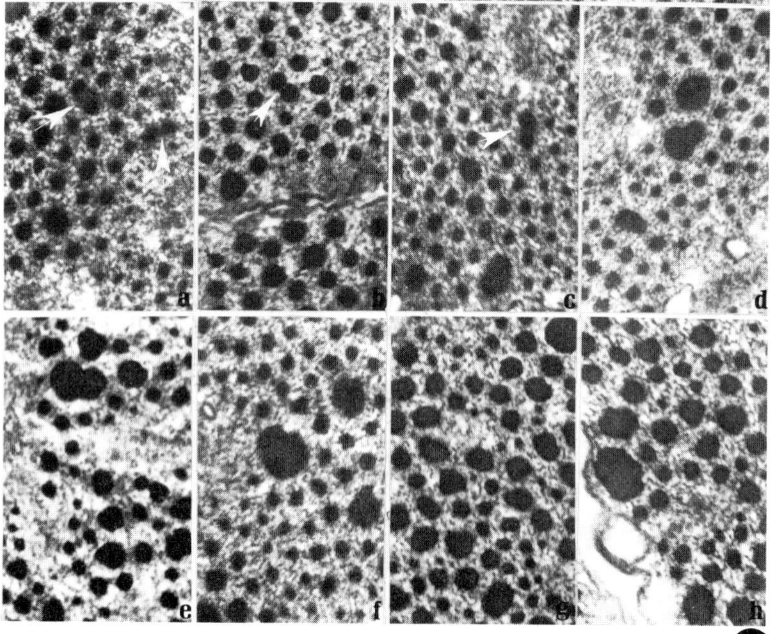

8

for the catch mechanism (Achazi, 1982; Cohen, 1982) the same model of fiber could fulfill a range of different functional needs, including powerful action, by providing itself with the means to strengthen its mechanical support. The thicker filaments thus obtained would be more suitable than the thinner ones for maintaining tension and resisting stretch, with no great reduction of contraction speed during the active contractions because they are not too long.

The fusion hypothesis has been tested by classifying the filaments in ranks of 20-nm cross-section diameter. The frequency counts give types characteristic monomodal distributions, whose modal classes (30 nm) significantly differ for the three fiber types ($P_{(X^2)} < 5\%$) (Fig. 59.9).

The relationship between thick filament distribution patterns and the functional features of the fiber types can be better analyzed considering the total and class-related filament cross-section coverage of fiber unit areas (Fig. 59.10). The most characteristic features are the inversion of gradient between 30 and 50 nm filament classes (Figs. 59.10c,d,g,h), and the wider variability within B I-type filament classes.

The distribution pattern can be shown better assuming that a certain degree of "noise" can be introduced by different section diameters along the thicker filament classes; the last

Fig. 59.5. In longitudinal section the oversized thick filaments look tapered, and in their order in the sarcomere and their regular oblique stagger they look like myosin filaments. ×35,000

Fig. 59.6. Magnified longitudinal section of an extra-thick filament. Probably these filaments have a central paramyosin core, even though the typical longitudinal period is not recognizable in the sections. ×95,000

Fig. 59.7. Longitudinal detail showing possible fusion of myosin filaments to form a thicker one. ×65,000

Fig. 59.8. Highly magnified cross-sections show different organizational levels of the thick filaments; the details can be interpreted as the possible sequence of events leading to formation of the oversized filaments, by some sort of fusion mechanism between filaments. Starting from the first stages (a,b,c), we can reach further steps of association of myosin filaments, and finally the extra-thick filaments (100-110 nm in diameter), often showing very irregular outlines (d,e,f,g,h). ×95,000

9

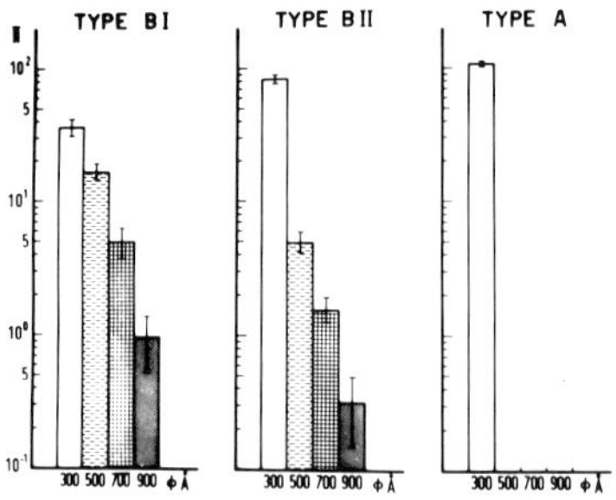
TYPE BI
TYPE BII
TYPE A
300 500 700 900 φ Å

10

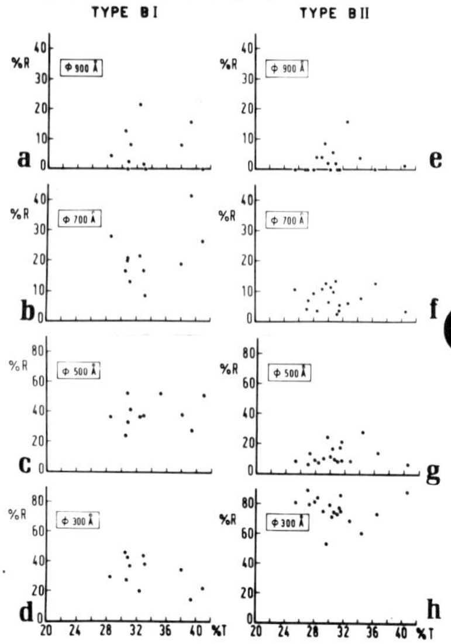
TYPE BI
TYPE BII
%R
φ 900 Å
φ 700 Å
φ 500 Å
φ 300 Å
a
b
c
d
e
f
g
h
%T

11

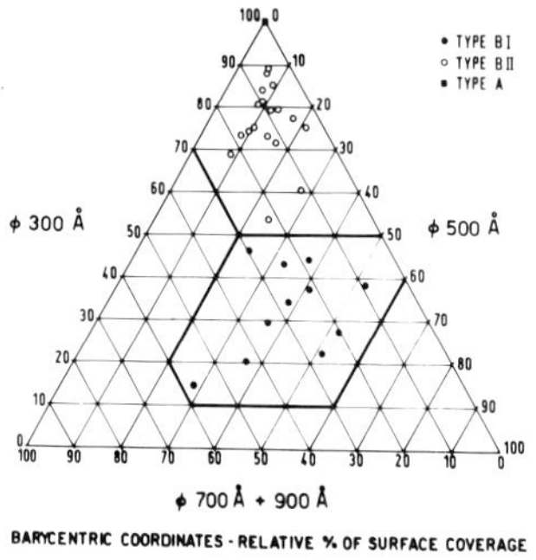
• TYPE BI
○ TYPE BII
▪ TYPE A
φ 300 Å
φ 500 Å
φ 700 Å + 900 Å
BARYCENTRIC COORDINATES - RELATIVE % OF SURFACE COVERAGE

12

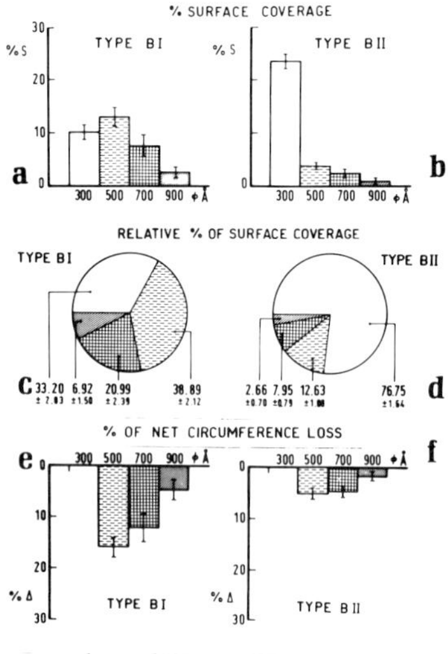
% SURFACE COVERAGE
TYPE BI
TYPE BII
%S
300 500 700 900 φ Å
a
b
RELATIVE % OF SURFACE COVERAGE
c
33.20 6.92 20.99 38.89
± 7.03 ±1.50 ± 2.39 ± 2.12
d
2.66 7.95 12.63 76.75
±0.70 ±0.79 ±1.00 ±1.64
% OF NET CIRCUMFERENCE LOSS
e
f
%Δ

13

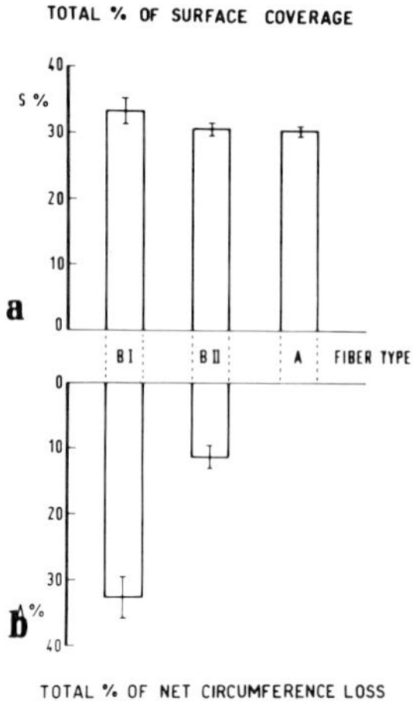
TOTAL % OF SURFACE COVERAGE
S %
a
BI
BII
A
FIBER TYPE
b
Δ%
TOTAL % OF NET CIRCUMFERENCE LOSS

two (70 and 90 nm) can be tentatively merged and the relative percentage surface coverage plotted on a barycentric graph (Fig. 59.11). There is a clustering of different section size micrographs. Figures 59.12a and b summarize the results of absolute (histograms) and relative (pie-charts) percentages of surface coverage per class per fiber type. The nonrandom clustering suggested by the barycentric plot seems to be confirmed.

Fig. 59.9. Mean frequencies ($\bar{N}$) of thick filaments per 1/4 μm^2 unit area of fiber cross-section and the corresponding standard errors are shown on a logarithmic scale for each fiber type and filament cross-section class.

Fig. 59.10. Surface coverage for each filament cross-section class as a percentage of the total surface coverage (%T) per fiber type (relative percentage: %R). The relative percentage surface coverage was calculated as the mean ratio between the surface of the thick filament cross-sections per unit area and the total sum of the thick filament cross-section surfaces per unit area. The total percentage surface coverage was calculated as the sum of the absolute percentage surface coverage of each filament class.

Fig. 59.11. The plot on barycentric coordinates of the relative percentage surface coverage per thick filament class (70 nm and 90 nm filaments have been pooled) shows a characteristic pattern of distribution. Each point represents a micrograph.

Fig. 59.12. Histograms and pie-charts of the relevant amounts discussed through the text are superimposed to emphasize the most striking features of the structure. (a) %S: mean absolute percentage surface coverage and the corresponding standard errors. (b) The relative percentage surface coverage was calculated as in Fig. 59.10. (c) the percentage net surface loss (%Δ) was calculated as the ratio between the area and the circumference corresponding to a given class of thick filament diameter, assuming 30 nm as the basic unit and weighting for the absolute percentage surface coverage.

Fig. 59.13. The total % surface coverage and the total percentage net circumference loss are obtained by adding the quantities shown in Fig. 59.12a and 12c.

These results led us to investigate the functional significance of the filament types and of the hypothetical fusion mechanism. The possibility of myosin filaments merging into thicker ones is also suggested on examining the total percentage of filament cross area coverage (Fig. 59.13), which does not significantly differ fro the three fiber types ($P_F < 1\text{‰}$). The mechanical strength achieved by this merging of filaments into thicker ones should entail a loss of filament lateral surface (considering that the circumference is linearly related to the radius, while area is related to the square of the radius) with a possible loss of actomyosin bridges (Figs. 59.12e and 59.13b). This might be true if the density of the bridges had to remain constant.

According to Cohen's recent model (1982), however, the structure of the very thick myosin filament of the catch muscles is different from that of the other muscles (Squire, 1975), namely, there is a particular myosin-paramyosin reciprocal arrangement allowing complete actomyosin and myosin-paramyosin interactions. The stretch resistance characterizing the catch state depends directly on this special organization. Therefore the apparent loss of filament lateral surface consequent to fusion between filaments is not inconsistent with increased strength of the fiber, because it is not inevitably related to a loss of actomyosin cross bridges. In any case, this would be compensated by other interactions and possible linkages between the matching molecules (Twarog and Mumola, 1972; Tameyasu and Sugi, 1976; Gilloteaux and Baguet, 1977; Pfitzer and Ruegg, 1982; Achazi, 1982; Cohen, 1982).

ACKNOWLEDGMENT

This work has been supported by a CNR grant no. CTO2431.

REFERENCES

Achazi R.K.: Catch muscle, in: "Basic Biology of Muscles: A comparative approach." (B. M. Twarog, R. J. C. Levine, M. M. Dewey, eds.), Raven Press, New York, pp. 291-308 (1982).

Candia Carnevali M.D. and A. Saita: Different myofilament patterns in the muscles of a comatulid (Echinodermata-Crinoidea). Biol. of the Cell, 45, 273 (1982).

Candia Carnevali M.D. and A. Saita: Muscle system organization in the Echinoderms—II. Microscopic anatomy and functional meaning of the muscle-ligament-skeleton system in the arm of the comatulidis (Antedon rosacea), J. Mar. Biol. Ass. U.K., in press (1983a).

Candia Carnevali M.D. and A. Saita: Muscle system organization in the Echinoderms—III. Fine structure of the contractile apparatus in the arm flexor muscles of the comatulids (Antedon rosacea), J. Mar. Biol. Ass. U.K., in press (1983b).

Cohen C.: Matching molecules in the catch mechanism, Proc. Nat. Acad. Sci. U.S.A., 79, 3176-78 (1982).

Gilloteaux J. and F. Baguet: Contractile filament organization in functional states of the anterior byssus retractor muscle (ABRM) of Mytilus edulis L., Eur. J. Cell Biol., 15, 192-220 (1977).

Lanzavecchia G.: Studi sulla muscolatura elicoidale e paramiosinica—IV. La muscolatura longitudinale e circolare della parete corporea degli Anellidi Tubificidi, Accad. Naz. Lincei (Rend. Sci. fis. mat. e nat.), 50, 6-12 (1971).

Lanzavecchia G.: Morphological modulations in helical muscles (Aschelminthes and Annelida), Int. Rev. Cytol., 51, 133-86 (1977).

Morrison C.M. and P.H. Odense: Ultrastructure of some pelecypod adductor muscles, J. Ultrastruc. Res., 49, 228-51 (1974).

Nie N.H., C. H. Hull, J.G. Jenkins, K. Steinbrenner and D.M. Brent: SPSS-Statistical package for the social sciences, McGraw-Hill, New York, Toronto (1975).

Nonomura Y.: Fine structure of the thick filament in molluscan catch muscle, J. Mol. Biol., 88, 445-55 (1974).

Pfitze G. and J.C. Ruegg: Molluscan catch muscle: regulation and mechanics in living and skinned anterior byssus retractor muscle of Mytilus edulis, J. Comp. Physiol., 147, 137-42 (1982).

Saita A. and M.D. Candia Carnevali: Myofilament arrangement in some muscles of Crinoidea (Echinodermata), Caryologia, 35, 112-13 (1982).

Saita A., M.D. Candia Carnevali and M. Canonaco: Muscle system organization in the Echinoderms—I. Intervertebral muscles of Ophioderma Longicaudum (Ophiuroidea), J. Submicro. Cytol., 14, 291-304 (1982).

Smith D.S., S.A. Wainwright, J. Baker and M.L. Cayer: Structural features associated with movement and "catch" of sea-urchin spines, Tissue & Cell, 13 299-320 (1981).

Sobieszek A.: The fine structure of the contractile apparatus of the anterior byssus retractor muscle of Mytilus edulis, J. Ultrastruc. Res., 43, 313-43 (1973).

Squire J.M.: General model of myosin filament structure. III. Molecular packing arrangements in myosin filaments, J. Mol. Biol. 77, 291-323 (1973).

Tameyasu T. and H. Sugi: The semielastic component and the force-velocity relation in the anterior byssal retractor muscle of Mytilus edulis during active and catch contraction, J. Exp. Biol., 64, 497-510 (1976).

Twarog B.M. and Y. Mumoka: Calcium and the control of contraction and relaxation in a molluscan catch muscle, Cold Spring Harbor Symp. Quant. Biol., 37 489-503 (1972).

60 Mode of Filament Assembly of Myosins Extracted from Gravid and Nongravid Human Uterus

F. Cavaille and I. Pinset-Harstrom

Our first investigations concerning human uterine myosin led us to the conclusion that the protein found in the nongravid organ (NG) is not strikingly different from those found in the gravid one (G): we had not found any difference in the K-EDTA-ATPase activity and in the peptide maps obtained after limited chymotryptic proteolysis of the heavy chains (F Cavaille, J. J. Leger, 1983 in press). These results are contradictory with those of Huszar and Bailey (1980) who reported higher ATPase activity for G Myosin and differences in the amino acid composition of a fragment of the heavy chains.

The purpose of this work was to study the ability of uterine myosin to form filaments in solution, comparing G and NG proteins, for it has been shown in skeletal muscle that the different isoenzymes of myosin appearing during the muscle development have not the same polymerization properties. (Whalen et al., 1981).

Recent works seem to indicate that LC_{20} phosphorylation plays an important role in filament formation of smooth muscle myosin, so we have compared G and NG uterine myosin in their phosphorylated (P) and nonphosphorylated (nP) forms. Our results provide evidence that NG myosin is more soluble than G myosin, and that the phosphorylation is necessary to obtain long filaments.

MATERIAL AND METHODS

NP myosin was rapidly extracted from gravid and nongravid uteri as previously described (Cavaille, Leger, 1983). P myosin is obtained by the same procedure including a phosphorylating step before column chromatography as in Chacko et al. (1977).

Two dimensional gel electrophoresis (Isoelectrofocusing followed by SDS polyacrylamide gel electrophoresis) was used to assess the level of LC_{20} phosphorylation.

Formation of myosin filaments was obtained by controlled dilution to a final KCl concentration of 0.12 M in the presence of 3 mM $MgCl_2$ and 1 mM EGTA. Various pH, ranging from 6.3 to 8.0 were tested. When ATP was added, its concentration was 2 mM. Electron microscopy observation of the filaments was performed as in Pinset-Harstrom, Truffy (1979).

One aliquot of diluted myosin solutions was centrifuged in a Beckman Airfuge (130,000 g, 15 min.) and resulting supernatant was assayed for protein to estimate the percentage of aggregated myosin and of soluble products.

RESULTS

Phosphorylation. LC_{20} phosphorylation is very low (<10%) for the myosin extracted without a step of phosphorylation, while phosphorylation level is higher than 80% in the preparations including this step. Myosin is free of proteolysis.

Filaments. Three main types of filaments were obtained: bipolar with a bare zone of 1,500-1,900 Å, with a mixed polarity or side polar. The length of the side polar filaments varied from 0.3 μm to 2.1 μm. They have no central bare zone. The cross bridges pointed out all along their length, except at the ends. Some very thin and short minifilaments were also seen.

NP myosins aggregated in bipolar or mixed polar filaments, but soluble material is important (40%) for G; nP myosin and very few filaments were seen with NG, nP myosin (99% soluble).

In the presence of ATP, G nP myosin was completely solubilized at pH higher than 6.5 P-myosins aggregated in side polar filaments, but also in mixed polarity filaments for nG P myosin. Some minifilaments were seen in these preparations. These filaments are sedimented by ultracentrifugation. GP myosin gave side polar filaments of mean length 1.2 ± 0.6 μm (n = 30) while nGP myosin gave mixed polar and side polar filaments shorter (0.6 ± 0.2 μm, n = 76) than GP myosin filaments.

When P myosin was diluted with a solution containing ATP, only bipolar or mixed polarity filaments were obtained and soluble material increased (80%). This effect was more important at pH higher than 6.5. When ATP was added to still formed long side polar filaments of GP myosin, the filaments were not disrupted.

DISCUSSION

The filaments obtained in this study with human uterine myosin look like those described by Craig and Megerman (1977), Kendrick-Jones et al. (1983) who studied aorta and gizzard myosins. The higher solubility of nP myosin and the lability of the bipolar filaments in the presence of ATP we observed with uterine myosin agree well with the results of Suzuki et al. (1978).

The long side polar filaments we obtained with G-P myosin are similar in their aspect and length to those seen by Kendrick-Jones et al. (1983) studying P-gizzard myosin. Although the level of LC_{20} phosphorylation is similar in the preparations of G and nG myosins, the nG myosin led to shorter side polar filaments and to mixed polarity filaments. This difference in their aggregation mode reflects probably structural differences between the two molecules, the nG myosin, phosphorylated or not, being more soluble than the G myosin.

The presence of ATP in diluting solution modify the aggregation mode of P uterine myosin; this effect was not reported by Kendrick-Jones (1983), but it is important to notice that the mode of filament assembly of smooth muscle myosin depends dramatically on the experimental conditions, and it is hazardous to compare results not obtained exactly with the same protocol.

In conclusion, it appears from this work that human uterine myosin aggregates in filaments of size and type still reported for other smooth muscles and that LC_{20} phosphorylation permits the formation of side polar filaments. Myosins from gravid and nongravid human uterus differ in their degree of association in a manner that is independent of the phosphorylation level. The higher solubility of the myosin of nongravid uterus can reflect structural differences between the two molecules.

REFERENCES

F. Cavaille, J. J. Leger, Obstet. Gynec. Invest. (1983) in press.

S. Chacko, M. A. Conti, R. S. Adelstein. Proc. Natl. Acad. Sci. USA (1977), 74, 129-93.

R. Craig, J. Megerman, J. Cell. Biol. (1977), 75, 990-96.

G. Huszar, P. Bailey, Proc. Ann. Met. Amer. Soc. Biol. Chem. (1980), Abstract no. 2960.

J. Kendrick-Jones, W. Z. Cande, P. J. Tooth, R. C. Smith, J. M. Scholey, J. Mol. Biol. (1983), 165:139-62.

I. Pinset-Harstrom, J. Truffy, J. Mol. Biol. (1979), 134, 173-88.

H. Suzuki, H. Onishi, K. Takahashi, S. Watanabe, J. Biochem., Tokyo (1978) 84, 1529-42.

R. G. Whalen, S. M. Sell, G. S. Buttler-Browne, K. Schwartz, P. Bouveret, I. Pinset-Harstrom, Nature (1981), 292, 805-09.

61 Organizing Centers for Intermediate Filaments

Barry S. Eckert and Susan E. Caputi

INTRODUCTION

Intermediate filaments have been studied extensively in recent years with particular interest to their organization in the cytoskeleton. Several reports (Eckert and Daley, 1981; Klymkowsky, 1981; Gawlitta et al., 1982) demonstrated that microinjection of antibody specific for keratin or vimentin resulted in formation of a juxtanuclear aggregate of intermediate filaments. It has been suggested (Eckert et al., 1982a,b) that this response indicates the presence of an intermediate filament organizing center (IFOC).

In this paper, we present further observations of the behavior of living PtK1 cells that increase our understanding of the IFOC.

MATERIALS AND METHODS

Cell Culture. PtK1 cells are cultured in Ham's F-12 media supplemented with 10% Fetal Bovine Serum. Cells to be used for immunofluorescence are grown on 18 mm square coverslips.

Wounding of Cultures. Confluent monolayers of PtK1 cells were wounded by scratching the surface of the coverglass with a sterile plastic cell scraper.

Immunofluorescence. Details of preparation and characterization of antibody specific for keratin are given in Eckert and Daley (1981). Staining procedure using 37% formaldehyde containing 0.5% Triton X-100 is given in detail in Parysek and Eckert (1983).

Video Microscopy. Time lapse video recordings are made on a panasonic NV 8030 time lapse recorder. A panasonic low light video camera is used. Cells are kept at 37°C with a Sage air curtain incubator and pH is maintained at 7.2 with HEPES added to 10 mM.

Random Photomicroscopy. Random fluorescent micrographs were made by moving the mechanical stage on a Zeiss Ultraphot without looking in the oculars. The resulting field of cells was focused and photographed regardless of content.

RESULTS

Random observation of PtK1 cells prepared for immunofluorescence reveal that subconfluent cultures contain a significant number of cells showing aggregation of keratin near the nucleus (Fig. 61.1a). These aggregates are quite similar in appearance to those observed after microinjection (Eckert and Daley, 1981). We have confirmed through time lapse video that sparse cultures contain a high proportion of motile cells, roughly equal to the percentage of cells showing aggregation of filaments (69%). In contrast, confluent cultures contain relatively few motile cells. Approximately 85% of confluent cells show well-spread keratin networks (Fib. 61.1b) and few keratin aggregates.

We also noticed in the video images that motile cells contained a phase dark spot close to the nucleus. This was not seen in nonmotile cells. To identify this spot we prepared cells for

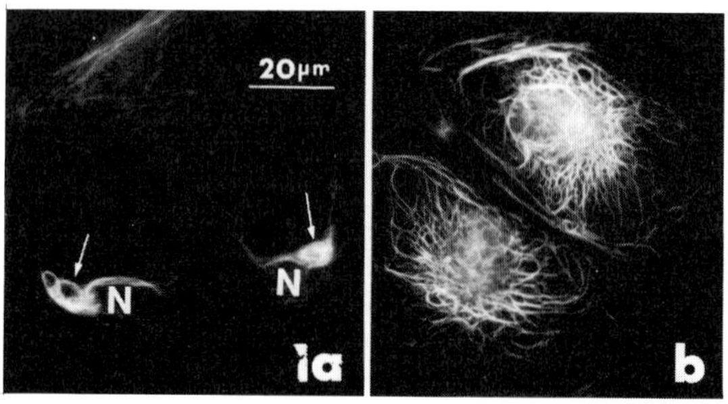

Fig. 61.1. Comparison of PtK1 cells from sparse and dense cultures by immunofluorescent labeling with antikeratin. (a) Cells from sparse cultures showing prominent aggregates of keratin (arrows) near the nucleus (N). Cells of this morphology accounted for 68% of the cells in sparse cultures. (b) Cells from dense cultures of PtK1 showing well-spread networks of keratin filaments. Cells with this morphology accounted for 85% of cells in dense cultures.

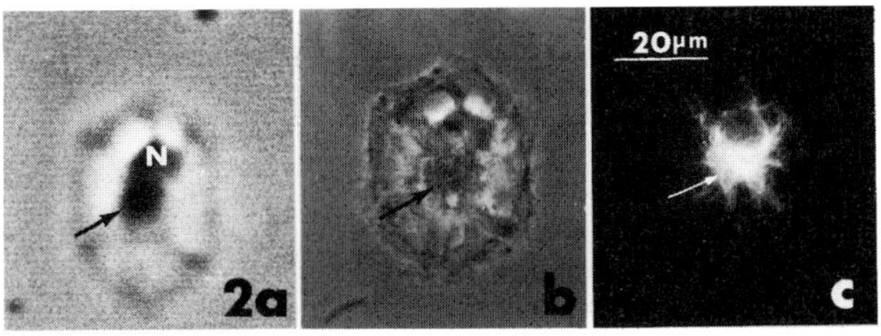

Fig. 61.2. Correlation between video image of living cells and immunofluorescence of those same cells. (a) An image from video of a PtK1 cell. The nucleus (N) and phase-dark spot (arrow) are indicated. (b) Phase contrast image (directly from the microscope) of the same cell. (c) Immunofluorescence of the same cell showing the distribution of keratin. The phase-dark spot is revealed as a keratin aggregate.

immunofluorescence after video recording and located the same cells in the fluorescence microscope. In all cases the phase dark spot observed in motile cells corresponded to an aggregate of keratin filaments (Fig. 61.2). Furthermore, this aggregate was observed on the side of the nucleus closest to the leading edge of migrating cells.

To further demonstrate the relationship of juxtanuclear keratin filament aggregation to motility, confluent cultures were wounded. PtK1 cells begin migrating into the wounded area of the coverglass within 90 minutes. At the same time, those cells migrating showeed keratin filament aggregates close to their nuclei(Fig. 61.3).

DISCUSSION

The results presented have shown that when PtK1 cells become motile or change shape, their keratin filaments form an aggregate close to the nucleus. In motile cells, this aggregate is on the leading side of the nucleus. We believe that this consistent location is the site of the IFOC.

These observations allow us to speculate on the function of the IFOC. It is clear from work of other laboratories that intermediate filaments are essentially completely assembled in the cytoplasm. It also appears that intermediate filament proteins assemble quite soon after they are synthesized (Bilkstead and Lazarides, 1983). This would tend to rule out assembly as a function of the IFOC. This leaves control of distribution of

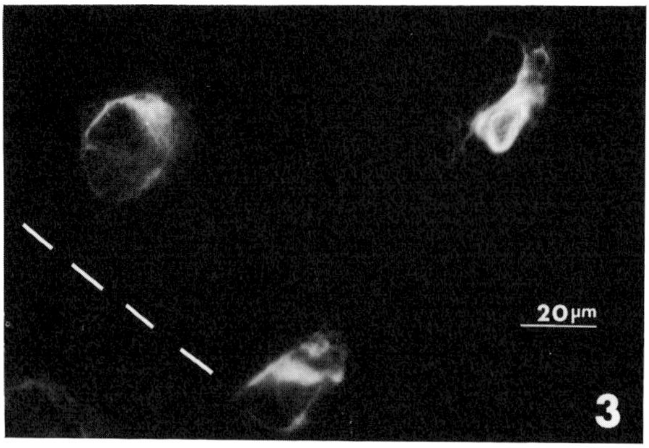

Fig. 61.3. An immunofluorescence micrograph of cells that are beginning to move into a wound in a confluent culture of PtK1 cells. The broken white line indicates the edge of the wound. The cells that have begun to migrate show prominent keratin aggregates on the leading side of their nuclei.

filaments as the primary function. We are, therefore, renaming this region of the cell the Intermediate Filament Distribution Center (IFDC).

We believe that the function of the IFDC is the redistribution of the intermediate filament network in dynamic cells. Clearly it would be difficult for a cell that is changing shape to move a complex, intact network of filaments through its cytoplasm. We suggest that more efficient mechanisms would be the collapse of the network into a compact aggregate, movement of the aggregate, and redistribution of filaments into a new network. We have not observed spheroid keratin bodies (Franke et al., 1982) in our cells. This leaves the IFOC as the only distribution point in PtK1 cells.

The components and ultrastructure of the IFOC are unknown. Although the centriole is not in this area, dense amorphous material is frequently observed in the electron microscope to be associated with intermediate filament aggregates. The significance of this material is unknown but is currently being investigated. It is clear, however, that the IFOC is an important functional component of the cytoskeleton.

ACKNOWLEDGMENTS

Supported by Grant PCM 81-10876 from the National Science Foundation. This work was done during the tenure of an Established Investigatorship of the American Heart Association with funds contributed in part by the Western New York Chapter.

REFERENCES

Bilkstad, I. and Lazarides, E. 1983 Vimentin filaments are assembled from a soluble precursor in avian erythroid cells. J. Cell Biol. 96:1803-08.

Eckert, B. S. and Daley, R. A. 1981 Response of cultured epithelioid cells (PtK1) to microinjection of antibody specific for keratin. Biol. Cell 41:227-30.

Eckert, B. S., Daley, R. A. and Parysek, L. M. 1982a Assembly of keratin onto PtK1 cytoskeletons: Evidence for an intermediate filament organizing center. J. Cell Biol. 92: 575-78.

Eckert, B. S., Daley, R. A. and Parysek, L. M. 1982b In vivo disruption of the cytokeratin cytoskeleton in cultured epithelial cells by microinjection of antikeratin: Evidence for the presence of an intermediate filament organizing center. Cold Spr. Harbor Symp. Quant. Biol. 46:403-12.

Franke, W. W., Schmid, E., Grund, C. and Geiger, B. 1982 Intermediate filament proteins in non-filamentous structures: Transient disintegration and inclusion of subunit proteins in granular aggregates. Cell 30:103-13.

Gawlitta, W., Osborn, M. and Weber, K. 1981 Coiling of intermediate filaments induced by microinjection of a vimentin-specific antibody does not interfere with locomotion and mitosis. Eur. J. Cell Biol. 26:83-90.

Klymkowski, M. W. 1981 Intermediate filaments in 3T3 cells collapse after intracellular injection of a monoclonal anti-intermediate filament antibody. Nature 291:249-51.

PART V

DISORDERS OF CONTRACTILE AND CYTOSKELETAL COMPONENTS

62 Pathophsiology of the Myoephithelial Cells

G. Bussolati and P.Gugliotta

THE MYOEPITHELIAL CELLS

Epithelial cells with contractile activity, spindle-shaped and flattened against the basement membrane, have been described in the salivary, the lacrimal, the bronchial and esophageal sero-mucosal, the sweat and the mammary glands (Hamperl, 1970).

These cells have been termed myoepithelial cells: their best known and common function is contraction leading to the ejection of the gland product under chemical and nervous or, in the case of the mammary glands, hormonal stimulation.

These cells might, however, be involved also in other functions, such as the transfer of various substances from the capillaries to the secretory cells and the production of type IV collagen for building up the basement membrane (Sanger, 1975).

The morphological identification of the myoepithelial cells on tissue sections, while easy in normal conditions, can be difficult or even impossible with any certainty in pathological conditions: these cells can undergo processes of degeneration and even atrophy, dislocation, and hyperplasia, thus losing their typical morphology and distribution (Hamperl, 1970). Electron microscopical investigations have been employed to identify in normal and pathological conditions the myoepithelial cells, which are mainly characterized by bundles of microfilaments (Ahmed, 1974; Haguenau, 1959; Ozzello, 1971).

The ultrastructural approach may, however, be often unrewarding or too time consuming. Thus several studies have been conducted for the selective identification of these cell types by means of special staining procedures (Macartney et al., 1979; Van Bogaert et al., 1977; Van Bogaert et al., 1977). Van Bogaert and coworkers tried to apply Putchler's staining methods for smooth muscle cells to the identification of the myoepithelial cells of the breast (Putchtler et al., 1969). These methods, however, proved unreliable in pathological conditions. Given the high content in contractile proteins of the myoepithelial secretory cells, the immunocytochemical approach would represent a practical staining method for the identification of these cells.

IMMUNOCYTOCHEMICAL PROCEDURES

Cytoskeletal Protein

Actin (Archer et al., 1971; Archer and Kao, 1968; Gabbiani et al., 1973), actinin, myosin, tropomyosin (Archer et al., 1971; Gabbiani et al., 1976; Macartney et al., 1979) and prekeratin (Franke et al., 1980; Gusterson et al., 1982) have been localized in the myoepithelial cells by immunofluorescence procedures. The antigenicity of these proteins can be affected by routine formalin fixation and paraffin embedding; the application of these techniques to pathological studies and to differential histological diagnosis is therefore not straightforward. To localize actin we have tested the following procedures:

- Fluoresceinated phalloidin kindly supplied by Prof. Th. Wieland (Max Planck Institute for Medical Research, Heidelberg) and used at a concentration of 0.1 mg/ml according to the procedure described by Wulf et al. (1979).
- DNAse type 1 (from Boehringer, Mannheim), diluted in PBS at a concentration of 1 mg/ml, followed by rabbit anti-DNAse antiserum employed at a 1:10 dilution in PBS according to the procedure described by Wang and Goldberg (1978). A standard indirect immunofluorescence procedure with antirabbit fluoresceinated antibodies or alternatively an immunoperoxidase method with the PAP procedure (Sternberger et al., 1970) was used to detect the bound DNAse-anti-DNAse complexes.
- Human smooth muscle autoantibodies, from patients with chronic hepatitis. These were employed in a standard indirect immunofluorescence procedure.
- Rabbit antiactin antibodies, prepared against chicken gizzard F actin. Affinity purified antibodies were kindly supplied by Dr. Klaus Weber (Max Planck Institute for Biophised Research, Goettinger) and from Dr. P. Cappuccinelli (Institute of Microbiology of the University of Sassari).

These were employed diluted 1:10 in PBS in indirect immunofluorescence and 1:100 up to 1:1,000 in immunoperoxidase. Fluoresceinated goat antirabbit IgG antisera (from Behringwerke, Marburg) diluted 1:5 in PBS were used for immunofluorescence. For immunoperoxidase, the peroxidase antiperoxidase (PAP) procedure according to Sternberger et al. (1970) and the avidin-biotin-peroxidase complex (ABC) procedure (Hsu et al., 1981) with a slight modification (Bussolati and Gugliotta, 1983) to avoid nonspecific staining of mast cells were employed. In addition, a recently proposed immunogalactosidase procedure (Bondi et al., 1982) has been used in alternative.

Rabbit antimyosin antisera, either kindly supplied by Dr. Vincenzo Eusebi, of the University of Bologna, or purchased from Miles (Elkhart, USA) have been tested on formalin or methacharn-fixed tissues, with or without prior proteinase treatment according to Gusterson et al. (1982). Immunofluorescence and immunoperoxidase procedures have been employed.

Antikeratin antisera have obtained in rabbits using human callous keratin as antigen. A commercially available antiserum from DAKO (Santa Barbara, CA, USA) has also been tested in standard immunocytochemistry procedures.

S-100 Protein

The presence of the nervous system protein S-100, or of related antigens, has been reported in the myoepithelial cells of the breast (Nakajima et al., 1982; Kahn et al., 1983) and salivary glands (Nakazato et al., 1982). To localize this protein in formalin fixed paraffin embedded tissues by immunoperoxidase procedures we have employed rabbit anti-S-100 protein from Dako (Santa Barbara, CA, USA) diluted 1:300.

STAINING RESULTS

Of the different cytoskeletal proteins investigated, we have been able to obtain a reproducible staining in routinely fixed and embedded tissues only with rabbit antiactin sera (Fig. 62.1), while other methods to visualize actin, although positive on fresh frozen sections, gave negative results.

Staining for myosin, although slightly improved by prior trypsin treatment and on methacharn fixed tissues did not allow a selective and reproducible identification of these cells. Anti-

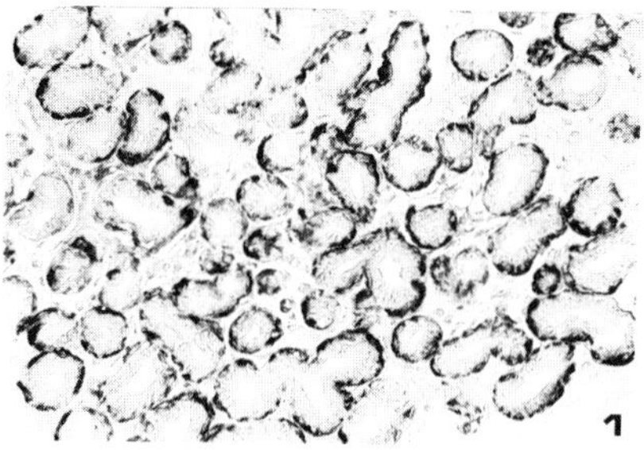

Fig. 62.1. Normal human mammary gland. Formalin fixation, paraffin embedding. The myoepithelial cells of the ductules forming a continuous outer layer are heavily stained by rabbit antiactin antibodies and the immunoperoxidase ABC procedure. The epithelial cells are unstained. Nuclei counterstained with Haemalum. (250×)

keratin sera gave a very weak staining in formalin-fixed tissues, while instead allowing a definite staining of both epithelial and myoepithelial cells in alcohol fixed tissues. Differentiation of these two cell types was not permitted by this procedure.

Staining for S-100 protein was positive in myoepithelial cell cytoplasm and apparently in the nuclei. However, for unknown reasons (Figs. 62.2 and 62.3) not all the cells of this type were revealed on tissue sections; thus, the staining appeared neither specific nor selective. We have, therefore, adopted staining for actin with rabbit antibodies on routinely fixed and embedded tissue sections as the staining of choice for the myoepithelial cells. This staining should not be regarded as specific: actin-rich epithelial cells in the exocrine glands, in normal and pathological conditions, should be interpreted as possibly myoepithelial in nature; the identification would reach certainty if the positive reaction was matched by the classical morphology and distribution. Vice versa, the absence of actin would allow to negate such identification. This holds true also in neoplastic processes, where a positive reaction for actin in any epithelial cell should be retained as evidence, but not proof, of its myoepithelial differentiation.

PATHOLOGY OF THE MYOEPITHELIAL CELLS

Tumors of myoepithelial cells (s.c. myoepithelioma, a variant of pleomorphic adenoma) of the salivary glands have been described (Chaudhry et al., 1982; Lomax-Smith and Azzo-

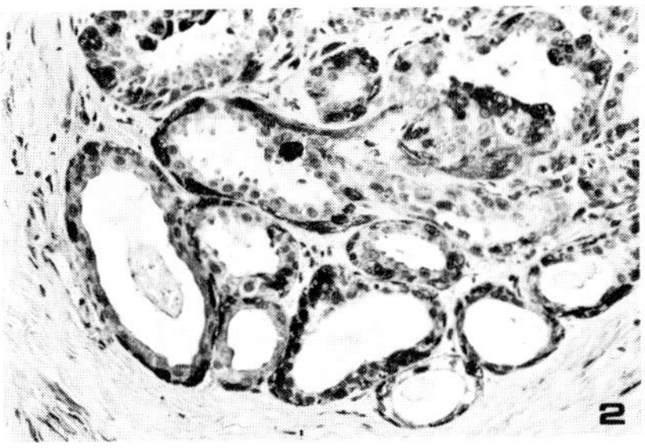

Fig. 62.2. Normal human mammary gland. Some myoepithelial cells are stained by anti-S-100 protein antiserum and the immunoperoxidase procedure. The staining is, however, neither specific nor selective. (250×)

pardi, 1978; Sciubba and Brannon, 1982). Ultrastructural investigations on the two cell types which can be present in these tumors, either the spindle-shaped or the hyaline globoid cells, have revealed a rich array of microfilaments. A positive reaction for actin would therefore reach a diagnostic significance in the identification of these cells. This has been confirmed in this laboratory—in a collaborative study with Drs. I. Bearzi and E. Fulcheri of the Universities of Ancona and Genova—even by the serial semithin-ultrathin sections procedures: a marked staining for actin in thick sections matched with the presence of bundles and aggregates of microfilaments visible, in the same cells, in electron microscopy. These same cells were also positive for the S-100 protein; however, in agreement with the observation of Nakazato et al. (1982), we found this marker present in a wide variety of cells of epithelial and myoepithelial nature.

In the breast, actin-rich myoepithelial cells are uniformly distributed along the basement membrane of ducts and lobular ductules. Their presence and distribution characterizes various pathological processes. We have observed that these cells are usually absent from the epithelium lining large cysts in cystic disease, while very numerous, forming bundles and aggregates in sclerosing adenosis (Bussolati et al., 1982).

In this latter form and in fibroadenomas the ductular epithelial cells can atrophy with luminal obliteration: myoepithelial

Fig. 62.3. Benign intraductal papilloma of the breast. Same method as Fig. 62.2. Some basally-located myoepithelial cells are stained by anti-S-100 protein antibodies. (250×)

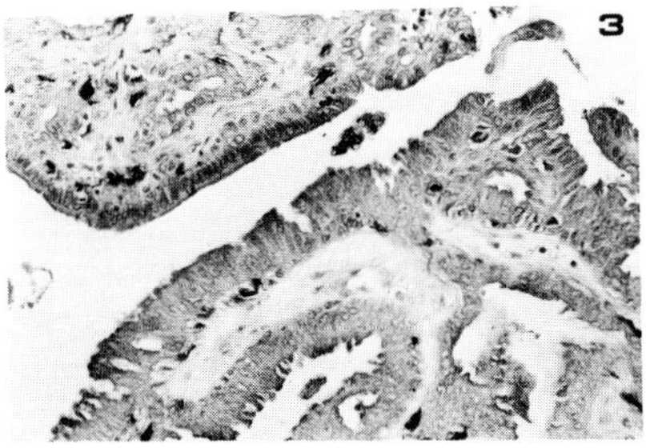

cells, hardly recognizable in routine histopathology, but positive in the immunocytochemical method for actin, appear entrapped in the sclerosing stroma (Bussolati et al., 1980).

The interest in the immunocytochemical identification of the myoepithelial cells is also linked to the observation that not all the basally located cells in the mammary ducts and ductules correspond to such cell type. Cases with hyperplasia of endocrinelike argyrophilic cells have been described by Eusebi and Azzopardi (1980). In addition, infiltrating foam cells (s.c. pseudopagetoid spreading) (Toker and Goldberg, 1977) and clear basal epithelial cells (Popotti et al., 1982) can be distinguished from the myoepithelial cells on the base of the lack of staining for actin.

The presence and behavior of the actin-rich myoepithelial cells in the different histological types of breast carcinoma can reach diagnostic significance. In in situ ductal carcinomas of the comedo, solid and cribriform variety the actin-rich myoepithelial cells are mostly absent (Bussolati et al., 1980; Gabbiani et al., 1976), as it is also confirmed by the immunocytochemical investigation of Macartney and coworkers (1979) using anti-myosin antibodies and by ultrastructural observations (Ozzello, 1971; Ozzello and Sanpitak, 1970). While this process of myoepithelial cell atrophy or destruction is quite obvious in advanced lesions, in less prominent, probably earlier lesions a few actin-rich cells can still be appreciated between the neoplastic epithelial cells and the surrounding stroma (Fig. 62.4). During the process of lobular cancerization the disappearance of myoepithelial cells closely follows the progression of the cancer cells. The pathogenesis of the above described phenomenon is not clear; it can, however, present a diagnostic interest in the differential diagnosis between ductal in situ carcinoma and hyperplasia (s.c. epitheliosis) (Bussolati et al., 1980). In this latter condition, the myoepithelial cells are uniformly preserved and in some cases even hyperplastic. Since it has been suggested that one of the

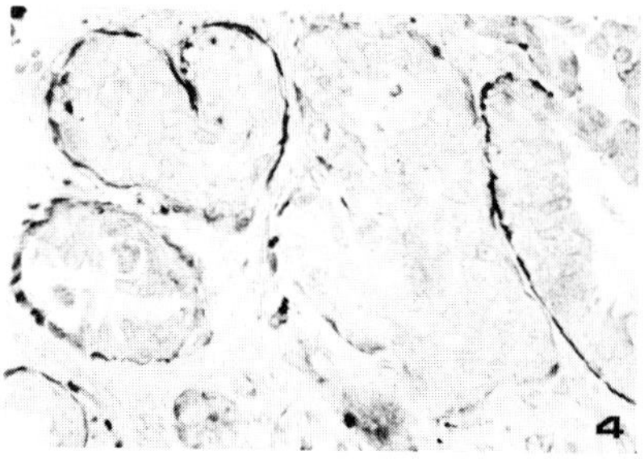

Fig. 62.4. Ductal carcinoma in situ of the breast. The ducts are partly filled by the malignant proliferations; some residual myoepithelial cells (stained by antiactin antibodies) are present at the periphery of the ducts. (250×)

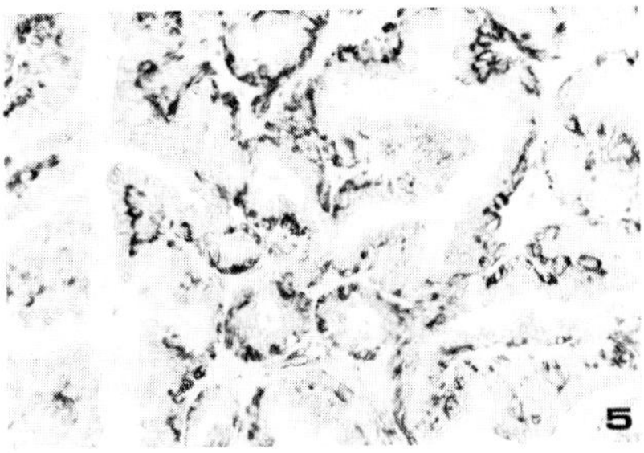

Fig. 62.5. Benign intraductal papilloma of the breast. The immunoperoxidase method for actin reveals the myoepithelial cells evenly present at the base of the papillary fronds. (250×)

functions of the myoepithelial cells is to produce type IV collagen for the building up of the basement membrane (Warburton et al., 1982), it is possible that the lack of myoepithelial cells might explain the well-known infiltrative and aggressive behavior of ductal carcinomas.

Papillary neoplasms of the breast constitute an interesting aggregate of benign and malignant intraductal neoplasm: various histological characteristics have been tabulated to differentiate benign papillomas and papillary carcinomas (Kraus and Neubecker, 1962). However, according to Azzopardi (1979), the most helpful and valuable feature is the absence of the myoepithelial cells in carcinomas. Since the epithelium lining the stromal branches is often arranged in multiple layers both in benign and malignant lesions, these cells cannot easily be identified by histology alone. Actin immunocytochemical detection can therefore play an important role in differential diagnosis. In a recent study on 56 cases of papillary tumors of the breast (15 benign papillomas; 41 papillary carcinomas in situ and/or invasive) (Papotti et al., 1983a), we have shown that while all the other histological features are not discriminative, the presence or absence of the myoepithelial cells is a reliable characteristic easily appreciable by the immunocytochemical staining for actin (Figs. 62.5 and 62.6). We have in fact observed that clear basal cells are often

Fig. 62.6. Papillary carcinoma of the breast. The general pattern is similar to that of the papilloma of Fig. 62.4; the staining for actin is, however, negative, since no myoepithelial cells are present at the base of the epithelium. (250×)

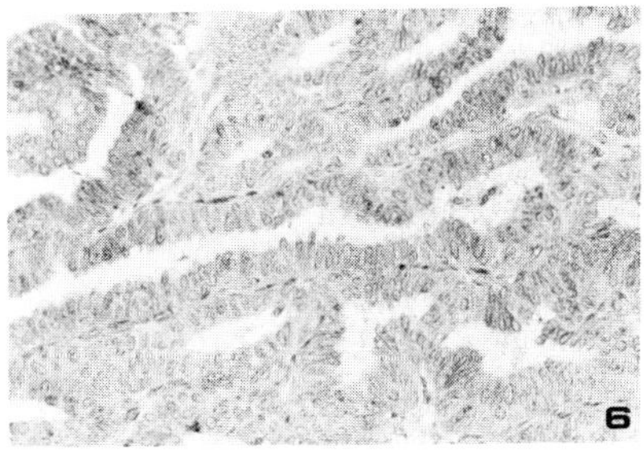

present in papillary carcinomas. These cells, being negative for actin but positive for epithelial cell markers (Bussolati et al., 1982) such as the Epithelial Membrane Antigen (EMA) appear to be of epithelial rather than of myoepithelial nature. In addition, we have recognized cases where areas of multiple papillomas (with the myoepithelial cells component still present) are admixed with areas of in situ ductal carcinomas. These cases, albeit rare, are indicative of the preneoplastic potential of multiple papillomas (Haagensen et al., 1981).

In a recent study on 18 cases of multiple papillomas associated with ductal carcinoma we have shown (Papotti et al., 1983b) that myoepithelial cell-free foci of carcino-embryonic antigen (CEA) positive epithelial cells can originate inside the stalk of multiple papillomas.

In situ lobular carcinomas (CLIS) show a distribution of the actin-rich myoepithelial cells quite different from that of ductal carcinomas, in agreement with the different natural history of this condition. Ultrastructural investigations (Ozzello, 1971; Tobon and Price, 1972) had already revealed the presence of a residual layer of myoepithelial cells, pressed against the basement membrane by the proliferating neoplastic cells. In a study of 13 cases of CLIS this observation has partly been confirmed by immunoperoxidase investigations for the detection of the actin-rich cells (Bussolati, 1980). In some cases, mostly associated with areas of infiltrative carcinoma, cells rich in actin are absent from the periphery of the lobular lesions, being apparently dislodged by the neoplastic cells. This peculiar feature has been confirmed by parallel immunocytochemical-ultrastructural investigations on serial semithin-ultrathin sections (Bussolati et al., 1981). In selected cases, this procedure has, in fact, revealed the presence of an extensive nest of intraductular dendriform cells. These actin-rich cells had abundant cytoplasmic microfilaments and were admixed among neoplastic cells with evidences of epithelial (secretory) differentiation. The exact significance of the above described feature is not clear, nor can we be certain of the hyperplastic or neoplastic nature of these dendriform cells. These patterns might identify cases of CLIS with different histogenesis and evolution—and possibly with a different prognosis.

Infiltrative breast carcinomas, of the ductal or lobular type, were, in our experience, mostly negative for actin. This observation is in contrast with immunofluorescence studies of Gabbiani and coworkers (1979, 1976) who found that most cases of infiltrating lobular and ductal carcinomas of the breast and most cases of mucoid carcinomas showed cytoplasmic areas, mostly peripheral, positive for actin. Whether this feature identifies cases with myoepithelial differentiation or if it is instead related to the contractile capacity of cancer cells cannot presently be decided.

The latter interpretation seems to find support in the immunoperoxidase investigations of Gusterson et al. (1982) who found myoepithelial-like cells, positive both for myosin and keratin, present in only a small minority of infiltrating carcinomas of the breast.

In tubular carcinomas, the infiltrating neoplastic ductlike structures are formed of epithelial cells with secretory activity and are devoid of the myoepithelial cell layer, as confirmed also by electron microscopical investigations (Eusebi et al., 1979; McDiviett et al., 1982).

In difficult cases, the immunoperoxidase staining can allow to differentiate tubular carcinomas from areas of sclerosing adenosis, since the actin-rich myoepithelial cells are uniformly present in this condition (Eusebi et al., 1979).

CONCLUSIONS

The myoepithelial cells are present in exocrine glands, such as the sweat, salivary, and mammary glands, being characterized by an epithelial nature and contractile functions. Their behavior and significance in different pathological and neoplastic conditions is unclear, mainly because of the lack of selective staining procedures. Being characterized by a rich network of cytoskeletal filaments, mainly of actin, selective immunocytochemical procedures can be employed to identify these cells in formalin-fixed paraffin embedded tissue sections and to follow their behavior in pathological lesions. An immunoperoxidase procedure with rabbit antiactin antibodies has allowed us to trace the actin-rich myoepithelial cells in retrospective studies on salivary gland tumors (s.c. myoepithelioma) and on different dysplasic and neoplastic lesions of the breast. These studies indicate that the identification of the cytoskeletal components, richly expressed in myoepithelial cells, can be useful in the differential diagnosis between benign and malignant lesions of the breast (such as papillomas versus papillary carcinomas, or sclerosing adenosis versus tubular carcinoma), to identify hitherto unknown pathological entities and to demonstrate a morphological and functional differentiation of human neoplastic cells.

ACKNOWLEDGMENTS

Work supported by grants from the M.P.I., Rome, the C.N.R. (grant no. 82.00248.96) and the A.I.R.C., Milan. One of us (P.G.) is recipient of a fellowship from the Regione Piemonte.

REFERENCES

Ahmed A.: The myoepithelium in human breast carcinoma. J. Pathol. 113:129-35 (1974).

Archer F.L., Beck J.S. and Melvin J.M.O.: Localization of smooth muscle protein in myoepithelium by immunofluorescence. Am. J. Pathol. 63:109-118 (1971).

Archer F.L. and Kao V.C.: Immunohistochemical identification of actomyosin in myoepithelium of human tissues. Lab. Invest. 18:669-74 (1968).

Azzopardi J.G.: Problems in breast pathology. W. B. Saunders, London, Philadelphia, Toronto (1979).

Bondi A., Chieregatti G., Eusebi V., Fulcheri E. and Bussolati G.: The use of beta-galactosidase as a tracer in immunocytochemistry. Histochemistry 76:153-58 (1982).

Bussolati G.: Actin-rich (myoepithelial) cells in lobular carcinoma in situ of the breast. Virchows Arch. (Cell Pathol.) 32:165-76 (1980).

Bussolati G., Alfani V., Weber K. and Osborn M.: Immunocytochemical detection of actin on fixed and embedded tissues: its potential use in routine pathology. J. Histochem. Cytochem. 28:169-73 (1980).

Bussolati G., Botta G. and Gugliotta P.: Actin-rich (myoepithelial) cells in ductal carcinoma in situ of the breast. Virchows Arch. (Cell Pathol.) 34:251-59 (1980).

Bussolati G., Botto Micca F., Eusebi V. and Betts C.M.: Myoepithelial cells in lobular carcinoma in situ of the breast: a parallel immunocytochemical and ultrastructural study. Ultrastruct. Pathol. 2:219-30 (1981).

Bussolati G., Gugliotta P. and Papotti M.: Detection and significance of epithelial and myoepithelial cell markers in carcinoma of the breast. In "New frontiers in mammary patholoty" Hollmann K.H., De Brus J. and Verley J.M. Editors, Plenum Publish Corp., New York Vol. 2:249-64 (1983).

Bussolati G., Papotti M. and Gugliotta P.: Histology and histochemistry of cystic breast disease. In "Endocrinology of cystic breast disease: A. Angeli, H.L. Bradlow and L. Dogliotti Editors, Raven Press, New York 2:7-18 (1983).

Chaudhry A.P., Satchidanand S., Peer R. and Cutler L.S.: Myoepithelial cell adenoma of the parotid gland: A light and ultrastructural study. Cancer 49:288-93 (1982).

Erlandson R.A. and Rosen P.P.: Infiltrating myoepithelioma of the breast. Am. J. Surg. Pathol. 6:785-93 (1982).

Eusebi V. and Azzopardi J.G.: Lobular endocrine neoplasia in fibroadenoma of the breast. Histopathology 4:413-28 (1980).

Eusebi V., Betts C.M. and Bussolati G.: Tubular carcinoma: a variant of secretory breast carcinoma. Histopathology 3: 407-19 (1979).

Franke W.W., Weber K., Osborn M., Schmid E. and Freudenstein C.: Antibody to prekeratin: decoration of tonofilament-like arrays in various cells of epithelial character. Exp. Cell Res. 116:429-45 (1978).

Gabbiani G.: The cytoskeleton in cancer cells in animals and humans. Methods Archiev. Exp. Pathol. 9:231-43 (1979).

Gabbiani G., Csank-Brassert J., Schneeberger J.C., Kapanci Y., Trenchev P. and Holborow E.J.: Contractile proteins in human cancer cells. Immunofluorescent and electron microscopic study. Am. J. Pathol. 83:457-74 (1976).

Gabbiani G., Ryan G.B., Lamelin J.-P., Vassalli P., Majno G., Bouvier C.A., Cruchaud A. and Lüscher E.F.: Human smooth muscle autoantibody. Its identification as antiactin antibody and a study of its binding to "nonmuscular" cells. Am. J. Pathol. 72:473-88 (1973).

Gusterson B.A., Warburton M.J., Mitchell D., Ellison M., Munro Neville A. and Rudland P.S.: Distribution of myoepithelial cells and basement membrane proteins in the normal breast and in benign and malignant breast diseases. Cancer Res. 42:4763-70 (1982).

Haagensen C.D., Bodian C. and Haagensen D.E. Jr.: Breast carcinoma. Risk and detection. W. B. Saunders Comp., Philadelphia, London, Toronto, Mexico City, Sidney, Tokyo pp 197-237 (1981).

Haguenau F.: Les myofilaments de la cellule myoépithéliale. Etude au microscope électronique. C. R. Seances Acad. Sci. 249:182-84 (1959).

Hamperl H.: The myothelia (myoepithelial cells): normal state; regressive changes; hyperplasia; tumours. Curr. Top. Pathol. 53:161-220 (1970).

Hsu S.M., Raine L. and Fanger H.: Use of avidin-biotin-peroxidase complex (ABC) in immunoperoxidase technique—A comparison between ABC and unlabeled antibody (PAP) procedures. J. Histochem. Cytochem. 29:577-80 (1981).

Kahn H. J., Marks A., Thom H. and Baumal R.: Role of antibody to S100 protein in diagnostic pathology. Am. J. Clin. Pathol. 79:341-47 (1983).

Kraus F.T. and Neubecker R.D.: The differential diagnosis of papillary tumours of the breast. Cancer 15:444-55 (1962).

Lomax-Smith J.D. and Azzopardi J.G.: The hyaline cell: a distinctive feature of "mixed" salivary tumours. Histopathology 2:77-92 (1978).

Macartney J.C., Roxburgh J. and Curran R.C.: Intracellular filaments in human cancer cells: a histological study. J. Pathol. 129:13-20 (1979).

Macartney J.C., Trevithick M.A., Kricka L. and Curran R.C.: Identification of myosin in human epithelial cancers with immunofluorescence. Lab. Invest. 41:437-45 (1979).

McDivitt R.W., Boyce W. and Gersell D.: Tubular carcinoma of the breast. Clinical and pathological observations concerning 135 cases. Am. J. Surg. Pathol. 6:401-411 (1982).

Nakajima T., Kameya T., Watanabe S., Hirota T., Sato Y. and Shimosato Y.: An immunoperoxidase study of S-100 protein distribution in normal and neoplastic tissues. Am. J. Surg. Pathol. 6:715-27 (1982).

Nakazato Y., Ishizeki J., Takahashi K., Yamaguchi H., Kamei T. and Mori T.: Localization of S-100 protein and glial fibrillary acidic protein-related antigen in pleomorphic adenoma of the salivary glands. Lab. Invest 46:621-26 (1982).

Ozzello L.: Ultrastructures of the human mammary gland. Pathol. Annu. 6:1-79 (1971).

Ozzello L. and Sanpitak P.: Epithelial-stromal junction of intraductal carcinoma of the breast. Cancer 26:1186-98 (1970).

Papotti M., Gugliotta P., Eusebi V. and Bussolati G.: Immunohistochemical analysis of benign and malignant papillary lesions of the breast. Am. J. Surg. Pathol. 7:451-61 (1983a).

Papotti M., Gugliotta P., Ghiringhello A. and Bussolati G.: Origin of breast carcinoma in cases with multiple intraductal papillomas: an histological and immunohistochemical investigation. (1983b submitted for publication).

Puchtler H., Sweat-Waldrop F., Terry M.S. and Conner H.M.: Investigation of staining, polarization and fluorescence microscopic properties of myoendothelial cells. J. Microsc. 89:95-104 (1969).

Sanger J.W.: Intracellular localization of actin with fluorescently labelled heavy meromyosin. Cell Tissue Res. 161:431-44 (1975).

Sciubba J.J. and Brannon R.B.: Myoepithelioma of salivary glands: report of 23 cases. Cancer 49:562-72 (1982).

Sternberger L.A., Hardy P.H. Jr., Cuculis J.J. and Meyer H.C.: The unlabelled antibody enzyme method of immunohistochemistry: preparation and properties of soluble antigen-antibody complex (horseradish peroxidase antihorseradish peroxidase) and its use in identification of spirochaets. J. Histochem. Cytochem. 18:315-33 (1970).

Tobon H. and Price H.M.: Lobular carcinoma in situ. Some ultrastructural observations. Cancer 30:1082-91 (1972).

Toker C. and Goldberg J.D.: The small cell lesion of mammary ducts and lobules. Pathol. Annu. 12:217-49 (1977).

Van Bogaert L.J., Abarka J. and Maldague P.: Appraisal and pitfalls of myoepithelial cell staining by Levanol Fast Cyanine 5RN. Histochemistry 54:251-58 (1977).

Van Bogaert L.J., Maldague P., Abarka J. and Colette J.M.: Etude histochimique comparative des cellules myoépithéliales de la glande mammaire humaine. Acta. Histochem. 59:8-14 (1977).

Wang E. and Goldberg A.R.: Binding of deoxyribonuclease I to actin: a new way to visualize microfilament bundles in non muscle cells. J. Histochem. Cytochem. 26:745-49 (1978).

Warburton M.J., Ferns S.A. and Rudland P.S.: Enhanced synthesis of basement membrane proteins during the differentiation of rat mammary tumour epithelial cells into myoepithelial like cells in vitro. Exp. Cell Res. 137:373-80 (1982).

Wulf E., Deboben A., Bautz F.A., Faulstich H. and Wieland Th.: Fluorescent phallotoxin, a tool for the visualization of cellular actin. Proc. Natl. Acad. Sci. USA 76:4498-4502 (1979).

63 Red Cell Cytoskeletal Filaments: Alterations with Physiologic Aging

Jerome M. Loew, Yehuda Marikovsky,
Henry D. Tazelaar, Ronald S. Weinstein

INTRODUCTION

Throughout its life span, the normal human red blood cell remains highly deformable and is able to return to a basic biconcave disc shape in the absence of shear stress. It is widely held that the elasticity of the membrane results from interaction between the plasma membrane and the underlying cytoskeleton. The model of the arrangement of red cell membrane proteins proposed by Lux and coworkers, suggests that membrane deformability, flexibility, and morphology are primarily dependent on and controlled by a submembranous cytoskeletal protein scaffolding composed of spectrin bands 1 and 2, actin, and band 4.1 interacting with other proteins such as bands 3 and 4.2 on the cytoplasmic surface (Lux, 1979). The role and contribution of the membrane skeleton to the major structural and rheological features of the red cell has been noted frequently. Reduced deformability of the red cell is an age-related change, although the loss of shear deformability per se is not generally considered the major factor limiting cell survival. Reduced deformability may be related to alterations in the cytoskeleton. Changes in the cytoskeleton may also play a major role in the topographic distribution of integral glycoproteins, and thus may influence recognition and sequestration of aged red cells. We recently observed (Marikovsky, Levy, and Weinstein 1983) ultrastructural alterations of red cell cytoskeletons with physiologic aging. These include evidence of changes in dimensions of short filaments, tentatively identified as oligomeric spectrin, and a decrease in the overall thickness of the cytoskeletal assembly. In this report we expand our preliminary findings through the use of several different techniques of specimen preparation for electron microscopy in an attempt to clarify the role of the cytoskeleton in determining the fate of the aged red cell.

MATERIAL AND METHODS

Human erythrocytes were obtained from adult volunteer donors and anticoagulated with heparin. Cells were separated into young and old cell populations by differential density separation (Danon and Marikovsky, 1964). Red cell ghosts were prepared from washed red cells by gradual hemolysis (Danon, Nevo, and Marikovsky, 1956) or by rapid hypotonic lysis (Dodge, Mitchell, and Hanahan, 1963).

Thin sectioning: Red cell membranes were fixed with 2% tannic acid and 2.5% glutaraldehyde in 0.1 M sodium cacodylate buffer, pH 7.4 at 4°C overnight, as previously described (Tsukita, Tsukita, and Ishihara, 1980). The fixative was removed by rinsing in the buffer and the pellet postfixed with ice-cold 1% Os)4 in 0.1 M sodium cacodylate buffer pH 7.4 for two hours. After rinsing in distilled water, pellets of fixed membranes were embedded in 1% agar, stained en bloc with 0.5% uranyl acetate for two hours at room temperature, dehydrated, and embedded in Epon 812 for thin sectioning.

Freeze-drying: The method is a minor modification of that described by Nermut (1981). Freeze-dried membranes were prepared by replication of monolayers of red cells, mounted on Alcian blue coated freshly cleaved mica, lysed in 5 mM sodijm phosphate, pH 8 (5P8), squirted with a syringe, and fixed with 8% formaldehyde for 10 minutes. Optional treatment of squirted membranes with Triton X-100 or EDTA was performed before fixation. Fixation was followed by washing and then freezing, first in Freon for 10 seconds and finally in liquid nitrogen before mounting on a cold stage of a Balzers BAF 301 freeze-etch unit.

Negative staining: For the negatively stained preparations erythrocytes were washed in phosphate buffered saline (PBS) and ghosts prepared by hypotonic lysis in 5P8 containing, for some experiments, 50 micromolar phenylmethylsulfonylfluoride. Ghosts were allowed to settle onto formvar-carbon coated copper grids. Lipid was extracted by incubation with 1% Triton X-100 in 5P8 at room temperature for 1 to 4 minutes. Excess Triton was removed with a rapid dip into high purity water, and ghosts were fixed with 0.1% glutaraldehyde. The grid was again washed and negatively stained with 0.9% uranyl acetate containing 0.25% dextrose (weight per volume).

Actin filaments were prepared from rabbit skeletal muscle actin purified by the method of Spudich and Watt (1971).

Electron micrographs were taken on a Phillips 301 at 80 kv.

RESULTS

Thin sections: Tangentially sectioned red cell membranes after tannic acid fixation show a network as illustrated in Figure 63.1. The network is composed of units or segments of varying length. It cannot be determined if one segment represents one specific macromolecular complex or if one such complex can span several segments, with changes in direction at points where filaments cross. The length of the segments is variable between and within preparations. In cross section (Fig. 63.2), the granulofibrillar network extends short distance into the cytoplasmic compartment and the remainder of the cytoplasmic compartment appears empty. Prominent in the sections are small electron-dense granules which represent the short filaments or elongated particles seen in tangential sections. The alteration in membrane morphology observed with erythrocyte aging has been recently reported (Marikovsky, Levy, and Weinstein, 1983). These consist of thickening of the elements of the reticulum as seen in tangential sections, and a decrease in the overall thickness of the submembranous network from 15-20 nm to 10-15 nm.

Treatment with EDTA under conditions which deplete membranes of spectrin and actin dramatically alters the appearance (Fig. 63.2). The trilaminar unit membrane is unchanged, but the submembranous components are removed by EDTA extraction.

Freeze-dried membrane replicas: These preparations appear similar to the illustrations in the original Nermut paper describing the method. Again, filaments of varying thickness are evident. In our preparations, some networks were nonuniform. There was a definite isotropy to the network at the periphery of the adherent membrane compared to the areas in the center of the cell. This indicates that there is elastic deformation of the cell during removal of the top portion of the membrane.

In replicas of freeze-dried membranes the age-related alterations are also apparent. In the young cells the reticulum appears to consist of filamentous aggregates formed from finer filaments. In many areas the finer structures extend from the aggregates along the plane of the membrane (Fig. 63.3). The thicker filaments appear flattened, and also vary in thickness. Some of these thicker filaments appear slightly twisted, as if spun from finer filamentous units. In aged red cells the branch-

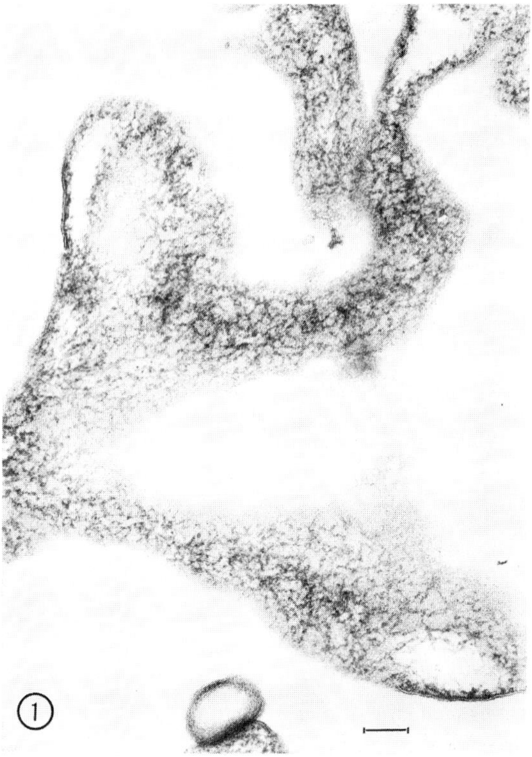

Fig. 63.1. Tangential section of red cell membrane after tannic acid-glutaraldehyde fixation. The interrelationship of multiple small subunits forming the network can be seen, but differences in thicknesses of different subunits are less apparent than with other methods. Bar = 0.1 micron.

ing into finer filaments is less apparent. Most of the reticulum appears composed of nearly cylindrical elements with a nearly clean background (Fig. 63.4), compared to the fine filaments on the membrane in the younger cells. These appear more uniform in diameter than the filaments in younger cells. However, in the aged red cells there are more gaps in the network where the membrane appears devoid of this reticulum.

Negatively stained Triton extracted membranes: In these preparations, the presence of two or more types of filaments and/or elongated particles was most clearly appreciable. The networks were again isotropic, but a clear distinction could be seen between thin filaments (arrows) and shorter thicker parti-

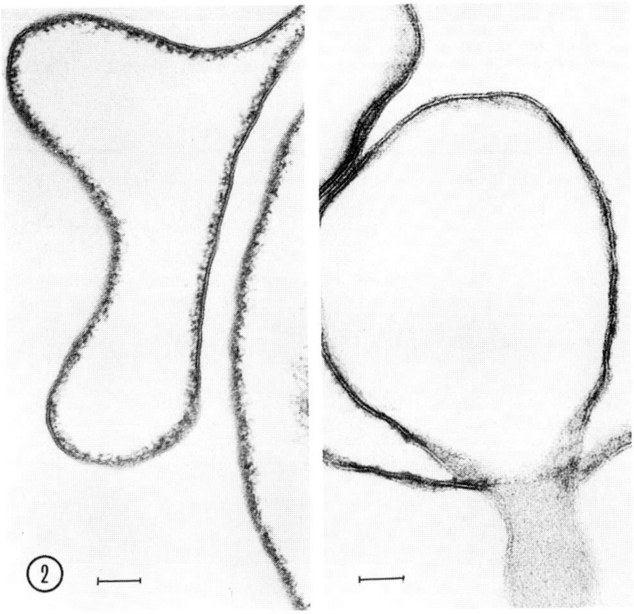

Fig. 63.2. Left: Cross section of red cell membranes after tannic acid-glutaraldehyde fixation. The granulofibrilar network extends short distance into the cytoplasmic compartment. Bar = 0.1 micron. Right: Cross section of EDTA treated red cells after tannic acid-glutaraldehyde fixation. The granulofibrilar components are extracted by EDTA treatment. Bar = 0.1 micron.

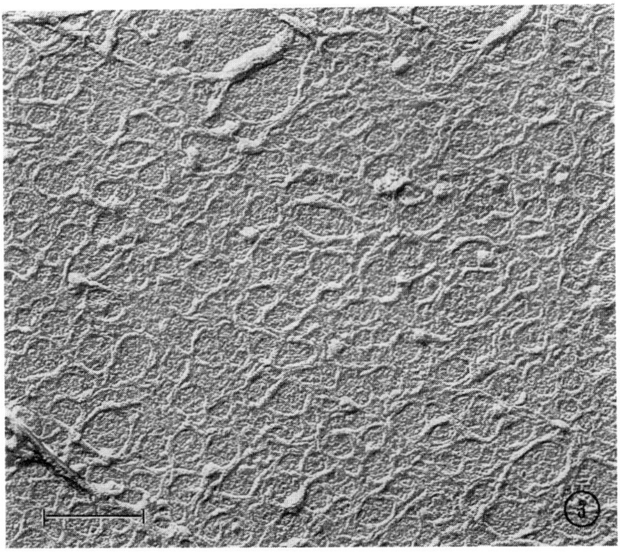

Fig. 63.3. Replica of freeze-dried membrane from young red cells. Note the multiple projections or attachments of the filamentous components onto the underlying membrane. Bar = 0.2 microns.

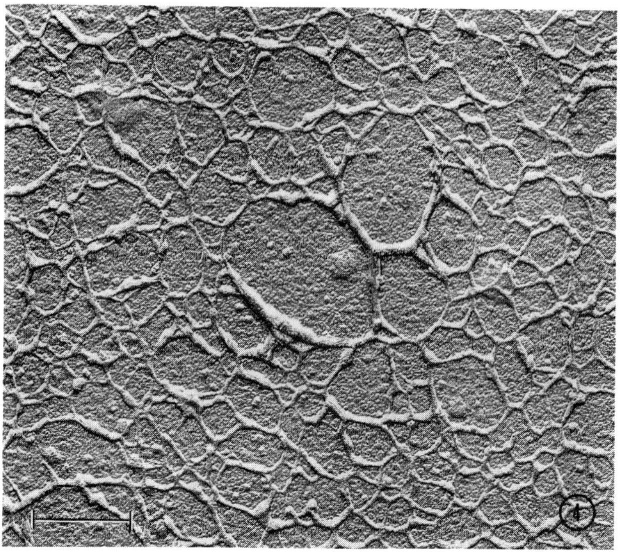

Fig. 63.4. Replica of freeze-dried membrane from old red cells. Bar = 0.2 microns.

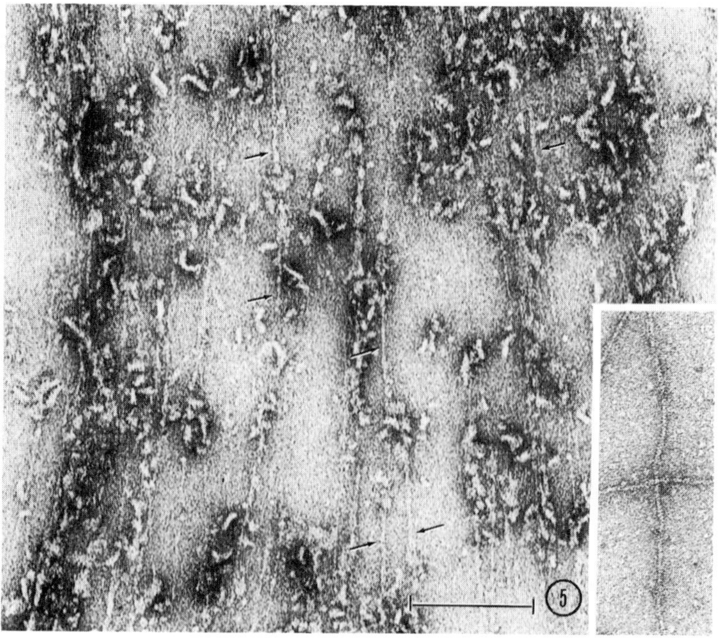

Fig. 63.5. Negatively-stained, Triton extracted red cell membrane. Inset shows negatively stained actin filaments. Bar = 0.2 microns

cles wrapped around or bridging between two or more of the longer thin filaments (Fig. 63.5). There is a similarity in thickness between the thin filaments and actin thin filaments (Fig. 63.5, inset).

DISCUSSION

The survival of the red cell is uniform. In normal humans the cells survive approximately 120 days in the circulation and undergo nonrandom removal. The removal must in some way be related to development of an abnormality in the membrane or cytoskeleton, but controversy still exists as to whether the destruction of the cell is mechanical, due to increased rigidity and decreased resistance to shear stress (Weed and Reed, 1966), or immunologic (Lutz and Kay, 1981; Kay et al. 1983). It is also possible that different mechanisms prevail in abnormal compared to normal cells.

Specimens were examined with multiple electron microscopic methods to minimize the possibility that the structures observed were merely artifacts of the specimen preparative technique. The freeze-dried preparation avoids artifacts of fixation, staining, and effects of the electron beam on the proteins, since replicas are viewed. Tannic acid fixed thin sections and negatively stained specimens are not subject to freezing. Similarly, only the embedded, thin section preparations are subject to dehydration in organic solvents and to embedding resin. It was not possible to entirely avoid the effects of ghost production, but both gradual lysis by dialysis and rapid osmotic lysis were used without apparent morphologic differences (not illustrated).

Our findings demonstrate alterations in the configuration of the filamentous units visualized in the cytoskeleton in aged as compared to younger cells. We assume that this represents a change in organization or association state of major cytoskeleton constituents. Assuming that the spectrin oligomers exist as cylinders with a density of 1.3 grams per ml, refolding a tetramer to decrease its length from 2,000 to 1,000 angstroms would only increase its diameter from 28 to 38 angstroms. Increasing a 1,000 angstrom cylinder from 38 to 53 angstroms would increase the molecular weight from that of a spectrin tetramer to that of an octomer. Thus, small changes in appearance may reflect large changes in the organization of the components of the cytoskeleton. This shortening and thickening must decrease the number of interconnections between different filamentous subunits and between filaments and other components of the membrane.

Even if the trigger for the removal of the senescent erythrocyte is recognition of an altered antigenic determinant at the outer surface of the cell membrane, altered organization of skeletal proteins could modify the properties of one of the integral membrane proteins of the cell in such a way as to lead to the recognition of the aged cell as an altered or damaged cell. The mobility and distribution of integral proteins of the red cell membrane are tightly restricted by interaction with skeletal proteins (Sheetz, 1983; Nigg and Cherry, 1980). These integral proteins bear numerous antigenic determinants and the glycophorins carry much of the surface charge of the cell through their sialic acid residues (Furthmayr, 1978). Altered skeletal organization could lead to altered distribution of charges on the surface of the aged cell. Such altered charge distribution has been demonstrated for sickle cell membranes (Hebbel et al., 1980) and for acid damaged membranes (Timme, 1981). This could lead to macrophage recognition of the cells and the removal of the aged red cell.

We cannot determine from these studies if the thicker filaments seen in our preparations are important structural elements in the intact circulating erythrocyte. Nevertheless, the structures seen and the differences between young and old cells can be interpreted in terms of an hypothesis which could explain several membrane phenomena associated with red cell aging. In the young cells the skeletal components appear as fine fibrils partially organized into thicker aggregates. These fine elements appear anchored to the plasma membrane at many points, presumably by interaction with integral proteins. Some of the elasticity of the membrane could result from the unraveling of the larger aggregates with mechanical stress. In the aged red cell the only components seen are thicker, more nearly cylindrical structures. These do not appear to have such numerous fine extensions along the surface of the membrane. If this is indeed the case, interactions between the cytoskeleton and integral components of the plasma membrane, both protein and lipid, change in their distribution and presumably in their number and nature with cell aging. The formation of apparently more tightly organized filamentous structures with aging also suggests that bonds between skeletal proteins become more numerous or stronger with red cell aging. Such a change in the aged cell could explain the altered surface charge distribution in such cells (Marikovsky and Danon, 1969; Danon, Goldstein, Marikovsky, and Skutelsky, 1972). These could be due to loss of interaction between integral and skeletal proteins with other forces regulating distribution of the charged integral proteins.

A recent paper (O'Connell and Swislocki, 1983) indicates that spectrin in aged red cells differs from that of young cells in its ability to undergo phosphorylation by endogenous protein kinase. We still do not know the organization of membrane proteins in the filaments we observe, but presumably spectrin is a major component. We feel that the finding of biochemical changes in one of the major skeletal proteins with aging supports our observation of altered cytoskeletal organization with erythrocyte aging.

ACKNOWLEDGMENT

The excellent technical help of Mrs. Ora Asher, Mr. H. Waks and Mr. S. Himmelhoch is greatly appreciated.

REFERENCES

Danon, D., Goldstein, L., Marikovsky, Y., J. Skutelsky, E. Use of cationized ferritin as a label of negative charges on cell surfaces. J. Ultrastruct. Res. 38:500-510, 1972.

Danon, D., and Marikovsky, Y. Determination of density distribution of red cell population. J. Lab. Clin. Med. 64:668, 1964.

Danon, D., Nevo, A., and Marikovsky, Y. Preparation of erythrocyte ghosts by gradual haemolysis in hypotonic aqueous solutions. Bull. Res. Council Israel, 6E:36, 1956.

Dodge, J. T., Mitchell, C., and Hanahan, D. J. The preparation and chemical characteristics of hemoglobin-free ghosts of human erythrocytes. Arch. Biochem. Biophys. 100:119, 1963.

Furthmayr, H. Glycophorins A, B, and C: A family of sialoglycoproteins. Isolation and preliminary characterization of trypsin derived peptides. J. Supramol. Struct. 9:79-95, 1978.

Hebbel, R. R., Yamada, O., Moldon, C. F., Jacob, H. S., White, J. G., Eaton, J. W. Abnormal adherence of sickle erythrocytes to cultured vascular endothelium: possible mechanism for microvascular occlusion in sickle-cell disease. J. Clin. Invest. 65:154-60, 1980.

Kay, M. M. B., Goodman, S. R., Sorensen, K., Whitfield, C. F., Wong, P., Zaki, L. and Rudloff, V. Senescent cell antigen is immunologically related to band 3. Proc. Nat. Acad. Sci. 80:1631-35, 1983.

Lutz, H. U., Kay, M. M. B. An age-specific cell antigen is present on senescent human red blood cell membranes. A brief note. Mechanisms of aging and development 15:65-75, 1981.

Lux, S. E. Dissecting the red cell membrane skeleton. Nature 281:426-29, 1979.

Marikovsky, Y. and Danon, D. Electron microscope analysis of young and old red blood cells stained with colloidal iron for surface charge evaluation. J. Cell. Biol. 43:1-7, 1969.

Marikovsky, Y., Levy, R. and Weinstein, R. S. Red cell cytoskeletal alterations with physiological aging, in: Oplatka, A., and Balaban, M. (eds). Biological Structure and Coupled Flows. New York, Academic Press, 1983, pp. 363-68.

Nermut, M. V. Visualization of the "membrane skeleton" in human erythrocytes by freeze-etching. Eur. J. Cell. Biol. 25:265-71, 1981.

Nigg, E. A. and Cherry, R. J. Anchorage of a band 3 population at the erythrocyte cytoplasmic membrane surface: Protein rotational diffusion measurements. Proc. Nat. Acad. Sci. USA 77:4702-06, 1980.

O'Connell, M. A. and Swislocki, N. I. Spectrin phosphorylation in senescent rat erytyrocytes. Mechanism of aging and development. 22:51-70, 1983.

Sheetz, M. P. Membrane skeletal dynamics: Role in modulation of red cell deformability, mobility of transmembrane proteins, and shape. Semin. Hematol. 20:175-88, 1983.

Spudich, J. A., and Watt, S. The regulation of rabbit skeletal muscle contraction. J. Biol. Chem. 246:4866-71, 1971.

Timme, A. H. The ultrastructure of the erythrocyte cytoskeleton at neutral and reduced pH. Ultrastruct. Res. 77:199-209, 1981.

Tsukita, S., Tsukita, S., and Ishikawa, H. Cytoskeletal network underlying the human erythrocyte membrane. J. Cell. Biol. 85:568, 1980.

Weed, R. I. and Reed, C. F. Membrane alterations leading to red cell destruction. The Amer. J. Medicine 41:681-98, 1966.

64 Light, Polarization, and Fluorescence Microscopic Demonstration of Myoid Structures in Various Types of Cells and Lesions

H. Putchtler, F. S. Waldrop,
S. N. Meloan, B. P. Barton

INTRODUCTION

Myosins in nonmuscle cells have been studied since the 1840s. Because convenient histochemical technics for myosins and related structures were lacking, the TP-Amidoblack 10B reaction was developed (Puchtler, 1956). This technic was based on X-ray diffraction studies by Astbury and coworkers and on polarization microscopic findings. Chemical data on dye binding led to the substitution of milling dyes for demonstration of myofibrils and myoid material (Puchtler et al., 1969a, b, 1974, 1975a, b; Waldrop et al., 1972a, b). These technically facile and inexpensive reactions are suitable for general hospital laboratories. Observations with TP-dye technics will be discussed in context with classical literature and recent immunological data. Infrared fluorescence microscopic studies of diseases of muscle (Puchtler et al., 1980; Meloan et al., 1983) will be reviewed briefly.

LESIONS OF MUSCLE

According to textbooks, conventional technics do not show myocardial infarcts within six hours after onset of ischemia. The TP-Levanol Fast Cyanine 5RN (TPL) reaction visualized alterations of A bands in experimental myocardial infarcts within one hour after ligation of a coronary artery (Hollingsworth et al., 1967). This technic proved convenient also for studies of skeletal muscle and can be combined with the PAS reaction (Puchtler et al., 1969a). Slight changes, not detectable by light microscopy, can be demonstrated by polarization microscopy of sections treated with TP-Thiazine Red R (Meloan et al., 1975).

Correlation with chemical and electron microscopic data indicated that these changes are due to structural alterations of myosin.

MYOENDOTHELIAL AND MYOINTIMAL CELLS

Koelliker (1854) regarded endothelial cells as contractile elements similar to smooth muscle. Histochemical (Puchtler et al., 1969b) and immunological studies (Becker and Murphey, 1969) confirmed the presence of myosin. Polarization microscopy showed alignment of myofibrils parallel to the long axis of the vessel (Puchtler et al., 1969b). Myointimal cells were described by Remak (1850) and studied extensively in arteriosclerosis (Trompetter, 1876; Westphalen, 1886). Derivation of intimal from endothelial cells was suggested by Henle (1841). Borst and Enderlen (1909) observed transformation of endothelial into smooth muscle cells and migration of such myoendothelial cells across sutures of vessels into transplants. TP-dye technics show in transplanted human kidneys subendothelial layers of myointimal cells that are separated from the media by a conspicuous band of collagen (Waldrop et al., 1972). Immunological studies are required to determine whether or not such cells are derived from the transplant.

MYOEPITHELIAL CELLS

In 1847 Koelliker discovered smooth muscle in sweat glands. Similar cells in salivary and lacrimal glands were termed basket cells (Boll, 1868, 1869). Engelmann (1871) regarded these cells as transformed glandular epithelium whose contraction expedited expulsion of secretory products. Drasch (1889) demonstrated contraction in vivo. The term myoepithelial cells was coined by Renaut (1897) and Kolossow (1898), who expressed surprise that they were lacking in liver, pancreas, and kidney, where they seemed as necessary as in salivary glands.

TP-dye technics readily demonstrate classical myoepithelial cells, e.g., in lingual and eccrine sweat glands (Puchtler et al., 1974). In a patient with acquired immune deficiency syndrome (AIDS) myoepithelial cells in sweat glands showed significant loss of myofibrils.

TERMINAL BAR-TERMINAL WEB SYSTEM

The layer now known as terminal web was described in intestine by Brettauer and Steinach (1857). Similar bands were seen in gall bladder epithelium (Virchow, 1857), ciliated cells (Billroth, 1858), Brunner's glands, bile canaliculi and stomach (Schwalbe, 1872). Histochemical studies indicated a myosinlike protein in these structures (Schwalbe, 1872; v. Ebner, 1872). This band was studied extensively. As stated by Prenant (1899): "Cette ligne a été si souvent constantée, qu'il est inutile de citer les observationes qu'on fait." A continuous layer extending from canaliculi through ducts was identified in liver (Krause, 1893) and pancreas (Renaut, 1903). Zimmermann (1898) ascribed the open lumina of canaliculi to the tonus of a contractile meshwork. Schmidt (1943) showed that fibrils of the intestinal terminal web were oriented at right angle to the long axis of the cells.

TP-dye reactions visualized the terminal web in intestine (Puchtler, 1957, 1958), lingual and sweat glands (Puchtler et al., 1974), bronchi, biliary, and pancreatic pathways (Puchtler et al., 1975a). The pattern of dye binding is very similar to immunofluorescence pictures obtained with antibodies to myosin, e.g., by Armstrong and MacSween (1973), Drenckhahn et al., 1977, 1980), Bretscher and Weber (1978), Mooseker et al. (1978), but differs strikingly from immunohistochemical reactions for prekeratins. Considerable thickening of the pericanalicular layer was found in liver and pancreas with obstruction of ducts (Puchtler et al., 1975a). In some diseases the myoid material was barely recognizable (Badaruddin et al., 1976). In alcoholic liver disease Mallory bodies reacted strongly with TPL (Meloan and Puchtler, 1982). Hepatocytes of a patient with AIDS contained conspicuous myoid deposits that resembled Mallory bodies. Skin of this patient showed loss of myoid material from the terminal web of sweat glands.

KIDNEY

Basal filaments (Basalreifen) of renal epithelium were seen by Wedel in 1850 and known to early French histologists as figures—festonnées; contractility of renal epithelium was demonstrated by Ranvier (Hortoles, 1881). Abram (1900) stressed the similarity of basal fibrils and muscle fibers. Benda ascribed "un role moteur" to filaments in renal epithelium (Policard, 1905). Zimmermann (1929) found fine fibrils in mesangium but could not determine their nature. In electron microscopic studies Pease (1968) demonstrated myoid systems in tubules and glomeruli. Harper et al. (1969, 1970) visualized the terminal web and basal fibrils with TPL. Rostgaard showed that basal fibrils are contractile systems composed of 6-8 and 13-17 nm filaments. In normal kidneys myoid fibrils in glomeruli were often below the resolution of the light microscope, but became prominent in various diseases; in some cases of systemic lupus erythematosus and scleroderma, podocytes resembled smooth muscle cells (Harper, 1970). Immunological studies confirmed the presence of actomyosin in glomeruli (Becker, 1972; Trenchev et al., 1974, 1976; Scheinman et al., 1976). Myointerstitial cells occurred in glomerular crescents and between tubules in early stages of fibrosis (Harper et al., 1970, 1971).

THYMUS

In 1888 Mayer identified striated muscle cells in thymus of amphibia. Pensa (1902, 1904) observed transformation of striated into smooth myoid cells in birds. Myoid cells in mammalian thymus were studied in detail by Hammar (1905, 1909). Maximow (1909) confirmed differentiation of epithelial into myoid cells; Pappenheimer (1912/13) demonstrated transformation into contractile myoid cells in tissue cultures. Severe alterations of myoid cells and Hassall's corpuscles were observed in infectious diseases (Wiesel, 1912; Hart, 1913). In immunological investigations smooth myoid cells of thymus reacted with antibodies against striated muscle but not smooth muscle (Strauss et al., 1964, 1966; Vetters, 1966, 1967).

The TPL reaction visualized myoid cells at the boundary and throughout the thymic cortex. Hassall's corpuscles were formed from myoid cells; these findings confirm earlier observations by Hammary (1909). Polarization microscopy showed highly oriented myoid fibrils in these structures. Fluorescence microscopy visualized different stages of formation and degeneration of Hassall's corpuscles (Puchtler et al., 1975b). Comparison of consecutive sections treated with TPL and the peroxidase-antiperoxidase (PAP) reaction for prekeratin showed significant differences in the distribution of myoid and prekeratin material; the patterns often looked like positives and negatives of the same structures. Numerous Hassall's corpuscles and myoid cells forming extensive meshworks were found in infants with congenital malformations of the heart. The clinical significance of these alterations is not yet clear.

CENTRAL NERVOUS SYSTEM

The earliest stains for glia fibers were Weigert's fibrin stain and Heidenhain's iron hematein; both methods also colored muscle (Müller, 1899). Glia fibers yield the same x-ray diffraction pictures as fibrin (Wilke, 1951; Wilke and Kircher, 1952). Since x-ray diffraction patterns of fibrin and myosin are virtually identical (Low, 1953), the diagrams of glia fibers also resemble those of myosin. In 1968 Puszkin and coworkers isolated from brain an actomyosin that was not derived from the vascular system. Comparative studies of actomyosin from striated muscle, smooth muscle, and brain revealed distinct differences (Berl and Puszkin, 1970).

The TPL reaction visualized meshworks of myoid fibers in the marginal layer, along blood vessels and in white matter; spiderlike cells containing myoid fibers were found in some areas of grey matter. The strong anisotropy of myoid fibers treated with TP-Sirius Red 4BA indicated a highly oriented filamentous structure at the submicroscopic level (Waldrop et al., 1972). Correlation with electron microscopic data indicated similarity of the distribution of myoid and glia fibers. Toh et al. (1976) demonstrated binding of antibodies to smooth muscle by astrocyte fibrils. Duplicated sections stained with TPL and with antibodies to glial fibrillar acidic protein (GFAP) showed striking differences. With the former, glia fibers stood out clearly against the unreactive cytoplasm and could often be traced over considerable distances. The GFAP reaction colored material in the cytoplasm of glia cells the same shade as fibrils; this background stain renders correlation of myoid and GFAP fibrils difficult.

INFRARED (IR) FLUORESCENCE OF STAINED MYOFIBRILS

Quantitative studies demonstrated wide variations in IR fluorescence of dyes with the chemical nature of the substrate (Ramsley, 1968). In human tissues stained with suitable triarylmethane dyes myofibrils, e.g., in myoepithelium, smooth and striated muscle emitted strong IR fluorescence; adjoining collagen showed little or no fluorescence (Meloan and Puchtler, 1974; Puchtler et al., 1980). Pilot studies indicated impairment of IR fluorescence in various lestions (Paschal et al., 1978). Comparison of photomicrographs taken with Kodachrome 25 and Ektachrome IR film indicated great sensitivity of IR fluorescence. Muscle fibers that seemed virtually normal in visible light often showed decrease or partial loss of IR fluorescence. Considerable alterations of IR fluorescence patterns were found in diverse diseases, e.g., Duchenne-type and Werdnig-Hofmann muscular dystrophy, myasthenia gravis, Parkinson's disease, amyotrophic lateral sclerosis, scleroderma, osteogenesis imperfecta, lead poisoning, tetanus, eclampsia, anorexia nervosa, hepatic cirrhosis (Meloan et al., 1983). The abnormal IR fluorescence patterns differed significantly from disease to disease. However, owing to the heterogeneity of the available material, it is not yet possible to draw conclusions concerning the potential diagnostic significance of different IR fluorescence patterns.

Limitations of space preclude inclusion of a bibliography. However, references can be obtained from the authors.

REFERENCES

Abram, J. H.: Carcinoma of the kidney arising in the glomeruli. J. Path. Bact. 6:384-86, 1900.

Armstrong, E. M., MacSween, R. N. M.: Demonctration of bile canaliculi by an immunofluorescent staining technic. Anat. Rec. 177:317-20, 1973.

Badaruddin, R. H., Waldrop, F. S., Puchtler, H.: Alterations of the myoid pericanalicular layer: a light microscopic pilot study of human autopsy material. Arch. Path. 100:616-19, 1976.

Becker, C. G.: Demonstration of actomyosin in mesangial cells of the renal glomerulus. Amer. J. Path. 66:97-110, 1972.

Becker, C. G., Murphey, G. E.: Demonstration of contractile protein in endothelium and cells of heart valves, endocardium, intima, arteriosclerotic plaques and Aschoff bodies of rheumatic heart disease. Am. J. Path. 55:1-37, 1969.

Berl, S., Puszkin, S.: Mg^{2+}-Ca^{2+}-activated adenosine triphosphatase system isolated from mammalian brain. Biochemistry 9:2058-67, 1970.

Billroth, T.: Über die Epithelialzellen der Froschzunge, sowie über den Bau der Cylinder- und Flimmerepithelien und ihr Verhältnis zum Bindegewebe. Arch. f. Anat. Physiol. 1858: 159-77.

Boll, F.: Ueber den Bau der Thränendrüse. Arch. mikrosk. Anat. 4:146-53, 1868.

Boll, F.: Die Bindesubstanz der Drüsen. Arch. mikrosk. Anat. 5:334-56, 1869.

Borst & Enderlen: Die Transplantation von Gefässen und ganzen Organen. Dtsch. Zschr. f. Chirurgie 99:54-163, 1909.

Bretscher, A., Weber, K.: Localization of actin and microfilament associated proteins in the microvilli terminal web of the intestinal brush border by immunofluorescence microscopy. J. Cell Biol. 79:839-45, 1978.

Brettauer, J., Steinach, S.: Untersuchungen über das Cylinderepithelium der Darmzotten und seine Beziehung zur Fettresorption. Sitzungsber. Akad. Wiss. Wien, Math.-Naturwiss. Kl. 23:303-13, 1857.

Drasch, O.: Beobachtungen an lebenden Drüsen mit und ohne Reizung der Nerven derselben. Arch. f. Anat. u. Physiol. physiol. Abt. 1889: Quoted from Zimmermann, K. W., 1927.

Drenckhahn, D., Groeschel-Stewart, U., Unsicker, K.: Immunofluorescence microscopic demonstration of myosin and actin in salivary glands and exocrine pancreas of the rat. Cell Tissue Res. 183:273-80, 1977.

Drenckhahn, D., Steffens, R., Groeschel-Stewart, U.: Immunocytochemical localization of myosin in the brush border region of the intestinal epithelium. Cell Tissue Res. 205:163-66, 1980.

Ebner, V. v.: Über die Anfänge der Speichelgänge in den Alveolen der Speicheldrüsen. Arch. mikrosk. Anat. 8: 481-513, 1872.

Engelmann, Th. W., 1871: Quoted from Zimmermann, K. W., 1927.

Hammar, J. A.: Zur Histogenese und Involution der Thymusdrüse. Anat. Anz. 27:23-30, 41-89, 1905.

Hammar, J. A.: Fünfzig Jahre Thymusforschung: Kritische Übersicht der normalen Morphologie. Ergebn. Anat. Entwickl. gesch. 19:1-274, 1909.

Harper, J. T., Puchtler, H., Meloan, S. N., Terry, M. S.: Histochemical demonstration of myoepithelial filaments (myofibrils) in tubular and glomerular epithelium of human kidney. Lab. Invest. 20:585-86, 1969.

Harper, J. T., Puchtler, H., Meloan, S. N., Terry, M. S.: Light microscopic demonstration of myoid fibrils in renal epithelial, mesangial, and interstitial cells. J. Microsc. 91: 71-85, 1970.

Hart, C.: Thymusstudien: Die Pathologie der Thymus. Virchows Arch. path. Anat. 21:41-83, 1913.

Henle, J.: Allgemeine Anatomie. Lehre von den Mischungs- und Formbestandteilen des menschlichen Körpes. L. Voss, Leipzig, 1841.

Hollingsworth, A. S., Teabeaut, J. R., Puchtler, H.: Application of newer histochemical stains to myocardial infarcts. Lab. Invest. 16:653, 1967.

Hortoles, C.: Recherches histologiques sur le glomérule et les épithéliums du rein. Arch. Anat. Physiol. (Norm. et Path. sér. 2) 8:861, 1881.

Koelliker, A.: Manual of Human Microscopical Anatomy (Transl. by G. Busk and T. Huxley; DaCosta, ed.). Lippincott, Grambo & Co., Philadelphia, 1854.

Kolossow, A.: Eine Untersuchungsmethode des Epithelgewebes besonders der Drüsenepithelien und die erhaltenen Resultate. Arch. mikrosk. Anat. 52:1-43, 1898.

Krause, R.: Beiträge zur Histologie der Wirbeltierleber. I. Über den Bau der Gallencapillaren. Arch. mikrosk. Anat. 42: 53-82, 1893.

Low, B. W.: The structure and configuration of amino acids, peptides and proteins. In: The Proteins, vol I, part A. H. Neurath, K. Bailey, eds. Academic Press, New York, 1953. pp. 235-391.

Maximow, R.: Untersuchungen über Blut und Bindegewebe. II. Über die Histogenese der Thymus bei Säugetieren. Arch. mikrosk. Anat. 74:525-621, 1909.

Mayer, S.: Zur Lehre von der Schilddrüse und Der Thymus bei den Amphibien. Anat. Anz. 3:97-103, 1888.

Meloan, S. N., Puchtler, H.: Observations on infrared fluorescence of stained sections. J. South Carolina Med. Assoc. 70:81, 1974.

Meloan, S. N., Puchtler, H.: Mallory bodies: lesions of hepatocytes containing proteins of the keratin-myosin-epidermin group. Histochemistry 75:445-60, 1982.

Meloan, S. N., Puchtler, H., Branch, B. W.: Staining, polarization and fluorescence microscopic studies of myoid cells in human thymus. Bull. Beorgia Acad. Sci. 33:93, 1975.

Meloan, S. N., Puchtler, H., Paschal, L. D.: Demonstration of early lesions of muscle by infrared fluorescence microscopy. Acta histochemica Suppl. 28:253-63, 1983.

Mooseker, M. S., Pollard, T. D., Fujiwara, K.: Characterization and localization of myosin in the brush border of intestinal epithelial cells. J. Cell Biol. 79:444-53, 1978.

Müller, E.: Studien über Neuroglia. Arch. mikrosk. Anat. 55: 11-62, 1899.

Pappenheimer, A. M.: Further studies in histology of the thymus. Amer. J. Anat. 14:299-332, 1912/13.

Paschal, L. D., Puchtler, H., Meloan, S. N.: Demonstration of lesions of striated muscle by infrared fluorescence microscopy. Lab. Invest. 38:359, 1978.

Pease, D. C.: Myoid features of renal corpuscles and tubules. J. Ultrastruct. Res. 23:304-320, 1968.

Pensa, A.: Osservazioni a proposito di una particolarita di structura del timo. Boll. Soc. Med. Chir. Pavia 1902:288-202.

Pensa, A.: Ancora a proposito di una particolarita di struttura del timo ed osservazioni sullo sviluppo del timo negli anfibii anuri. Boll. Soc. Med. Chir. Pavia 1904:65-79.

Policard, A.: Sur la striation basale des cellules du canalicule contournée du rein des mammiferes. Compt. rend. Seanc. Soc. Biol. 59:568-69, 1905.

Prenant, A.: Cellules vibratiles et cellules à plateau. Bibliogr. Anat. 7:21-38, 1899.

Puchtler, H.: Histochemical analysis of terminal bars. J. Histochem. Cytochem. 4:439, 1956.

Puchtler, H.: Histochemical analysis of the tectal plate as seen in the intestinal epithelium. Rev. Can. de Biol. 16:537, 1957.

Puchtler, H.: Histochemical analysis of the "terminal web" in the epithelial cells of rat intestine. Anat. Rec. 130:360, 1958.

Puchtler, H., Waldrop, F. S., Terry, M. S., Conner, H. M.: A combined PAS-myofibril stain for demonstration of early lesions of striated muscle. J. Microsc. 89:329-38, 1969a.

Puchtler, H.: Sweat, F., Terry, M. S., Conner, H. M.: Investigation of staining, polarization and fluorescence microscopic properties of myoendothelial cells. J. Microsc. 89:95-104, 1969b.

Puchtler, H., Waldrop, F. S., Carter, M. G., Valentine, L. S.: Investigation of staining, polarization and fluorescence microscopic properties of myoepithelial cells. Histochemistry 40:281-89, 1974.

Puchtler, H., Waldrop, F. S., Meloan, S. N., Branch, B. W.: Myoid fibrils in epithelial cells: Studies of intestine, biliary and pancreatic pathways, trachea, bronchi, and testis. Histochemistry 44:105-118, 1975a.

Puchtler, H., Meloan, S. N., Branch, B. W., Gropp, S.: Myoepithelial cells in human thymus: staining, polarization and fluorescence microscopic studies. Histochemistry 45: 163-76, 1975b.

Puchtler, H., Meloan, S. N., Paschal, L. D.: Infrared fluorescence microscopy of stained tissues: Principles and technic. Histochemistry 68:211-30, 1980.

Puszkin, S., Berl, S., Puszkin, E., Clarke, D. D.: Actomyosin-like protein isolated from mammalian brain. Science 161: 170-71, 1968.

Remak, R.: Histologische Bemerkungen über die Blutgefässwände. Müller's Arch. f. Anat. und Physiol. Wiss. Med. 1850:79-101.

Ramsley, A. O.: Infrared fluorescence of dyes. Am. Dyestuff Reptr. 57:25-29, 1968.

Renaut, J.: Traite d'Histologie Pratique. Paris, 1897.

Renaut, J.: La cuticule tubuleuse des canaux et des canalicules pancréatiques intralobulaires. Compt. rend. Assoc. Anat. (Nancy) 5:23-27, 1903.

Rostgaard, J.: Electron microscopy filaments in basal part of rat kidney tubule cells and their in situ interaction with heavy meromyosin. Z. Zellforsch. 132:497-521, 1972.

Scheinman, J. I., Fish, A. J., Brown, D. M., Michael, A. J.: Human glomerular smooth muscle (mesangial) cells in culture. Lab. Invest. 34:150-58, 1976.

Schmidt, W. J.: Über Doppelbrechung und Feinbau der Darmepithelzellen der Kaulquappe. Zschr. Zellforsch. 33:1-4, 1943.

Schwalbe, G.: Beiträge zur Kenntnis der Drusen in den Darmwandungen, in's Besondere der Brunner'schen Drüsen. Arch. mikrosk. Anat. 8:92-140, 1872.

Strauss, A. J. L., van der Geld, H. W. R., Kemp, P. G., Exum, E. D., Goodman, H. C.: Immunological concomitants of myasthenia gravis. Ann. New York Acad. Sci. 124:744-66, 1964.

Strauss, A. J. L., Kemp, P. G., Douglas, S. D.: Myasthenia gravis. Lancet I: 772-73, 1966.

Toh, B. H., Muller, H. K., Elrick, W. L.: Smooth muscle associated antigen in astrocytes and astrocytomata. Br. J. Cancer 33:195-202, 1976.

Trenchev, P., Sneyd, P., Holborow, E. J.: Immunofluorescence tracing of smooth muscle contractile protein antigens in tissues other than smooth muscle. Clin. Exper. Immunol. 16:125-35, 1974.

Trenchev, P., Dorling, J., Webb, J., Holborow, E. J.: Localization of smooth musclelike contractile proteins in kidney by immunoelectron microscopy. J. Anat. 121:85-95, 1976.

Trompetter, J.: Ueber Endarteriitis. Georgi, Bonn, 1876.

Vetters, J. M.: Myasthenia gravis. Lancet 1:314, 1966.

Vetters, J. M.: Muscle antibodies in myasthenia gravis. Immunology 13:275-80, 1967.

Virchow, R.: Über das Epithel der Gallenblase und über einen intermediären Stoffwechel des Fettes. Virchows Arch. 11: 574-78, 1857.

Waldrop, F. S., Humphries, A. L., Puchtler, H., Valentine, L. S.: Histochemical, polarization and fluorescence microscopic studies of arteries in transplants. Bull. Georgia Acad. Sci. 30:93, 1972.

Waldrop, F. S., Puchtler, H., Valentine, L. S.: Light microscopic demonstration of myoid fibres in the nervous system. J. Microsc. 96:45-56, 1972.

Westphalen, H.: Histologische Untersuchungen über den Bau einiger Arterien. Mattiesen, Dorpat, 1886.

Wiesel, J.: Pathologie des Thymus. Ergeb. allg. Path. path. Anat. 15:416-782, 1912.

Wilke, G.: Ueber Gliafaserbildung als intercellulärer Vorgang. Dtsch. Z. f. Nervenh. 166:447-63, 1951.

Wilke, G., Kircher, H.: Ueber roentgenographische Untersuchunger zur Frage der Gliafaserbildung. Dtsch. Z. f. Nervenh. 167:391-406, 1952.

Zimmermann, K. W.: Beiträge zur Kenntnis einiger Drüsen und Epithelien. Arch. mikrosk. Anat. 52:552-706, 1898.

Zimmermann, K. W.: Handbuch der mikroskopischen Anatomie des Menschen, vol 5, part 1. M. v. Möllendorf (ed.). Springer, Berlin, 1927.

Zimmermann, K. W.: Über den Bau des Glomerulus der menschlichen Niere. Zschr. mikrosk.-anat. Forsch. 18:520-52, 1929.

65 Perturbation of Myosin Light Chain Phosphorylation Levels in Adherent Fibroblasts

Susan A. Bayley and D. A. Rees

INTRODUCTION

Studies on isolated molecular components of smooth muscle and nonmuscle cells would suggest that light chain phosphorylation of myosins have important regulatory functions in physiological activity. Such phosphorylation has been shown to be a prerequisite for actin activation of the myosin ATPase and to be regulated by enzymes under the control of Ca^{++}-calmodulin or cAMP—all of which has led to the proposal of a plausible overall scheme (Adelstein, 1982). Light chain phosphorylation has also been shown to stimulate myosin assembly into thick filaments—consistent with a second level of control of a possible sliding filament mechanism in smooth muscle and nonmuscle systems (Suzuki et al., 1978; Scholey et al., 1980). Confirmation of the relevance of such mechanisms through comparison of the levels of light chain phosphorylation with physiological parameters in cells and tissues, is encouraging but still incomplete. In tracheal smooth muscle, myosin light chain phosphorylation rises with tension development but some features of the correlation are still difficult to explain (Aksoy et al., 1982; de Lanerolle et al., 1982; Perry, 1982). In platelet stimulation, part of the motile response is a cytoplasmic contraction associated with and perhaps driven by a rise in light chain phosphorylation with increased incorporation of myosin as its light chain phosphate into a detergent-insoluble residue to suggest its assembly with the cytoskeleton (Fix and Phillips, 1982; Carroll et al., 1982).

Following our previous interest in mechanisms of spreading and rounding of substrate-attached fibroblasts (Badley et al., 1980a, 1980b; Couchman et al., 1982) we have now examined the changes in myosin light chain phosphorylation in an attempt to relate these to the architecture of the cytoskeleton and shape changes of the whole cell.

METHODS

Balb/c 3T3 fibroblasts were grown for 24 hours on glass coverslips (76 mm × 64 mm) to subconfluency in glass petri dishes containing 10 ml Dulbecco's minimum essential medium (Gibco Biocults, special formulation) supplemented with 10% foetal calf serum and 0.1 mCi ^{32}P (Orthophosphate, Carrier free).

Control coverslips were then attached to Cu blocks rinsed with serum-free medium and immediately plunged into liquid nitrogen. The cell layers were then freeze dried. Prior to this freeze-drying process the remaining coverslips were subjected to various treatments as described in Table 65.1. All cell layers were rehydrated and detached from the substrate with 0.25 M sucrose, 0.5% NP4O, 50 mM NaF, 10 mM EDTA, 1 mM PMSF pH 8.3. NaF and EDTA inhibit the action of phosphatase and kinase respectively so that the phosphorylation levels in the cell extracts should be maintained at the same level as in the original cells. The samples were then briefly treated with RNAase A, clarified by ultracentrifugation and analyzed by two-dimensional electrophoresis (O'Farrell, 1975). Phosphorylated proteins were detected by autoradiography of the gels after drying down.

RESULTS

Assay of Phosphorylation

A typical autoradiogram prepared after incubation of resting Balb/c 3T3 cells with $^{32}PO_4^{3-}$ followed by processing and two-dimensional electrophoresis as described under Methods, is shown in Figure 65.1a. A similar autoradiogram from an identical batch of cells which, however, had been incubated prior to processing for 1 minute with 0.05% EGTA in Ca^{++} and Mg^{++} free phosphate buffered saline at 37°C, is shown in Figure 65.1b. The comparison shows clearly the disappearance of a spot that runs in the position expected for myosin light chain phosphate as a result of the EGTA treatment.

Table 65.1. Level of Myosin Light Chain Phosphorylation in Balb/c 3T3 Fibroblasts after Treatment with a Variety of Agents

Treatment	Phosphorylation	Cell Shape
Serum free medium rinse	++	Spread
PBS- 5 minutes	++	Spread
PBS- rinse then 1 minute in EGTA	-	Rounding
PBS- rinse then 4 minutes in EGTA	-	Rounded
PBS+ or - rinse then 1 to 4 minutes in Trypsin in PBS+ or -	-	Rounding
PBS+ rinse then 1 minute in Azide	++	Spread
PBS+ rinse then 20 minutes in Azide	+	Spread
PBS+ rinse, 1 minute Azide, PBS- rinse, 1 minute EGTA	++	Spread
PBS+ rinse, 20 minutes Azide, PBS- rinse, 20 minutes EGTA	+	Spread
PBS- rinse, 1 minute Azide + EGTA	++	Spread
PBS- rinse, 20 minutes Azide + EGTA	+	Spread

Abbreviations and Concentrations:

PBS+	Phosphate Buffered Saline
PBS-	Ca^{++} and Mg^{++} free Phosphate Buffered Saline
EGTA	0.05% in PBS-
Trypsin	10 μg/ml
Azide	10 mM Sodium Azide in PBS+
Azide + EGTA	10 mM Sodium Azide + 0.05% EGTA in PBS-

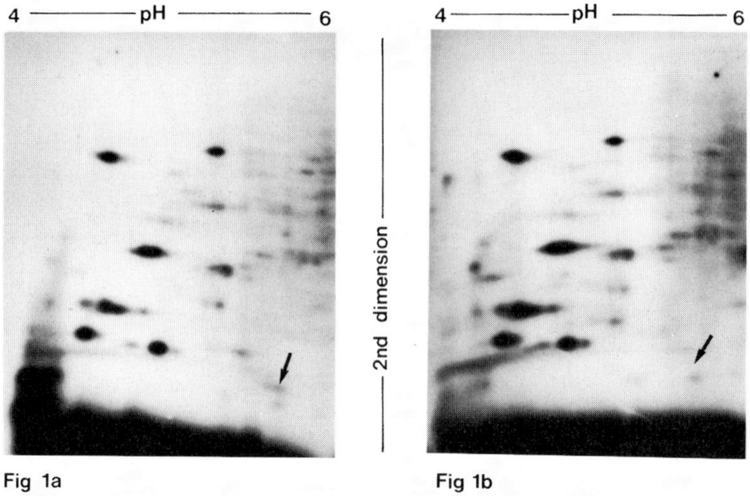

Fig. 65.1. Autoradiograms of ^{32}P labelled proteins from (a) untreated Balb/c 3T3 fibroblasts (b) after 1 minute in EGTA, then separation by two-dimensional electrophoresis. Samples were prepared for electrophoresis as described in the text. The second dimension is 5-20% PAGE. The positions expected for myosin light chain phosphate are indicated with arrows.

Phosphorylation Levels during Perturbation of Cell Shape and Cytoskeleton

Quantitative estimation of the levels of light chain phosphorylation were not possible with our method but clear cut qualitative changes could be observed (e.g., Fig. 65.1a and 1b) and classified. By comparison of autoradiograms prepared from the same number of cells which had been exposed in the same ways to radioisotope and using the spots made by other phosphorylated proteins as internal standards the level of phosphorylation was classified as high (++), intermediate (+), or low (-). The absence of film blackening by low levels suggests that the high levels may exceed the low levels by an order of magnitude. The results of such measurements after a variety of pretreatments of living cells, are shown in Table 65.1 together with the corresponding cell shape.

The simultaneous addition of azide and EGTA is still at an early stage of study, but, with this exception, the effect of each pretreatment was confirmed in at least 4 independent experiments.

DISCUSSION

We report several new observations on levels of myosin light chain phosphate in substrate attached fibroblasts. A marked reduction in myosin light chain phosphate is associated with cell rounding, whether induced by EGTA or trypsin. When rounding is prevented by pretreatment of cells with azide as previously described (Rees et al., 1980), the reduction in phosphorylation level is likewise prevented in the short term. This blocking influence of azide acts very rapidly, as shown by the observation that simultaneous addition of azide and EGTA to the cells caused no detectable change in the level of light chain phosphorylation.

The influence of EGTA in diminishing the level of light chain phosphorylation is readily understood in terms of the requirement of the appropriate kinase for Ca^{++}-calmodulin (Adelstein, 1982); removal of Ca^{++} would inactivate this enzyme and allow the phosphatase to prevail. The cell rounding which is energy-dependent (Bershadsky et al., 1980), follows as a distinct second step after the level of phosphorylation has decayed as is consistent with previous observations in other cell types that actomyosin redistribution as a result of such treatment occurs in advance of shape changes (Pollack and Rifkin, 1975; Badley et al., 1980a).

Stress fibers disintegrate as a consequence of Ca^{++} depletion and can be restored by returning the cells to Ca^{++}-containing medium (Badley et al., 1980a)—in interesting contrast to microfilament bundles in gut microvilli where the converse is the case (Matsudaira and Burgess, 1982). Since stress fibers are also disrupted by long-term (45 minutes) energy depletion of the cell (Bershadsky et al., 1980), it is possible that they could require continuing active contraction for their maintenance (compare Isenberg et al, 1976, Kreis and Birchmeier, 1980) and indeed we do observe a gradual decline in light chain phosphorylation over this period. On the other hand, it has been reported that treatment of other cell types to increase their content of cAMP causes—as expected from present understanding of relevant control mechanisms (Adelstein, 1982)—a reduction in light chain phosphorylation (Lockwood et al., 1982), but an enhancement in stress fiber formation (Willingham and Pastan, 1975; Lloyd et al., 1977; Lockwood et al., 1982).

At first sight and in principle, the influence of azide in maintaining phosphorylation levels in the presence of EGTA could be explained in terms of known mechanisms by an inactivation of either the ATP-dependent Ca^{++} pump or the cAMP-dependent kinase acting on the myosin light chain kinase, before other relevant ATP-required mechanisms are significantly affected. Other explanations may also be possible, however. The continued protection against rounding by azide as phosphorylation levels decay would reflect a parallel inhibition of the energy-dependent, phosphorylation-independent, rounding activity.

Our overall conclusion is that, while any correlation between phosphorylation and stress fiber maintenance remains unclear, cell rounding does not correlate with high levels of phosphorylation of myosin light chain within substrate-attached fibroblasts.

REFERENCES

Adelstein, R. S. Calmodulin and the regulation of the Actin-Myosin interaction in smooth muscle and non-muscle cells. Cell, 30, 349-50 (1982).

Aksoy, M. O., Murphy, R. A. and Kamm, K. E. Role of Ca^{++} and myosin light chain phosphorylation in regulation of smooth muscle. Am. J. Physiol., 242, C109-C119 (1982).

Badley, R. A., Woods, A., Carruthers, L. and Rees, D. A. Cytoskeleton changes in fibroblast adhesion and detachment. J. Cell Sci., 43, 379-90 (1980a).

Badley, R. A., Woods, A., Smith, C. G., Carruthers, L. and Rees, D. A. Actomyosin relationships with surface features in fibroblast adhesion. Exp. Cell Res., 126, 263-72 (1980b).

Bershadsky, A., Gelfand, V. I., Svitkina, T. M. and Tint, I. S. Destruction of microfilament bundles in mouse embryo fibroblasts treated with inhibitors of energy metabolism. Exp. Cell Res., 127, 421-29 (1980).

Carroll, R. C., Butler, R. G., Morris, P. A. and Gervard, J. M. Separable assembly of platelet pseudopodal and contractile cytoskeletons. Cell, 30, 385-93 (1982).

Couchman, J. R., Rees, D. A., Green, M. R. and Smith, C. G. Fibronectin has a dual role in locomotion and anchorage of primary chick fibroblasts and can promote entry into the growth cycle. J. Cell Biol., 93, 402-10 (1982).

Fox, J. E. B. and Phillips, D. R. Role of phosphorylation in mediating the association of myosin with the cytoskeletal structures of human platelets. J. Biol. Chem., 257, 4120-26 (1982).

Isenberg, G., Rathke, P., Hulsmann, N., Franke, W. W. and Wohlfarth-Botterman, K. E. Cytoplasmic actomyosin fibrils in tissue culture cells. Direct proof of contractility by visualisation of ATP-induced contraction in fibrils isolated by laser micro-beam dissection. Cell Tissue Res., 166, 427-43 (1976).

Kreis, T. E. and Birchmeier, W. Stress fibre sarcomeres of fibroblasts are contractile. Cell, 22, 555-61 (1980).

de Lanerolle, P., Condit, J. R., Tanenbaum, M. and Adelstein, R. S. Myosin phosphorylation, agonist concentration and contraction of tracheal smooth muscle. Nature, Lond., 298, 871-72 (1982).

Lloyd, C. W., Smith, C. G., Woods, A. and Rees, D. A. Mechanisms of cellular adhesion. III. The interplay between adhesion, the cytoskeleton and morphology of substrate-attached cells. Exp. Cell Res. 110, 427-37 (1977).

Lockwood, A. H., Trivette, D. D. and Pendergast, M. Molecular events in cAMP mediated reverse transformation. Cold Spring Harbor Symp. Quant. Biol., 46, 909-19 (1982).

Matsudaira, P. T. and Burgess, D. R. Partial reconstruction of microvillus cone bundle: Characterisation of villin as a Ca^{++}-dependent, actin bundling/depolymerising protein. J. Cell Biol. 92, 648-56 (1982).

O'Farrell, P. H. High resolution two dimensional electrophoresis of proteins. J. Biol. Chem., 250, 4007-4021 (1975).

Perry, S. V. Phosphorylation of the myofibrillar proteins and the regulation of contractile activity in muscle. Phil. Trans. R. Soc. London, B302, 58-71 (1983).

Pollack, R. and Rifkin, D. Actin-containing cables within anchorage dependent rat embryo cells are dissociated by plasmin and trypsin. Cell, 6, 495-506 (1975).

Rees, D. A., Badley, R. A. and Woods, A. Relationships between actomyosin stress fibres and some cell surface receptors in fibroblast adhesion. In, Cell Adhesion and Motility (Curtis, A. S. G. and Pitts, J. D., eds.): Cambridge University Press (1980).

Scholey, J. M., Taylor, K. A. and Kendrick-Jones, J. Regulation of non-muscle myosin assembly by calmodulin-dependent light chain kinase. Nature, Lond., 287, 233-35 (1980).

Silver, P. J. and Stull, J. T. Regulation of myosin light chain and phosphorylase phosphorylation in tracheal smooth muscle. J. Biol. Chem., 257, 6145-50 (1982).

Suzuki, H., Onishi, H., Takanishi, K. and Watanabe, S. Structure and function of chicken gizzard myosin. J. Biochem. Tokyo, 84, 1529-42 (1978).

Willingham, M. C. and Pastan, I. Cyclic AMP and cell morphology in cultured fibroblasts. Effects on cell shape, microfilament and microtubule distribution, and orientation to substratum. J. Cell Biol., 67, 146-59 (1975).

66 Morphological and Immunohistochemical Studies on Heart in Cardiomyopathic Hamsters

Zeng-hong Tu and Larry F. Lemanski

INTRODUCTION

Homberger et al. (1962) discovered an autosomal recessive condition in an inbred strain (B10 14.6) of Syrian hamster which resulted in a dystrophylike disease of the skeletal muscle, heart failure and premature death of the animal. Strain UM-X 7.1, a derivative of the original strain (B10 14.6), shows similar traits and has the advantage of showing very good consistency in the temporal onset of disease symptoms (Bajusz, 1973). The heart is severely affected and homozygous animals die of heart failure after only one-third to one-half the normal life span. Numerous papers have appeared on cardiomyopathic hamsters in recent years. A majority of these reports are descriptive morphological, physiological, or biochemical studies on late juvenile and adult animals. To date, few investigations have been directed toward analyzing the heart myocytes of young cardiomyopathic hamsters. A problem with studies on animals at advanced stages of the disease is that secondary effects probably result from the heart failure. To understand the primary underlying defect(s) of the cardiomyopathy, it seems necessary to study young animals before the experimental results become clouded by secondary abnormalities. Moreover, it would obviously be desirable to examine normal and cardiomyopathic (CM) heart myocytes under similar growth conditions (i.e., cell culture). Bajusz (1969) observed aberrant myofibril structure and arrangement in in vivo hearts of cardiomyopathic animals 30 days of age or older. Jasmin and Bajusz (1975) have suggested that the myofibril abnormalities may result from hypoxia due to heart failure in vivo. In the present study, we examined isolated normal and CM adult ventricular heart cells to more precisely evaluate their overall morphology. In addition, we prepared primary cultures of ventricular heart cells from newborn normal and CM hamsters and analyzed their differentiation by morphological and immunohistochemical methods.

METHODS

For the preparation of isolated adult heart cells, normal and CM hamster hearts were dissociated into single cells by retrograde aortic perfusion of the coronary vasculature. The perfusion solution consisted of a Ca^{++} and Mg^{++}-free Krebs-Ringer solution containing 0.05% collagenase, 0.1% hyaluronidase, and 0.5% bovine serum albumin, which was kept at 37°C and gassed continuously with 95% O_2 and 5% CO_2. After the heart softened, the atria were discarded and the ventricles were dissociated by gentle agitation and filtering through nylon mesh. The cells were collected by centrifugation resuspended in a glutaraldehyde-formaldehyde and picric acid fixative, osmicated, critically point dried, and viewed in a scanning electron microscope.

To prepare primary heart cell cultures, three-day-old normal and cardiomyopathic hamsters were used. The animals were killed by cervical dislocation and the hearts were extirpated. A solution containing 0.08% trypsin and 0.02% collagenase was used to dissociate the cells. To enrich the cultures for myocytes, the dissociated heart cell suspensions were preincubated in an Erhlenmeyer flask for 60 min at 37°C. Most of the fibroblasts attached to the bottom of the flask during this period. The remaining unattached cells (now about 90% myocytes) were diluted to an average density of 2×10^5 dispersed cells/ml of medium. Two ml aliquots of the myocyte-enriched cell suspensions were added to each tissue culture dish with collagen-coated glass coverslips on the bottom. The culture medium consisted of Eagle's MEM containing 15% fetal calf serum, penicillin 100 μ/ml, streptomycin 0.1 mg/ml and fungizone 0.25 μg/ml. The cultures were incubated at 37°C in an atmosphere of 5% CO_2 and 95% air. The culture medium was changed every 3 days.

Immunofluorescent staining for the contractile proteins myosin, tropomyosin, actin, and α-actinin were performed. The antibodies used and the staining protocol have been detailed in an earlier paper (Lemanski and Tu, 1983). Briefly, heart cells which had been grown on glass coverslips were fixed in a periodate-lysine-paraformaldehyde solution (McLean and Nakane, 1974), rinsed and stained by an indirect method using FITC-labelled antibodies.

RESULTS AND DISCUSSION

Scanning electron microscopic examination of the dissociated adult normal and CM cardiomyocytes proved very revealing. Normal cells, for the most part, exhibited straight cylindrical shapes with straight parallel myofibrils (Fig. 66.1). In contrast, the dissociated mutant cells often appeared to have aberrant shapes, the most common abnormality being a twisted or helical morphology (Fig. 66.2). Fluorescent staining of the myofibrils within the dissociated cells suggested that the myofibrils were twisted and abnormally arranged as well (Fig. 66.3). This abnormal myofibril morphology in itself would seem to provide an explanation for the decreased efficiency of heart contractions in CM hamsters and presumably is related to the observed heart failure in this genetically-based cardiomyopathy.

Primary cell cultures derived from the heart ventricles of normal and CM newborn hamsters after three and five days in culture were stained by an indirect method with FITC-labelled antibodies against the contractile proteins. Intense fluorescence was apparent in the myofibrils of both normal and CM cardiomyocytes. Antimyosin stained the A bands of the myofibrils in cultured cells. Normal cardiomyocytes after three days in culture contained well formed myofibrils which were, for the most part, arranged in parallel arrays (Fig. 66.4). By contrast, antimyosin staining revealed that the myofibrils in many of the 3-day cultured CM cells were disoriented with respect to each other (Fig. 66.5) and by 5 days the disarray seemed even more pronounced. In spite of the disoriented arrangements, the antimyosin staining of individual myofibrils appeared similar in normal and CM myocytes. Antiactin and antitropomyosin antibodies stained the I band regions of myofibrils. Again, although the myofibrils often appeared disoriented, there was no obvious difference in the staining of individual myofibrils. α-actinin staining again revealed the myofibril disarray. In addition, the Z line staining appeared wider in some cells and showed a somewhat irregular "zigzag" arrangement. Also, some of the CM cardiomyocyte contained collections of α-actin "spots" suggestive of Z bodies. The α-actinin staining in normal cells was suggestive of thin distinct Z lines with few Z body collections.

Thus, our results make it clear that myofibril formation is abnormal in CM hamster heart cells in culture and indicates that this condition in in vivo hearts is not caused solely by the hypoxic conditions resulting from decreased microcirculation, as has been suggested by previous studies. A more likely possibility might be that the observed disarray is a primary

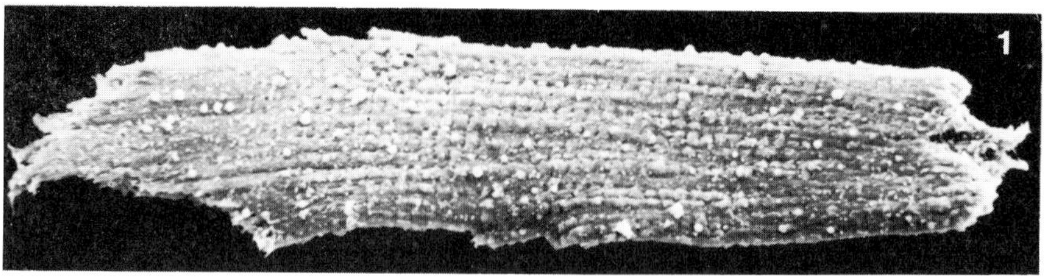

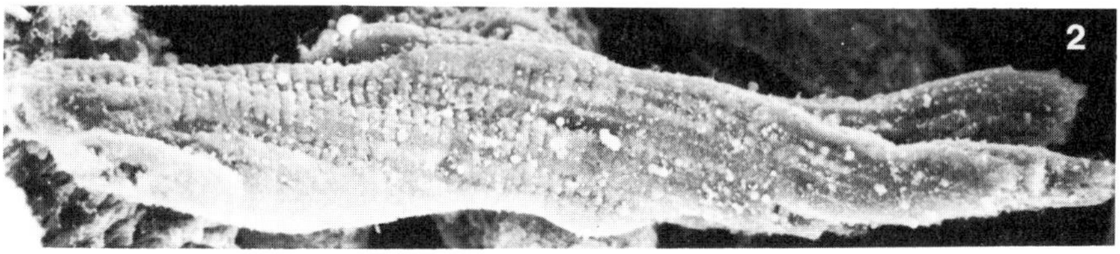

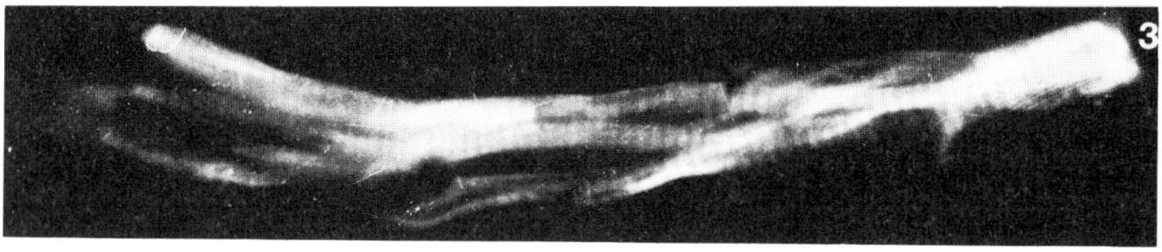

Fig. 66.1. Scanning electron micrograph of isolated adult normal cardiomyocyte. The cell shows a straight cylindrical morphology.

Fig. 66.2. Scanning electron micrograph of isolated adult CM cardiomyocyte. The cell shows a twisted or helical configuration.

Fig. 66.3. Fluorescent micrograph of isolated adult CM heart cell. The myofibrils appear to have a disoriented arrangement.

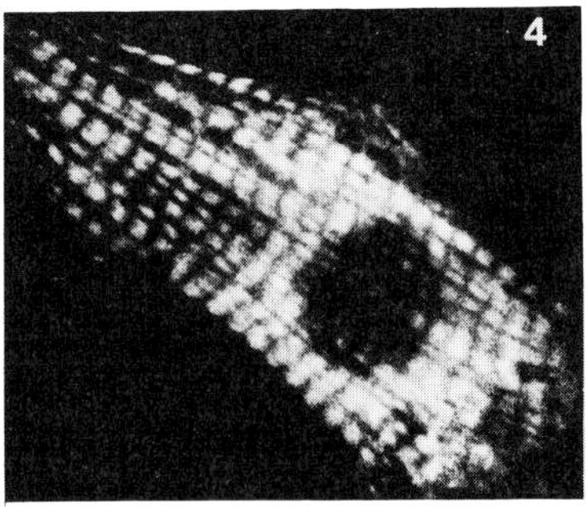

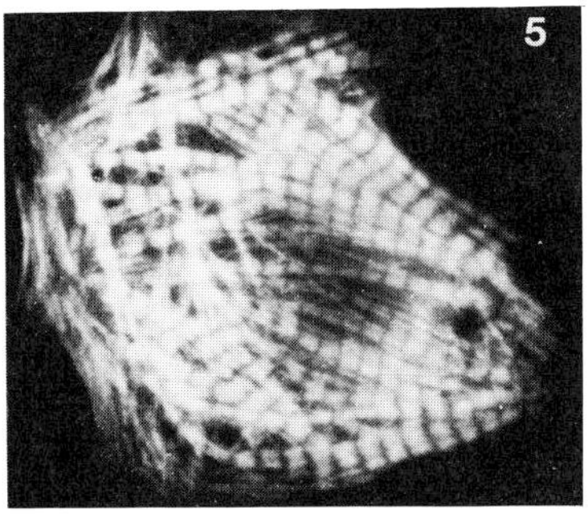

Fig. 66.4. Cultured normal hamster cardiomyocyte immunostained for myosin. The A-bands are stained and reveal that most of the myofibrils are arranged in parallel.

Fig. 66.5. Cultured cardiomyopathic hamster heart cell stained for myosin. There is obvious myofibril disarray.

defect of the mutation or, at least, a defect that is expressed very early in development. This would presumably result in decreased contraction efficiency for the CM heart cells, which could lead to heart failure and, eventually, to many of the other abnormalities at advanced stages of the disease. It is obvious that further studies will be required to fully understand the genetically-based cardiomyopathy in UM-X 7.1 Syrian hamsters.

REFERENCES

Bajusz, E. (1969). Hereditary cardiomyopathy: A new disease model. Amer. Heart J., 77:686-96.

Bajusz, E. (1973). A disease model of hereditary cardiomyopathy: Its usefulness and limitations. Rec. Advan. Studies Card. Struct. Metabol., 2:291-92.

Homberger, F., J. Baker, C. Nixon and G. Wilgram (1962). New hereditary disease of Syrian hamster. Primary, generalized polymyopathy and cardiac necrosis. Arch. Intern. Med., 110:660-62.

Jasmin, G. and E. Bajusz (1975). Prevention of myocardial degeneration in hamsters with hereditary cardiomyopathy. Rec. Advan. Studies Card. Struct. Metabol., 6:219-29.

Lemanski, L. F. and Z. H. Tu (1983). Immunofluorescent studies for myosin, actin, tropomyosin and α-actinin in cultured cardiomyopathic hamster heart cells. Develop. Biol., 97:338-48.

McLean, I. and P. K. Nakane (1974). Periodate-lysine-paraformaldehyde fixative. A new fixative for immunoelectron microscopy. J. Histochem. Cytochem., 22:1077-83.

67 Studies on Heart Development and Inductive Processes in Cardiac Mutant Axolotls: Ambystoma Mexicanum

Larry F. Lemanski, Lynn A. Davis, Craig S. Hill, Daniel J. Paulson

INTRODUCTION

A naturally occurring recessive genetic mutation which alters normal embryonic heart development was discovered in the Mexican axolotl, Ambystoma mexican by Humphrey (1972). The mutation has been designated c for cardiac nonfunction. Mutant (c/c) embryos are first distinguishable from their normal siblings (+/+ or +/c) at stage 34 when the normals develop contracting hearts. Mutant embryos at this stage appear similar to normals upon gross inspection, except that the mutant hearts do not beat properly since only the conus region contracts. No vascular circulation is established. During later stages in development, the mutant embryos acquire ascites and their hearts appear distended and thin-walled. In spite of the lack of cardiovascular function, mutant embryos survive approximately three weeks beyond the time when the heart normally should have begun to beat, presumably by simple diffusion of oxygen directly into the tissues. Skeletal muscle does not appear directly affected by gene c.

Some very revealing information concerning a possible mode of action for gene c was provided by Humphrey (1972). In one series of experiments, he transplanted mutant (c/c) heart primordia into the heart regions of normal (+/c or +/+) recipients at stages 29-30 and observed that the mutant hearts beat in this new "normal" environment. Conversely, when normal hearts were placed into mutant hosts, no heartbeat developed. In further studies, normal and mutant siblings were linked parabiotically at stage 25; the cardiac deficiency was not corrected in mutants and the normals were not adversely affected. Such conjoined animals could live indefinitely, and the mutant twins, except for their nonfunctional hearts, appeared normal. These studies suggested that gene c specifically affected the heart and indicated that the failure of normal heart development in cardiac nonfunction embryos might stem from abnormal inductive (or inhibitory) effects in the heart region.

METHODS

Cardiac mutant (c/c) embryos and their phenotypically normal (+/+ or +/c) siblings were obtained by matings of heterozygous adults (+/c × +/c). Additional homozygous normal (+/+) embryos were obtained from wild-type (+/+ × +/+) matings. The embryos were reared in either 25% or 50% Holtfreter's solution and were staged according to the system of Schreckenberg and Jacobson (1975). Tissues for routine ultrastructural examination were fixed in a glutaraldehyde-formaldehyde-picric acid combination, postfixed in osmium tetroxide and embedded in Epon (Lemanski, 1973a). For immunohistochemical studies, the tissues were fixed in a periodate-lysine-paraformaldehyde solution (McLean and Nakane, 1974), frozen sectioned and immunostained by an indirect method using FITC-labelled antibodies (Lemanski, Fuldner, and Paulson, 1980). Organ cultures were performed in sterile Holtfreter's solution (Hill and Lemanski, 1979; Lemanski, Paulson, and Hill, 1979).

RESULTS AND DISCUSSION

Morphological studies comparing normal and mutant heart development from stage 34 (heart beat stage) through 41 (when mutant embryos die) have been reported (Lemanski, 1973b). Electron microscopy reveals that normal ventricular heart myocytes contain organized sarcomeric myofibrils at stage 34-35. By stage 41, the normal ventricular myocardium shows trabeculae formation and contains well differentiated muscle cells. The mutant myocardium does not trabeculate and remains a single cell-layer in thickness. Mutant heart ventricular cells contain a few scattered thin (6 nm) and thick 15 nm) filaments and occasional Z bodies. Some mutant heart cells at stages 34 through 41 show a partial organization of myofilaments; however, distinct sarcomeric myofibrils are not observed. Mutant cells instead contain amorphous proteinaceous collections in their peripheral cytoplasm where myofibrils initially organize in normal cells.

Immunofluorescent studies to stain myosin, α-actinin, actin, and tropomyosin in normal and mutant heart cells were performed.

By stage 41, the normal cells, as expected, showed intense sarcomeric myofibril staining patterns with the various antibodies; antimyosin stained the A-bands, antiactin and antitropomyosin the I band regions, and anti-α-actinin the Z-lines. The mutant cells also stained intensely for myosin, actin, and α-actinin; however, there was no well-defined sarcomeric pattern of staining as in normal myocytes. Furthermore, the immunofluorescent staining for tropomyosin in mutant hearts appeared less than normal (Lemanski, Fuldner, and Paulson, 1980). Radioimmunoassay studies also suggested a reduced quantity of antigenically-detectable tropomyosin in mutant hearts (Moore and Lemanski, 1983).

Thus, although mutant hearts contain almost normal complements of the myofibrillar proteins myosin, actin, and α-actinin, tropomyosin appears deficient. Apparently tropomyosin is either quantitatively reduced or antigenically (biochemically) altered in some way so as to escape recognition by the antibodies used in experiments. In any event, normal myofibrillogenesis fails. In view of the failure of much of the actin in mutant hearts to form into filaments (Lemanski, et al., 1976) and considering the close molecular association between actin and tropomyosin in organized myofibrils of muscle (Gergeley, 1976), studies were undertaken to determine whether there might be a relationship between the deficiency of normal muscle tropomyosin in mutant hearts and their failure to form filamentous actin. Our results suggested that there very well may be a correlation (Lemanski, 1979). When mutant hearts were glycerinated and incubated in a solution containing purified tropomyosin from chicken or rabbit skeletal muscle, large numbers of 6 nm actin filaments formed from the amorphous collections in the cells. Thus, it appears that insufficient or abnormal tropomyosin in mutant heart cells may be a key in their failure to form normally organized myofibrils. Based on these observations, it is inviting to speculate that the presence of normal muscle-type tropomyosin in developing embryonic hearts may be a necessary prerequisite for actin to become filamentous and for myofibrillogenesis to take place.

With regard to heart induction studies on cardiac mutant axolotls, it is germane to point out that the heart, as with most vertebrate organs, requires inductive interactions for normal development. These interactions control and direct heart formation. The anterior endoderm (gut) in vertebrate embryos is the most potent heart inductive tissue. This has been established experimentally in amphibians (Jacobson and Duncan, 1968; Fullilove, 1970) and is probable for chicks and mammals as well.

Since the inductive role of anterior endoderm in amphibians is well established and since Humphrey's (1972) transplantation experiments were suggestive of abnormal inductive processes in cardiac mutant embryos, we performed experiments to determine whether the cardiac defect could be corrected by organ culturing mutant (c/c) hearts with normal (+/+) anterior endoderm (Lemanski, Paulson, and Hill, 1979). Controls included culturing mutant hearts alone or in the presence of epidermis, somites or posterior endoderm. An additional control included culturing stage 35 normal sibling hearts by themselves or with epidermis. When stage 35 mutant (c/c) hearts were cultured with stage 30 normal (+/+) anterior endoderm, 81% beat throughout their lengths by 24 hours in culture. None of the control mutant hearts showed contractions propagated throughout their lengths although contractions were visible in the conus regions of some mutant hearts. Ninety-one percent of the normal hearts exhibited propagated contractions. After 48 hours in culture, the ventricular portions of the hearts were examined by electron microscopy. The normal hearts, as expected, were composed of myocytes containing numerous well-organized sarcomeric myofibrils. The stage 35 mutant hearts cultured with stage 30 normal anterior endoderm also contained myofibrils of normal morphology (Fig. 67.1). By contrast, no organized sarcomeres could be found in mutant heart controls cultured without anterior endoderm. The results of these experiments suggest that normal anterior endoderm corrects the heart defect in cardiac lethal embryos and in essence transforms mutant hearts into normal ones. We believe that this takes place because the normal anterior endoderm provides a necessary final inductive influence to cause the mutant hearts to form organized sarcomeric myofibrils and begin beating.

Very recently, organ culture experiments in our laboratory have been performed on mutant axolotl hearts using conditioned medium prepared by growing normal anterior endoderm in Holtfreter's solution (Davis and Lemanski, 1983). These studies demonstrate that a factor(s) produced by the normal endoderm and released into the medium is capable of inducing myofibrillogenesis in the mutant hearts. Experiments are currently underway to characterize this inducing factor(s). Preliminary work has shown that the ability of normal endoderm to produce the factor(s) decreases after stage 33. Furthermore, it appears that conditioned media from +/+ endoderm is a more potent inducer of cardiac differentiation than conditioned media from +/c endoderm. The factor(s) retains its activity after being boiled for five minutes; however, the activity appears to be

Fig. 67.1. Electron micrograph illustrating a portion of a mutant heart ventricle which has been cultured for 48 hours with normal (+/+) anterior endoderm. A well-organized sarcomeric myofibril of normal morphology can be seen. A, A-band; I, I-band; Z, Z-line.

lost after storage at -70°C for one week. Passage over a Sephadex G100 column has permitted the separation of conditioned media into active and inactive fractions. Thus, we are hopeful that further analysis of the active fraction of conditioned media will lead to identification and characterization of the factor(s) produced by normal anterior endoderm which is (are) necessary for normal heart differentiation.

REFERENCES

Davis, L. A. and L. F. Lemanski (1983). Inductive properties of a factor produced by endoderm. J. Cell Bio. (Abstract), In press.

Fullilove, S. L. (1970). Heart inductor: Distribution of active factors in newt endoderm. J. Esp. Zool., 75:323-26.

Gergeley, J. (1976). Troponin-tropomyosin dependent regulation of muscle contraction by calcium. In Cell Motility: Cold Spring Harbor Conferences on Cell Proliferation (ed. Goldman, Pollard, and Rosenbaum), 3:137-49.

Hill, C. S. and L. F. Lemanski (1979). Morphological studies on cardiac lethal mutant salamander hearts in organ cultures. J. Exp. Zool., 290:1-20.

Humphrey, R. R. (1972). Genetic and experimental studies on a mutant gene (c) determining absence of heart action in embryos of the Mexican axolotl (Ambrystoma mexicanum). Develop. Biol., 27:365-75.

Jacobson, A. G. and T. T. Duncan (1968). Heart induction in salamanders. J. Exp. Zool., 167:79-103.

Lemanski, L. F. (1973a). Heart development in the Mexican salamander, Ambystoma mexicanum. II. Ultrastructure. Amer. J. Anat., 136:487-526.

Lemanski, L. F. (1973b). Morphology of developing heart in cardiac lethal mutant Mexican axolotls, Ambystoma mexicanum. Develop. Biol. 33:312-33.

Lemanski, L. F. (1979). Role of tropomyosin in actin filament formation in embryonic salamander heart cells. J. Cell Biol., 82:227-38.

Lemanski, L. F., R. A. Fuldner and D. J. Paulson (1980). Immunofluorescence studies for myosin, α-actinin and tropomyosin in developing hearts of normal and cardiac lethal mutant Mexican axolotls, Ambystoma mexicanum. J. Embryol. Exp. Morph., 55:1-15.

Lemanski, L. F., M. S. Mooseker, L. D. Peachey, and M. R. Iyengar (1976). Studies of muscle proteins in embryonic myocardial cells of cardiac lethal mutant Mexican axolotls (Ambystoma mexicanum) by use of heavy meromyosin binding and sodium dodecyl sulfate polyacrylamide gel electrophoresis. J. Cell Biol., 68:375-88.

Lemanski, L. F., D. J. Paulson and C. S. Hill. (1979). Normal anterior endoderm corrects the heart defect in cardiac lethal mutant salamander embryos. Science 294:860-62.

McLean, I. W. and P. K. Nakane (1974). Periodate-lysine-paraformaldehyde fixative. A new fixative for immunoelectron microscopy. J. Histochem. Cytochem., 22:1077-83.

Moore, P. B. and L. F. Lemanski (1982). Quantification of tropomyosin by radioimmunoassay in developing hearts of cardiac mutant axolotls, Ambystoma mexicanum. J. Mus. Res. and Cell Motil., 3:161-67.

Schreckenberg, G. M. and A. G. Jacobson (1975). Normal stages of development of the axolotl, Ambystoma mexicanum. Develop. Biol., 42:391-408.

68 Calcium Dependent Disorganization of Thrombin Aggregated Platelets (platelet strip)

Matteo A. Russo and Leon Salganicoff

INTRODUCTION

Cell disorganization occurs when cytoplasmic Ca^{++} exceeds certain levels, either due to membrane lesions or internal failures in Ca^{++} homeostasis (1,2,3). It has been shown that extracellular influx of Ca^{++} is a common mediator of the in vivo cell necrosis (1,2,3). Different mechanisms have been proposed to explain this phenomenon such as the activation of lysosomal acid proteases (4), cytoplasmic neutral proteases (5), or superoxide radical production (for a review see 6). While the above-mentioned hypothesis may be correct, ultrastructural kinetic studies suggest an alternate mechanism based on a <u>mechanical lesion of</u> the membrane, which results from the uncontrolled contraction of the cytoskeleton in cells that are rich in 6 nm microfilaments. This mechanism has been called "supercontraction" to evidentiate the abnormal mechanical forces that are generated by the unmodulated or uncontrolled forces of contracting proteins when cytoplasmic Ca^{++} increases over a certain level. Evidence supporting this mechanism has been presented for dissociated myocardiocytes (7), for hepatocytes in primary cultures in the presence of phalloidin (8), and in normal and tumor T lymphocytes (9). To further explore the "supercontraction" hypothesis, we present now a study of the ultrastructural changes in platelets when their cytoplasmic Ca^{++} is increased by external Ca^{++} using the ionophore A23187 to facilitate the translocation. For this purpose we use a mechanochemical model of thrombin activated and irreversibly aggregated platelets (the platelet strip) (10,11,12). This model, similar to a white thrombus, is naturally rich in contractile microfilaments. The preparation can be depleted of the excess of cytoplasmic Ca^{++}, by incubation in an EGTA-Mg saline, leading to relaxation. External Ca^{++} can be increased afterwards without inducing a contractile response and the preparation is able to display contraction-relaxation cycles on addition of agonists like ADP, epinephrine, and TxA_2 analogs, demonstrating in this manner an intact stimulus-contraction coupling (13). Such contractile cycles are controlled by the increase in cytoplasmic Ca^{++} due to release from receptor operated pools. It has also been shown that the Ca^{++} ionophore A23187 is able to permeabilize the membrane toward extracellular Ca^{++}. To verify that a Ca^{++}-dependent supercontraction occurs in the platelet strip, we have correlated the contractile responses of the preparation with ultrastructural changes occurring at selected points of their contractile cycle under experimental conditions allowing a normal Ca^{++} homeostasis or when cytoplasmic Ca^{++} is increased by permeabilizing the membrane to external Ca^{++} with the ionophore A23187. The preparations have been fixed in situ to maintain the special relationships of the cytoskeleton.

MATERIAL AND METHODS

Preparation and conditioning of the platelet strip have been described in detail previously. Briefly: cold recalcified human platelet rich plasma is centrifuged in a special flat bottom tube over a high compliance nylon mesh. The platelet poor plasma is decanted and the flat homogeneous pellet of packed platelets is activated by heating it at 37° for 20 min. The tube is disassembled and the resulting giant aggregate (22 mm diameter θ) containing the mesh as a holding skeleton is cut in strips of 4×10 mm. The strips are hung in organ baths and handled in a similar manner to vascular smooth muscle strips. Salines: modified K-H: NaCl: 119.0 mM; KCl 4.7 mM; $NaHCO_3$:25.0 mM; Mg Cl_2:2.0 mM; KH_2PO_4:1.2 mM; EGTA:0.05 mM glucose 5 mM. The pCa^{++} of this saline is 8.00, the EGTA being added to chelate the contaminating Ca^{++} ($\sim$ 25 μM). Ca^{++} is purposely omitted so it can be adjusted afterwards to any desired concentration in the organ bath.

Chemicals: A23187 was a generous gift from Lilly Labs, cytochalasin D and ADP were obtained from Sigma Corp., St. Louis, Missouri. Epinephrine was in ampoules 1/1,000 (Parke Davis).

All strips were fixed in situ under stretching by changing the bath medium with a 2% glutaraldehyde solution in phosphate buffer 0.1 M pH 7.3 containing 0.2 mM Ca^{++}. After several washings in the same buffer, the samples were postfixed in a 1.33% O_sO_4 solution in the same buffer and processed by standard techniques for infiltration in Epon 812. They were embedded in a silicone rubber flat mold with an orientation adequate to obtain precise transverse sections along the vector of the stretching. Ultrathin sections obtained with a diamond knife were stained with uranyl acetate and lead hydroxide.

For optical microscopy thick sections from the same samples used for electronic microscopy were stained with Azur II in borax.

RESULTS

Control

Preincubation of the platelet strips for 4 hours in a Krebs-Henseleit, where Ca^{++} has been substituted for 2 mM EGTA, relaxes the platelet strip. During relaxation, the preparation responds to agonists, the contraction being reversible on wash-out of the agonist (Fig. 68.1a). This behavior demonstrates that the pretreatment depletes the excess cytoplasmic Ca^{++} which causes the contracted state. The receptor operated pools of Ca^{++} are maintained, as shown by the intact stimulus-receptor coupling. When the preparation is relaxed, stepwise addition of Ca^{++} does not recontract the preparation. After Ca^{++} addition, its behavior is similar to that of the chelant treated preparation suggesting that release of internal Ca^{++} pools is the most important response to agonist stimulation (Fig. 68.1b).

At the optical microscopy, this strip appears uniform when observed in a cross section, a very thin layer of disorganized platelets being visible on both sides (top and bottom) of the strip (Fig. 68.2a).

At the electron microscope, platelets are aggregated, showing well-organized cell-to-cell contacts, irregular shape with long pseudopods and protrusions, and in the cytosol, numerous bundles of microfilaments, mostly oriented along the direction of the stretching. The amount of granules (alpha-granules and lysosomes) appears to decrease rapidly from the center to the periphery of a transverse section of the strip (Fig. 68.2a,b), being the total amount very low. No differences are found ultrastructurally between the relaxed and contracted preparations whether in the presence or absence of external Ca^{++}.

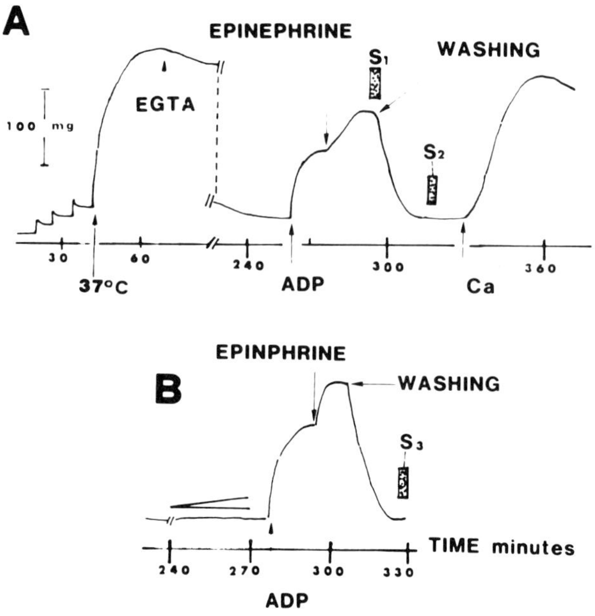

Fig. 68.1. Platelet strips are cooled for 1 hour in KH at 1° to inhibit metabolism and adjusted to an initial tension of 100 mg. Then the KH is changed to a KH-EGTA and left to relax until a stable tension is achieved (∿ 4 hours). The medium is changed to KH. The responses shown are to ADP (40 μM) and epinephrine (10 μM). Ca ions are added stepwise to the KH following a doubling geometric progression (25 μM initial concentration, 50 μM, 100 μM, 1 mM final) i different samples (S1, S2, S3) for optical and electron microscopy.

Effect of the Ionophore A23187 on the Platelet Strip Behavior

Addition of 10 μM ionophore A23187 to the preparation in the absence of external Ca^{++} does not change noticeably the behavior of the preparation. The preparation is able to respond to agonist and the only difference seen is a limited decrease in the maintenance of the tonic response. This is probably due to the outwardly directed facilitated diffusion of the cytoplasmic Ca^{++} through the ionophore channel (Fig. 68.3a).

When Ca^{++} is added to the preparation after incubation with the ionophore, the preparation starts to contract but immediately loses tension and its capacity to respond to agonists.

When Ca^{++} addition is done as a bolus, the preparation contracts and maintains its contraction in the absence of the ionophore but immediately loses tension if the ionophore is present (Fig. 68.3b).

At the optical microscope strips treated with the ionophore in the presence of Ca^{+} appear to be increased in thickness; in a cross section the width of the strip is more than doubled suggesting that the preparation has swollen due to the isometric state contraction. Three layers can be seen, each of one is about one-third of the total thickness. The central layer appears to be quite uniform, being organized as the control strip. In the peripheral layers numerous small lacunae are scattered all over the strip, which in not being stained by Azur II, may be identified as wide extracellular space or holes among inside platelets (Fig. 68.4a,b).

Three distinctive features characterize the disorganized regions of the strip, at the electron microscope level (Fig. 68.4a,b):

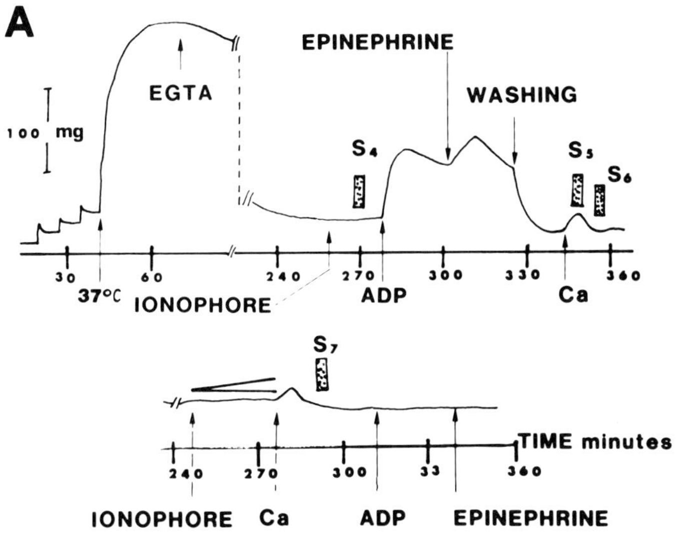

Fig. 68.2. The initial treatment is as in Fig. 68.1. When maximal relaxation is reached, A23187 Ca-ionophore and/or bolus of 1 MM Ca^{++} (Ca) are added. Samples for optical and electron microscopy (S4, S5, S6) were taken in different conditions and different times (30* -45* -60* -180* -600* sec) from the addition of Ca^{++}, 600 sec. is also the time needed for the complete diffusion of drugs and ions inside the strip (Salganicoff in prep).

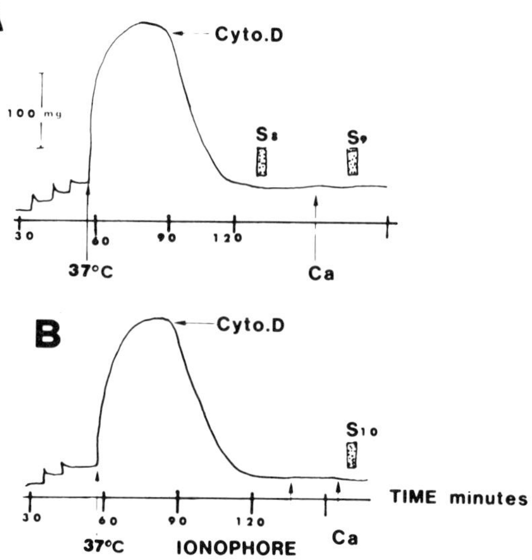

Fig. 68.3. The initial treatment is as in Fig. 68.1. Then 15 μM cytochalasin D (CyD) is added, then 10 μM A23187 Ca-ionophore and a bolus of 1 mM Ca^{++} are added. Samples (S7, S8, S9, S10) were taken for optical and electron microscopy.

1. Electron-dense clumps of supercontracted filaments inside the electron-lucent (holes) cytosol of necrotic platelets.
2. Disruption of plasmamembrane at the level of cell-to-cell junctions among platelets.
3. Wide and apparently empty extracellular spaces which otherwise contain platelet debris.

Platelets which do not undergo supercontraction are still well preserved with irregular profiles, long pseudopods, electron-dense cytosol and cell-to-cell junctions still well organized (Fig. 68.4b,c,d). No differences are seen ultrastructurally, whether the Ca^{++} is added stepwise or as a bolus suggesting that the lesions occur as a result of cytoplasmic Ca^{++} increase.

Effect of Pretreatment with Cytochalasin D on the Platelets Treated with Ca^{++} and Ionophore

To test the role of actin depolymerization on the genesis of the lesions the platelet strip is pretreated with 15 μM cyto-

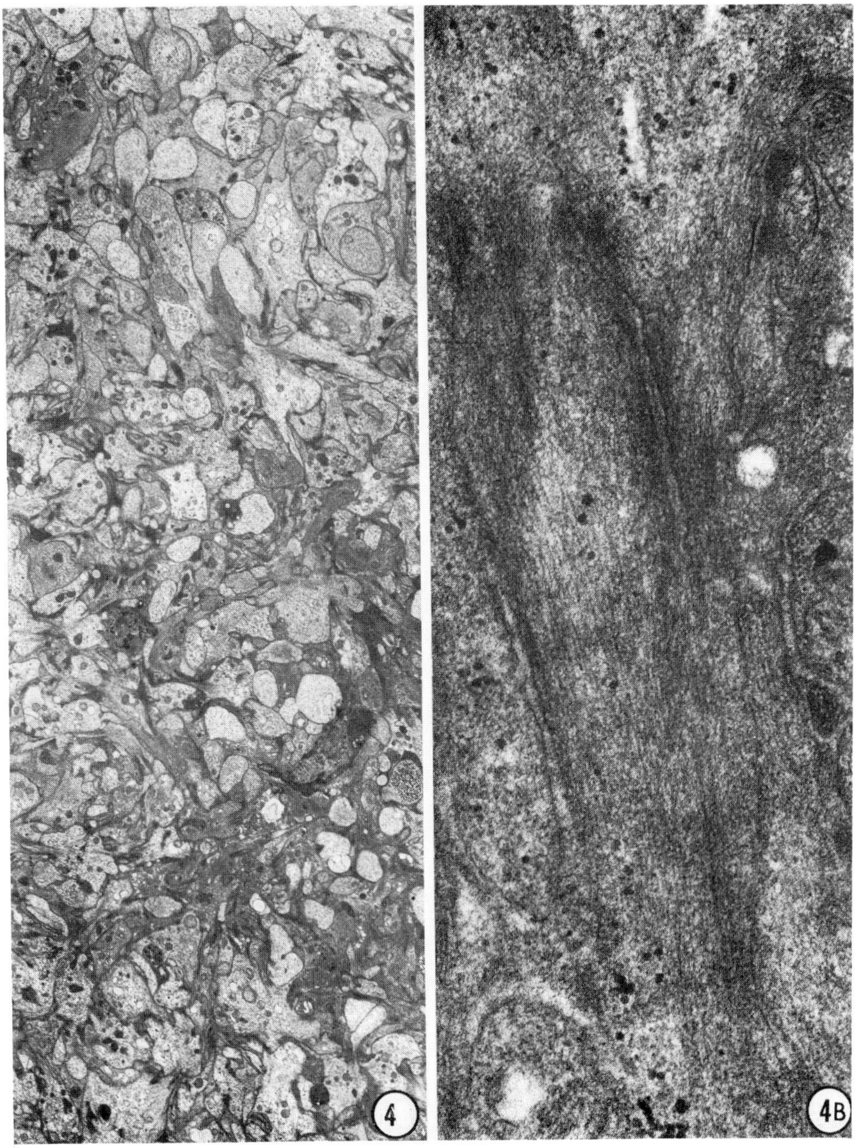

Fig. 68.4. Control strip (S1 – S2 – S3 – S4). Ultrathin section taken along the direction of the stretching. (a) General view. (b) Detail. Uniformly well organized aggregated platelets. Extracellular spaces are not visible or filled with electron-dense material. Well-preserved cell-to-cell contacts. Bundles of microfilaments are oriented mainly along the direction of the stretching.

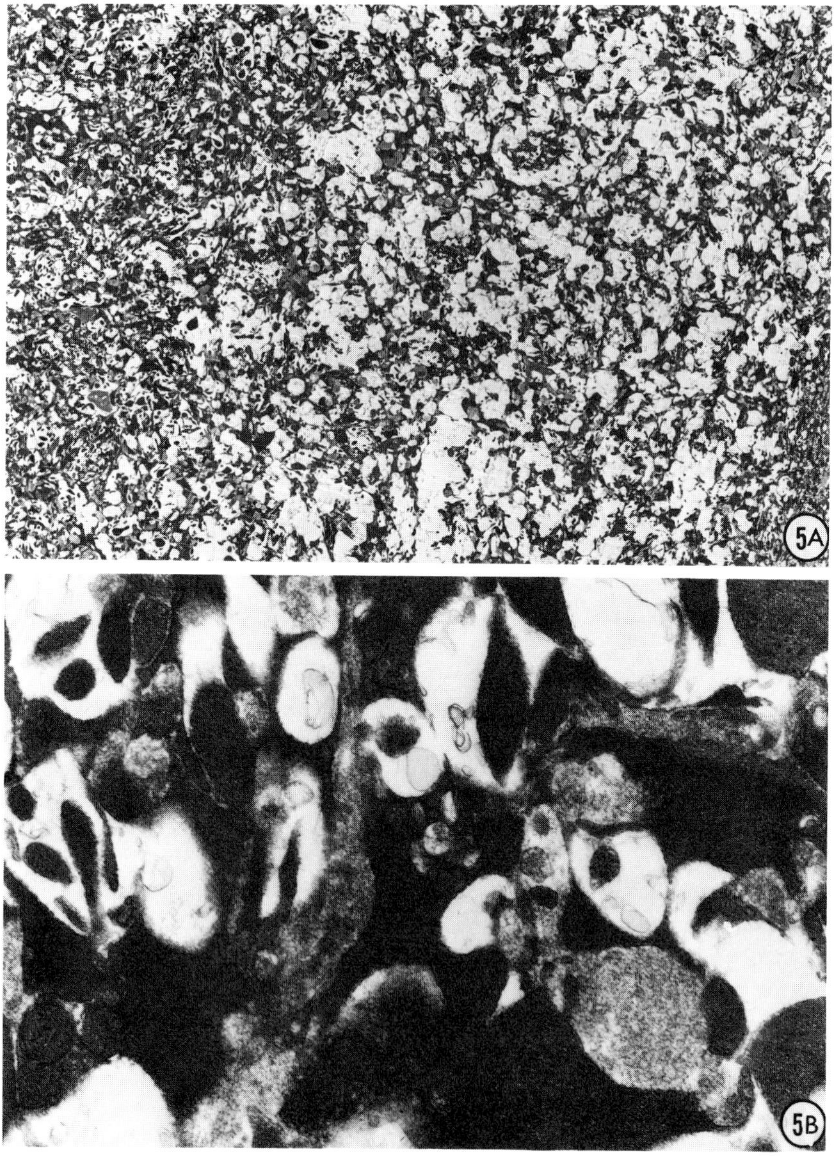

Fig. 68.5. (S5 – S6 – S7). Strip incubated in the presence of A23187 Ca-Ionophore and then in the presence of extracellular 1.8 mM Ca^{++}. (a) General view. (b) Detail. Platelets are disorganized; wide extracellular spaces are present. In higher magnification micrograph (b) most of the times these spaces are shown to be disorganized platelets containing clumps of supercontracted microfilaments (see also Fig. 68.6). Note that plasmamembranes are fragmented at level of cell-to-cell contacts.

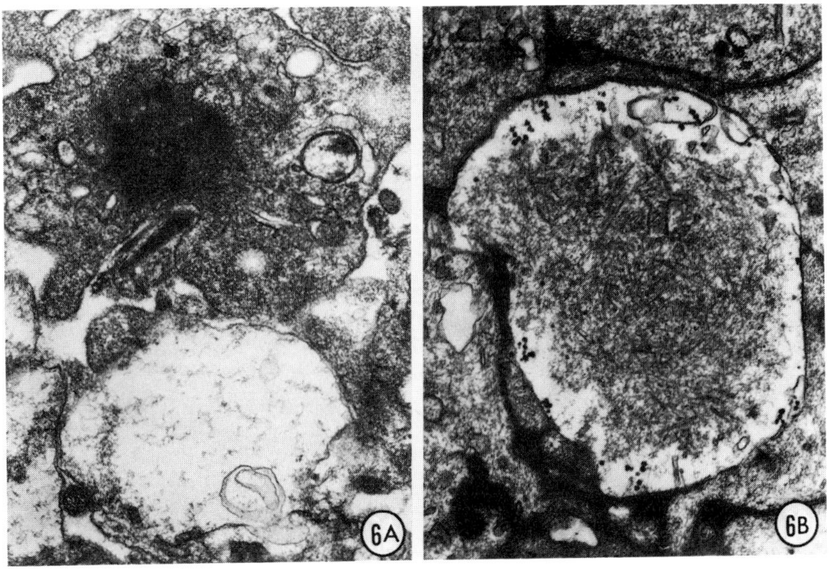

Fig. 68.6. Details of supercontracted platelets. Micrograph (a) shows an apparently empty platelet (left) with membrane and organelle debris; another platelet (right) presents in the cytoplasm an electron-dense clump of supercontracted microfilaments. Micrograph (b) shows a platelet which plasmamembrane is fragmented at level of cell-to-cell contacts. In the cytoplasm is present a mass of filaments; some of them are thin (6 nm) filaments, others are thicker and, probably, represent "tactoid bodies" of myosin.

chalasin D. This results in total relaxation. Afterwards the platelets are treated with 10 μM ionophore and 1.3 mM Ca^{++}. No contraction or apparent change in behavior is seen (Fig. 68.5). At the optical microscope, the strips which were treated with cytochalasin D alone have the same organization of the strips treated with A23187 plus Ca^{++}; three layers were present in a cross section through the thickness of the strip; the internal one appeared still organized in a fashion similar to that displayed by the control strip while the external ones shown lacunae, holes, and wide extracellular spaces; furthermore, the round and regular profile of individual platelets were recognizable. The external layers represent the 70-75% of the total thickness of the platelet strip (Fig. 68.6a).

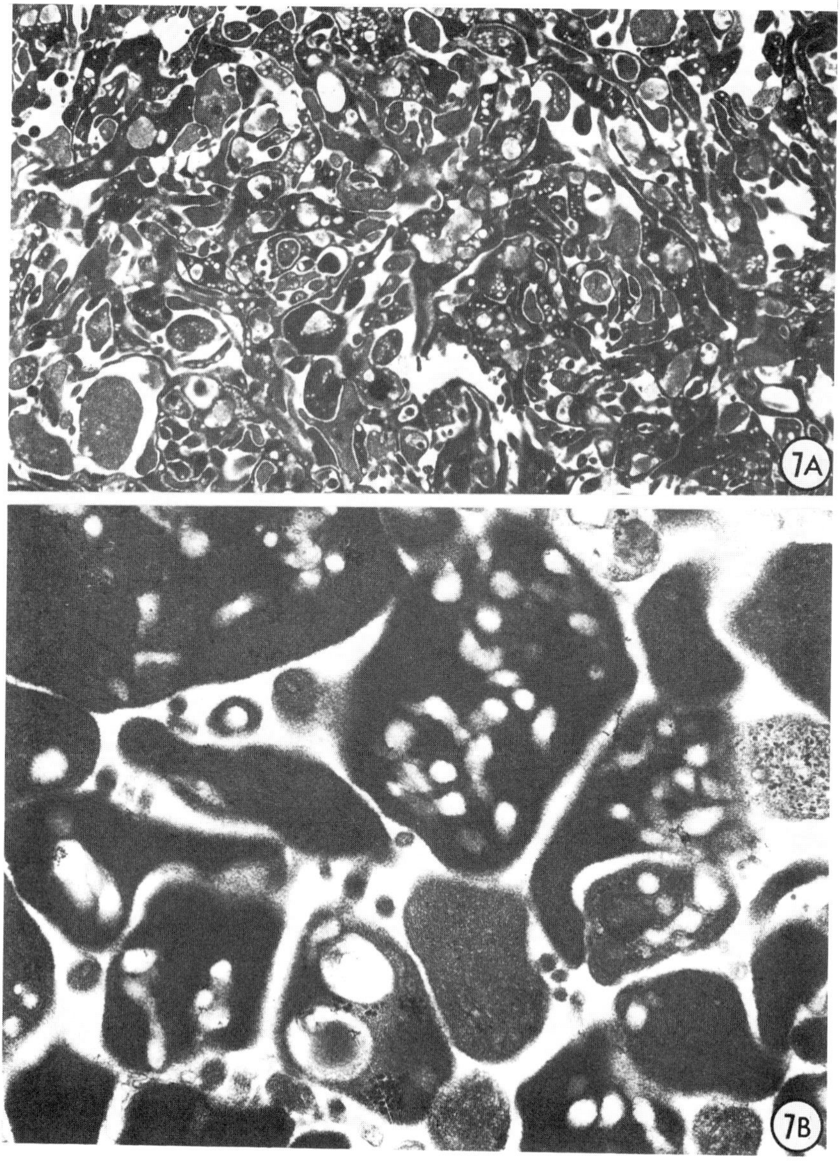

Fig. 68.7. Effect of preincubation with cytochalasin D on strips treated with ionophore and Ca^{++} at mM concentrations. By comparing Fig. 68.7a to 68.5a and 68.7b to 68.5b, it is evident that Cyt. D. is able to completely prevent the disorganization by supercontraction. When compared to the control strip (compare 68.7a,b to 68.4a,b) the only differences are some increase of extracellular spaces and the change in the shape of platelets, which appears mostly spheroidal, short, and with a few pseudopods after if treated with Cyt. D.

At the electron microscope it is evident that pretreatment with cytochalasin D completely prevents the disorganization by supercontraction of the strip. The unstained spaces seen at the optical microscope, were shown to be extracellular spaces among platelets which had lost their cell-to-cell contact and had assumed spherical shape. The organization of the individual platelets appeared well preserved but the pseudopods tended to be less numerous, shorter, and with filaments difficult to visualize. Sometimes the surface connecting system and the dense tubular system are extremely dilated (Fig. 68.6).

DISCUSSION

The untrastructural lesions appearing in the platelet strip are similar to that previously described for myocardiocytes, filament rich hepatocytes, and thymoma ascites cells. Inhibition of contraction by actin depolymerization avoids the ultrastructural changes suggesting therefore a mechanical origin of the lesion. It is of interest that the levels of cytoplasmic Ca^{++} needed for full contraction are not enough to disrupt the platelet ultrastructure. Only when external Ca^{++} is high enough to allow appreciable influx in the presence of the ionophore the lesions occur. Considering that the lesions occur in the preparation fixed at 38 sec. after the Ca^{++} addition, it is unlikely that proteases or any other of the previously mentioned mechanisms had time to act. The only difference seen as a function of time is the increase in the disorganized layers at the periphery of the strip. In conclusion, we suggest that disorganization occurs when the external Ca^{++} is high enough to allow appreciable influx through the ionophore induced translocation. This event leads to the increase of cytoplasmic Ca^{++} to abnormally high levels, an event followed by an irreversible contraction and finally disorganization of the cytoskeleton.

ACKNOWLEDGMENTS

Partially supported by W. W. Smith Charitable Foundation and CNR (Italy)—Joint-grant between Italy and U.S.A.—1983.

REFERENCES

1. Schanne F. A., Kane, A. B., Young, E. E. and Farber, J. L. Calcium dependence of toxic cell death: A final common pathway. Science 206:700, 1979.

2. Farber, J. L. The role of calcium in liver cell death. In "Progress in liver disease," 8:347; Popper & Schaffner Edts. Greene & Stratton, Inc., New York, 1982.
3. Farber, J. L. Membrane injury and Ca^{++} homeostasis in the pathogenesis of coagulative necrosis. Lab. Inves. 47: 114, 1982.
4. Kane, A. B., Stanton, R. P., Raymond, E. G., Dobson, M. E., Knafell, M. E. and Farber, J. L. Dissociation of intracellular lysosomol rupture from the cell death due to silica. J. Cell Biol. 87:643, 1980.
5. Toyo-Oka, T., Shimizu, T. and Masaki, T. Inhibition of proteolytic activity of calcium activated neutral protease by leupeptin and antipain. Biochem. Biophys. Res. Comm. 82:484-91, 1978.
6. Rubin, R. and Farber, J. L. Mechanisms of the killing of cultured hepatocytes by hydrogen peroxide. Archives Biochem. Biophys. In press.
7. Russo, M. A., Cittadini, A., Dani, A. M., Inesi, G. and Terranova, T. An ultrastructural study of calcium induced degenerative changes in dissociated heart cells. J. Mol. Cell Cardiol. 13:265-79, 1981.
8. Russo, Matteo A., Kane, Agnes B. and Farber, John L. Ultrastructural Pathology of Phalloidin-Intoxicated Hepatocytes in the Presence or Absence of Extracellular Calcium. Amer. J. Pathol. 109:133-44, 1982.
9. Russo, Matteo A., Mercorella, I., Dani, A. M., Wolf, F. and Cittadini, A. Role of Ca^{++} and Cytoskeleton in the Control of Cell Shape and Architecture of normal and Tumor Lymphocytes. In: "Membranes in Tumor Growth." T. Galeotti, G. Neri and S. Papa Edts. Elsevier/North-Holland Biomedical Press, Amsterdam, pp. 105-14, 1982.
10. L. Salganicoff, Matteo Russo and M. Loughane. The Platelet Strip: A New Model for the Study of the Mechanochemical Properties of a Platelet Aggregate. VIth International Congress on Thrombosis and Haemostasis, Philadelphia, Pennsylvania, 1977.
11. L. Salganicoff. A deficient Ca^{++} pump as the cause of the irreversibility of clot retraction. Thromb. Haemostasis 42:215, 1979.
12. Daniel, J. L., Molisch, I. R., Holmsen, H. and Salganicoff, L. Phosphorylation of myosin light chain in intact platelets: Possible role in platelet aggregation and clot retraction. Cold Spring Harbor Conf. on Cell Proliferation. 1981, Vol. 8, 913-27.

13. Salganicoff, L. and Sevy, R. Reversible relaxation of a contracted platelet aggregate with sulphate. Fed. Proc. (1980) 38, 582.

69 Estrogen Stimulation Induces Cytoskeletal Changes in Breast Cancer: MCF-7 Cells

A. Sapino, L. Guidoni,
P. C. Marchisio, G. Bussolati

INTRODUCTION

The MCF-7 human breast carcinoma cell line (11), has specific receptors for steroid and polypeptide hormones (1-5). This cell line, responds to estrogen stimulation by increase in: mitotic index (8), metabolic activity, and synthesis of various proteins, such as progesterone receptor (6), plasminogen activator (2), and other proteins (4-13), whose function and significance have not yet been defined.

The effects of estrogen stimulation on the ultrastructure and the cell surface of these cells have been studied by scanning and transmission electron microscopy: physiological concentrations of estradiol have been observed to change these tumor cells into secretory cells, and to induce an increase in the number and length of microvilli; a parallel change of the cellular shape takes place, since they become more rounded and less tightly attached to culture dishes (12).

These data suggest that estrogens could induce changes of cytoskeletal structures. In the present study, MCF-7 cells have been cultured in the presence of different concentrations of 17-β-estradiol, and the architectural components of the cytoplasmic matrix have been investigated using immunofluorescence procedures and specific antibodies to prekeratin, vimentin, actin, and tubulin.

MATERIALS AND METHODS

MCF-7 human breast carcinoma cells have been grown in three different media:

1. Maintenance medium, composed of RPMI medium supplemented with 10% fetal calf serum, having a final E (estrogen) concentration of about 10^{-9} M;
2. E-deprived medium, composed of RPMI medium with 10% charcoal absorbed fetal calf serum, with a final E concentration of about 10^{-10} M;
3. E-supplemented medium, composed as the previous one, but supplemented with 17-β-estradiol in order to obtain a final concentration of 10^{-8} M.

The cells, grown serially in flasks, were transferred for the experiments onto glass coverslips in 60 mm Petri dishes (5×10^5 cells/dish).

Cells have been examined by immunofluorescence microscopy using antibodies raised against a number of cytoskeletal antigens (9).

RESULTS

MCF-7 Cells Grown in E-Deprived Medium

Cells cultured for three days in E-deprived medium appeared either scattered and roundish or arranged in clumps with occasional intercellular contacts.

In IRM (Interference Reflection Microscopy) they appeared flattened and rich in adhesion areas.

The staining for actin revealed that microfilaments occurred in small bundles, commonly located at the cell periphery often at intercellular contacts. Stress fibers, typical of cells in culture that exhibit little or no locomotion, were absent.

Prekeratin antibodies decorated MCF-7 cells very strongly, and specifically revealed numerous convoluted fibers (tonofilaments) evenly distributed in the cytoplasm and forming a more compact network in the perinuclear zone. Endings of tonofilaments were occasionally observed at sites of cell-cell contact.

Intermediate filaments of the vimentin type were also detected by the appropriate antibodies, but they formed a looser network as compared to tonofilaments. This is in agreement with the epithelial origin of these cells and adds to the problem of vimentin significance in cultured cells (7).

The staining for tubulin revealed a complex array of microtubules, similar to that observed in other cultured epithelial cells.

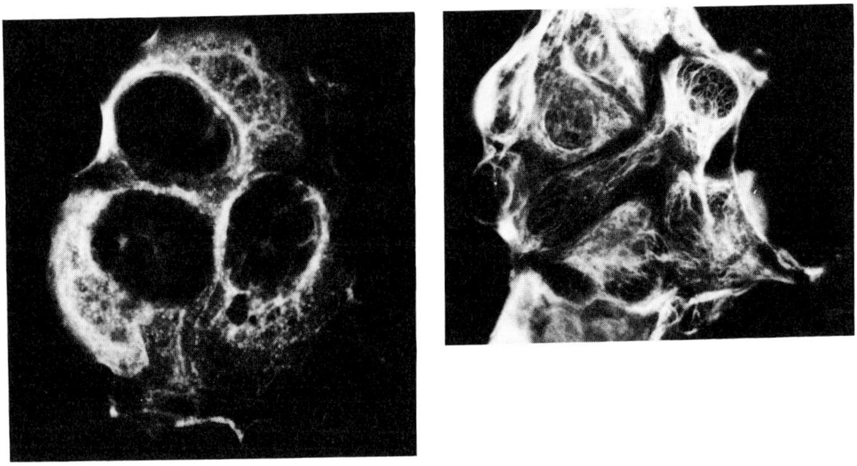

Fig. 69.1. Prekeratin Fibers (revealed by specific antibodies). (a) Cells cultured in estrogen-deprived medium: prekeratin fibers are mostly distributed in the perinuclear cytoplasm. (b) Cells cultured in estrogen supplemented medium: increase in prekeratin fibers, mainly present inside the projections sprouting from the cell body.

MCF-7 Cells Grown in E-Supplemented Medium

MCF-7 cells responded to estrogen stimulation by increasing their mitotic index; increased proliferation led to the appearance of clusters and irregular glandlike figures. These cells, particularly those located at the periphery of the clumps, displayed pseudopodial cytoplasmic protrusions, containing a well developed cytoskeletal network.

As seen in IRM, adhesion areas were mostly localized in these projections.

The periphery of the same cells and their cytoplasmic processes were strongly positive for actin. Stress fibers were absent as well.

Estrogen stimulation induced an apparent increase in tonofilaments that joined the perinuclear zone to the projections sprouting from the cell body.

Staining for vimentin intermediate filaments as well as for microtubules were not modified by estrogen stimulation.

DISCUSSION

Several studies have been reported on the MCF-7 cell line, which took origin from a human breast carcinoma producing metastases and pleural effusion (11). The interest in this cell line stems from its well-proved estrogen dependency: receptors for 17-β-estradiol have been demonstrated (1-5), and addition of estrogen in the culture medium induces a striking increase in mitotic index and in protein turnover (4-8-13).

In the present study, we have investigated these cells in order to clarify whether their response to estrogen stimulation was matched by changes in the overall cytoskeletal structure.

Previous ultrastructural investigation (12) showed that estrogen stimulation of MCF-7 cells induces an increase in number and length of microvilli at the cell surface; this effect was accompanied by cell detachment. In agreement with these investigations, our IRM data have shown marked changes in adhesion, while cells grown in E-deprived medium appeared markedly flattened over the dish, E-stimulation induced the appearance of tight adhesion areas only at the base of the long projections and processes sprouting from the cell body. Loss of adhesion and the appearance of similar projections have also been observed in Shionogi 115 mouse mammary tumor cells stimulated by testosterone (14) and might therefore correspond to a general specific response to steroid stimulation in hormone dependent tumors. Further immunofluorescence studies in these androgen-dependent cells (3) showed steroid-modulated changes of actin microfilaments and microtubules, while intermediate filaments were not investigated. Changes of the cytoskeletal components have not yet been reported on the specific response of MCF-7 or other estrogen dependent tumor cell lines. In our study some filaments (e.g., microtubules and vimentin intermediate filaments) did not appear significantly affected by the hormone conditioning; conversely, a rearrangement of tonofilaments in the cel projections and the formation of a rich perinuclear network represented instead the main effect of E-stimulation. Microfilaments underwent also a marked redistribution and appeared in a thick meshwork at the cell periphery in connection with the apparent increase of surface activity. These changes are unlikely to be related to the increase in mitotic index while it might be a primary effect of estrogen-stimulation. Specific high affinity sites for estradiol receptor have in fact been identified in cytoskeletal components (10). A direct effect of estrogen stimulation on the rearrangement of intermediate filaments and actin accordingly of the cell shape of cancer cells can be suggested; this might be independent of the well-known effect on cell proliferation. Whether such a hormone-induced change in the shape of tumor cells might affect their invasive and metastatic potential remains to be clarified.

ACKNOWLEDGMENTS

Work supported by grants from the M.P.I., Rome, the C.N.R. (grant n. 82.00248.96) and the A.I.R.C., Milan.

REFERENCES

1. Brooks S.C., Locke E.R. and Soule H.D.: Estrogen receptor in a human cell line (MCF-7) from breast carcinoma. J. Biol. Chem. 248:6251-53 (1973).
2. Butler W.B., Kirkland W.L. and Jorgensen T.L.: Induction of plasminogen activator by estrogen in a human breast cancer cell line (MCF-7). Biochem. Byophys. Res. Commun. 90:1328-34 (1979).
3. Couchman J.R., Yates J., King R.J.B. and Badley R.A.: Changes in microfilament and focal adhesion distribution with loss of androgen responsiveness in cultured mammary tumor cells. Cancer Res. 41:263-69 (1981).
4. Edwards D.P., Adams D.J., Savage N. and McGuire W.L.: Estrogen induced synthesis of specific proteins in human breast cancer cells. Biochem. Byophys. Res. Commun. 93: 804-812 (1980).
5. Horwitz K.B., Costlow M.E. and McGuire W.L.: MCF-7: A human breast cancer cell line with estrogen, androgen, progesterone and glucocorticoid receptors. Steroids 26: 785-95 (1975).
6. Horwitz K.B. and McGuire W.L.: Estrogen control of progesterone receptor in human breast cancer. J. Biol. Chem. 253:2223-28 (1978).
7. Lazarides E.: Intermediate filaments: a chemically heterogeneous, developmentally regulated class of proteins. Ann. Rev. Biochem. 51:219-50 (1982).
8. Lippman M.E., Monaco M.E. and Bolan G.: Effects of estrone, estradiol, and estriol on hormone responsive human breast cancer in long-term tissue culture. Cancer Res. 37:1901-07 (1977).
9. Naldini L., Di Renzo M.F., Tarone G., Giancotti F.G., Marchisio P.C. and Comoglio P.M.: Protein substrates of tyrosin phosphokinase encoded by VSRC oncogene are associated with cytoskeleton and adhesion structures in transformed cells. (this volume)
10. Puca G.A. and Sica V.: Identification of specific high affinity sites for the estradiol receptor in the erythrocyte cytoskeleton. Biochem. Biophys. Res. Commun. 103:682-89 (1981).

11. Soule H.D., Vazquez J., Long A., Albert S. and Brennan M.A.: A human cell line from a pleural effusion derived from a breast carcinoma. J. Natl. Cancer Inst. 51:1409-1412 (1973).
12. Vic P., Lignon F., Derocq D. and Rochefort H.: Effects of estradiol on the ultrastructure of the MCF-7 human breast cancer cells in culture. Cancer Res. 42:667-73 (1982).
13. Westley B. and Rochefort H.: A secreted glycoprotein induced by estrogen in human breast cancer cell line. Cell 20:353-62 (1980).
14. Yates J. and King R.J.B.: Correlation of growth properties and morphology with hormone responsiveness of mammary tumor cells in culture. Cancer Res. 41:258-69 (1981).

70 Dynamic Changes of Myofibrillar Protein Isoforms in Human Diseases

Giovanni Salviati, Romeo Betto,
Massimo Zeviani, Daniela Danieli-Betto

INTRODUCTION

In recent years biochemical and immunological evidences have been presented for a coordinated expression of type-specific isoforms of several myofibrillar proteins (myosin, tropomyosin, and troponin) in fast and slow skeletal muscles. For one of these proteins, i.e., myosin, isoforms characteristic of the stage of the ontogenic development (embryonic, neonatal, and adult) have also been found.

In human myopathies alteration of this coordinated gene expression has been suggested in order to explain the discrepancies between the histochemical fiber type composition of the muscle and the immunological reactivity and the light chain composition of the isolated myosin. Furthermore, the large number of cells containing both fast and slow isoforms of myosin light chains and of troponin T and troponin I in DMD has been interpreted as evidence for an incomplete differentiation of the muscle cells.

In this study, the myofibrillar protein composition of human single muscle fibers was compared in biopsies from different myopathies. The biopsies studied have as common feature an increase of the intermediate fibers, irrespective of the type of myopathy, i.e., DMD, myotonic dystrophy, hyper- and hypothyroidism. Furthermore, in DMD fibers no evidence for a fetal isozyme could be detected suggesting that DMD muscle fibers are still able to undergo developmental transitions from the fetal to the adult stage of differentiation.

MATERIAL AND METHODS

Normal muscles were obtained from informed volunteers (biceps m.) and from radical mastectomy (pectoralis minor m.).

Pathological muscles were from 1 patient with DMD, 3 patients with myotonic dystrophy, 4 patients with hyperthyroidism, and 2 patients with hypothyroidism. Bundles of fibers, 3 mm thick, were tied on a wooden stick at the resting length and were chemically skinned as already described (Salviati et al., 1982). Single fibers were isolated under dissecting microscope and were solubilized overnight with 20-30 μl of 10% glycerol, 5% 2-mercaptoethanol, 2.3% SDS, 62.5 mM Tris-HCl, pH 6.8. Myofibrillar protein were analyzed by SDS-gel electrophoresis according to Laemmli on 10-20% polyacrylamide linear gradients. For myosin heavy chains analysis the concentration of polyacrylamide gel was decreased to 5%. Histochemical typing of single fibers was carried out as previously described (Salviati et al., 1982).

RESULTS AND DISCUSSION

Normal Muscles

Since human muscles such as the biceps brachialis and the pectoralis minor contain a mixture of type I (slow) and type II (fast) fibers, segments of single fibers isolated from these muscles were first typed histochemically according to the alkaline and acid sensitivity of myosin ATPase. The remaining fiber segments were subsequently analyzed by SDS gel electrophoresis.

Fast and slow human muscle fibers have type-specific forms of myosin light chains, tropomyosin, and troponin that can be very well resolved on the basis of the apparent molecular weight. With regard to myosin light chain composition, uncertainties are still present in the literature about the number of chains of fast and slow human myosin. Our results indicate that fast (type II) fiber myosin always contains three light chains (LC1F, LC2F, and LC3F), although the LC3F content is variable. On the other hand, two populations of slow fibers type I can be identified.

In one population myosin contains only two light chains (LC1S and LC2S), in accordance with the results of Billeter et al., (1981) on single fibers, and of Volpe et al., (1982) and Pons et al., (1983) on isolated myosin. In the other population, three light chains are identified either in one dimension (Fig. 70.1) or in two dimension gels (not shown). This light chain has been described in human myosin preparations by Fitzsimmons and Hoh (1981), and it has been called LC1Sa in analogy with

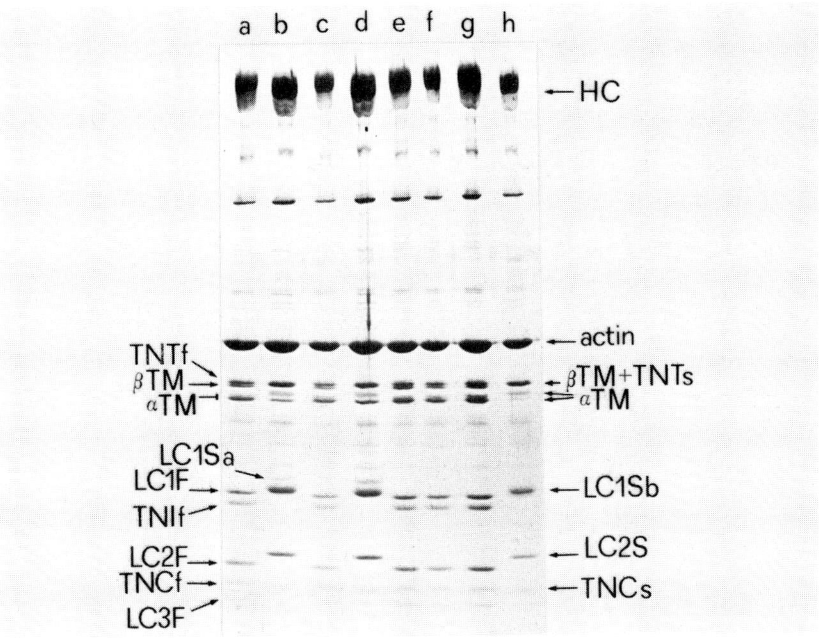

Fig. 70.1. Myofibrillar protein patterns of single fibers from normal human muscle (pectoralis minor). Segments of single fibers chemically skinned were incubated for histochemical myosin ATPase. The remaining segments were analyzed by one dimensional SDS gel electrophoresis on a 10-20% polyacrylamide linear gradient according to Laemmli (1970). The gel was stained with Coomassie Blue. Protein were identified by comparison with rabbit fast and slow myofibrillar proteins, a,c,e,f,g, type II (fast) fibers; b,d,h, type I (slow) fibers. Key words: F, fast; S, slow; HC, myosin heavy chains; LC, myosin light chains; TM, tropomyosin; TN, troponin T, I, and C.

the chain present in slow myosin of the cat, rat, and rabbit. It is present in low amount as compared to LC1Sb and in a percentage of fibers which varies in different human muscles (50% of slow fibers in bicipes brachialis; 75% in vastus lateralis, and 25% in pectoralis major).

Regulatory proteins (and tropomyosins, troponin T, I, and C) also show specific isoforms according to fiber type (Fig. 70.1). In particular tropomyosin is spitted into two subunits both in fast and in slow fibers, but it is present in different amounts and with different apparent molecular weight. These results confirm previous reports suggesting that in normal human muscles the expression of fast and slow isoforms of myosin, tropomyosin, and troponin is under a very coordinated control.

Pathological Muscles

The analysis of myofibrillar protein pattern of single fibers from a DMD muscle biopsy as well as from two DMD carriers shows a number of muscle cells where the segregation of fast and slow isoforms is lost and coexistence of fast and slow myosin light chains can be demonstrated.

As shown in Table 70.1, the presence of fibers with coexistence of both types of myosin light chains is not a peculiar feature of DMD muscles, since it can be demonstrated in normal muscles also. Nevertheless, in DMD muscles this population of fibers is greatly expanded. Several types of myosin light chain patterns are observed: fiber with all six fast and slow light chains (LC1Sa-LC1Sb-LC2S-LC1F-LC2F-LC3F); fibers missing fibers missing the LC1Sa only, or fibers with the fast set of light chains plus a slow light chain, more frequently the LC2S.

The increase of fibers with coexistence of fast and slow myosin light chains fits well with the increase of type IIC fibers which has been described in DMD muscle. Type IIC fibers have been interpreted as immature fibers, since they are found in fetal muscles (Brooke et al., 1971). Since DMD muscles are characterized by conspicuous processes of cell necrosis and regeneration, it is possible that type IIC fibers represent immature regenerating fibers.

A fetal form of myosin light chain is expressed in human fetal muscles as well as in culture of fetal muscle cells (Strohman et al., 1983). This light chain differs from the adult isoforms by the apparent molecular weight so that it can be identified on electrophoretic gels. However, the search of such a chain in single fibers form the DMD patient was unsuccessful. This negative result can be explained by the fact that for technical reasons we are unable to isolate from the biopsy very small fibers. The smallest fibers we can isolate are in fact not below 15 μ in diameter. Since the fetal form of myosin is found in myotube it is possible that we miss in our sampling muscle cells expressing the fetal form.

On the other hand, these results suggest that the hypothesis of a block of maturation of muscle fibers in DMD muscles is not correct. The transition from fetal to adult isoform of myosin is in fact not impaired, since small young muscle fibers are able to accumulate adult isoforms.

A second observation that supports this conclusion is that the increased number of type IIC fibers or fibers with coexistence of fast and slow myosin light chains can be demonstrated also in other human myopathies where degenerative and regenerative processes are negligible. Table 70.1 shows that in myotonic dystrophy and in hyper- and hypothyroid human myopathies the proportion of fibers with coexistence of fast and slow myosin light chains is even higher than in DMD muscles. Such a coexistence is found in experimental muscles undergoing transformation from fast to slow type and vice versa by chronic electrical stimulation or cross-innervation. Thus fibers with coexistence of fast and slow myosin could also be interpreted as intermediate fibers that are changing from a mature type to the other.

Histochemical reactivity for myosin ATPase shows that fibers with coexistence of fast and slow myosin light chains are not only type IIC fibers, but also type I and type II. The discrepancy between histochemical results and the pattern of myosin light chains can be explained by assuming that the sensitivity of myosin ATPase to acid and alkali, which is the basis for the histochemical classification, is a property of myosin heavy chains. Thus, independently of the pattern of light chains, type IIC fibers should contain both types of myosin heavy chain. This assumption was checked by SDS-gel electrophoresis using a 5% polyacrylamide gel. Figure 70.2 shows that in fast (type II) and slow (type I) human fibers myosin heavy chains have different electrophoretic mobilities and that type IIC fibers have indeed both types of myosin heavy chains. This is confirmed by the results of peptide mapping of myosin heavy chains according to Cleveland et al. (1977) which demonstrate in type IIC fibers peptides specific for fast and slow myosin heavy chains (not shown).

On the other hand, the presence in other fibers of both fast and slow types of myosin light chains and only one type of heavy chains (Fig. 70.2j) is indicative of the existence of molecular hybrids of myosin. These fibers are histochemically classified either type I or type II.

The demonstration of molecular hybrids of myosin explains why there can be no correlation between the histochemical fiber typing of a pathological human muscle and the composition of light chains of the isolated myosin.

All together these results show that several human myopathies, not only DMD, have as a common feature an increased number of intermediate fibers.

Table 7.1. Increased Incidence of Intermediate Muscle Fibers in Human Myopathies

	(n. fibers)	LC1F LC2F LC3F	LC1Sb LC2S	LC1Sa LC1Sb LC2S	LC1S LC2S LC1F	LC1S LC2S LC2F	LC1S LC2S LC3F	LC2S LC1F LC2F LC3F	LC1S LC2S LC1F LC2F LC3F
normal[1]	(122)	57.4	19.7	19.7	–	–	–	1.6	1.6
normal[2]	(262)	45.8	24.0	25.6	–	–	–	0.4	1.5
DMD carrier[1]	(66)	51.5	16.7	13.6	–	–	–	13.6	4.6
DMD[1]	(75)	73.3	13.4	1.3	–	–	–	5.3	6.7
Myotonic Dystrophy[3]	(81)	3.7	29.6	12.3	–	2.5	3.7	2.5	45.7
Hyperthyroidism[1]	(95)	70.5	5.3	2.1	–	–	–	3.2	18.9
Hypothyroidism[1]	(90)	41.1	16.7	23.3	–	–	10.0	4.4	4.5

Note: Single fibers chemically skinned were analyzed by SDS gel electrophoresis. The percentage distribution was calculated on the basis of the composition of myosin light chains. 1, biceps brachialis muscle, 2, pectoralis minor muscle, 3, gastrocnemious muscle.

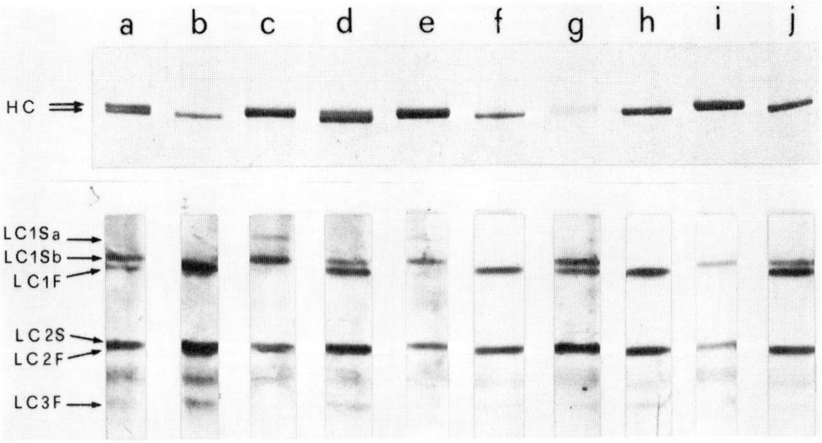

Fig. 70.2. Myosin hybrids and coexistence of fast and slow myosin in single human muscle fibers. Upper panel: SDS gel electrophoresis of myosin heavy chains on 5% polyacrylamide gel. Lower panel: SDS gel electrophoresis of myosin light chains of the same fibers on 10-20% polyacrylamide linear gradient. Single fibers were solubilized overnight on 15 μ of 10% glycerol, 5% 2-mercaptoethanol, 2.3% SDS, 62.5 mM Tris/HCl, pH 6.8 10 μl were used for the analysis of myosin light chain pattern, and the remaining 5 μl for that of myosin heavy chains. Gels were stained with silver (Salviati et al., 1982). Key words as in Fig. 70.1.

These intermediate fibers represent either immature young fibers that are specializing toward the adult slow type (DMD), or mature fast and slow fibers that are changing toward the opposite type because of the altered electrical activity of the surface membrane (myotonic dystrophy) or because of the altered hormone stimulation (hyper- and hypothyroid myopathy).

REFERENCES

Billeter, R., Heizmann, C. W., Howald, & Jenny, E. Analysis of myosin light and heavy chain types in single human skeletal muscle fibers. Eur. J. Biochem. 116, 389-95 (1981).

Brooke, M. H., Williamson, E. & Kaiser, K. K. The behavior of the four principal muscle fiber types in the developing rat and in reinnervated muscle. Arch. Neurol. 25, 360-66 (1971).

Cleveland, D. W., Fischer, S. G., Kirshner, M. W. & Laemmli, U. K. Peptide mapping by limited proteolysis in sodium dodecyl sulfate and analysis by gel electrophoresis. J. Biol. Chem. 252, 1102-06 (1977).

Laemmli, U. K. Cleavage of structural proteins during the assembly of the head of bacteriophage T4. Nature Lond. 227, 680-85 (1970).

Fitzsimons, R. B. & Hoh, J. F. Y. Isomyosins in human type 1 and type 2 skeletal muscle fibres. Biochem. J. 193, 229-33 (1981).

Pons, F., Leger, J., Georgesco, M., Bonnel, F. & Leger, J. Myosin light chains in normal and pathological human skeletal muscles. Muscle & Nerve 6, 40-47 (1983).

Salviati, G., Betto, R., Danieli-Betto, D. Polymorphism of myofibrillar proteins of rabbit skeletal-muscle fibres. An electrophoretic study of single fibres. Biochem. J. 207, 261-72 (1982).

Strohman, R. C., Micou-Eastwood, J., Glass, C. A. & Matsuda, R. Human fetal muscle and cultured myotubes derived from it contain a fetal-specific myosin light chain. Science 221, 955-57 (1983).

Volpe, P., Biral, D., Damiani, E. & Margreth, A. Characterization of human muscle myosins with respect to the light chains. Biochem. J. 195, 251-58 (1981).

71 Ontogeny of the Concerted Regulation of Mammalian Sperm Motility

C. Pariset, J. Feinberg, M. Loir, J. L. Dacheux,
J. Weinman, S. Weinman, J. Demaille

INTRODUCTION

Considerable information is now available concerning the Ca^{2+}- and cAMP-mediated regulation of the major biological processes characteristic of the spermatozoa. In contrast, little is known about the role of these two intracellular second messengers in determining the sperm cell development and the acquisition of flagellar motility. It was felt that a study of the ontogeny of calmodulin and cAMP-dependent protein kinase might provide some insight into the molecular events of sperm cell maturation. These two ubiquitous classes of protein have been characterized and localized within the sperm (see Garbers and Kopf, 1980, and Tash and Means, 1983 for reviews). To this end, the calmodulin content and localization as well as the cAMP-dependent protein kinase activity have been determined in homogenous sperm cells at different stages of maturation collected by centrifugal elutriation, and in spermatozoa collected from different regions of the ram epididymis.

MATERIAL AND METHODS

Preparation of cells. Ram testis spermatogenic cells were obtained by centrifugal elutriation, as described by Loir and Lanneau (1982).

Epididymal spermatozoa. The sperm samples from the corpus and caput epididymides were collected by cannulation of small areas of the organ by the technique described by Dacheux (1980).

Calmodulin assay. Calmodulin was assayed by the highly specific calcium calmodulin dependent activation of myosin light chain kinase according to Le Peuch et al. (1979).

cAMP-dependent protein kinase activity was assayed as described by Pariset et al. (1983).

Motility estimation. The motility of the epididymal spermatozoa was estimated as the percentage of motile spermatozoa after a 10 min incubation at 37°C before agglutination, using a SORO Doppler spectrokinesimeter (Dubois et al. 1975; Jouannet et al. 1977).

Indirect immunofluorescence studies. Calmodulin was localized by indirect immunofluorescence as described by Feinberg et al. (1981).

RESULTS

Evolution of the Calmodulin Level and cAMP-Dependent Protein Kinase Activity during Spermatogenesis

Cell preparation. The distribution of cell types in fractions of ram testis germ cells separated by centrifugal elutriation is given in the legend to Figure 71.1.

Calmodulin levels. As can be seen on Figure 71.1, the largest quantities of calmodulin were found in pachytene spermatocytes. In round and elongating spermatids, calmodulin levels were lower but they remained very high. In elongating spermatids, they progressively tended to the values previously reported for bull sperm.

Calmodulin localization. In round spermatids, the developing acrosome was highly labeled by the specific antibody and appeared as a bright cap with a brighter spot corresponding to the proacrosome granule. The rest of the cell displayed a diffuse level of fluorescence. In elongating spermatids, the developing acrosome was still highly fluorescent; the bright spot which could be seen in the earliest stages was attenuated in the latest stages where the staining seemed more homogeneous. Elongated spermatids exhibited two highly fluorescent areas: the anterior region of the head and a spot in the neck region at the base of the tail corresponding to the centriole and to the capitulum region. The rest of the tail displayed a lower level of fluorescence. In epididymal and ejaculated spermatozoa, calmodulin became present in the postacrosome, whereas the fluorescence in the anterior part of the head was notably missing except at the apical ridge where the acrosome reaches its maximum thickness.

cAMP-dependent protein kinase activity. As shown in Figure 71.1 the cAMP-dependent protein kinase activity was

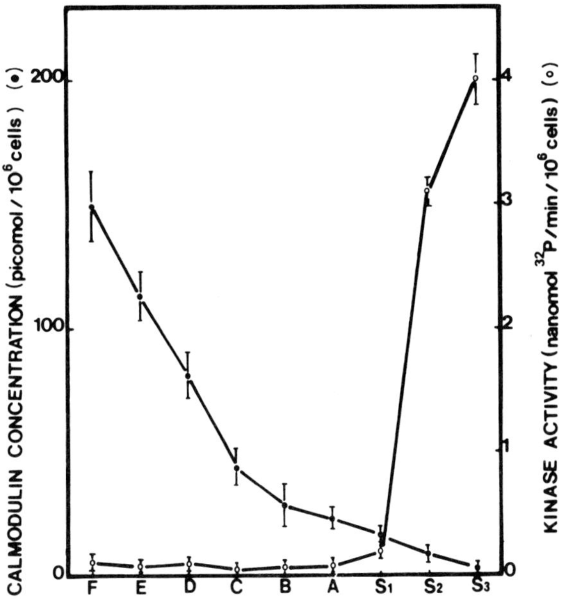

Fig. 71.1. Calmodulin concentration (•) and cAMP-dependent protein kinase activity (○) in ram germ cell fractions collected at different stages of spermatogenesis by centrifugal elutriation. Each value is the mean ± SEM of closely agreeing triplicate determinations made on six experiments throughout the year. Fraction F: pachytene spermatocytes, 60%; Fraction D: spermatids 1-8, 95%; Fraction B: spermatids 9-12, 86%; Fraction S_1: spermatids 13-15, 98%; S_2: epididymal spermatozoa; S_3: ejaculated spermatozoa. Fractions E, C, and A should be considered as transition fractions.

approximately the same in spermatocytes and spermatids but increased dramatically in epididymal and ejaculated spermatozoa.

Evolution of the Calmodulin Level and cAMP-Dependent Protein Kinase Activity during the Epididymal Transit

Calmodulin levels. As can be seen in Figure 71.1, the calmodulin level decreases during the epididymal transit. Yet, no statistically significant differences could be detected in the spermatozoa collected in different regions of the epididymis.

cAMP-dependent protein kinase activity. As shown in Figure 71.2, the acquisition of the flagellar motility which occurs

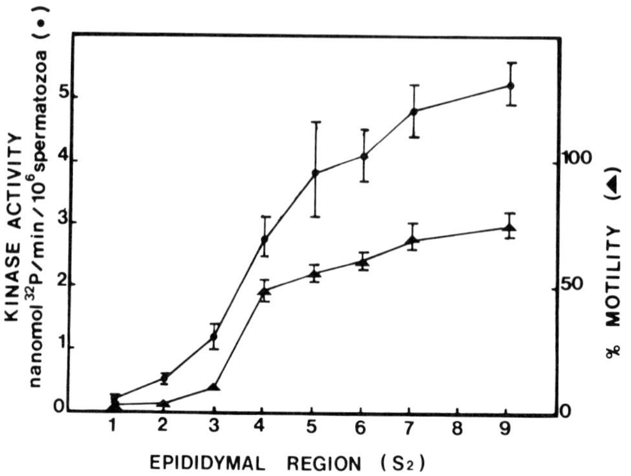

Fig. 71.2. cAMP-dependent protein kinase activity (●) and percentage of motile cells (▲) in spermatozoa collected from different regions of the epididymis.

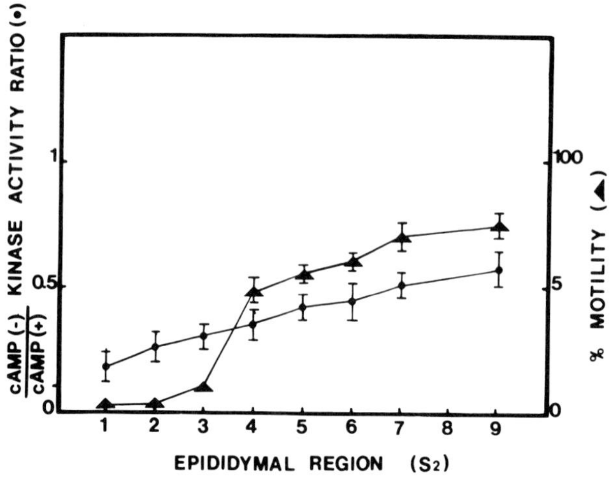

Fig. 71.3. cAMP (-) vs cAMP (+) protein kinase activity ratio (●) and percentage of motile cells (▲) in spermatozoa collected from different regions of the epididymis; about 2×10^7 spermatozoa were used for this determination.

in the epididymal regions 3 and 4 is associated with a major increase in the cAMP-dependent protein kinase activity. In contrast, the acquisition of progressive motility which occurred in zones 8 to 10 could not be correlated to any modification of the cAMP-dependent protein kinase activity. As shown in Figure 71.3, a regular increase in the cAMP (-) vs cAMP (+) protein kinase activity ratio could be detected during the epididymal transit.

DISCUSSION

Calmodulin has been found to be abundant in testis and in spermatozoa (Jones et al., 1979). For the first time, we demonstrated a marked decrease in the calmodulin levels occurring from spermatocytes to spermatozoa. Moreover, using a monospecific antibody against calmodulin, we demonstrated that despite this continuous decline, the distribution pattern changes from the developing acrosome to the postacrosomal and flagellar structures. Thus, the role of calmodulin in the sperm cell maturation might be the Ca^{2+} regulation of the developing cell metabolism (meiosis, enzyme activities, and calcium fluxes) and the Ca^{2+} regulation of the morphological maturation of the gamete. This point of view is substantiated by the fact that in the case of human spermatozoa, an abnormal development of the sperm is frequently associated to a decrease in the calmodulin intracellular levels (Pariset et al., 1983).

The present results provide further data confirming the role of the cAMP-dependent protein kinase activity in initiating flagellar motility. Moreover, the acquisition of forward motility occurs without any modification of the cAMP-dependent protein kinase which would suggest that the cAMP-dependent protein kinase is involved in the generation of the flagellar beat exclusively. In addition, the enhancement of the cAMP-dependent protein kinase activity is associated with a progressive increase in the cAMP (-) vs cAMP (+) protein kinase activity ratio. This fact suggests that the regulation of this enzyme activity is at least partly mediated by the cAMP-dependent dissociation of preexistent RC complexes rather than by an increased biosynthesis of enzyme molecules. The next step in the elucidation of these regulatory mechanisms involving cAMP-dependent protein kinase and calmodulin would be the identification and quantitation of the flagellar ligands or substrates of these two proteins.

REFERENCES

Dacheux, J. L. An in vitro luminal perfusion technique to study epididymal secretion. IRCS Med. Sci. 8, 137 (1980).

Dubois M., Jouannet P., Berge P., Vologhine B., Serres C. and David G. Méthode et appareillage de mesure objective de la motilité des spermatozoides humains. Ann. Phys. Biol. Med. 9, 19-41 (1975).

Feinberg J., Weinman J., Weinman S., Walsh M., Harricane M., Gabrion J. and Demaille J. Immunocytochemical and biochemical evidence for the presence of calmodulin in bull sperm calmodulin. Biochim. Biophys. Acta, 673 3030-31 (1981).

Feinberg J., Pariset C., Rondard M., Loir M., Lanneau M., Weinman S. and Demaille J. Evolution of Ca^{2+}- and cAMP-dependent regulatory mechanisms during ram spermatogenesis. Developmental Biology, in press (1983).

Garbers D. and Kopf D. The regulation of spermatozoa by calcium and cyclic nucleotides. In: Advances in Cyclic Nucleotide Res. Vol. 13. Edited by P. Greengard and G. A. Robison. Raven Press, New York (1980).

Jones H., Lenz R., Palewitz B., Cormier M. Calmodulin localization in human spermatozoa. Proc. Natl. Acad. Sci. USA 77, 2772-76 (1980).

Jouannet P., Volochine B., Deguent P., Serres C. and David G. Light scattering determination of various characteristic parameters of spermatozoa motility in a series of human sperm. Andrologia 9, 36-49 (1977).

Loir M. and Lanneau M. A strategy for an improved separation of mammalian spermatids. Gamete Res. 6, 179-88 (1982).

Pariset C., Roussel C., Weinman S. and Demaille J. Calmodulin intracellular concentration and cAMP-dependent protein kinase in human sperm samples in relation to sperm morphology and motility. Gameter Res. 8, in press (1983).

Tash J. and Means A. Cyclic adenosine 3'5' monophosphate calcium and protein phosphorylation in flagellar motility. Biol. Reprod. 28, 75-104 (1983).

72 Myosin Isoenzymic Modification at Critical Stages of Development of Mice and Urodelan Amphibians

A. d'Albis, C. Janmot, J. Weinman, Cl. L. Gallien

INTRODUCTION

The existence of various specific myosins, each characteristic of different muscle types, striated skeletal, cardiac and smooth, is recognized in adult organisms. Besides, in the course of ontogeny and along with differentiation, the appearance, disappearance, and even reappearance of different myosin isoforms has been observed. An extensive study (Whalen et al., 1981) carried out on a rat fast-twitch contracting muscle has clearly demonstrated the sequential appearance from the end of gestation to about three weeks after birth, of three different heavy chain isomyosins. These changes appear to be mainly controlled by the increase of thyroxine secretion in perinatal period (Butler-Browne et al., 1982; Gambke et al., 1983).

In this paper the modifications in the myosin isoforms pattern in development are investigated in the mouse, Mus musculus, for which recent studies (Chanoine, 1983) indicate an important increase in plasmatic thyroid hormones (T_3-T_4) during the 20 first postnatal days. We also analyze myosin isoenzymic modifications in two urodelan amphibian species: Pleurodeles waltlii and Ambystoma mexicanum, the second one not naturally undergoing metamorphosis. Their comparison is of particular interest since thyroid hormone level is known to markedly increase during amphibian metamorphosis (Etkin, 1968; Larras-Regard et al., 1981).

MATERIALS AND METHODS

Animals and Muscles

Tongues and diaphragms from mice of various ages were generous gifts of F. Rieger and M. Pinçon-Raymond. Dorsal axial muscles were dissected from Pleurodeles waltlii and Ambystoma mexicanum at successive steps of development, defined from the staging series by Gallien and Durocher (1957).

Myosin Extracts

Mouse muscles were first washed with 5 vol. of 20 mM NaCl, 5 mM sodium phosphate, 1 mM EGTA pH 6.5, and crude extraction of myosin was then performed by gentle agitation with 3 vol. of 100 mM $Na_4P_2O_7$, 5 mM EGTA, 1 mM DTT pH 8.5. After 20 mn, the mixture was clarified using a bench centrifuge and the myosin containing supernatant either used right away for electrophoretic analysis or diluted twice with glycerol for conservation at -20°C (d'Albis et al., 1982).

Myosin from amphibian species was extracted as already described (Pliszka et al., 1981) in the case of Rana esculenta. Small pieces of muscle were washed 5 mm with 4 vol. of 600 mM KCl, 40 mM $NaHCO_3$, 10 mM Na_2CO_3, 1 mM $MgCl_2$, 10 mM $Na_4P_2O_7$ pH 8.8; the mixture was centrifuged and the pellet treated for 90 mn with 4 vol. of the same buffer. After centrifugation, the supernatant was kept as above.

Gel Electrophoresis

Analysis of the native myosin molecule was performed by electrophoresis under nondissociating conditions (Hoh et al., 1976; d'Albis et al., 1979). Running buffer was 20 mM $Na_4P_2O_7$ pH 8.5, 10% glycerol, 0.01 2-mercaptoethanol, 2 mM $MgCl_2$, to which was sometimes added 1 mM ATP. Cylindrical gels (6 × 0.5 cm) were 3.88% in acrylamide and 0.12% in Bis. Electrophoresis was carried out at a constant current of 5 mA per tube for 20 to 22 hours at 2°C.

RESULTS

The polymorphism of myosin has been analyzed by electrophoresis in nondissociating conditions.

Myosin Isoenzymes in Mouse Muscles

Crude extracts of tongue muscle (fast-twitch muscle type) from foetal, young (1 to 13 days after birth), and adult mice were analyzed (Fig. 72.1a). Each band on the gels represents at least one species of native myosin. Three bands are clearly distinguished at the end of gestation (18 days foetuses). They persist until after the 13th day of postnatal life, but are no more detected in adults. The "adult" type bands, with lower migration than the neonatal ones, become visible on the gels as soon as the second day after birth. The two kinds of myosin are in approximately equal amounts at about the 12th day.

This description of the isomyosin transition around birth is probably oversimplified, as it may well be that some of the myosin forms are so similar that the electrophoretic procedure would not allow their clearcut separation. As already observed in the case of rabbit (Hoh and Yeoh, 1979), comigration of myosins actually demonstrates the close migration of the two isoforms having respectively the lowest mobility in foetus 18 days of gestation and the highest one in adult (Fig. 72.1b).

Diaphragm, another striated muscle, appears to undergo the same myosin modifications, showing in addition the progressive appearance of a slow-type myosin, of much lower electrophoretic mobility (Fig. 72.1b).

Myosin Isoenzymes in Urodelan Muscles

Myosins extracted from the dorsal axial muscle of Pleurodeles waltlii were analyzed at successive steps of development (Fig. 72.2a). During larval life, and till stage 54, three main isoenzymes are present; at precocious stage 38 (feeding), myosin migrates at the same level, but as a blur, not being clearly resolved. From stage 55, when anatomical signs of metamorphosis are observed, bands of lower mobilities gradually appear. At stage 55c, up to nine bands can be detected. No difference is observed using as donors either normal individuals or giant atypical larvae. At stage 56, when amphibian life totally replaces aquatic conditions, the isoenzymes of intermediate mobilities disappear first, followed by the "larval" type myosin forms. Adults display only three myosin bands; the isoenzymic pattern is identical for both anterior and posterior (caudal) samples of the dorsal axial muscle. These adult type bands have specially low electrophoretic mobilities, when compared both to the larval forms and to adult myosins from other species, such as avians, mammalians or even anuran amphibians (Fig. 72.2b).

Ambystoma mexicanum is of special interest since animals of this species will not undertake metamorphosis under usual laboratory conditions; they reach adulthood and sexual maturity

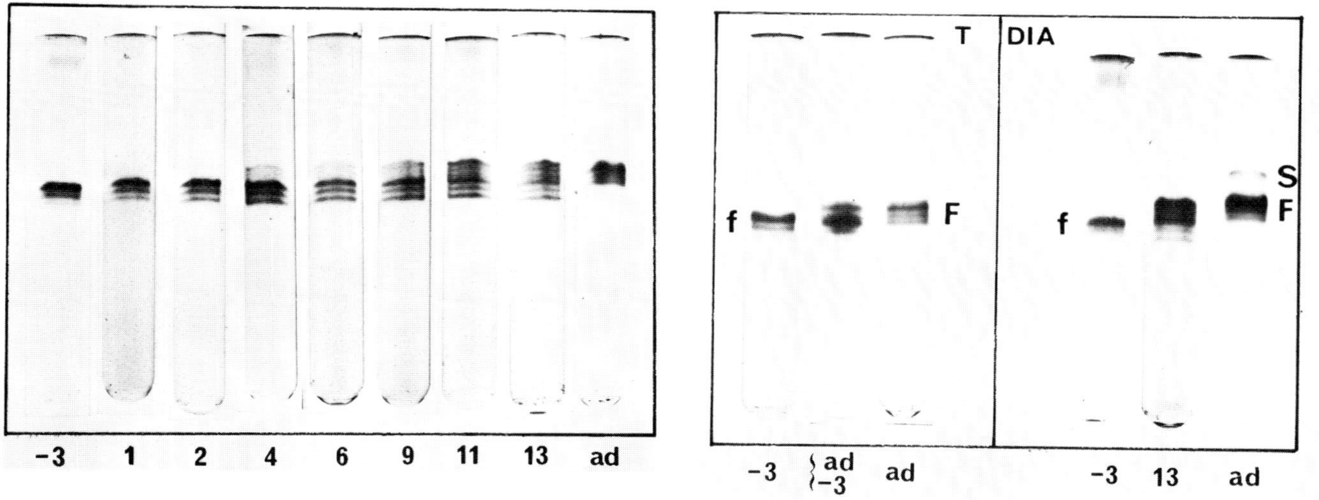

Fig. 72.1. Polymorphism of intact native myosin. Electrophoresis in nondissociating conditions of various mouse muscle extracts. (a) Tongue myosins from mice of several ages, between 3 days before birth (-3) up to 13 days postnatal life and to adult (ad). (b) Tongue (T) and diaphragm (DIA) myosins. (f and F represent neonatal and adult fast myosins, and S, slow myosin).

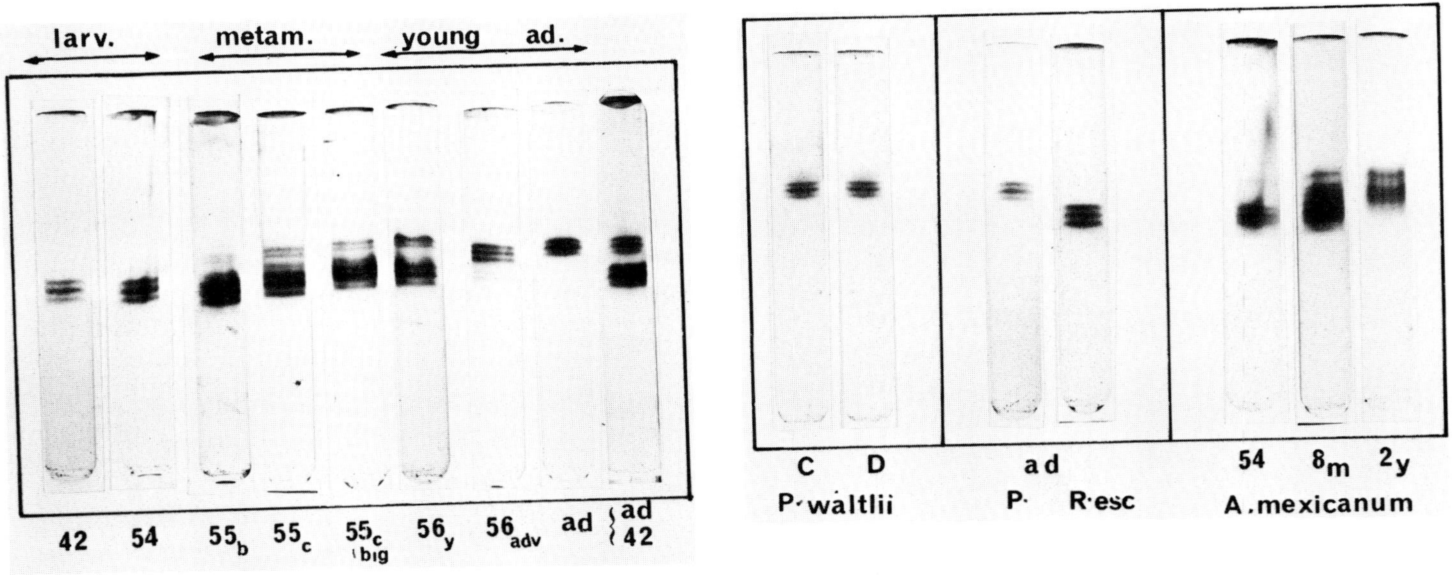

Fig. 72.2. Polymorphism of intact native myosin. Electrophoresis in nondissociating conditions of various amphibian muscle extracts. (a) Dorsal muscle myosins from Pleurodeles waltlii. Numbers correspond to successive stages in the development (Gallien and Durocher, 1957) and cover periods of approximately 90 days until the end of the larval life and 20 days during metamorphosis. (55_{big} means big animal; 56_y and 56_{adv} represent respectively young and advanced states of this stage). (b) Myosins from caudal (C) and dorsal (D) muscles from adult Pleurodeles waltlii. Myosins from adult Pleurodeles waltlii (P) and Rana esculenta (R.esc). Myosins from a young larva (54), an 8 months larva (8m), and 2 years old adult (2y) Ambystoma mexicanum.

as giant neotenic larvae. Myosins from Ambystoma mexicanum dorsal axial muscle are not well resolved (Fig. 72.2b). It is clear however that young larvae exhibit three myosin isoenzymes with electrophoretic mobilities similar to those observed for Pleurodeles waltlii at stage 54. Adult type myosins are quite delayed; they appear only late in the animal life.

DISCUSSION

Transition between various types of myosin during development appears to be a general phenomenon which takes place in all kinds of muscles in a variety of animal species. It has been demonstrated for both striated and smooth muscles in the case of avians and mammalians (Hoh and Yeoh, 1979; Lompré et al., 1981; Bandman et al., 1982; Takano-Ohmuro et al., 1982, 1983). In this paper we confirm the observation for mammalians (mouse) and extend it to skeletal striated muscles of amphibians.

The chronology may change from one species to another. We observe that adult myosins appear a few days earlier in mouse, as compared with rat. This may be due to the type of muscle, which is not the same in the two studies; it may also reflect a real difference in the rates of development of the two animals. As a matter of fact, myosin transition in mouse striated muscle appears to be closely correlated with an increase in plasmatic thyroid hormones level. This correlation between T_3-T_4 level in plasma and myosin transition is clearly confirmed by our observations on urodelan amphibians. Transition occurs during metamorphosis in Pleurodeles waltlii; it is delayed in Ambystoma mexicanum which does not undertake metamorphosis, due to its low thyroidian activity.

REFERENCES

Bandman E., Matsuda R. and Strohman R.C. (1982). Developmental appearance of myosin heavy and light chain isoforms in vivo and in Dev. Biol. 93, 508-18.

Butler-Browne G.S., Bugaisky L.B., Cuénoud S., Schwartz K. and Whalen R.G. (1982). Denervation of newborn rat muscles does not block the appearance of adult fast myosin heavy chain. Nature (London) 299, 830-33.

Chanoine C. (1983). Evolution du taux de calmoduline dans les tissus musculaires chez la souris (Mus musculus) au cours de la période périnatale. Corrélation avec l'évolution du taux d'hormones thyroidiennes dans le plasma. Mémoire de DEA (Embryologie) Université Pierre et Marie Curie—Paris.

d'Albis A., Pantaloni C. and Béchet J.J. (1979). An electrophoretic study of native myosin isozymes and of their subunit content. Eur. J. Biochem. 99, 261-72.

d'Albis A., Pantaloni C. and Béchet J.J. (1982). Myosin isoenzymes in rat skeletal muscles. Biochimie 64, 399-404.

Etkin W. (1968). Hormonal control of Amphibian metamorphosis. In: Metamorphosis, a problem in developmental biology (W. Etkimaud, L. I. Gilberts, eds) pp. 313-48. Appleton-Century-Crofts—New York.

Gallien L. & Durocher M. (1957). Table chronologique du développement chez Pleurodeles waltlii Michah. Bull. Biol. France et Belgique 19, 97-114.

Gambke B., Lyons G.E., Haselgrove J., Kelly A.M. and Rubinstein N.A. (1983). Thyroidal and neural control of myosin transitions during development of rat fast and slow muscles. FEBS Lett., 156, 335-39.

Hoh J.F.Y., McGrath P.A. and White R.I. (1976). Electrophoretic analysis of multiple forms of myosin in fast-twitch and slow-twitch muscles of the chick. Biochem. J. 157, 87-95.

Hoh J.F.Y. and Yeoh G.P.S. (1979). Rabbit skeletal myosin isoenzymes from fetal, fast-twitch and slow-twitch muscles. Nature (London) 280, 321-23.

Lompré A.M., Mercadier J.J., Wisnewsky C., Bouveret P., Pantaloni C., d'Albis A. and Schwartz K. (1981). Species and age-dependent changes in the relative amounts of cardiac myosin isoenzymes in mammals. Dev. Biol. 84, 286-90.

Larras-Regard E., Tavrog A. and Dorris M. (1981). Plasma T_4 and T_3 levels in Ambystoma tigrinum at various stages of metamorphosis. Gen. Comp. Endocrinology 43, 443-50.

Pliszka B., Strzelecka-Golaszewska H., Pantaloni C. and d'Albis A. (1981). Comparison of myosin isoenzymes from slow-tonic and fast-twitch fibers of frog muscle. Eur. J. Cell. Biol. 25, 144-49.

Takano-Ohmuro H., Obinata T., Masaki T. and Mikawa T. (1982). Changes in myosin isozymes during development of chicken breast muscle. J. Biochem. 91, 1305-1311.

Takano-Ohmuro H., Obinata T., Mikawa T. and Masaki T. (1983). Changes in myosin isozymes during development of chicken gizzard muscle. J. Biochem. 93, 903-08.

Whalen R.G., Sell S.M., Butler-Browne G.S., Schwartz K., Bouveret P. and Pinset-Härström I. (1981). Three myosin heavy-chain isozymes appear sequentially in rat muscle development. Nature (London) 292, 805-09.

73 Cytoskeletal Components of Human Hairy Cell Leukemia Cells

Alois B. Lang, Bernhard F. Odermatt,
Jacques R. Rüttner

INTRODUCTION

Hairy cell leukemia (HCL) is a neoplasm of disputed origin. The small lymphoid cells display several surface antigens, including immunoglobulins, characteristic of B cells (Janckila et al., 1983; Jansen et al., 1982, Golomb et al., 1980) while the hairy surface and the content of Fc-receptors place them closer to the monocytic lineage (Golomb et al., 1975; Alptuna et al., 1978). To study such prominent surface structures, a homogeneous cell population would be advantageous. Several cell lines from HCL have been established; however, all of them contained Epstein-Barr virus (EBV) and therefore show features typical of lymphoblastoid cell lines (Miyoshi et al., 1981; Saxon et al., 1978). In vivo, hairy cells are EBV-negative.

Here we report the establishment and the morphology of HCLZ-1, the first cell line which maintains the typical features of hairy cells and lacks EBV, as well as of three EBV containing lines HCLZ-2-4. A preliminary biochemical and immunological comparison of the cytoskeletal proteins of these cells is also presented.

MATERIAL AND METHODS

Cell cultures: Tissue from four patients with clinically and hematologically confirmed HCL was obtained at splenectomy. A spleen cell suspension was prepared by teasing the tissue through a wire mesh in Ca, Mg free phosphate buffered saline (PBS). Cells were separated by Ficoll-Hypaque centrifugation and the isolated cells were seeded in tissue culture flasks in enriched RPMI-1640 medium (manuscript in preparation).

Transmission electron microscopy (TEM): Cell were fixed in 2.5% glutaraldehyde in 0.1 M cacodylate buffer (pH 7.4) and postfixed in 1% osmium tetroxide. They were dehydrated in graded concentrations of ethanol and embedded in epoxy resin. Ultrathin sections were stained with uranyl acetate and lead citrate and examined with a Philips 201 EM.

Scanning electron microscopy (SEM): Cells were allowed to adhere on tissue culture coverslips (Lux. Sci. Corp.) and then fixed in 2.5% glutaraldehyde in cacodylate buffer and in 1% osmium tetroxide. After dehydration, the cells were dried by the critical point method and coated with gold. The specimens were examined with a Jeol JSM-T20 scanning electron microscope.

Antibodies: Two monoclonal antibodies which react specifically with vimentin and actin were used for immunofluorescence studies. Briefly, a cytoskeletal preparation of human fibroblasts was used to immunize the mice. Hybridoma cultures were screened for antibody production by ELISA, indirect immunofluorescence on cultured cells and frozen tissue sections and immunoblotting following SDS-polyacrylamide gel electrophoresis (Laemmli, 1970; Towbin et al., 1979). Cytoskeletal proteins from HCL cells, normal human lymphocytes and human fibroblasts were prepared by high salt buffer and Triton X-100 extraction (Franke et al., 1979). Tests for the Epstein-Barr virus nuclear antigen (EBNA) were performed as described (Reedman, 1973), using known EBNA-positive and negative cell lines as controls.

RESULTS AND DISCUSSION

Immediately after seeding, most of the spleen cells began to adhere to, and spread on, the plastic tissue culture flasks. The rest of the cells remained freely floating. In the early stages of the culture the proliferation was low. After 1-2 months with periodic media changed, growing colonies were detectable among the attached cells in the culture from one patient. The cells reached confluence after about 4 months and could then be cultured every week by trypsinization. Although piling up, all cells remain tightly attached. Ultrastructural examination of these cells revealed that, even after 12 months in culture, the cells possessed typical features of hairy cells (Fig. 73.1A): numerous cytoplasmic projections, somewhat irregular nuclei, large amounts of small fragments of rough endoplasmic reticulum (RER), and free ribosomes. Similarly, SEM studies of these cells (12 months in culture) revealed numerous surface projections with microvilli and ruffled mem-

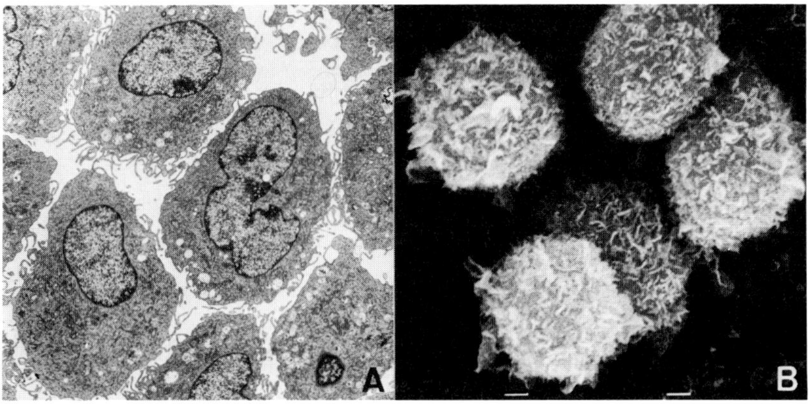

Fig. 73.1. Transmission (A) and scanning (B) electron microscopy of EBNA-negative HCL cells (HCLZ-1) after 12 months of in vitro cultures showing: (A) the characteristic cytoplasmic projections, irregular nuclei, striking amounts of small fragments of RER and free ribosomes; (B) ruffled membranes and microvilli.

branes (Fig. 73.1B). Tests for EBNA with uncultured HCL cells of the patient and with the derived cell line after 6 month of in vitro culture were negative. To our knowledge this line, designated HCLZ-1, represents the first EBNA-negative permanent cell line (now over 2 years in culture) with typical hairy cell features (Fig. 73.1; see also Pilan, 1982; Alptuna et al., 1978).

In the cultures of the other three patients, with advancing culture age, most cells detached and degenerated soon after seeding. But some remaining cells grew in suspension and formed loose clumps, reminiscent of lymphoblastoid cells. Also, all these cells were EBNA-positive and, furthermore, secreted immunoglobulin (IgM/k). Ultrastructural studies of the lymphoblastoid cell lines, designated HCLZ-2-4, revealed further differences to HCLZ-1 cultures (Fig. 73.2A). The cells showed considerable heterogeneity in nuclear size and shape and in cytoplasmic organelles. The nuclei contained very prominent nucleoli and in the cytoplasm long strands of RER were apparent. SEM studies revealed a cell population heterogeneous with respect to the surface morphology. The coarse, long microvilli as visible in Figure 73.2B are present only on some cells, while others show a smooth surface.

To elucidate further the nature of the surface structures of these cell lines, we started to analyze their cytoskeletal pro-

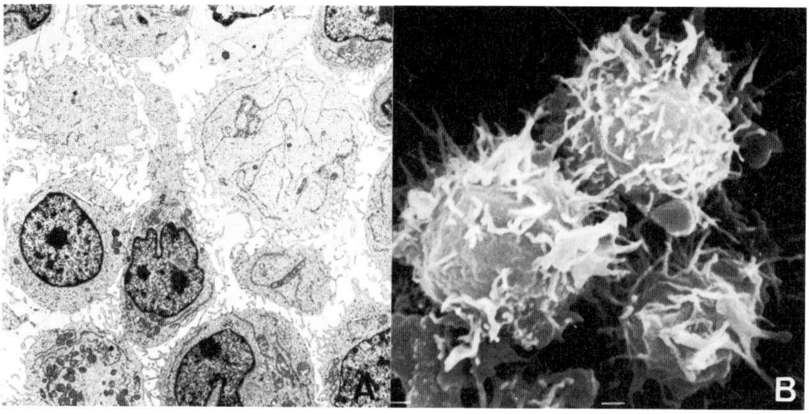

Fig. 73.2. TEM (A) and SEM (B) examination of the lymphoblastoid (EBNA-positive) hairy cell line HCLZ-3 after 6 months of in vitro cultures showing: (A) heterogeneity in nuclear size and shape, prominent nucleoli and long strands of RER; (B) very coarse long microvilli.

teins. Cytocentrifuge preparations of uncultured HCL cells from the patients as well as cells of HCLZ-1 and HCLZ-3 were labeled with monoclonal antibodies to vimentin. In all cells, fluorescence of vimentin filaments in the cytoplasm was evident, with little fibrillar staining.

A rather weak uniform staining was seen when uncultured hairy cells as well as cells of HCLZ-1 were stained with monoclonal antibodies to actin (data not shown). In contrast among lymphoblastoid cells of HCLZ-3 many cells had intensely stained surface structures. These data imply that actin is a constituent of the surface structures of lymphoblastoid cells seen in REM (Fig. 73.2B). The appearance of cell surface cytoskeletal components on lymphocytes transformed by mitogens or EBV has been reported (Bachvaroff et al., 1980; Rubin et al., 1982).

The polypeptide composition of HCLZ-1, HCLZ-3, and of normal lymphocytes, and the cytoskeleton of HCLZ-1 and human fibroblasts for comparison is shown in Figure 73.3A. The major component of the cytoskeleton of HCLZ-1 is vimentin, identified by the monoclonal antibody to vimentin in the immunoblotting experiment. As is evident, the actin polypeptide is a minor component of HCLZ-1 cytoskeleton and of the total protein preparation. In normal lymphocyte and HCLZ-3 actin is a major component. The HCLZ-1 cytoskeleton reveals two further polypeptides larger than vimentin, which are of considerable interest

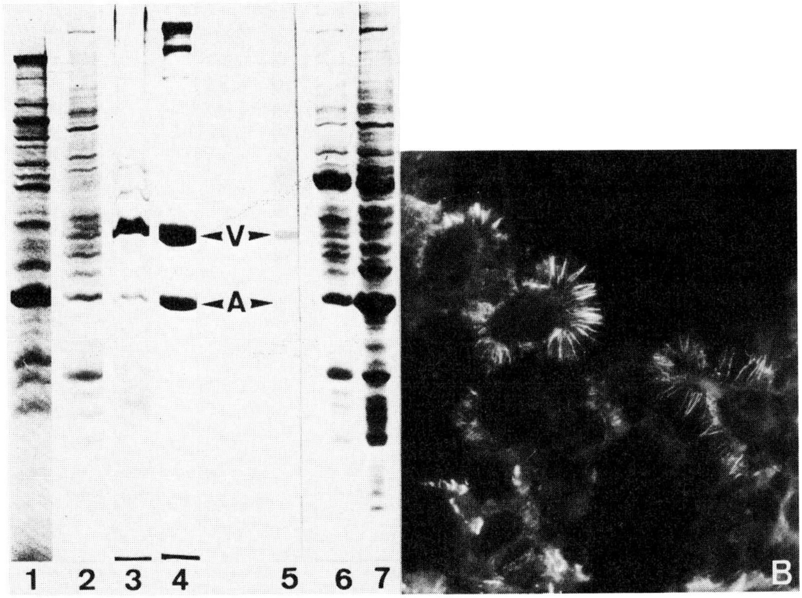

Fig. 73.3. (A) SDS-polyacrylamide gel electrophoresis of (1) soluble proteins from normal peripheral blood lymphocytes, (2) soluble proteins and (3) cytoskeletal proteins from the cells of the HCL-line HCLZ-1, (4) cytoskeletal proteins from cultured human fibroblasts, (5) immunoblotting assay of total protein of HCLZ-1 with the monoclonal antibody to vimentin, (6) total protein of HCLZ-1 and (7) total protein of the lymphoblastoid cell line HCLZ-3. A: actin; V: vimentin. (B) Indirect immunofluorescence staining of HCLZ-3 with a monoclonal antibody to actin; note the prominent actin stain of the surface structures.

in elucidating the protein components of the ruffles and microvilli of the hairy cells (Bretscher, 1983; Burridge, 1983).

In conclusion, we have established a permanent growing EBNA-negative cell line (HCLZ-1) with typical features of hairy cells. This cell line differs morphologically and biochemically from the lymphoblastoid cell lines established from other HCL patients. A preliminary study of cytoskeletal proteins showed that lymphoblastoid cells share common cell surface components different from HCL cells. Therefore, HCLZ-1 should facilitate a detailed biochemical and immunological study of the nature of the surface structure of hairy cells. Efforts to produce and characterize monoclonal antibodies against these structures of HCLZ-1 are in progress in our laboratory.

ACKNOWLEDGMENTS

We wish to thank Dr. C. Wyss for critically reading the manuscript; Ms. D. R. Gut for excellent EM examination; Ms. S. Homberger, Ms. G. Waelle and Ms. L. Kuhn for skilled technical assistance and Ms. R. Felchlin for typing the manuscript. This work was supported by a grant from the Krebsliga des Kantons Zürich.

REFERENCES

Alptuna, E., Anteunis, A., Krulik, M., Astesano, A., Robineaux, R., Debray, J. Hairy cell leukemia: A clinical, immunological and ultrastructural study. Virchows Arch. B Cell Path. 28, 135-49 (1978).

Bachvaroff, R. J., Miller, F., Rapaport, F. T. Appearance of cytoskeletal components on the surface of leukemia cells and lymphocytes transformed by mitogens and Epstein-Barr virus. Proc. Natl. Acad. Sci. 77, 4979-83 (1980).

Bretscher, A. Purification of an 80,000-dalton protein that is a component of the isolated microvillus cytoskeleton, and its localization in nonmuscle cells. J. Cell Biol. 97, 425-32 (1983).

Burridge, K., Connell, L. A new protein of adhesion plaques and ruffling membranes. J. Cell Biol. 97, 359-67 (1983).

Franke, W. W., Schmid, K., Weber, K., Osborn, M. HeLa cells contain intermediate-sized filaments of the prekeratin type. Exp. Cell Res. 118, 95-109 (1979).

Golomb, H. M., Braylan, R., Polliack, A. Hairy cell leukemia (leukemic reticuloendotheliosis): a scanning electron microscopic study of eight cases. Br. J. Haem. 29, 455-60 (1975).

Golomb, H. M., Mintz, U., Vardiman, J., Wilson, C., Rosner, M. C. Surface immunoglobulin, lectin-induced cap formation, and phagocytic function in five patients with the leukemic phase of hairy cell leukemia. Cancer 46, 50-55 (1980).

Janckila, A. J., Stelzer, G. T., Wallace, J. H., Yam, L. T. Phenotype of the hairy cells of leukemic reticuloendotheliosis defined by monoclonal antibodies. Am. J. Clin. Pathol. 79, 431-37 (1983).

Jansen, J., LeBien, T. W., Kersey, J. H. The phenotype of the neoplastic cells of hairy cell leukemia studied with monoclonal antibodies. Blood 59, 609-14 (1982).

Laemmli, U. K. Cleavage of structural proteins during the assembly of the head of bacteriophage T4. Nature 227, 680-85 (1970).

Miyoshi, I., Hiraki, S., Tsubota, T., Hikita, T., Masuji, H., Nishi, Y., Kimura, I. Hairy cell leukemia: Establishment of a cell line and its characteristics. Cancer 47, 60-65 (1981).

Reedman, B. M., Klein, G. Cellular localization of an Epstein-Barr virus associated complement-fixing antigen in producer and non-producer lymphoblastoid cell lines. Int. J. Cancer 11, 499-520 (1973).

Rubin, R. W., Quillen, M., Corcoran, J. J., Ganapathi, R., Krishan, A. Tubulin as a major cell surface protein in human lymphoid cells of leukemic origin. Cancer Res. 42, 1384-89 (1982).

Saxon, A., Stevens, R. H., Quan, S. G., Golde, D. W. Immunological characterization of hairy cell leukemias in continuous culture. J. Immunol. 120, 777-82 (1978).

Towbin, H., Staehelin, T., Gordon, J. Electrophoretic transfer of proteins from polyacrylamide gels to nitrocellulose sheets: Procedure and some applications. Proc. Natl. Acad. Sci. 76, 4350-54 (1979).

74 Changes of Contractile Structure of Muscle Maintained in a Shortened Position

Anna Jakubiec-Puka with the technical assistance of Hanna Chomontowska

INTRODUCTION

It is known that maintenance of skeletal muscle in a shortened state brings about a reduction of sarcomere number of the myofibrils. It appears that this alteration adjusts the sarcomere length to that optimal for muscle function (Tabary et al. 1972; Williams and Goldspink, 1978; Huet de la Tour et al. 1979). However, the way in which this alteration occurs remains obscure. Therefore, in the present work the ultrastructure of the contractile apparatus was observed in muscle maintained in shortened position to gain insights into the cellular process involved.

METHODS

The rat soleus muscle was removed on the 3-7th day after immobilization of the ankle joint in extension. All procedures for electron microscopy were performed as described previously (Jakubiec-Puka et al. 1982a).

RESULTS

Within a few days after immobilization of the ankle joint in extension the length of the soleus muscle diminished to 80-90% of the contralateral control. Simultaneously, several changes developed in the ultrastructure of the contractile apparatus. The most characteristic changes were evaluated quantitatively (Table 74.1). Variable size of sarcomeres, "extra sarcomeres," small diameter myofibrils and disappearance of the Z-line (Figs. 74.1, 74.2) were very frequent. Actin filaments were often

Table 74.1. Quantitative Evaluation of Contractile Structure Changes in the Muscle Maintained in Shortened Position for 3-7 Days

	Day of Experiment	
	3-4th	7th
Muscle length % of contralateral	83-85	83
Number of fibers studied	70	56
Percent of fiber containing		
"Extra sarcomeres"	74	75
Z-line disappearance	66	73
Postmortemlike changes	39	12
Foci of Z-line "streaming"	21	45
Rods	–	9
Whirls or total disorganization	11	3

Note: Fibers were taken randomly.

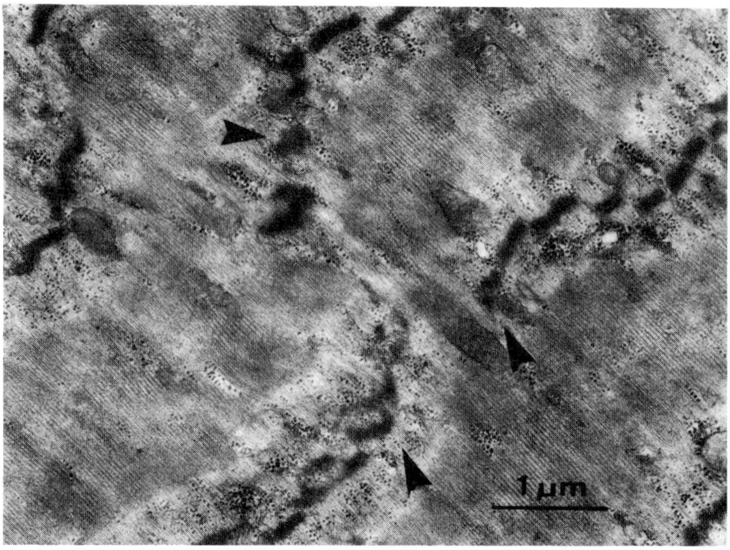

Fig. 74.1. Irregularity and variable size of sarcomeres, disappearance of Z-line (arrows).

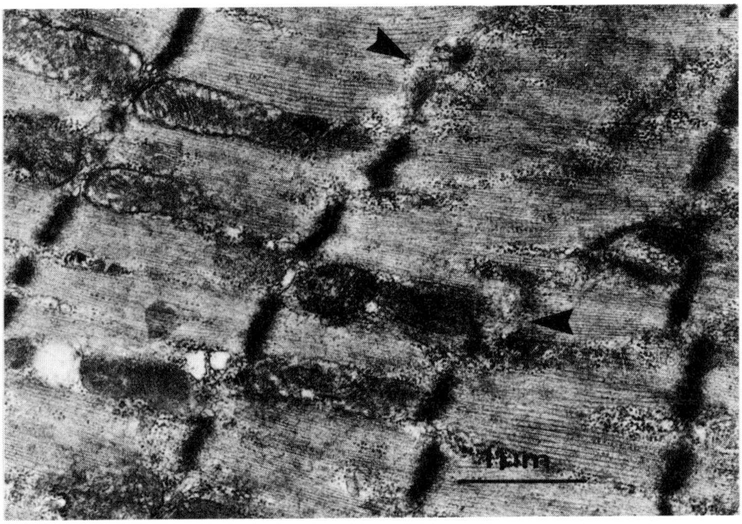

Fig. 74.2. Disappearance of Z-line (arrows), two neighboring sarcomeres seem to transform to one.

preserved in spite of a loss of the Z-line. These filaments were also frequently present in the central part of sarcomeres. Disorganization of the contractile structure with whirls of myofibrils, sometimes around deep folds of sarcolemma (Fig. 74.3), as well as destructions, like those appearing in postmortem muscle (MacNaughtan, 1978), were also frequently observed. Small foci of Z-line "streaming" were present in many fibers. Total disorganization of the contractile structure and postmortem-like changes were more intensive on the 3-4th day of the experiment than on the 7th day. On the 7th day rodlike changes were occasionally found. Many lysosomes and myelinlike figures were present between myofibrils and in the subsarcolemmal region.

DISCUSSION

The present data clearly show a disintegration of the sarcomere structure for the soleus muscle maintained at shortened length. The observed changes were very intense on the 3-7th day of the experiment (Table 74.1) and could be transient forms in the reconstruction of the contractile apparatus to reduce the

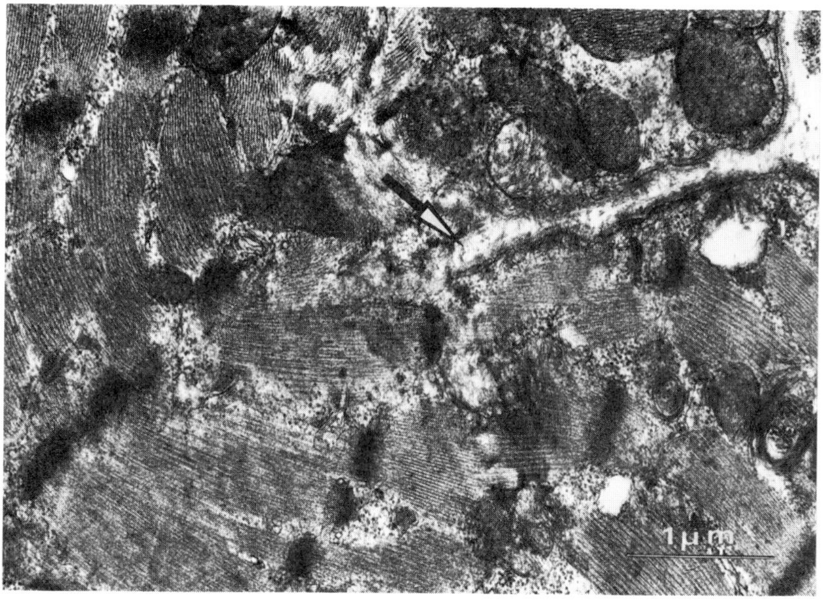

Fig. 74.3. Deep invagination of sarcolemma (arrow) and whirls of myofibrils.

number of sarcomeres. The process of the reduction of sarcomere number seems to begin with the disappearance of the Z-line (Fig. 74.1, 74.2). The next step, after loss of the Z-sheets, could be displacement of myofilaments from two neighboring sarcomeres to give one new sarcomer, as Figure 74.2 suggests. Another possible way of reduction of the sarcomere number and decrease of muscle fiber length, would be elimination of part of the contractile structure. Disorganization and whirls of myofibrils (Fig. 74.3), as well as postmortemlike changes of the contractile structure, could signal the beginning of this process, which leads to total sarcomere destruction and disappearance of part of the muscle fiber. The frequency of these changes (Table 74.1) shows that they could be the transient forms during the process of reduction of the sarcomere number and of atrophy in shortened muscle. Interestingly, the changes are considerably different from those found in the contractile structure of the soleus muscle maintained in an extend position (Jakubiec-Puka et al. 1982b), which may lead to new sarcomere production.

REFERENCES

Huet de la Tour E., Tabary J.C., Tabary C., Tardieu C., The respective roles of muscle length and muscle tension in sarcomere number adaptation of guinea-pig soleus muscle, J. Physiol (Paris), 75:589-92, 1979.

Jakubiec-Puka A., Kulesza-Lipka D., Kordowska J., The contractile apparatus of striated muscle in the course of atrophy and regeneration. II. Myosin and actin filaments in mature rat soleus muscle regenerating after reinnervation, Cell Tiss. Res. 227:641-50, 1982a.

Jakubiec-Puka A., Kordowska J., Chomontowska H., Z-line changes in the rat muscle maintained in lengthened position, J. Muscle Res. Cell Motil., 3:482-83, 1982b.

MacNaughtan A.F., A histological study of post mortem changes in the skeletal muscle of the fowl (Gallus domesticus). II. The cytoarchitecture, J. Anat., 126:7-20, 1978.

Tabary J.C., Tabary C., Tardieu C., Tardieu G., Goldspink G., Physiological and structural changes in the cat's soleus muscle due to immobilization at different lengths by plaster cats, J. Physiol. (London), 224:231-44, 1972.

Williams P.E., Goldspink G., Changes in sarcomere length and physiological properties in immobilized muscle, J. Anat., 127:459-68, 1978.

75 Electrophoretic Analysis of Human Uterine Structural Proteins in Normal and Myomatous Tissues

Thota Srihari, Marcus C. Schaub, Kaspar A. Büchi

INTRODUCTION

The uterus is composed of an outer layer of connective tissue (perimetrium), the middle layer of smooth muscle cells (myometrium), and an innermost layer consisting of several glands lined with epithelial cells supported by connective tissue elements (endometrium). It is striking that about 20% of all women around 35 years of age develop uterine tumors (mainly benign leiomyomas) of various size which arise primarily from the muscle cells of myometrium. About 0.5% of these benign tumors may even become sarcomatous (see Wynn, 1977). We asked the question whether these benign leiomyomas affect cell metabolism resulting in distinct changes in the molecular morphology. We therefore collected tissue samples from normal and affected myometria and also from pregnant and older aged women. Tissues were analyzed by two-dimensional acrylamide gel electrophoresis and the results suggest that during tumorous transformation distinct changes indeed occur in the polypeptide pattern.

MATERIALS AND METHODS

The uterine samples analyzed in this study were obtained from 27 female patients at the Dept. of Obstetrics and Gynecology of the University Hospital, Zürich. Out of these 27 cases, 10 were affected by leiomyomas, 2 were in their 39th week of pregnancy, and the remaining 15 were normal, 3 of these women were around 60 years old. Immediately after surgery myometrial samples were collected, placed on ice, and stored at -60°C until analyzed. Total tissue extracts were prepared by addition of ampholine solution (1:3; wt/vol) containing 9.5 M urea, 1.75% ampholines in the pH range 3.5-10 and 5% beta-mercaptoethanol; then they were homogenized and subjected directly to isoelectric focusing (IEF). Additional samples were glycerinated and washed twice in order to remove membranes and soluble proteins. The fine-cut tissue was mixed (1:10; wt/vol) with a solution containing 20 mM Tris-Cl, pH 7.6 100 mM KCl, 5 mM EDTA, 14 mM beta-mercaptoethanol and 50% glycerol and left at -20°C for at least one week. Then samples were washed twice with 20 mM Tris-Cl buffer, pH 7.6, containing 5 mM $MgCl_2$. The remaining sediment was prepared for IEF as described for total tissue extracts. IEF in the first dimension was carried out on acrylamide gels (11.5 cm × 3 mm) according to O'Farrell (1975). Electrophoresis in the second dimension was run in the presence of sodium dodecyl sulfate (SDS-PAGE) according to Laemmli and Favre (1973) on slab gels as described in detail elsewhere (Srihari et al., 1981). Proteins were stained in 0.25% Coomassie Brilliant Blue R-250.

The following proteins were used for reference and molecular weight estimation of polypeptides in uterine samples: serum albumin (68 kDalton), actin (42 kDa), alpha-tropomyosin (34 kDa), beta-tropomyosin (36 kDa), myosin light chains from fast-twitch muscles LC1f (25 kDa), LC2f (18 kDa), LC3f (16 kDa) and myosin light chains from slow-twitch skeletal muscles LC1s (27 kDa), LC2s (19kDa). Myosin was prepared and purified from uterine tissue by ammonium sulfate fractionation between 38-53% saturation and alpha-actinin (100 kDa) as described by Craig et al. (1982). Commercial alpha-actinin was from Sigma (St. Louis, USA) and bovine serum albumin from Boehringer (Mannheim, West Germany). Purified calmodulin was a gift from Dr. C. Heizmann (Dept. of Pharmacology and Biochemistry, Vet. Med. Faculty, University of Zürich). Myofibrils were prepared from fast-twitch psoas and slow-twitch soleus rabbit skeletal muscles. Protein concentrations were estimated according to Lowry et al. (1951). Quantitative evaluation of polypeptides in two-dimensional gel electrophoresis after staining with Coomassie Blue was done in each gel relative to tropomyosin (TM) as described recently for heart muscle proteins (Schaub et al., 1983).

RESULTS

Two protein patterns from total tissue extracts of uterine smooth muscle are shown in Figure 75.1. 60-70 distinct polypeptides can be distinguished. In normal uterine tissue (Fig. 75.1a) actin and tropomyosin are the most abundant proteins. Coelectrophoresis of uterine preparations with myofibrillar proteins of slow-twitch and fast-twitch skeletal muscles of rabbit revealed identical electrophoretic mobility of actin in all 3 tissues (42 kDa) whereas uterine TM is similar to the beta-type TM (36 kDa) found in slow-twitch muscles. Coelectrophoresis of purified uterine myosin with myofibrillar proteins of slow-twitch muscle (Fig. 75.2) indicates that the myosin light chains ULC1 and ULC2 have lower apparent molecular weights of around 18 kDa and 16 kDa, respectively. ULC1 and ULC2 comigrate almost entirely with myosin LC2f and LC3f from fast-twitch muscle. Similar results have been reported for bovine uterine myosin light chains by Leger and Focant (1973). Since in uterine smooth muscle myosin constitutes only around 5-8% of total cell protein (Cohen and Murphy, 1978; Huszar and Roberts, 1982) which is 4-6 times less than in skeletal muscles, its light chains could hardly be detected in two-dimensional electrophoresis unless the runs were grossly overloaded with protein. In addition to the above proteins the following polypeptides with their apparent molecular weights in brackets could be identified tentatively in uterine extracts by comparison of their electrophoretic properties with those of isolated proteins and of known proteins in the myofibrillar preparations of skeletal muscles: serum albumin (68 kDa), desmin (51 kDa), vimentin (54 kDa) and calmodulin (16.5 kDa). Albumin and desmin are clearly visible in glycerinated and washed myofibrillar samples of slow-twitch soleus muscle (Fig. 75.2).

Comparison of tissue samples from normal and leiomyomatous tissue revealed a large number of consistent differences in their protein patterns. The relation between actin and TM did not vary much, whereby the molar ratio of actin to TM was between 5-10 per mole of TM. In contrast a number of proteins are significantly increased in relative amounts in total tissue extracts from leiomyomas. Above all, albumin increased 20-100 times, desmin 5-40 times, calmodulin 5-10 times, and a number of unidentified polypeptides located closely together between actin, desmin, and TM in the molecular weight range of 36-51 kDa as well as 6 proteins scattering from the basic to the acidic pH range with molecular weights lying between 14-27 kDa were variably elevated (indicated by arrowheads in Fig. 75.1b). After glycerination and washing the soluble proteins calmodulin, the scattered proteins in the small molecular weight region as well as a large fraction of the albumin disappeared in the samples from myomatous tissues. The increase in desmin and the poly-

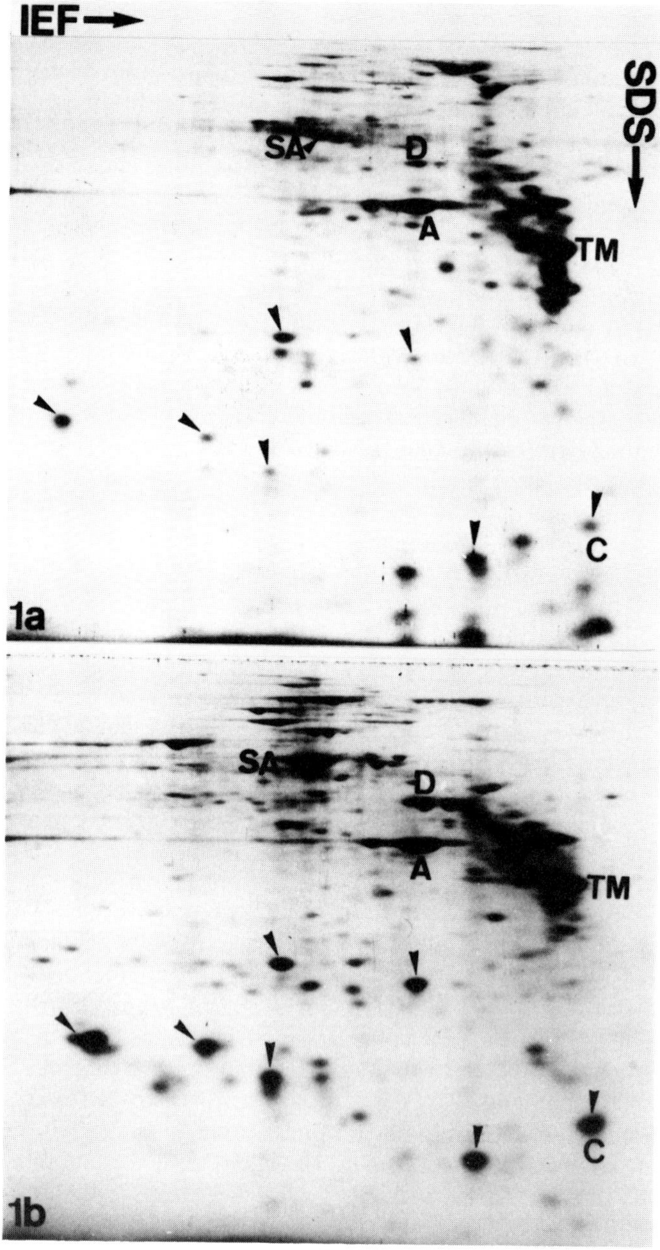

Fig. 75.1. Two-dimensional electrophoresis of proteins from total tissue extracts of myometrium from normal uterus (1a) and of nonaffected myometrium from a uterus containing leiomyomas (1b). About 120 μg of protein per each sample were applied. Gels are presented with the pH range 7.1-4.4 from left to right and decreasing molecular weight from top to bottom. A, actin; C, calmodulin; D, desmin; SA, albumin; TM, tropomyosin.

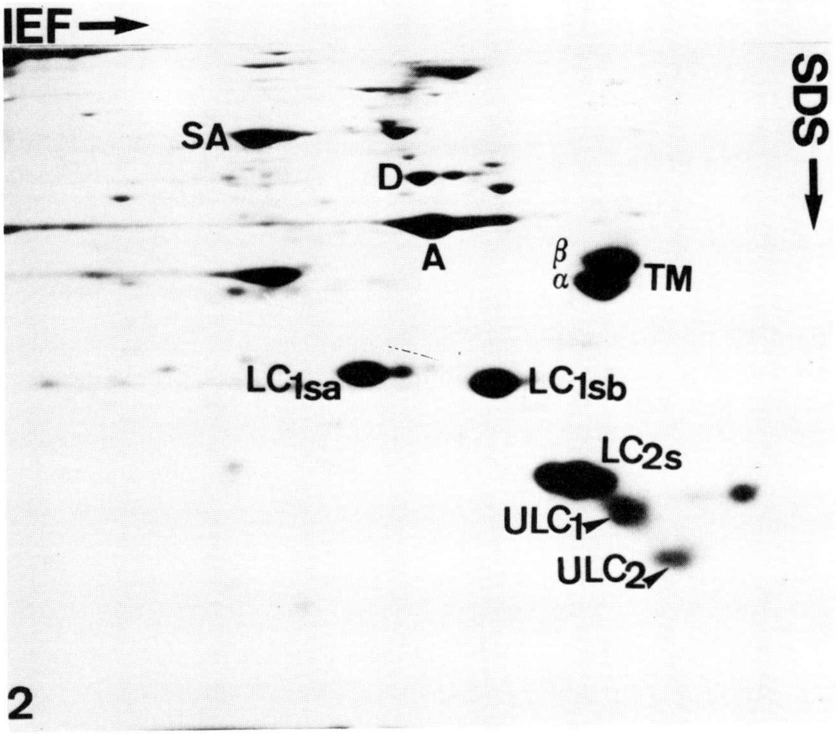

Fig. 75.2. Two-dimensional coelectrophoresis of myofibrillar proteins of rabbit slow-twitch soleus muscle (150 μg of protein applied) and purified myosin from normal human myometrium (30 μg of protein applied). The myosin light chains from slow-twitch soleus muscle (LC1sa, LC1sb, LC2s) and human uterus muscle (ULC1, ULC2) are indicated. Presentation and other abbreviations are the same as in Fig. 75.1.

peptides between actin, desmin, and TM persisted and albumin was still 3-10 times more abundant when compared to healthy control myometrium. The most interesting finding was that in all cases samples from nonaffected areas collected far from the myomatous foci exhibited almost identical protein patterns to those of the tumors with all the increased protein levels (Fig. 75.1b). The myometrial samples from women pregnant in the 39th week displayed a protein pattern very similar to that of normal uteri except for a 3-10-fold increase in desmin (Fig. 75.3a). The protein pattern of myometrium of the women of

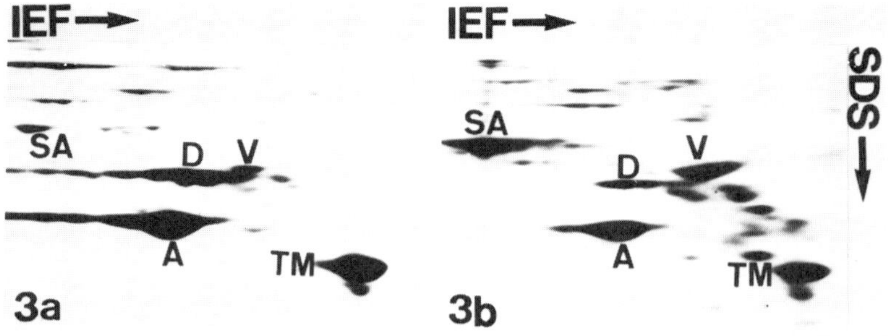

Fig. 75.3. Two-dimensional electrophoresis of total myometrial tissue extracts of samples obtained from 39 week pregnant (3a) and 61 years old (3b) women. About 120 μg of protein per each sample were applied. Only the upper part of the gels are shown where distinct differences as compared to normal uterine samples were found. V, vimentin resembling protein. Presentation and all other abbreviations are the same as in Fig. 75.1.

an age of around 60 years approached the appearance found in myomatous uteri, i.e., an increase in albumin, desmin, and in particular, a large increase in a protein suspected to be vimentin (Fig. 75.3b). In most normal cases vimentin was hardly detectable and in myomatous tissues only little increase occurred. It is known, however, that this protein is very susceptible to proteolytic degradation (Lazarides and Granger, 1982) and some of the polypeptides in the region below vimentin between desmin, actin, and TM may represent such degradation products.

DISCUSSION

Our findings concern an increase relative to the content of TM of the intracellular structural proteins desmin and possibly also vimentin. Both are constituents of the intermediate filament system. Previously, immunocytochemical studies have demonstrated an increase in desmin in uteri affected by leiomyomas (Gabbiani et al., 1981; Evans et al., 1983). In the case of tumor cells the suggestion was made that an accumulation of intermediate filament proteins is a sign of degeneration or of impending necrosis (see Ghadially, 1980). Contrary to such an assumption our results indicate an increase of these proteins during pregnancy as well as in the nonaffected parts of myomatous uteri.

In addition to the structural proteins also soluble cell proteins increase in myomatous tissues including calmodulin and albumin. Calmodulin is believed to be involved in cell proliferation and is found increased in many tumorous tissues (Veigl et al., 1982). The dramatic increase in albumin in total tissue extracts may be due to the large portion of extracellular space in smooth muscle where it may be deposited. However, after glycerination and washing albumin was still 3-10 times increased over control samples indicating its partial association with intracellular structures. Recently the identity of serum albumin and intracellularly located albumin in chicken skeletal muscles has been demonstrated (Heizmann et al., 1981). Albumin was found to be structurally integrated in the myofibrils in larger quantities in slow-twitch than in fast-twitch muscle fibers (Müller and Heizmann, 1982). Similarly, our analysis revealed a much higher amount of albumin in rabbit slow-twitch soleus myofibrils (Fig. 75.2) than in those from fast-twitch muscles (not shown).

No changes in the contractile proteins proper, myosin and actin, were found in relation to TM. In invasive cancerous cells an increase in contractile proteins was demonstrated immunocytochemically (Gabbiani et al., 1976). Since leiomyomatous cells are noninvasive they may not require an increase in contractile proteins or else, all three proteins, myosin, actin, and TM, may have increased in their content in a fixed stoichiometric relationship. Whatever the case, albumin, intermediate filament proteins, calmodulin, and further unidentified polypeptides increase disproportionately to TM not only in the transformed leiomyomatous tissue but also in the large portions of the myometrium which are not affected by the tumor directly. In late pregnancy desmin is increased selectively to a large extent, while in older age the myometrial protein pattern starts resembling that found in myomatous uteri (Fig. 75.3b). These biochemical findings underlying the mechanism of myomatous transformation as compared to growth during pregnancy and to aging could be of diagnostic value.

ACKNOWLEDGMENTS

This study was supported by the Swiss National Science Foundation Grant No. 3.152.81.

REFERENCES

Cohen, D. M. and Murphy, R. A.: Differences in cellular contractile protein contents among porcine smooth muscles. Evidence for variation in the contractile system. J. Gen. Physiol. 72. 369-80. 1978.

Craig, S. W., Lancashire, C. L. and Cooper, J. A.: Preparation of smooth muscle alpha-actinin. Methods in Enzymology, 85 B. 316-21. 1982.

Evans, D. J., Lampert, J. A. and Jacobs, M.: Intermediate filaments in smooth muscle. J. Clin. Pathol. 36. 57-61. 1983.

Gabbiani, G., Csank-Brassert, J., Schneeberger, J. C., Kapanci, Y., Trenchev, P. and Holborow, J.: Contractile proteins in human cancer cells. Amer. J. Pathol. 83. 457-74. 1976.

Gabbiani, G., Kapanci, Y., Barazzone, P. and Frank, W. W.: Immunochemical identification of intermediate-sized filaments in human neoplastic cells. Amer. J. Pathol. 104. 206-16. 1981.

Ghadially, F. N.: Diagnostic electron microscopy of tumors. Butterworths, London, 1980.

Heizmann, C. W., Muller, G., Jenny, E., Wilson, K. J., Landon, F. and Olomucki, A.: Muscle beta-actinin and serum albumin of the chicken are indistinguishable by physicochemical and immunological criteria. Proc. Nat. Acad. Sci. USA, 78. 74-77. 1981.

Huszar, G. and Roberts, J. M.: Biochemistry and pharmacology of the myometrium and labor; regulation at the cellular and molecular levels. Amer. J. Obstet. Gynecol. 142. 225-37. 1982.

Laemmli, U. K. and Favre, M.: Maturation of the head of bacteriophage T4. I. DNA packaging events. J. Mol. Biol. 80. 575-99. 1973.

Lazarides, E. and Granger, B. L.: Preparation and assay of the intermediate filament proteins desmin and vimentin. Methods in Enzymology 85 B. 488-508. 1982.

Leger, J. J. and Focant, B.: Low molecular weight components of cow smooth muscle myosins; characterisation and comparison with those of striated muscle. Biochim. Biophys. Acta, 328. 166-72. 1973.

Lowry, O. H., Rosebrough, N. J., Farr, A. L. and Randall, R. J.: Protein measurement with the Folin phenol reagent. J. Biol. Chem. 193. 265-75. 1951.

Müller, G. and Heizmann, C. W.: Albumin in chicken skeletal muscle. Eur. J. Biochem. 123. 577-82. 1982.

O'Farrell, P. H.: High resolution two-dimensional electrophoresis of proteins. J. Biol. Chem. 250. 4007-21. 1975.

Schaub, M. C., Tuchschmid, C. R., Srihari, T. and Hirzel, O. H.: Myosin isoenzymes in human hypertrophic hearts; shift in atrial myosin heavy chains and in ventricular myosin light chains. Eur. Heart J. 1983 (in press).

Srihari, T., Wiehrer, W., Pette, D. and Harris, B. G.: Electrophoretic analysis of myofibrillar proteins from the body wall muscle of Ascaris suum. Molec. Biochem. Parasitol. 3. 71-82. 1981.

Veigl, M. L., Sedwick, W. D. and Vanaman, T. C.: Calmodulin and Ca^{2+} in normal and transformed cells. Fed. Proc. 41. 2283-88. 1982.

Wynn, R. M.: Biology of the uterus. Plenum Press, New York, 1977.

76 Changes in the Cytoskeleton and the Mitochondria during Neurogensis

Jean-Luc Vayssiere, Francis Berthelot,
Marie-Madeleine Portier, Philippe Denoulet,
Bernard Croizat, François Gros

The mouse neuroblastoma cells, under appropriate conditions of culture, exhibit many biochemical, physiological, and morphological properties of the mature nerve cell. For this reason, they are considered to be a convenient model for studying the genetic expression during neuronal terminal differentiation. In particular, they have proved very useful in the study of formation of neurites, the appearance of enzymes that synthesize neurotransmitters and the acquisition of membrane excitability (Augusti-Tocco et al., 1964; Schubert et al., 1971).

In addition to these transitions, considered as specific of neuronal differentiation, we observed changes at the level of some cytoskeletal components and membrane proteins.

CHANGES IN THE CYTOSKELETON AND ASSOCIATED PROTEINS DURING NEUROBLASTOMA DIFFERENTIATION

There are reductions in the incorporation of methionine into α-isotubulins and actin (see Fig. 76.1) and changes in the microheterogeneity of β-isotubulins (Eddé et al., 1981). These analyses were made using high resolution two-dimensional electrophoresis (see legend of Fig. 76.2). Further examination of these modulations, with several clones in different conditions of differentiation, allowed us to suggest a sequence of the molecular events occurring during the morphological differentiation (Portier, 1982).

A new differentiation inducer, 1 methyl cyclohexane carboxylic acid (CCA), appeared very useful in this study. It promotes a strong stimulation of the methionine incorporation into some proteins, in particular vimentin and the insoluble membrane protein Y (Fig. 76.1). We detected protein Y in the

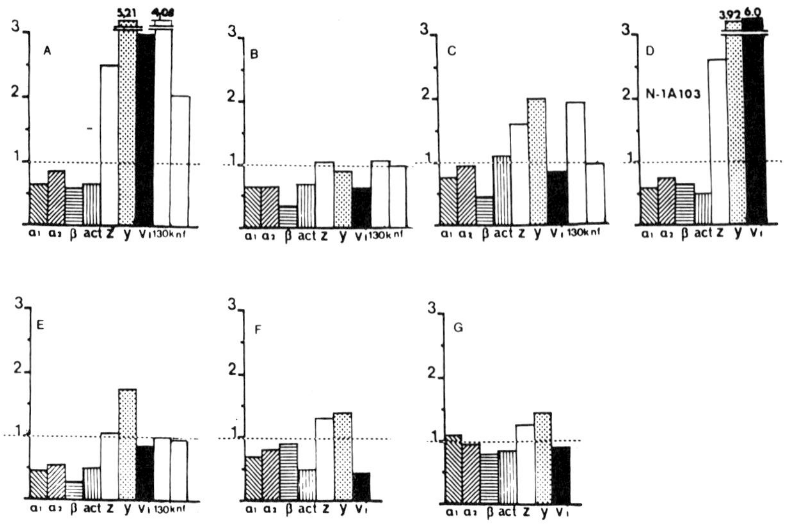

Fig. 76.1. Graphic representation of the methionine incorporation into some proteins in various culture conditions. The effects of various culture conditions are shown on the methionine incorporation into isotubulins, actin, and some other proteins. The data are expressed relative to the values found in control cultures which can then be considered = 1 (---). Control cultures are cells grown in complete medium and maintained in monolayer conditions. The experiments were made with clone N1E115 except when mentioned N1A103: (A) monolayer culture in a complete medium containing CCA; (B) monolayer culture in a complete medium containing DMSO; (C) monolayer culture in a medium without serum; (D) monolayer culture of N1A103 cells in a complete medium containing CCA (cells are blocked in a stationary phase without expressing any morphological differentiation); (E) suspension cells in a complete medium containing CCA; (F) suspension cells in a medium without serum; (G) suspension cells in a complete medium; α_1, α_2: α_1, α_2-iso-tubulins; β, β-isotubulin; act, actinin; Vi, vimentin; 130 k, vinculin; nf, 70 k neurofilament.

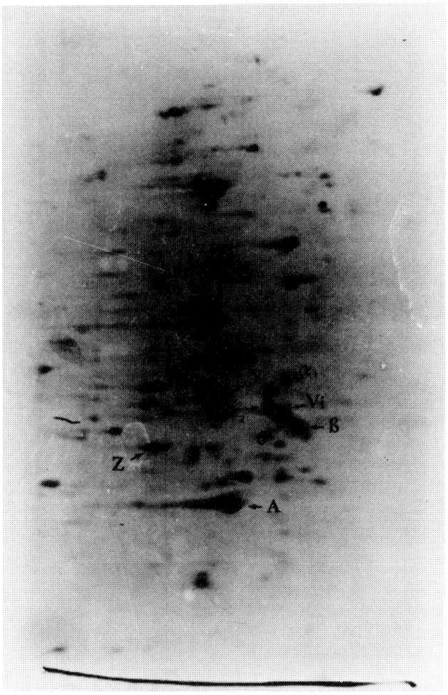

Fig. 76.2. Electrophoregram showing the proteins discussed. The proteins are shown on a two-dimensional gel loaded with a pellet from N1E115 neuroblastoma extract. Cells were labeled for 24 h with L-|^{35}S|methionine and lysed with Nonidet P40. The crude extract thus obtained was separated into pellet and supernatant, which were treated for isoelectric focusing according to O'Farrell (1976). Except for isotubulins, actin, and neurofilament, equally distributed in both pellet and supernatant, the proteins studied here are insoluble and thus exclusively recovered in the pellet. About 10^6 cpm corresponding to ^{35}S-labeled proteins are loaded on each gel. The cpm corresponding to each spot are counted and the percentages of each protein are calculated vs total cpm loaded on the first-dimensional gel. The data are expressed relative to the values found in a control culture (cells grown in a complete medium and maintained in monolayer conditions). Graphic representation of the data is shown in Fig. 76.1. α_1, α_2:α-isotubulins; β, 6:β-isotubulins, A: actin; Vi: vimentin.

neuroblastoma and identified it as a characteristic protein of the peripheral nervous system (Portier, 1983).

It can be postulated that CCA causes some important modifications of the neural surface since it ensures better adhesion to the substratum than other differentiation inducers. Concomitant with this, neurite extension proceeds more efficiently, neurites are more stable than after induction by other substances, and the synthesis of membrane-bound proteins is increased.

EFFECTS OF CCA ON CELLULAR ENERGETICS

Moreover, CCA promotes a significant increase in the level of cellular energetics as estimated by the cellular accumulation of isomerase-2-deoxy[^{14}C]D-glucose-6-phosphate which reflects glucose utilization, according to Sokoloff (1977).

A considerably higher amount of radioactivity—2.5 to 4 times—is found in CCA-treated cells, as compared to other types of cultures, corresponding to a higher rate of deoxyglucose penetration and utilization (Croizat, 1981). The burst in cellular energetics might play a role in triggering the neuronal differentiation. It is interesting to note that drastic changes in neuroblastoma metabolism induced by CCA are accompanied by modifications in the ultrastructure as observed by electron microscopy. For example, CCA-treated cells seem to contain more mitochondria. Furthermore, their resistance to oligomycin is also drastically increased (data not shown).

For these reasons, it seemed interesting to pay particular attention to the mitochondrial proteins and more generally to mitochondrial function at different stages of brain maturation and during neuroblastoma differentiation.

MITOCHONDRIAL ACTIVITY

Preliminary results reveal that CCA enhances the mitochondrial ATPase activity in mouse brain, whether the drug is injected intraperitoneally to the animal (+18%), or added to the reaction mixture, a standard ATPase assay, in the presence of an ATP regenerating system (+35%). Likewise, the ATPase activity of differentiated neuroblastoma is higher (+33%) when differentiation is induced by CCA than when it occurs in a medium without fetal calf serum.

The CCA effect on mitochondrial respiration is equally studied by oxygraphy: CCA seems to act as an uncoupler of the oxydative phosphorylations, at least in the brain mitochondria.

MITOCHONDRIAL PROTEINS

Figure 76.3 shows the two-dimensional electrophoretic pattern of mitochondrial proteins prepared from neuroblastoma cells. Three proteins (indicated by arrows) have been selected for further study, for the following reasons: (i) they are major spots; (ii) they exhibit quantitative modulations in the course of neuroblastoma differentiation and/or brain maturation. An examination of these proteins may elucidate the possible role of the mitochondria in neuronal differentiation.

Two experimental procedures provided information concerning the identity and processing of the mitochondrial proteins: (i) mitochondrial function was impaired using a specific inhibitor, the K^+ ionophore nonactin; (ii) total cell mRNA or polysomal RNA from mouse brain or neuroblastoma were translated in a

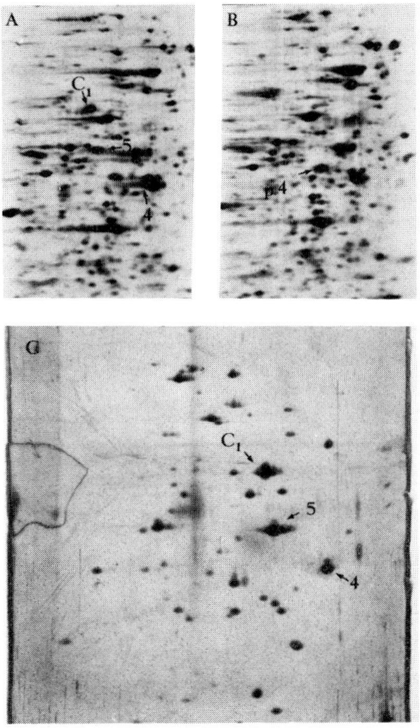

Fig. 76.3. Electrophoregram showing mitochondrial proteins. 1-A shows an autoradiogram of proteins from a soluble extract from N1E115 neuroblastoma cells in a complete medium (DMEM + 7.5% fetal calf serum), labeled during 6 hours before harvesting. 1-B. Same conditions, except for a 18 hours treatment with 5 μM nonactin before harvesting. Proteins C1, 5, and 4 disappear after nonactin treatment while P4 the precursor of protein 4 appears. 1-C is a two-dimensional silver stained gel, as Morrissey (1981), showing purified mitochondrial proteins. Purification was processed by successive centrifugations followed by a sucrose step gradient according to Anderson (1981).

cell-free reticulocyte system. The product of the translation was also analyzed on two-dimensional electrophoresis.

PROTEIN 4 (apparent molecular weight: 52 kda; pH_i 5.2)

Treatment of cultures with nonactin results in the loss of the spot corresponding to protein 4 from the two-dimensional pattern of labeled polypeptides. A new spot appears corresponding to the precursor (mitpro) of protein 4 (see insert in Fig. 75.3). The precursor is made by the nucleocytoplasmic genetic system. This is shown by the synthesis of the mitpro spot in a reticulocyte cell-free system fed with total mRNA from mouse brain. The final product, protein 4, is not synthesized in the same conditions.

PROTEIN C1 (apparent molecular weight: 68 kda; pH_i 5.8)

There is no synthesis of that protein when cultures have been treated with nonactin. We have not identified any spot corresponding to a mitpro. Nevertheless, it might be present on the gel with important modifications in the molecular weight and/or pH_i. In these conditions, we have not observed any product, mitpro or protein C1, in the cell-free system fed with mRNAs from neuroblastoma or mouse brain.

PROTEIN 5 (apparent molecular weight: 55 kda; pH_i 5.5)

This protein might correspond to the IEF24 protein reported by Larson et al. (1980) which might serve, according to these authors, as a linkage protein of the mitochondria to intermediate filaments structures. The identification of protein 5 is in progress. Its synthesis is also inhibited by nonactin in neuroblastoma cells and no candidate for the mitpro was identified on the electrophoregrams. There is no synthesis of protein 5 in the cell-free system fed with total mRNA from mouse brain or neuroblastoma. However, the protein is synthesized when the system is fed with bound polysomes from neuroblastoma. This means that the translation (probably dependent on the mitochondrial electrochemical potential) is not initiated in a cell-free system. However, the run-off of the messengers, attached to the polysomes, occurs.

The proteins mentioned above exhibit quantitative modulations during brain maturation as measured by methionine incorporation at fetal, neonatal, and adult stages. Preliminary results show that CCA stimulates methionine incorporation into proteins C1 and 5, in neuroblastoma cultures. DMSO, another differentiation inducer, stimulates the protein 4 (data not shown).

Our conclusions support the idea that, concomitant with neuroblastoma differentiation, CCA induces a large increase in cellular metabolic rate and this probably involves a rearrangement of the mitochondria and cytoskeletal proteins.

REFERENCES

Anderson, L., Identification of mitochondrial proteins and some of their precursors in two-dimensional electrophoretic maps of human cells. Proc. Natl. Acad. Sci. USA 78, 2407-11 (1981).

Augusti-Tocco, G., and Sato, G., Establishment of functional clonal lines of neurons from mouse neuroblastoma. Proc. Natl. Acad. Sci. USA 64, 311-15 (1969).

Croizat, B., Berthelot, F., Portier, M. M., Ohayon, H., and Gros, F. Effects of 1-methyl cyclohexane carboxylic acid (CCA) on cellular energetics in neuroblastoma cells. Biochem. Biophys. Res. Commun. 103, 1044-61 (1981).

Eddé, B., Jeantet, C., and Gros, F., One β-isotubulin subunit accumulates during neurite outgrowth in mouse neuroblastoma cells. Biochem. Biophys. Res. Commun. 103, 1035-43 (1981).

Moose-Larsen, P., Bravo, R., Fey, S. J., Small, J. V., and Celis, J. E., Putative association of mitochondria with a subpopulation of intermediate-sized filaments in cultured human skin fibroblasts. Cell 31, 681-92 (1982).

Morrissey, J., Silver stain for proteins in polyacrylamide gels: A modified procedure with enhanced uniform sensitivity. Anal. Biochem. 117, 307-10 (1981).

O'Farrel, P., High resolution two-dimensional electrophoresis of proteins. J. Biol. Chem. 250, 4007-10 (1975).

Portier, M. M., Croizat, B., and Gros, F., A sequence of changes in cytoskeletal components during neuroblastoma differentiation. FEBS Lett. 146, 283-88 (1982).

Portier, M. M., Croizat, B., de Néchaud, B., Gumpel, M., and Gros, F., La protéine Y, un nouveau marqueur potentiel du système nerveux périphérique. Compt. Rend. Acad. Sc., in press (1983).

Schubert, D., Humphreys, D., de Vitry, F. and Jacob, F., Induced differentiation of neuroblastoma. Develop. Biol. 25, 514-46 (1971).

Sokoloff, L., Rewick, M., Kennedy, C., Des Rosiers, M. H., Patlak, C. S., Pettigrew, K. D., Sakurada, O., and Shinohara, M., The $|^{14}C|$deoxyglucose method for the measurement of local cerebral glucose utilization: theory, procedure and normal values in the conscious and anaesthetized albino rat. J. Neurochem. 28, 897-916 (1977).

77 Altered Patterns of Expression of Myoisin Subunits in Diseased Human Muscle

Alfredo Margreth, Donatella Biral,
Ernesto Damiani

INTRODUCTION

Skeletal muscle myosins exist in several isoforms during early ontogenesis, according to the differentiation stage of myoblasts and myotubes and, possibly, depending on their future destiny as fast-twitch and slow-twitch fibers. The establishment of stable nerve-muscle interactions appears, however, to be critical for the molecular transition of fetal and neonatal isomyosins to adult fast and slow isomyosins and their segregation in the corresponding types of fibers during post-natal development (see, Caplan et al., 1983). Early skeletal muscle contains a unique type of myosin with respect to the heavy chains (HC emb and HC neonatal) and one type of light chain (LCemb), but two major light chain components appear to be identical to those of adult fast-twitch myosin (LC1F and LC2F). Species-specific differences in developmental transitions of myosin isoforms have been observed, such as the absence of LCemb in avian muscle (see, Caplan et al., 1983). An embryonic type of light chain has been found recently in humans (Biral et al., 1983; Strohman et al., 1983; Cummins, 1983a). When analyzed under native conditions, human fetal myosin resolved into fetal-specific electrophoretic components (Fitzsimmons and Hoh, 1981a), but its HC composition was not established, however.

Changes in myosin composition have been described in association with several forms of human myopathies (see Cummins, 1983b), and certain specific patterns of changes are now emerging, such as the persistence or resumption of synthesis of fetal myosin at the adult stage (Fitzsimmons and Hoh, 1982), the partial or complete suppression of synthesis of fast myosin (Fardeau et al., 1978; Volpe et al., 1981a), and the presence of a raised steady-state level of fibers containing hybrid myosin molecules or truly intermediate fibers (Takagi et al., 1982). We report here on these particular aspects on account of recent results obtained in our laboratory.

MATERIALS AND METHODS

Myosin was prepared from fetal muscles (15-18 wks gestation), neonatal muscle (1 day postnatal), adult and pathological muscle (sporadic congenital nemaline myopathy (CNM), congenital myopathy with type I fiber predominance and mitochondrial abnormalities (CM); myotonic dystrophy; Duchenne muscular dystrophy (DMD), using a procedure described previously (Volpe et al., 1981). SDS-polyacrylamide gel electrophoresis of myosins was performed by Laemmli's method (1970) and two-dimensional electrophoresis, according to O'Farrel (1971), under conditions described by Volpe et al. (1981). Peptide maps of myosin heavy chains were obtained by a modification (Carraro et al., 1981) of the method of Cleveland (1977), using S. aureus V8 protease (25 μg total). A 16-19% polyacrylamide linear gradient gel, 1 mm thick, was used in the digestion gel. Myosin heavy chains were digested during the run in the stacking gel (about 90 min). Antibodies against human fetal myosin were obtained by injecting an adult hen intramuscularly with 1 mg of immunogen emulsified with an equal volume of complete Freund's adjuvant, at two week intervals. The animal was bled after 35 days from the beginning of immunization and antibodies were purified by affinity chromatography. One-step ELISA and two-step competitive ELISA were carried out as described by Biral et al. (1982).

RESULTS AND DISCUSSION

One- and two-dimensional polyacrylamide gel electrophoresis of adult human myosin has shown the presence of a main slow light chain (LC1Sb) with isoelectric point (pI) and molecular weight properties closer to that of LC1F than in any other species so far investigated, LC2F and LC2S components and low amounts of both LC1Sa and LC3F (Volpe et al., 1981; Fitzsimmons and Hoh, 1981b). Myosin prepared from fetal 15-18 wks gestation muscle, when analyzed by SDS gel electrophoresis, exhibits LC1F and LC2F as the major light chain components (see Volpe et al., 1981), low amounts of a molecular weight species of 19,000, probably representing LC2S, and a fetal-specific molecular species of 26,500 (Fig. 77.1). Two-dimensional electrophoresis of both adult and fetal myosin indicates that the fetal-specific peptide has pI properties similar to those of LC1F and, by these properties and its higher weight, can be distinguished from all adult types of LC1. This result is in agreement with the recent observations of Strohman et al. (1983) and of Cummins (1983a).

In most muscle specimens from patients with DMD, LCemb is present in low amounts in the purified myosin, in marked contrast with several forms of congenital myopathies, as well

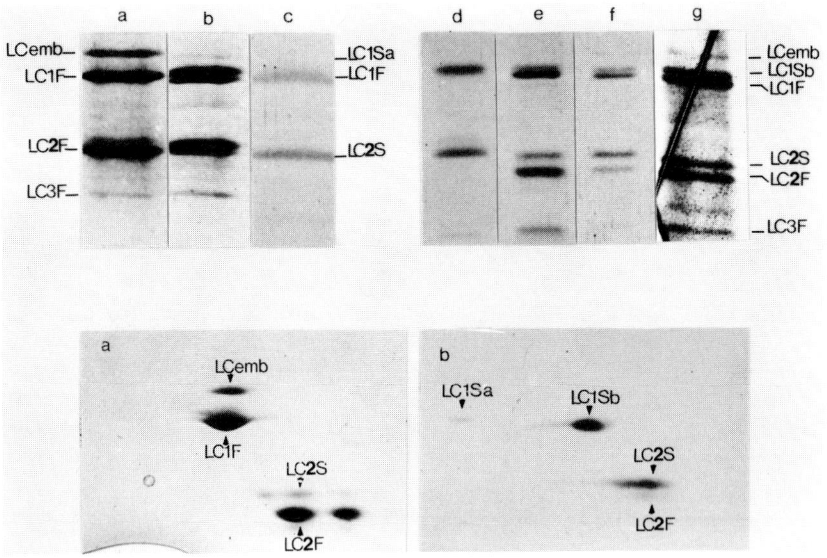

Fig. 77.1. Light chain pattern of human muscle myosins. Upper panel: 1-D electrophoresis; lower panel: 2-D electrophoresis (isoelectrofocussing, first dimension and SDS-PAGE, second dimension). Only the light chain region is shown. SDS-PAGE was carried out according to Laemmli in 15% polyacrylamide gels. Isoelectrofocussing was performed with 1% ampholines (Pharmacia) of pH range 3-10. About 30-40 µg of myosin were loaded per gel. The slabs were stained with CBB or silver (g). Key: (a) fetal myosin (limb muscles, 15-18 wks); (b) normal adult myosin (pect. minor, 40 y); (c) biopsy from patient with sporadic DNM gastrocnemius 3y); (d) biopsy from patient with congenital myopathy (gastrocnemius 8y); (e,f) biopsies from patients with myotonic cystrophy (gastocnemius 29y, 27y); (g) biopsy from patient with DMD (forearm, 16y).

as myotonic dystrophy (Fig. 77.1). The expression of LCemb in DMD appears to occur concomitantly with that of the adult light chains both of the fast and slow type, a result compatible with the existence of promiscuous fibers, as shown by Takagi et al. (1982).

In myotonic dystrophy, which is almost paradigmatic of type II fiber predominance, the fibers contain a myosin whose light chains are those characteristic of adult fast-twitch myosin (LC1F, LC2F, and LC3F), as expected from theory, or the light chains are virtually LC1F associated with LC2S and LC2F (Fig. 77.1). In these cases, the calculated LC2S:LC1S molar ratio was as high as 2, as compared to a normal stoichiometric value of about 1. This finding which agrees with the observation of a decreased reactivity of myosin from myotonic dystrophy muscle with antibodies specific for LC1F (Volpe et al., 1982b), is reminiscent of earlier findings on the LC composition of chick embryonic myosin (Caplan et al., 1983). However, a similarly abnormal LC composition has also been found by us recently for the myosin of chronically denervated rabbit slow-twitch muscle. Consequently, the formation of these particular myosin LC hybrids may be interpreted in several ways, though a likely explanation, for myotonic dystrophy, at least, is an on-going type-transition of muscle fibers. In congenital myopathies associated with type I fiber predominance, e.g., nemaline myopathy, myosin changes are somewhat the mirror image of those in myotonic dystrophy, in that the myosin LC pattern tends to be essentially of the pure slow type (Fardeau et al., 1978; Volpe et al., 1982a and Fig. 77.1).

The nature of myosin HC in adult skeletal muscle has been investigated recently in humans by peptide map analysis techniques in single fibers and was shown to differ between type I, type IIa and type IIb fibers (Billeter et al., 1981). By similar techniques (see Methods), we have been able to demonstrate a unique pattern for fetal myosin HC, i.e., distinct from that obtained with adult skeletal muscle, though closely similar to that of newborn muscle. Thus, two isoforms of myosin HC, at least, appear to occur sequentially during muscle development in humans.

We have investigated further the structural differences between fetal and adult isoforms of myosin HC, by a combined immunological approach, using antibodies raised in the chicken against fetal myosin, and which could be shown to be directed essentially to the HC by Western blot techniques. These antibodies are able to distinguish between fetal and adult myosin when tested by ELISA techniques (Fig. 77.2, 77.3), and there-

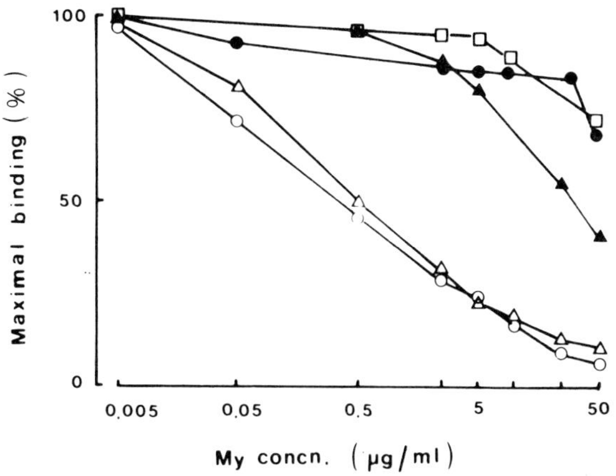

Fig. 77.2. Two-step competitive ELISA of human muscle myosins. In the first step, a solution of affinity purified anti-(human fetal myosin)antibody (10 µg/ml) was preincubated with an equal volume of antigen solution at the final concentrations indicated on the abscissa. Key: ○: fetal myosin; △: neonatal myosin (V. lateralis, 1d postnatal); ●: normal adult myosin (V. lateralis, 60y); ▲: myosin from biopsy muscle of three patients (average values) with DMD; □: myosin from biopsy muscle of two patients (average values) with myotonic dystrophy.

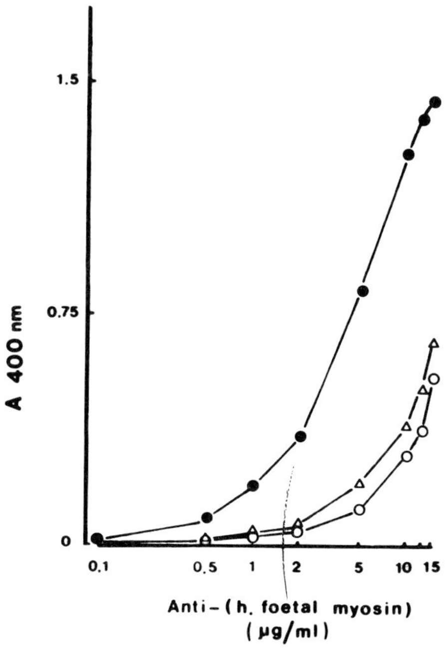

Fig. 77.3. One-step ELISA of human muscle myosins. Anti-(human fetal myosin)-antibody solution, at the concentrations indicated on the abscissa, was tested in microtiter wells coated with antigen (5 µg/ml). Key: ●: fetal myosin; △: adult myosin; ○: sporadic CNM.

fore afford a very sensitive tool to detect the presence of fetal myosin in samples of pathological muscle material.

As shown by the present results, myosin HC can be classified of the adult type both in myotonic dystrophy and in congenital myopathy. The latter finding, therefore, provides additional evidence that the basic abnormality in this group of muscle diseases does not involve a blocked fetal-adult switch, as implied in earlier theories (Volpe et al., 1982a). On the other hand, the results obtained with myosin samples of several DMD patients, showing a moderately increased immunological reactivity with anti-(fetal myosin)antibody, agree well with LC changes. As estimated from competitive ELISA (Fig. 77.2), the percentage of fetal myosin is about 2-3% of total myosin, a figure in the range of values reported by Fitzsimmons and Hoh (1981a) on account of their electrophoretic studies of native myosin in DMD. These relatively low figures are compatible with the interpretation (Fitzsimmons and Hoh, 1981a) that the resumption of synthesis of fetal myosin in DMD reflects the well-known occurrence of regenerative phenomena in this disease, since it has been shown (Carraro et al., 1983) that muscle regeneration in the rat recapitulates muscle ontogenesis.

ACKNOWLEDGMENTS

This work was supported by institutional funds from the Consiglio Nazionale delle Ricerche and, in part, by a grant from Legato Dino Ferrari. E.D. is a fellow of the Fondazione Anna Villa Rusconi.

REFERENCES

Billeter R., C. W. Heizmann, H. Howald and E. Jenny, Eur. J. Biochem., 116, 389, 1981.

Biral D., E. Damiani, P. Volpe, G. Salviati and A. Margreth, Biochem. J., 203, 529, 1982.

Biral D., E. Damiani, A. Margreth, E. Scarpini, G. Scarlato and M. Buscaglia, XIX Riunione della Sezione di Neuropatologia, St. Vincent, 27-28 Maggio, 1983.

Caplan A.I., M.Y. Fiszman, H.M. Eppenberger, Science, 221, 921, 1983.

Carraro U., C. Catani and L. Dalla Libera, Exp. Neur., 72, 401, 1981.

Carraro U., L. Dalla Libera and C. Catani, Exp. Neur., 79, 106, 1983.

Cleveland D.W., S.G. Fischer, M.W. Kirschner ans U.K. Laemmli, J.B.C., 252, 1102, 1977.

Cummins P., in: Perspectives in cardiovascular research, vol. 7, Myocardial hypertrophy and failure, N. R. Alpert ed, Raven Press, New York, pp 417, 1983a.

Cummins P., J. Muscle Res. Cell Mot., 4, 5, 1983b.

Fardeau M., J. Godet-Guillain, F.M.S. Tomè, S. Carson and R.G. Whalen in: Current topics in Nerve and Muscle Research (Agnayo A.J. and Karpati G. eds) pp. 164, Excerpta Medica, Amsterday, Oxford, 1978.

Fitzsimmons R.B. and J.F.Y. Hoh, J. Neurol. Sci., 52, 367, 1981a.

Fitzsimmons R.B. and J.F.Y. Hoh, Biochem. J., 193, 229, 1981b.

Fizsimmons R.B. and J.F.Y. Hoh, The Lancet, 1, 480, 1982.

Laemmli U.K., Nature (London), 227, 680, 1970.

O'Farrel P.H., J.B.C., 250, 4007, 1975.

Strohman R.C., J. Micou-Eastwood, C.A. Glass and R. Matsuda, Science, 221, 955, 1983.

Takagi A., S. Ishiura, I. Nonaka and H. Sugita, Muscle & Nerve, 5, 399, 1982.

Volpe P., D. Biral, E. Damiani and A. Margreth, Biochem. J., 195, 251, 1981.

Volpe P., E. Damiani, A. Margreth, G. Pellegrini and G. Scarlato, Neurology, 32, 37, 1982a.

Volpe P., E. Damiani, A. Margreth, G. Meola and G. Scarlato, Acta Neurol. Belgica, 82, 145, 1982b.

78 Pressure Induced Cytoskeletal Changes in Human Erythrocytes

J. A. Paciorek

INTRODUCTION

In this long-term study of mens' reaction to and living with a simulated but equivalent pressure of 300, 420, 500, or 540 meters of seawater (msw), we monitored the changes of many haematological parameters with preparations at pressure and after various decompression procedures.

The hyperbaric chamber-complexes of these dives were sited on shore. The breathing gas was a mixture of helium and oxygen, but with an increase in the partial pressure of oxygen, e.g., from 0.2 bar on surface to between 0.4 and 0.5 bar at depth. At 500 msw this is an equivalent O_2-gas percentage of only 0.85% balance 99.15% helium in the breathing gas. Helium gas has a cooling effect when breathed and so the chambers have to maintain a constant temperature of 30°C (±2°). The humidity within the chambers was between 50% and 80%.

Compressions to depths at this magnitude usually are achieved within 30 hours. Decompression takes up to 20 days from 540 meters of seawater. Blood sampling at pressure requires training of the divers in venepuncture and careful treatment of haemoglobin and haematocrit assays, light microscopy slides and in glutaraldehyde fixation. Due to the presence of irreversible morphological changes within the erythrocyte population of divers (Carlyle et al., 1979), a more rigorous study of red cell morphology and membrane protein components was performed during the NUTEC 500 msw dive in October 1981. To our knowledge this was the first full study of cytoskeletal proteins in men during hyperbaric exposure.

METHODS

Blood was sampled from each subject in the fasting state before the dive for periods up to three weeks. This established an individual baseline at 1 bar and control levels for each man. Routine haematological parameters such as maemoglobin (Hb), haematocrit (Hct) and red cell numbers were performed to monitor each subject's medical welfare. Additionally, enzymes such as glucose-6-phosphate dehydrogenase (G6PD), superoxide dismutase (SOD) and carbonic anhydrase activities (CA) with red cell morphology were analyzed. Oxyhaemoglobin dissociation curves were obtained by the Aminco Hemoscan and measurement of 2,3-DPG levels by the UV-Sigma kit method. Monitoring of Hb and Hct at pressure were done by portable instrumentation and compared with hospital values.

Upon commencement of hyperbaric exposure subjects were bled at 48 hour intervals. Blood for cytoskeletal protein analysis was collected into lithium-heparin-tubes and decompressed in a small pressure vessel. The pressure vessel was passed out of the chamber and brought to surface by a slow gas release at a rate of 1 meter/min.

All cells used in these experiments were kept at room temperature to attain equal in vitro states. Rapid decompression of heparinized blood, without a "Champagne-effect," was achieved by decompression within the medical lock in 1 min. from 500 msw, i.e., 500 m/min. This method was also used to check for decompression artifacts. Erythrocyte morphology of cells fixed at pressure in 8% glutaraldehyde and slowly or rapidly decompressed (Paciorek et al., 1982), was carried out by SEM.

Postdive samples were obtained up to one month after each dive. Heparinized red cells (3 mls) collected for cytoskeletal analysis after decompression were separated from the plasma by bench centrifugation at 3,000 rpm. These were washed three times in isotonic 5 mM phosphate buffered saline (PBS). Erythrocytes were then stored for 4 hours in acid-citrate-dextrose (ACD), 700 µl per one ml of red cells at 4°C until erythrocyte membrane extraction. Separation of membrane ghosts from the diving subjects' blood samples and five surface controls were in a lysis buffer at pH 8.0 of cold 5 mM phosphate, 0.1 mM EDTA. Ghosts were washed three times with ice-cold 0.3 mM phosphate, 0.1 mM EDTA, pH 7.5, centrifuged at 15,000 rpm in a Sorval SS 34 and finally resuspended in one-third of the starting blood volume in 0.3 mM phosphate buffer (pH 7.5). Each step took ten minutes. Aliquotes of the ghost preparation, either in 50 or 100 µl volumes were added to an equal volume of 2% SDS running buffer, either Tris bicine or glycine. Each sample was mixed with 20 µl of glycerol containing Bromophenol Blue and mercaptoethanol.

After a 10 min boiling, the samples were stored frozen at -20°C until required for electrophoresis; on either small vertical plates (3 × 10 cm) of polyacrylamide Tris bicine gel in 0.1% SDS at 130 volts for 30 to 45 min; or by Tris glycine 15-16% gradient gels in 0.1% SDS for 10 hours at 90 volts at 4°C. Gels were stained with Coomassie Brilliant Blue and destained in 30% ethanol-5% acetic acid. A limited supply of each subject's blood (15 ml/48 hours) was available. Packed red cells containing the buffy coat were obtained at surface (pre- and postdive) and at 200, 300, and 400 msw and were mixed with 5 ml Sigma Histopaque No. 1077 in 5 ml PBS; which was centrifuged at 10,000 rpm for 30 min. The red cell fraction (1 ml) was then placed in 6 ml Percoll, dissolved in 3 ml sodium diatrizoate to 4 ml distilled water. Centrifugation of this red cell solution (in 3 mls vols.) was carried out at a rate of 45,000 × g for 30 min which was modified from Vettore et al. 1980).

Three layers were taken from each tube, top, middle, and bottom. These were then washed with 5 mM phosphate buffered saline, pH 7.4. The three centrifuged fractions of the original diluted starting samples were pooled from the layers and analyzed for Hb-content, reticulocyte count, G6PD, SOD, CA, IgG-bindings, Spectrin antibody and FITC, morphological findings and deviations and SDS-electrophoresis of ghost preparations. Additionally, cells from the subjects were incubated at different pressure levels with either of the following per ml of blood:

1. 300 µl of anti IgG,
2. 100 µl of anti-complement $C(C_3)$,
3. 700 µl of ACD,
4. 1 ml of 10 mM glucose, in isotonic PBS,

and then decompressed at 1 m/min in the pressure vessel. Dried blood-films of all subjects in the 500 msw dive were made for light microscopy at pressure and also decompressed in the pressure vessel. All protein estimates for each sample used in electrophoresis was done by the method of Lowry.

RESULTS

Figures 78.1-78.6 reveal the morphological changes that occurred during the NUTEC 500 msw dive when cells were fixed

Figs. 78.1-78.6. SEM sequence of erythrocyte morphology changes in blood collected from one man in a hyperbaric chamber during a 500 msw dive; samples were fixed at pressure.

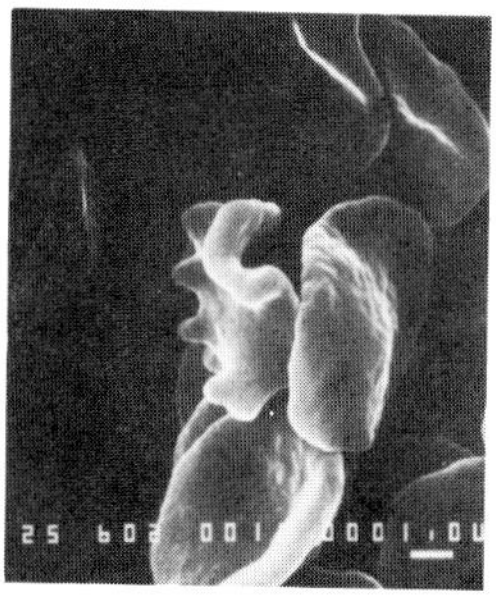

Fig. 78.1. Compression to 500 m, five days previously. Note loss of discoid form, enlarged acanthocytic cells, some crenation (1%) which is a normal percentage.

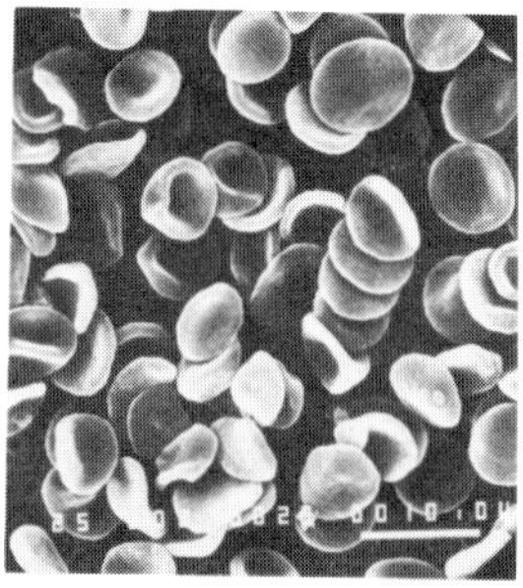

Fig. 78.2. Eight days following compression to 500 m, cells are sensitized and stressed by pressure, but few are acanthocytic; predominantly early echinocytes, stomatocytes, leptocytes and codocyte.

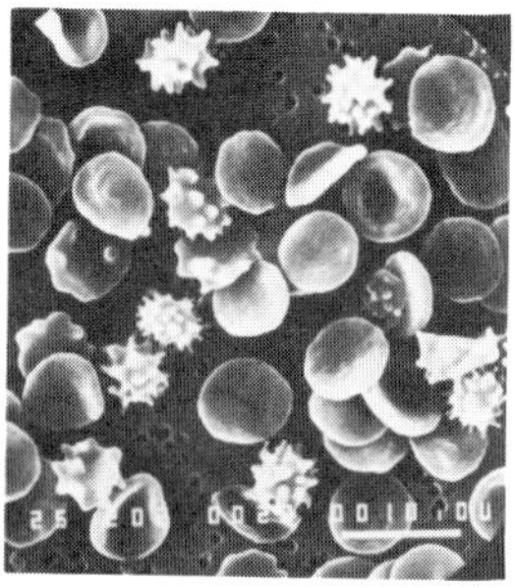

Fig. 78.3. Sixteen days following compression early in decompression at 300 msw. Note the appearance of echinocytes II, III, and IV stages and reappearance of a number of discocytes.

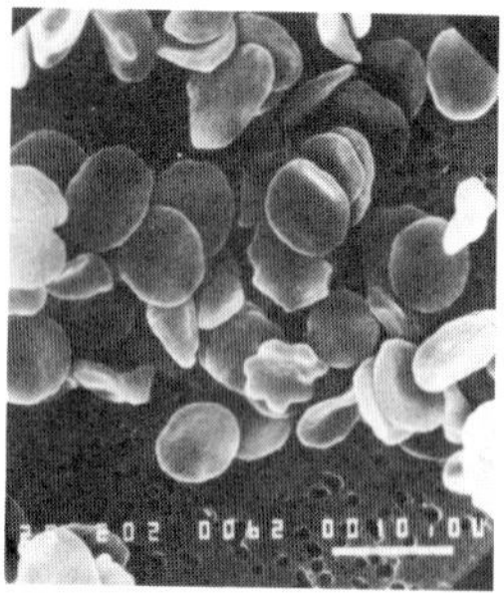

Fig. 78.4. During the decompression phase of the dive the number of echinocytes stage II and III diminished as the hemoglobin values decreased. (Table 78.1). Samples obtained at 200 msw.

Fig. 78.5. At same sampling as Fig. 78.4, but showing the presence of a spheroechinocyte (center). Other erythrocytes present have a distorted echinocyte I stage, but these are unable to undergo the discocyte-echinocyte transformation.

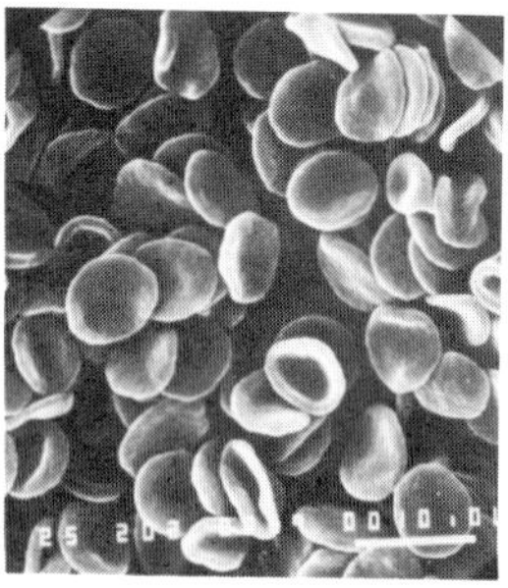

Fig. 78.6. Typical decompression SEM of blood samples obtained at 100 m from a man living at pressures greater than 200 msw. Samples Obtained after 24 days at pressure with the reappearance of distorted discocytes, but there is an absence of echinocyte III and IV.

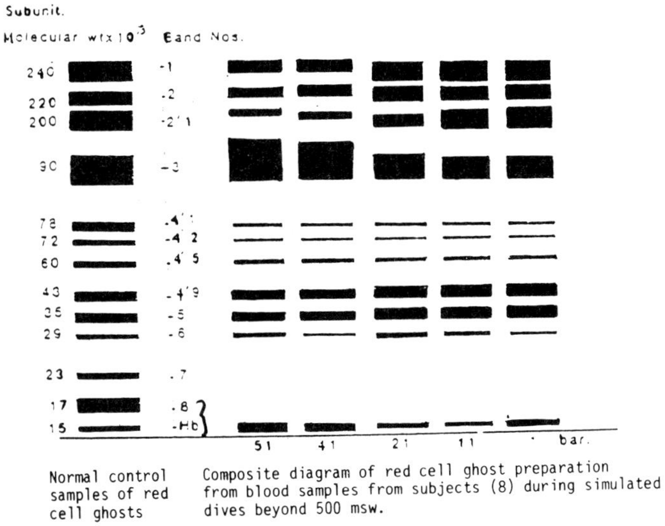

Fig. 78.7. Electrophoretic separation of ghost preparations.

at pressure. Based on the combined results of four dives beyond 400 msw at AMTE (UK) and NUTEC, the protein separation pattern on SDS has yielded the following; the Spectrin subunit alpha is declined by some 20% when compared to the starting amount of Band 1 in the predive phase. Replacement of this decrement is slow and not complete until after the dive. Actin or Band 5 is similarly reduced (8%) but at a later stage of the dive (Fig. 78.7); a composite representation of all the divers (12) individual SDS-electrophoresic separation of ghost proteins on the 15-16% gradient gels. An increased band of protein at 92,000 to 110,000 daltons was recorded, which may be the degradation complex from Band 1. Protein mapping of the degradation complex requires further analysis. Ghosts prepared from samples at pressure are usually pink and show an increase of Hb at Band 8 by electrophoresis. A cytosolic protein of 15,000 daltons is removed on the second harsh washing of ghosts prepared by hypotonic lysis. Resealing the ghosts were also carried out at 37°C, to eradicate the possibility that Spectrin may have been lost in preparation (Kansu et al., 1980). The separated fraction of the erythrocytes collected from samples at surface, 400, 300, 200 msw and postdive, showed the following, which is summarized in Table 78.1.

Cells fixed after being brought to surface had far too large numbers of nondiscoid cells, up to 40%. Recent studies with

Table 78.1. Fraction of Separated Cells and Depth

Procedure	Top					Middle					Bottom				
	S	400	300	200	P	S	400	300	200	P	S	400	300	200	P
Hb g/100 ml	15.	15.	14.5	14.	15.	15.	15.	14.	14.	14.	13.	13.	13.	12.	12.
Hct %			50%					49%				44%		42%	
G6PD activity			normal					normal					reduced		
SOD activity			normal					normal					reduced		
CA activity			normal					normal					reduced		
Retic. count	2%	–	1%	–	3%	1%	–	0.1%	–	0.1%	0%	–	–	–	–
	Antispectrin-FITC treatment of intact red cells (Paciorek et al. 1982) demonstrated an uninterrupted fluorescence on the inner surface of the membrane; whereas echinocyte IV and Spheroechinocytes showed areas of discontinuous fluorescence.														
SDS-Electrophoresis															
Band 1x)	100	100	100	100	110	100	90	90	90	80	100	80	80	–	90
Band 5	100	100	90	80	120	90	100	90	90	80	100	70	80	80	90
Band 3	100	120	130	100	100	100	120	130	140	90	100	170	110	110	80
Band 8	110	150	120	100	100	100	140	110	100	100	140	130	140	120	120

Code: S denotes surface predive values, x) equals percentage of surface value, P -"-"- postdive values.

respect to this very impaired metabolic cellular facility during slow decompression, has revealed that the incubated cells pre- and postdive in either glucose or ACD did not decrease the numbers of deformed cells.

Electrophoresis of ghost protein also confirmed that the quantity of the morphologic transformation was not reduced by the predecompressed metabolic treatment. However, whole blood procedures with either anti IgG or anti C_3 kept the number of irreversibly deformed cells at a level of 15%. The morphological study showed no binding of the cytosolic protein band of 15,000 daltons. Nor did the ghost preparation display the typical pink coloration due to the Hb-binding. Rapid decompression of samples caused 15% discocyte-echinocyte transformation only.

DISCUSSION

When men are exposed to depths greater than 200 msw for more than ten days, they suffer a reduction in Hb-content and red cell number. Initially there is a supression of the erythropoiesis due to the high partial pressure of oxygen in the breathing gas. This is reflected by the small and reduced reticulocyte counts.

Several of our observed changes in the erythrocytes structure and function have been implicated in the hypothesis of the aging process and disposal of senescent red cells. Changes in immunological recognition in the removal of erythrocytes by the monocyte-macrophage system is thought to be due to exposure of certain membrane surface antigens (Gattegno et al., 1976) and a concomitant increase in membrane surface IgG (Kay, 1975). During slow decompression, the cells became increasingly deformed, thus exposing antigen binding sites for IgG or C3. However, by the presence of added anti IgG and anti C3 into the cell suspension medium, the cellular transformation was significantly decreased.

Thus, as slow decompression proceeds, the binding of free IgG and C3 increases on to the further stressed membrane. The presence of antiserum to either IgG or C3 will then either (1) form antigenic-antibody complexes within the suspension medium, so reducing available free IgG or C3 for future binding, or (2) alternatively anti IgG or anti C3 acting as second but indirectly bound antibody to the cell membrane, which may induce a membrane conformation that prevents large numbers of cells from undergoing the discocyte-echinocyte transformation.

Metabolic impairment is well documented in the erythrocyte deformability of transfusion blood stored at 4°C. However, the origins of these deformations were always believed to be by ATP depletion. Feo and Mohandras (1976) clearly demonstrated that ATP per se is necessary for red cell shape maintenance. Cellular transformation during the in vitro decompression, in the NUTEC dive, did not have its basis in metabolic impairment. Incubation of the cells during decompression with glucose media did not reduce the number of abnormal cells. Measurement of cellular 2.3-DPG and ATP/ADP ratios showed these values to be normal even after decompression at 1 m per min. Thus the only methods by which we could prevent 40% of the erythrocytes underoing morphological changes was either by fixing the cells at pressure or by the presence of excess anti IgG or anti C3 in the blood suspension medium. Blood films made and dried at pressure for light microscopy also showed a reduced number of nondiscoid cells.

A reduction of SOD activity within the oldest fraction would permit the build-up of superoxide radicals, which would cause peroxidation within the red cell membrane and Hb molecule. This would agree with the report of Carell et al. (1975), that free radical reactions are a primary cause of red cell aging and exitus. In addition, sensitivity to exogenous compounds such as H_2O_2 produces irreducible protein crosslinkage in the 215- to 255,000 dalton range on SDS PAGE separation (Sauberman et al., 1981), whereas Snyder et al., 1983, have demonstrated that binding of Hbα-chains to the Spectrin alpha-subunit occurs during the in vivo oxidative damage to the membrane cytoskeleton. In our irreversibly transferred cells we observed discontinuity of stained Spectrin in the cytoskeleton, which may be considered as another marker of increased red cell senescence after pressure exposure.

ACKNOWLEDGMENTS

With thanks to Admiralty Marine Technology Establishment, Gosport, UK, and the Norwegian Underwater Technology Centre, Bergen, Norway, for extending their research facilities for this project, and to D.W. for help with the manuscript.

REFERENCES

Carrell, R. W., C. C. Winterbourn, and E. A. Rachmilewitz, Brit. J. Haematol. 30 259 (1975).

Carlyle, R. F., G. Nichols, J. A. Paciorek, P. M. Rowles, and N. Spencer, J. Physiol., 292, 34 (1979).

Feo, C. and N. Mohandas, P. 153 in "Red Cell Reology" (1976), ed. M. Bessis, S. B. Shohet, and N. Mohandas, Pub. Springer Verlag, Berlin-Heidelberg-New York, 1978.

Gattegno, L., D. Bladier, M. Garnier, and P. Cornillot, Carbohydrate Res., 52, 197 (1976).

Kansu, E., S. H. Krasnow, and K. Ballas, Biochim. Biophys. Acta, 596, 18 (1980).

Kay, M. M. B., Proc. Natl. Acad. Sci., U.S.A., 72, 3521 (1975).

Paciorek, J. A., E. Morild, and J. Onarheim, Norw. Underwater Technology Rept., 15b-82 (1982).

Sauberman, N., J. Piotrowski, W. Joshi, N. Fortier, and L. M. Snyder, Blood, 58, 34a (1981).

Snyder, L. M., N. Sauberman, H. Condara, J. Dolan, J. R. Jacobs, I. Szymanski, and N. L. Fortier, Brit. J. Haematol., 48, 435 (1981).

Vettore, L., M. C. DeMatteis, and P. Zampini, Am. J. Hematol., 8, 291 (1980).

79 A Novel Physiological Role of Microvilli: The Effects of Calcium, Zinc, and a Calcium-activated Protease on Vesiculation of Chick Intestinal Microvilli

Gillian M. Sainsbury and Barbara M. Luke

INTRODUCTION

If isolated brush borders, prepared from chicken small intestine, are exposed to a free calcium concentration > 10^{-6}M, the actin cores of the microvilli depolymerize and the microvillar membranes twist off to form vesicles which contain microvillar core proteins (see Burgess & Prum, 1982). It is likely that the protein, villin (M_r 95,000), is involved in this mechanism since at calcium concentrations > 10^{-6}M villin severs F-actin. The question arises as to whether or not vesicle formation from microvilli is a physiological process.

The reason why the absorptive surface of tissues as diverse as the photoreceptor of arthropods, the proximal tubule of the kidney, and the intestinal mucosa of vertebrates is covered with densely packed fingerlike microvilli is not completely explained. It has been suggested that these processes increase the surface area of the cell by as much as fortyfold. However, this large surface area may not be more efficient for transport than a smaller area for at least three reasons. Microvilli are closely packed with their glycocalyces intermeshed so that there are areas of relatively inaccessible surface. The cores of the microvilli are densely packed with filaments which must impede downward solute flow. Furthermore, as Parsons (1983) points out, the transported solutes must pass through the base of the microvillus which has only one-fiftieth of the area of the surface of the microvillus and is filled with fibers. I suggest that microvilli have a function additional to that of transport. Namely, it is that of buffering the cell from wear and tear damage, for example, in the case of the intestinal brush border, from abrasion by food particles and in the case of the arthropod photoreceptor, from damage by ultraviolet light. Wear and tear of microvillar membranes will tend to make them leady to extracellular solutes, which could lead to the destruction of the whole cell. However, since calcium in the gut lumen is in relatively high concentrations, inflow of calcium into the leaky microvillus will lead to vesiculation via the Ca^{2+} sensitive villin mechanism and sealing of the microvillar stump. The rest of the cell can continue to function normally with only a small loss of surface membrane area. Some evidence which supports microvillus vesiculation as a physiological mechanism comes from electron micrographs prepared from whole chicken small intestine. These clearly show vesicles and stumpy-ended microvilli adjacent to full length intact microvilli (Hobbs, 1980). Furthermore, Hobbs found that vesicles could be commonly found at the distal ends of the villi (where abrasion would be greatest) and much less often in the crypts of the villi. Vesicles were shown caught in the glycocalyx. Thus, it is possible that either vesicles are digested by exogenous enzymes and their components reabsorbed by adjacent microvilli, or they fuse with adjacent microvilli, and their membranes are salvaged by that route.

The first step in providing experimental evidence for the hypothesis that vesiculation is a physiological phenomenon designed to buffer the main body of the cell from wear and tear is to examine the factors which affect vesicle formation. In the experiments described below chicken small intestinal brush borders have been used to test the effects of calcium and zinc on vesicle formation. In addition, it seemed likely that intestinal brush borders contain a calcium-activated neutral protease (CAF) similar to that found in muscle where it is thought to be involved in myofibrillar turnover and where it acts on some muscle proteins. Preliminary experiments to test the presence of CAP in chicken small intestinal brush borders are described.

METHODS

Isolated brush borders were prepared from well-fed 18-month-old chicken small intestines by the method of Bretscher and Weber (1978) with the following modification: washed glass wood was used to filter the cells after homogenization. Brush borders (approximately 150 mg/tube) were suspended in KCl (75 mM), $MgCl_2$ (5 mM), EGTA (1 mM) and imidazole (10 mM, pH 7.0), and incubated in pyrex test tubes at 30°C for lengths of time as stated for each experiment. Calcium and zinc were added as solutions of $CaCl_2$ and $ZnCl_2$ made up in the above incubation medium. After incubation, reactions were stopped with EGTA (final added EGTA concentration of 4 mM). Samples were centrifuged for 2 min at 600 × g, to separate the brush borders from the suspension medium. Vesicles released from microvilli in the presence of calcium were found in the suspension medium and the degree of vesiculation was estimated by determination of the protein content of the suspension medium by the method of Lowry et al. (1951). Some samples were incubated at 30°C for 5 min in the presence of calcium and calcium with a calcium-activated protease prepared from turkey gizzard (CAF). For electron microscopy, samples of the brush border pellet fractions were fixed in glutaraldehyde (2.5%), sodium cacodylate buffer (0.15 M, pH 7.2), and sucrose (5% w/v) for 3 hr at 5°C. The pellet was then soaked in several changes of the same buffer overnight at 5°C. Samples were postfixed in osmium tetroxide (1%) for 1 hr, dehydrated in ethanol, and embedded in Araldite. The sections were stained with uranyl acetate and lead citrate and were viewed with a Philips EM 400 electron microscope. For the micrographs in Fig. 79.2a and b, the samples were applied to a carbon coated grid and negatively stained with aqueous uranyl acetate (2%).

RESULTS

The Effects of Calcium and Zinc on Isolated Brush Borders

Brush borders were incubated for 30 min in the absence (Fig. 79.1a) and presence of $CaCl_2$ (2.0 mM) (Fig. 79.1b), for 30 min. Microvilli on the borders incubated with calcium were short and stubbly and in some cases virtually absent, but rootlets could still be seen. The supernatant of the calcium treated brush borders was richly laden with vesicles (Fig. 79.2a). The protein concentration of the supernatant of calcium treated borders in the experiment shown in Figure 79.1 was 2.4 mg/ml as compared with 0.39 mg/ml in borders incubated without calcium. A six- to tenfold increase in supernatant protein was found in at least five other experiments. Electrophoresis of the supernatant proteins showed that they contained the full range of microvillar proteins (data not given). Brush borders from the experiment shown in Figure 74.1 were also incubated in the presence and absence of $ZnCl_2$ (2.0 mM) (Figs. 79.1d and 79.1a respectively). No obvious difference can be seen between microvilli in these borders. The protein concentration of the supernatant was slightly less in zinc treated borders

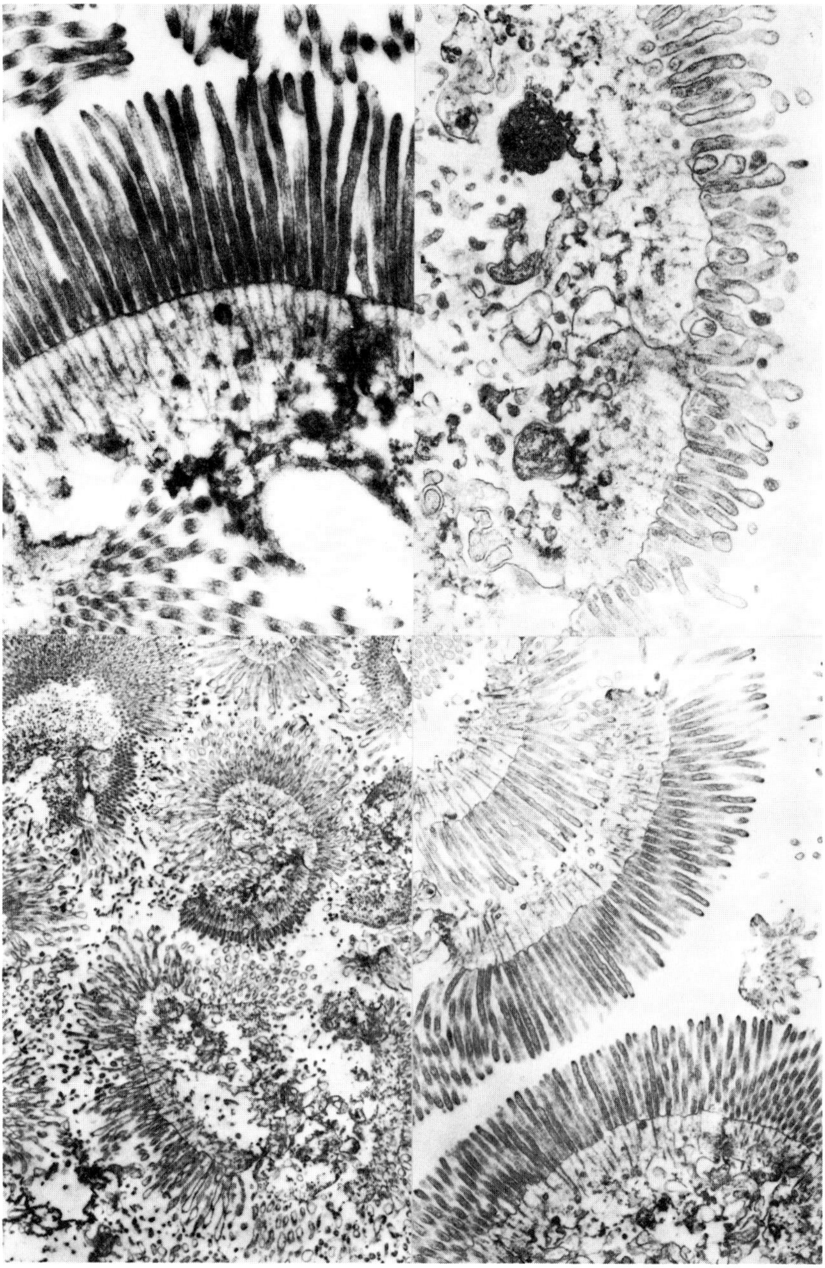

Fig. 79.1. Electron micrographs from one experiment in which isolated chick brush borders were prepared as described in the methods section and were incubated at 30°C for 30 min in the following conditions: (a) without additions, mag. × 25,000; (b) with $CaCl_2$, mag × 20,000; (c) with $CaCl_2$ and $ZnCl_2$, × 7,000; (d) with $ZnCl_2$, × 11,500.

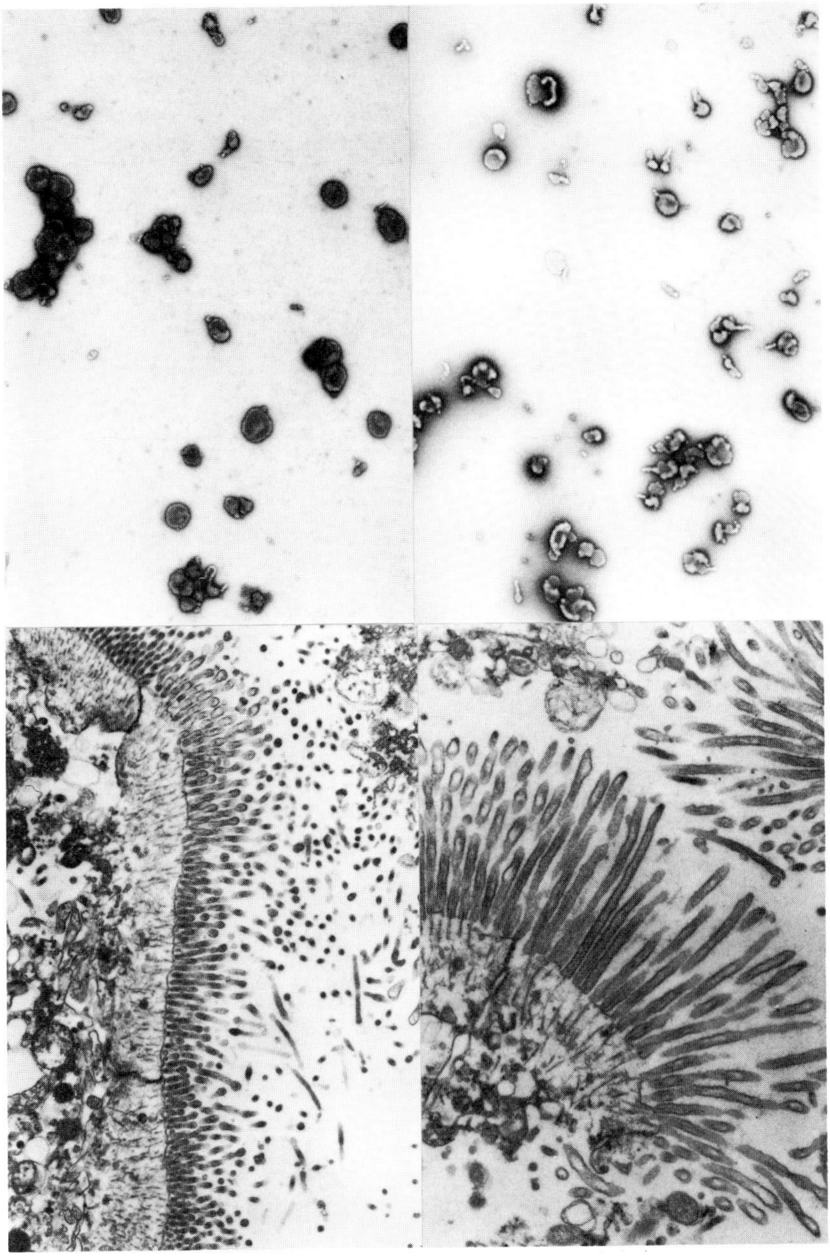

Fig. 79.2. Electron micrographs in (a) and (b) are of negatively stained vesicles from the supernatant of isolated brush borders incubated with (a) $CaCl_2$ mag × 25,000 and (b) $CaCl_2$ with $ZnCl_2$, mag × 25,000 for 30 min at 30°C. Electron micrographs in (c) and (d) are of isolated chick brush borders incubated for 5 min at 30°C in the presence of (c) $CaCl_2$ with calcium-activated protease mag × 15,000; and (d) $CaCl_2$ alone, mag × 20,000.

(0.30 mg/ml as compared with 0.39 mg/ml). This was found in other experiments also. When $ZnCl_2$ (2.0 mM) was added together with $CaCl_2$ (2.0 mM) (Fig. 79.1c) vesiculation of microvilli occurred but to a less extent (compare Figs. 79.1b with 79.1c). Protein in the supernatant of zinc-together-with calcium treated borders was 1.0 mg/ml as compared with 2.4 mg/ml in the calcium alone treated borders. From this it can be seen that zinc inhibits the vesiculation mechanism to some extent. Vesicles in the supernatant from the zinc-together-with calcium treated borders were frequently less regular in shape, many having short "tails" (Fig. 79.2b).

The Effect of Calcium-activated Protease (CAF) on Isolated Brush Borders

Isolated chick brush borders were incubated with $CaCl_2$ (0.17 mM) in the presence (Fig. 79.2c) and absence of CAF (Fig. 79.2d) for 5 min. From this and two similar experiments it appeared that in the presence of CAF, vesiculation of microvilli occurred at a more rapid rate. The calcium-alone treated brush borders were starting to vesiculate at 5 min whereas the CAF treated borders were in an advanced state of vesiculation.

CONCLUSIONS

This poster confirms the findings of other workers that free calcium concentrations $> 10^{-6}$M lead to vesicle formation in microvilli of brush borders from chicken small intestine (see Burgess & Prum, 1982). In addition, it is shown that zinc inhibits the effect of calcium to some extent. The mechanism of this is not yet known, but it may be relevant that for a number of years, zinc has been used to stabilize membrane preparations (see Warren et al., 1966). It remains to be seen whether zinc, which is in relatively high concentrations in small intestinal epithelium, plays any role in microvillar turnover. The other finding from this preliminary work is that microvilli appear to be a substrate for calcium-activated protease from muscle. This raises the question as to whether a calcium-activated protease is involved in microvillar turnover. The presence of a calcium-activated protease involved in turnover of blowfly photoreceptor microvilli was inferred by Blest et al. (1982). Relevant also to this question may be the findings that calcium-activated proteases have been implicated in other cytoskeletal systems, e.g., on filamin-actin interactions (Davies et al., 1978) and in pancreatic islets where they inhibit actin polymerization (MacDonald & Kowluru, 1982).

ACKNOWLEDGMENTS

This work was supported by the Muscular Dystrophy Group of Great Britain and St. Hilda's College, Oxford. I thank Prof. D. Goll (Arizona) for his generous gift of CAP, and Drs. C. A. R. Boyd and D. S. Parsons for contributing valuable discussion and criticism.

REFERENCES

Blest, A. D., Stowe, S. & Eddey, W. (1982), Cell Tissue Res. 223, 553-73. A labile Ca^{2+} dependent cytoskeleton in rhabdomeral microvilli of blowflies.

Bretscher, A. & Weber, K. (1978), Exptl. Cell Res. 116, 397-407. Purification of microvilli and an analysis of the protein components of the microfilament core bundle.

Burgess, D. R. & Prum, B. E. (1982), J. Cell Biol. 94, 97-107. Re-evaluation of brush border motility: calcium induces core filament solation and microvillar vesiculation.

Davies, P. J. A., Wallach, D., Willingham, M. C., Pastan, I., Yamaguchi, M. & Robson, R. M. (1978), J. Biol. Chem. 253, 4036-42. Filamin-Actin interaction. Dissociation of binding from gelation by Ca^{2+}-activated proteolysis.

Hobbs, D. G. (1980), J. Anat. 131, 635-42. The origin and distribution of membrane-bound vesicles associated with the brush border of chick intestinal mucosa.

Lowry, O. H., Rosebrough, N. J., Farr, A. L. & Randall, R. J. (1951), J. Biol. Chem. 193, 265-75. Protein measurement with the Folin phenol reagent.

MacDonald, M. J. & Kowluru, A. (1982), Arch. Biochem. Biophys. 219, 459-62. Calcium-activated factors in pancreatic islets that inhibit actin polymerization.